Nobel Laureate Contributions to 20th Century Chemistry

Nobel Laureate Contributions to 20th Century Chemistry

David Rogers
Danercon Ltd, Harrow, Middlesex, UK

RSC Publishing

ISBN-10: 0-85404-356-X
ISBN-13: 978-0-85404-356-9

A catalogue record for this book is available from the British Library

Published by The Royal Society of Chemistry,
Thomas Graham House, Science Park, Milton Road,
Cambridge CB4 0WF, UK

Registered Charity Number 207890

For further information see our web site at www.rsc.org

Typeset by Macmillan India Ltd, Bangalore, India
Printed by Henry Lings Ltd, Dorchester, Dorset, UK

Preface

Many years ago when I was an undergraduate chemistry student my lectures covered various aspects of the three classic branches of chemistry. It was sometimes apparent during the course of these lectures, particularly if we were being taught a named reaction, who originated a particular theorem, reaction or whatever. In most cases, however, the originators of the underlying science were not covered. Much of what is taught during an undergraduate degree builds on the previous work of earlier generations of scientists, and therefore remains anonymous. In some cases, however, some original work comes to light for which there are no precedents.

The award of the Nobel Prize for Chemistry addresses some of those scientists whose work is in a class of its own. Some of the scientists who have trained as chemists have achieved the distinction of being awarded a Nobel Prize in Physiology or Medicine, these are covered in the first section. Some scientists have been nominated for the Chemistry Nobel Prize on five or more occasions but for whatever reason failed to be awarded the ultimate accolade, these are covered in the last section – at least up to and including 1950. Yet further scientists sadly die before their work can be recognized by the award of a Nobel Prize. Of course many scientists are taken away from us when relatively young and who is to say that many might otherwise go on to join the august few who have received the award. Perhaps one young scientist, Rosalind Franklin, who died in the 1950's would have received the Nobel Prize had she lived a few more years. It is speculation but perhaps R. B. Woodward might have joined R. Hoffmann and K. Fukui in 1981 had he not died in 1979.

What is certain is that there are three Laureates who stand out above the group of other Laureates in these pages. Marie Curie is the only lady to have been awarded two Nobel Prizes one shared for Physics (1903) and the other awarded outright for Chemistry (1911). Linus Pauling is the only scientists to be awarded two undivided Nobel Prizes, one for Chemistry (1954) and one for Peace (1962). There is one chemist, however, who holds the unique distinction of being awarded two Chemistry Nobel Prizes. Fred Sanger was awarded an undivided prize for his work on insulin (1958) and shared the prize in 1980 for work that is known in the popular press as Genome science. There are pairs of Laureates, father and son, father-in-law and son-in-law who have both been awarded prizes in different categories, and there is the Curie family which between them have been awarded, or received on behalf of an organization (in the case of the Peace Prize) no less than six prizes (two for Physics, three for Chemistry and one for Peace).

In undertaking the research for this book I have used many reference sources and visited many libraries. There are a few members of various institutions who have freely given of their time in helping search out some useful information that proved otherwise elusive. They are Vicki Hammond of the Royal Society of Edinburgh, Lynn Saliba of the British Library, Martin Cadman of the Royal Society of Chemistry and various members of the library staff at the Royal Society. The Nobel Institute official website also deserves a mention for the data that it holds. I would also like to express my thanks to Maria Asp Romefors of the Royal Swedish Academy of Sciences for pointing me in a useful direction.

In reading accounts of Laureates from various sources I have come to realize that there are different methods of recording dates! In some cases learned societies have differed in the year a degree was obtained or an award given. There may be legitimate reasons for the slight discrepancies. For example a scientist may complete a degree in one year but be presented with the certificate the next year, or a scientist may receive notification of the award of a prize in one year and attend the institute the following year. In these cases the issue of dates is all in the wording. Finally, whilst I have been as thorough as I can be, there may still be errors in this book which are mine and mine alone.

Lastly, whilst writing this book those around me have come to know more than perhaps they wanted or needed to know concerning Nobel Chemistry Laureates. So, Carolyn, Adam and James, thank you for your patience.

Alfred Bernhard Nobel

21st October 1833 – 10th December 1896

Alfred Nobel never attended any school, with the exception that at the age of eight he was admitted to the first year of a parish elementary school. He remained there for one year only, after which the family moved to St. Petersburg. Alfred and his two brothers (eldest Robert, second brother Ludvig) received instruction from a private tutor. Alfred never attended university, nor did he ever obtain a degree. His formal education, such as it was, came to an end in 1850 when Alfred was 16 years old. Even at that age Alfred was scientifically trained as a chemist and a remarkable linguist who knew German, English, French, Russian and Swedish.

At the age of 16, the family finances allowed Alfred to travel abroad for two years. This trip included Italy, Germany and America, but most of the time he was in Paris where he continued his chemical studies in the laboratory of T. Jules Pelouze. Upon his return to Sweden he worked in his father's business until it went bankrupt in 1859. Soon after leaving his father's firm, Nobel started to undertake experiments with nitroglycerine, and made his first explosion in May or June of 1862. The explosion set the scene for the following activities:

1863	obtained a Swedish patent for the invention of a precussion detonator which he called the 'Nobel lighter'
1865	improves the blasting cap and sets up Alfred Nobel and Co., the factory being located in Krümmel near Hamburg
1866	established the United States Blasting oil company in the US

(continued)

(*continued*)

1866	invents dynamite
1871	founds the British Dynamite Company in Ardeer, Scotland
1873	starts the manufacture of nitroglycerine at Ardeer
1875	invents blasting gelatine in Paris and patents in 1876
1877	changes the name of the British Dynamite Company to Nobel's Explosives Company
1877	obtains the patent for blasting powder 'ballistite' in France
1891	moves to San Remo in Italy after a dispute with the French government over ballistite
1896	took out a patent for a pressure nozzle for glass.This technique was the basis for the manufacture of rayon after the second world war

Disaster struck in September 1864. Alfred had set up a small factory for the manufacture of nitroglycerine. Unfortunately there was an explosion in September 1864. Several lives were lost including Emil, Alfred's youngest brother. Undetered Alfred continued to set up factories for the production of nitroglycerine. And rapidly expanded his production from Sweden into Norway.

By his nature he was a melancholic, a dreamer and something of a recluse. He once wrote

> *...I want to live among trees and bushes, silent friends who respect the state of my nerves, and I escape when I can from both large cities and deserts...*

He was a lonely man, and with his sensitive nature he suffered keenly from the misfortune to be without a country that he could call home.

In 1875, he bought a house in Avenue Malakoff in Paris, but when the laboratory that he built within the house became too small he built a new one in Sevran on the outskirts of Paris. He was still unsettled and following a dispute with the French government moved to *Mio Nido* San Remo in Italy. He died there on 10th December 1896.

On his death, Nobel had assets in Sweden (5,796,140.00 Kr), Norway (94,472.28 Kr), Germany (6,152,250.09 Kr), Austria (228,754.20 Kr), France (7,280,817.23 Kr), Scotland (3,913,938.67 Kr), England (3,904,235.32 Kr), Italy (630,410.10 Kr) and Russia (5,232,773.45 Kr). A total of 33,233,792.20 Kr. Probating the will took sometime as there were some technical irregularities. Nevertheless, some considerable time after his death all of the obstacles were overcome and the first Nobel Prizes awarded in 1901. The tradition of awarding the prizes in Sweden on the anniversary of Nobel's death is carried out to this day as the ceremony takes place on the 10th December each year.

The Will of Alfred Bernhard Nobel

"The whole of my remaining realizable estate shall be dealt with in the following way: the capital, invested in safe securities by my executors, shall constitute a fund, the interest on which shall be annually distributed in the form of prizes to those who, during the preceding year, shall have conferred the greatest benefit on mankind. The said interest shall be divided into five equal parts, which shall be apportioned as follows: one part to the person who shall have made the most important discovery or invention within the field of physics; one part to the person who shall have made the most important chemical discovery or improvement; one part to the person who shall have made the most important discovery within the domain of physiology or medicine; one part to the person who shall have produced in the field of literature the most outstanding work in an ideal direction; and one part to the person who shall have done the most or the best work for fraternity between nations, for the abolition or reduction of standing armies and for the holding and promotion of peace congresses. The prizes for physics and chemistry shall be awarded by the Swedish Academy of Sciences; that for physiology or medical works by the Karolinska Institute in Stockholm; that for literature by the Academy in Stockholm, and that for champions of peace by a committee of five persons to be elected by the Norwegian Storting. It is my express wish that in awarding the prizes no consideration be given to the nationality of the candidates, but that the most worthy shall receive the prize, whether he be Scandinavian or not."

Paris, 27th November, 1895

The Royal Swedish Academy of Sciences

Nobel Prizes for Chemistry, and indeed Physics, have been awarded by the Academy since 1901 in line with the will of Alfred Nobel. The award ceremony takes place on 10th December of each year on the anniversary of Nobel's death. Laureates are announced in mid October.

Both the nominations and selection of the Nobel Laureates is to a large extent administered by the Chemistry Nobel Committee of the Academy. Each year this committee sends out invitations to thousands of scientists, members of academies and university professors asking them to nominate candidates. Former Laureates and Members of the Academy may also nominate.

Specially appointed experts help the Nobel Committee to review each nomination. Upon completion of these deliberations a vote is taken for the final choice of Laureates. Data of who has been nominated and who nominated the scientists is a closely guarded secret for 50 years following the year of the award. After that time the data may be obtained from the Academy. The Offices of the History of Science and Technology at the University of California compiled a list of nominees and nominators for the years 1901 to 1937 which was published in 1987. This data has been used where appropriate, and due reference made.

Contents

Dedicated to

Dick Staight

Section 1

Chemists Awarded the Nobel Prize for Physiology or Medicine

The Nobel Prize in Physiology or Medicine 1943

Henrik Carl Peter Dam (0.5)

"for his discovery of vitamin K"

Part of Dam's biographical abstract reads [2]:

...In 1925 Dam studied microchemistry in Graz (Austria) with F. Pregl. He became Assistant Professor at the Institute of Biochemistry, Copenhagen University, in 1928, and was promoted to Associate Professor at the same place in 1929, continuing to hold this post although he was working abroad part of the time, till 1941. On submitting a thesis Nogle Undersøgelser over Sterinernes Biologiske Betydning (Some investigations on the biological significance of the sterines) to the University of Copenhagen in 1934, Dam was awarded Ph.D. in Biochemistry.

To further his studies of the metabolism of sterols, Dam obtained a Rockefeller Fellowship and worked in Rudolph Schoenheimer's Laboratory in Freiburg, Germany, during 1932–1933, and later worked with P. Karrer, of Zurich, in 1935. He discovered vitamin K while studying the sterol metabolism of chicks in Copenhagen. He studied this vitamin further with respect to its occurrence and biological function in animals and plants, as well as its application in human medicine, its fundamental chemical and physical properties and its purification and isolation-the latter part of this research being carried out in collaboration with P. Karrer...

(i) Reproduced with permission.
(ii) Nobel Lectures, Physiology or Medicine 1942–1962, Elsevier Publishing Company, Amsterdam, 1964.

The Nobel Prize in Physiology or Medicine 1943

© Nobel Foundation [1]

Edward A. Doisy (0.5)

"for his discovery of the chemical nature of vitamin K"

Part of Doisy's biographical abstract reads [2]:

...He was educated at the University of Illinois, where he took his A.B. degree in 1914 and his M.S. degree in 1916. From there he went to Harvard University, where he took his Ph.D. degree in 1920.

From 1915 until 1917 he was assistant in biochemistry at Harvard Medical School and from 1917 until 1919 he did war service in the Sanitary Corps of the United States Army. From 1919 until 1923 he was Instructor, Associate, and Associate Professor at Washington University School of Medicine and in 1923 he became Professor of Biochemistry at St. Louis University School of Medicine. In 1924 he was appointed Director of the Department of Biochemistry.

Doisy has been concerned chiefly with biochemical studies of the sex hormones and vitamins K_1 and K_2. At the St. Louis School of Medicine he worked in collaboration with Edgar Allen on the refinement of the vaginal cytology (or smear) technique for the big-assay of the potency of oestrogenic hormones in ovariectomized rats...

(i) Reproduced with permission.
(ii) Nobel Lectures, Physiology or Medicine 1942–1962, Elsevier Publishing Company, Amsterdam, 1964.

The Nobel Prize in Physiology or Medicine 1945

Ernst Boris Chain (0.3)

"for the discovery of penicillin and its curative effect in various infectious diseases"

Part of Chain's biographical abstract reads [2]:

...He was from an early age interested in biochemistry and after graduation he worked for three years at the Charité Hospital, Berlin, on enzyme research. In 1933, after the access to power of the Nazi regime in Germany, he emigrated to England. Here, his first two years were spent working on phospholipids at the School of Biochemistry, Cambridge, under the direction of Sir Frederick Gowland Hopkins, for whose personality and scientific ability he came to have a great admiration.

In 1935 he was invited to Oxford University where he worked in the Sir William Dunn School of Pathology, becoming, in 1936, demonstrator and lecturer in chemical pathology. In 1948 he was appointed Scientific Director of the International Research Centre for Chemical Microbiology at the Istituto Superiore di Sanità, Rome. He became Professor of Biochemistry at Imperial College, University of London, in 1961, which position he still holds...

(i) Reproduced with permission.
(ii) Nobel Lectures, Physiology or Medicine 1942–1962, Elsevier Publishing Company, Amsterdam, 1964.

The Nobel Prize in Physiology or Medicine 1948

© Nobel Foundation [1]

Paul Hermann Müller

"for his discovery of the high efficiency of DDT as a contact poison against several arthropods"

Part of Müller's biographical abstract reads [2]:

...He commenced work in 1916 as a laboratory assistant at Dreyfus and Company and the following year he joined Lonza A.G. as an assistant chemist in the Scientific-Industrial Laboratory of their electrical plant, gaining a wealth of practical knowledge which later stood him in good stead in his career as an industrial chemist. He matriculated in 1918 and returned to school to obtain his diploma (1919) which entitled him to attend Basle University: he studied there under Professors Fichter and Rupe for his Doctorate which he received in 1925. He began his career with J. R. Geigy A.G. Basle, in May, 1925, to become Deputy Director of Scientific Research on Substances for Plant Protection in 1946.

Müller's first researches concerned the chemical and electrochemical oxidation of m-xylidine, and his early work at J. R. Geigy concerned vegetable dyes and natural tanning agents. He devoted some of his spare time to research on tanning agents and he invented synthetic agents which tanned hides pure white-they were, however, not fast to light. Later, in 1930, he developed the light-fast synthetic tanning agents Irgatan FL and Irgatan FLT. He worked on disinfectants for a short while, on moth-proofing agents for textiles, on pesticides in general, and he developed Graminone, a mercury-free seed disinfectant, before, in 1935, he started his researches on new synthetic contact insecticides.

Four years of intensive work led to the synthesis of dichlorodiphenyltrichloroethane (DDT) and the basic Swiss patent was granted in 1940. This compound was originally made in 1873 by an Austrian student, but had never received any particular attention...

(i) Reproduced with permission.
(ii) Nobel Lectures, Physiology or Medicine 1942–1962, Elsevier Publishing Company, Amsterdam, 1964.

The Nobel Prize in Physiology or Medicine 1950

Edward Calvin Kendall (0.3)

"for their discoveries relating to the hormones of the adrenal cortex, their structure and biological effects"

Part of Kendall's biographical abstract reads [2]:

...He was educated at Columbia University, where he obtained the degrees of Bachelor of Science in 1908 and Master of Science, specializing in Chemistry, in 1909. From 1909 until 1910 he was Goldschmidt Fellow of this University, and in 1910 he obtained his Ph.D. in Chemistry.

From 1910 until 1911 he was research chemist for Parke, Davis and Co., at Detroit, Michigan, U.S.A. Here he did research on the thyroid gland, and from 1911 until 1914 he continued this work at St. Luke's Hospital, New York.

In 1914 he was appointed Head of the Biochemistry Section in the Graduate School of the Mayo Foundation, Rochester which is part of the University of Minnesota, and in 1915 he was appointed Director of the Division of Biochemistry there and subsequently Professor of Physiological Chemistry. On April 1, 1951, Kendall reached the age of retirement from the Mayo Foundation and he accepted the position of Visiting Professor in the Department of Biochemistry at Princeton University...

(i) Reproduced with permission.
(ii) Nobel Lectures, Physiology or Medicine 1942–1962, Elsevier Publishing Company, Amsterdam, 1964.

The Nobel Prize in Physiology or Medicine 1950

Tadeus Reichstein (0.3)

"for their discoveries relating to the hormones of the adrenal cortex, their structure and biological effects"

Part of Reichstein's biographical abstract reads [2]:

...In 1916 Reichstein passed his school-leaving examination and began to study chemistry at the E.T.H. at Zurich, taking his diploma there in 1920. He then spent a year in industry and then began to work for his doctorate under Professor H. Staudinger. In 1922 he graduated and then began research under Professor Staudinger on the composition of the flavouring substances in roasted coffee.

After leaving Professor Staudinger, he continued to work for nine years on this subject, being financed for this purpose by an industrial firm, who provided him with an assistant. The aroma of coffee is, he found, composed of extremely complex substances, among which are derivatives of furan and pyrrole, and substances containing sulphur. Reichstein published during this period a series of papers on these substances and on new methods of demonstrating and making them, and also on the aromatic substances in chicory.

In 1929 he qualified as a lecturer at the E.T.H. Here he lectured on organic and physiological chemistry and in 1931, when his work on aromatic substances in coffee and chicory ended, he became assistant to Professor L. Ruzicka and was then able to devote himself exclusively to scientific research...

(i) Reproduced with permission.
(ii) Nobel Lectures, Physiology or Medicine 1942–1962, Elsevier Publishing Company, Amsterdam, 1964.

The Nobel Prize in Physiology or Medicine 1952

Selman Abraham Waksman

"for his discovery of streptomycin, the first antibiotic effective against tuberculosis"

Part of Waksman's biographical abstract reads [2]:

...In the autumn of 1911 he entered Rutgers College, having won a State Scholarship the previous spring. He received his B.Sc. degree in Agriculture from Rutgers in 1915. He was then appointed research assistant in soil bacteriology under Dr. J. G. Lipman at the New Jersey Agricultural Experiment Station, and was allowed to continue graduate work at Rutgers, obtaining his M.Sc. degree in 1916. In the same year, he became a naturalized United States citizen and was appointed a Research Fellow at the University of California where he received his Ph.D. in Biochemistry in 1918.

He was invited by Dr. Lipman to return to Rutgers, where he received an appointment as microbiologist at the Experiment Station and as Lecturer in Soil Microbiology at the University. He was appointed Associate Professor in 1925 and Professor in 1930. When the Department of Microbiology was organized in 1940, he became Professor of Microbiology and Head of the Department. In 1949, he was appointed Director of the Institute of Microbiology. He retired in 1958...

(i) Reproduced with permission.
(ii) Nobel Lectures, Physiology or Medicine 1942–1962, Elsevier Publishing Company, Amsterdam, 1964.

The Nobel Prize in Physiology or Medicine 1958

Edward Lawrie Tatum (0.25)

"for their discovery that genes act by regulating definite chemical events"

Part of Tatum's biographical abstract reads [2]:

...Tatum was educated at the University of Chicago and Wisconsin, taking his A.B. degree in Chemistry in 1931, his M.S. degree in Microbiology in 1932 and his Ph.D. degree in Biochemistry in 1934. For the Ph.D. degree his thesis was on work on the nutrition and metabolism of bacteria which he had done under the direction of Edwin Broun Fred and William Harold Peterson. This work no doubt laid the foundations of his later work with George Wells Beadle, which was to earn for their book, in 1958, the Nobel Prize in Physiology or Medicine.

After taking his doctor's degree, Tatum studied for a year at the University of Wisconsin and then was awarded a General Education Fellowship at the University of Utrecht, Holland. He then joined the Department of Biological Sciences at Stanford University, California, where he was Research Associate from 1937 until 1941, and Assistant Professor from 1941 until 1945. From 1945 until 1948 he was successively Assistant Professor of Botany and Professor of Microbiology at Yale University. In 1948 he returned to Stanford University as Professor of Biology and later became Professor of Biochemistry there...

(i) Reproduced with permission.
(ii) Nobel Lectures, Physiology or Medicine 1942–1962, Elsevier Publishing Company, Amsterdam, 1964.

The Nobel Prize in Physiology or Medicine 1962

Francis Harry Compton Crick (0.33)

"for their discoveries concerning the molecular structure of nucleic acids and its significance for information transfer in living material"

Part of Crick's biographical abstract reads [2]:

...Supported by a studentship from the Medical Research Council and with some financial help from his family, Crick went to Cambridge and worked at the Strangeways Research Laboratory. In 1949 he joined the Medical Research Council Unit headed by M. F. Perutz of which he has been a member ever since. This Unit was for many years housed in the Cavendish Laboratory Cambridge, but in 1962 moved into a large new building-the Medical Research Council Laboratory of Molecular Biology-on the New Hospital site. He became a research student for the second time in 1950, being accepted as a member of Caius College, Cambridge, and obtained a Ph.D. in 1954 on a thesis entitled «X-ray diffraction: polypeptides and proteins»...

(i) Reproduced with permission.
(ii) Nobel Lectures, Physiology or Medicine 1942–1962, Elsevier Publishing Company, Amsterdam, 1964.

The Nobel Prize in Physiology or Medicine 1962

© Nobel Foundation [1]

James Dewey Watson (0.33)

"for their discoveries concerning the molecular structure of nucleic acids and its significance for information transfer in living material"

Part of Watson's biographical abstract reads [2]:

...In 1947, he received a B.Sc. degree in Zoology. During these years his boyhood interest in bird-watching had matured into a serious desire to learn genetics. This became possible when he received a Fellowship for graduate study in Zoology at Indiana University in Bloomington, where he received his Ph.D. degree in Zoology in 1950. At Indiana, he was deeply influenced both by the geneticists H. J. Muller and T. M. Sonneborn, and by S. E. Luria, the Italian-born microbiologist then on the staff of Indiana's Bacteriology Department. Watson's Ph.D. thesis, done under Luria's able guidance, was a study of the effect of hard X-rays on bacteriophage multiplication.

From September 1950 to September 1951 he spent his first postdoctoral year in Copenhagen as a Merck Fellow of the National Research Council. Part of the year was spent with the biochemist Herman Kalckar, the remainder with the microbiologist Ole Maaløe. Again he worked with bacterial viruses, attempting to study the fate of DNA of infecting virus particles. During the spring of 1951, he went with Kalckar to the Zoological Station at Naples. There at a Symposium, late in May, he met Maurice Wilkins and saw for the first time the X-ray diffraction pattern of crystalline DNA...

(i) Reproduced with permission.
(ii) Nobel Lectures, Physiology or Medicine 1942–1962, Elsevier Publishing Company, Amsterdam, 1964.

The Nobel Prize in Physiology or Medicine 1964

Konrad E. Bloch (0.5)

"for their discoveries concerning the mechanism and regulation of the cholesterol and fatty acid metabolism"

Part of Bloch's biographical abstract reads [2]:

...In 1936 Bloch was able to immigrate to the United States as he had long hoped. On advice by the late Max Bergmann and with the generous support of the Wallerstein Foundation, he entered the Department of Biochemistry, College of Physicians and Surgeons, Columbia University, where he became a graduate student under Hans T. Clarke. Research leading to the Ph.D. degree was completed in 1938. Rudolf Schoenheimer then asked Bloch to join his research group. The first few years spent at Columbia with Schoenheimer and his associate David Rittenberg were certainly the most influential of his «Lehrjahre». This period more than any other developed his lasting interest in intermediary metabolism and problems of biosynthesis. During that time (in 1942) Bloch in collaboration with David Rittenberg initiated the work on the biological synthesis of cholesterol which was to occupy his research interests for nearly twenty years.

In 1946 Bloch moved to the University of Chicago as Assistant Professor of Biochemistry. Appointments to Associate Professor and Professor followed in 1948 and 1950, respectively...

(i) Reproduced with permission.
(ii) From Nobel Lectures, Physiology or Medicine 1963–1970, Elsevier Publishing Company, Amsterdam, 1972.

The Nobel Prize in Physiology or Medicine 1964

© Nobel Foundation [1]

Feodor Lynen (0.5)

"for their discoveries concerning the mechanism and regulation of the cholesterol and fatty acid metabolism"

Part of Lynen's biographical abstract reads [2]:

...Those who were to become responsible for his scientific training included Heinrich Wieland, Otto Hönigschmidt, Kasimir Fajans, and Walter Gerlach. The most enduring impression was left by Heinrich Wieland, who had won the Nobel Prize for Chemistry in 1927 and under whom Lynen graduated in March 1937 with the work: «On the Toxic Substances in Amanita». On completion of his doctoral thesis Lynen became acquainted under his guidance with the dynamic field of biochemistry, to which he has remained faithful to this day.

Lynen has also remained faithful to Munich University, where he became a chemistry lecturer in 1942, assistant professor in 1947, and biochemistry professor in 1953. In addition, in 1954 he became head of the Max-Planck-Institut für Zellchemie, newly created for him as a result of the initiative of Otto Warburg and Otto Hahn, then President of the Max-Planck-Gesellschaft zur Förderung der Wissenschaften. In this way Lynen acquired outstanding opportunities for scientific research...

(i) Reproduced with permission.
(ii) From Nobel Lectures, Physiology or Medicine 1963–1970, Elsevier Publishing Company, Amsterdam, 1972.

The Nobel Prize in Physiology or Medicine 1968

Robert W. Holley (0.33)

"for their interpretation of the genetic code and its function in protein synthesis"

Part of Holley's biographical abstract reads [2]:

...He studied chemistry at the University of Illinois and received his B. A. degree in 1942. Graduate work was at Cornell University, where the Ph.D. degree in organic chemistry, with Professor Alfred T. Blomquist, was awarded in 1947. Graduate work was interrupted during the war. He spent two years, 1944–1946, with Professor Vincent du Vigneaud at Cornell University Medical College, where he participated in the first chemical synthesis of penicillin.

After completing the Ph. D. degree, Holley spent 1947–1948 as an American Chemical Society Postdoctoral Fellow with Professor Carl M. Stevens at Washington State University. He then returned to Cornell University as Assistant Professor of Organic Chemistry at the Geneva Experiment Station in 1948. He was Associate Professor there from 1950–1957. During a sabbatical year, 1955–1956, he was a Guggenheim Memorial Fellow in the Division of Biology at the California Institute of Technology. In 1958, he returned to Ithaca, New York, as a Research Chemist at the U. S. Plant, Soil and Nutrition Laboratory, a U. S. Department of Agriculture Laboratory on the Cornell University campus. He had an appointment in the University throughout this period and became Professor of Biochemistry in 1962...

(i) Reproduced with permission.
(ii) From Nobel Lectures, Physiology or Medicine 1963–1970, Elsevier Publishing Company, Amsterdam, 1972.

The Nobel Prize in Physiology or Medicine 1968

© Nobel Foundation [1]

Marshall Warren Nirenberg (0.33)

"for their interpretation of the genetic code and its function in protein synthesis"

Part of Nirenberg's biographical abstract reads [2]:

...During this period he became interested in biochemistry. He continued studies in this field at the University of Michigan, Ann Arbor, and in 1957 received the Ph. D. degree from the Department of Biological Chemistry. Nirenberg's thesis, performed under the guidance of Dr. James Hogg, was a study of a permease for hexose transport in ascites tumor cells.

From 1957 to 1959 he obtained postdoctoral training with DeWitt Stetten Jr., and with William Jakoby at the National Institutes of Health as a fellow of the American Cancer society. During the next year he held a Public Health Service Fellowship and in 1960 became a research biochemist in the Section of Metabolic Enzymes, headed by Dr. Gordon Tompkins, at the National Institutes of Health.

In 1959 he began to study the steps that relate DNA, RNA and protein. These investigations led to the demonstration with H. Matthaei that messenger RNA is required for protein synthesis and that synthetic messenger RNA preparations can be used to decipher various aspects of the genetic code.

In 1962 he became head of the Section of Biochemical Genetics at the National Institutes of Health...

(i) Reproduced with permission.
(ii) From Nobel Lectures, Physiology or Medicine 1963–1970, Elsevier Publishing Company, Amsterdam, 1972.

The Nobel Prize in Physiology or Medicine 1988

Gertrude B. Elion (0.33)

"for their discoveries of important principles for drug treatment"

Part of Elion's biographical abstract reads [2]:

...I stayed there for a year and a half and was finally making the magnificient sum of $20 a week. By then I had saved some money and, with help from my parents, entered graduate school at New York University in the fall of 1939. I was the only female in my graduate chemistry class but no one seemed to mind, and I did not consider it at all strange.

After a year of graduate studies I had finished all the required courses but now needed to do the research work for my Master's degree. During this period, I took a job as a teacher-in-training and then as a substitute teacher in the New York City secondary schools, teaching chemistry, physics and general science for two years. In the meantime, I did my research work at night and on weekends at New York University, and obtained my Master of Science degree in chemistry in 1941...

...The work became fascinating almost from the very beginning. We were exploring new frontiers, since very little was known about nucleic acid biosynthesis or the enzymes involved with it. I had been assigned quite early to work on the purines and, with the exception of a few deviations into the pteridines and into some other condensed pyrimidine systems, the remainder of my work concentrated almost completely on the purines...

(i) Reproduced with permission.
(ii) From Les Prix Nobel. The Nobel Prizes 1988, Editor Tore Frängsmyr, [Nobel Foundation], Stockholm, 1989.

The Nobel Prize in Physiology or Medicine 1992

Edmond H. Fischer (0.5)

"for their discoveries concerning reversible protein phosphorylation as a biological regulatory mechanism"

Part of Fischer's biographical abstract reads [2]:

...It had always been my intention to go to the United States to pursue my studies in Biochemistry. In those days, that field was in its infancy in most European universities to such an extent that I was asked to present the very first course in Enzymology as a Privat Docent at the University of Geneva in 1950. Two events hastened my departure for the USA: the untimely death of Kurt Meyer following an asthma attack and my being abruptly issued a US immigration visa. Apparently, the US consulates abroad were clearing their files before the complicated McCarran Act would come into effect. I had decided to go to CalTech on a Swiss Post-doctoral Fellowship that Professor Paul Karrer succeeded in securing for me on a moment's notice. Some friends who knew of my arrival in New York had arranged for me to give some seminars on my way to Pasadena: Maria Fuld at Pittsburgh and Henry Lardy at Madison. To my utter surprise, I was offered a job in both places. Then, upon my arrival at CalTech I found a letter from Hans Neurath, Chairman of the Department of Biochemistry at the University of Washington, inviting me to come to Seattle, apparently for the same purpose. I thought that the Americans had to be crazy since at that time, academic positions in Europe were one-in-a-million. I visited Seattle with my wife and thought that the surrounding mountains, forests and lakes were beautiful, reminiscent of Switzerland. The Medical School was brand-new and when I was offered an Assistant Professorship, I accepted and have never regretted that decision...

(i) Reproduced with permission.
(ii) From Les Prix Nobel. The Nobel Prizes 1992, Editor Tore Frängsmyr, [Nobel Foundation], Stockholm, 1993.

The Nobel Prize in Physiology or Medicine 1992

© Nobel Foundation [1]

Edwin G. Krebs (0.5)

"for their discoveries concerning reversible protein phosphorylation as a biological regulatory mechanism"

Part of Krebs's biographical abstract reads [2]:

...The medical school years, 1940–1943, were war years, and although I did some research as a medical student, my main preoccupation was with becoming a physician who could serve in the armed forces. Nobody knew how long the war would last and our immediate concerns were with being a part of the war effort. After graduation from medical school I had eighteen months of residency training in internal medicine at Barnes Hospital in St. Louis, and then went on active duty as a medical office in the navy. The war ended and so did the period of my life in which I actively used my medical training in any practical sense. I believe I would have been happy practicing medicine but this was not to be.

After being discharged from the Navy in 1946, I returned to St. Louis with the idea of continuing residency and becoming an academic internist. However, it immediately became apparent that I would have to wait my turn to get back into hospital work, and I was advised by my professor of medicine, Dr. W. B. Wood, to study in a basic science department during the interim. Because of my background in chemistry, I chose biochemistry for this and was fortunate in being accepted by Dr. Carl and Gerty Cori as a postdoctoral fellow. After two years in their laboratory, during which time I studied the interaction of protamine with rabbit muscle phosphorylase, I became so enamored with biochemistry that I decided to remain in that field rather than returning to internal medicine. Again, I had found laboratory experience to be very satisfying just as it had been when I was a senior in college...

(i) Reproduced with permission.
(ii) From Les Prix Nobel. The Nobel Prizes 1992, Editor Tore Frängsmyr, [Nobel Foundation], Stockholm, 1993.

The Nobel Prize in Physiology or Medicine 1993

© Nobel Foundation [1]

Richard J. Roberts (0.5)

"for their discoveries of split genes"

Part of Roberts' biographical abstract reads [2]:

...As an undergraduate, David Ollis, the Professor of Organic Chemistry, really caught my imagination. His course emphasized problem solving, not memorization-more puzzles! Fortunately, he accepted me as his Ph.D. student and I began to explore the neoflavonoids found in a piece of heartwood from a Brazilian tree. Two pieces of luck followed. My tree contained more than its fair share of interesting new compounds and I was put in a lab with an exceptional postdoctoral fellow, Kazu Kurosawa, who proved a gifted teacher. Not only did he suggest the right experiments he explained why they should be done. Within one year I had essentially enough for my thesis and an understanding of how to do research. I had the luxury of spending the next two years following my nose, reading and experimenting.

During this time I came across a book, by John Kendrew, that was to change the course of my research career. It described the early history of crystallography and molecular biology focussing on the MRC Laboratory in Cambridge. It was my first exposure to "molecular biology" and I became hooked. For postdoctoral studies, I looked for a laboratory doing biochemistry that might accept an organic chemist and provide a pathway into molecular biology. Luckily, Jack Strominger offered me a position, not in Wisconsin as I had thought, but at Harvard where he had just been appointed Professor of Biochemistry and Molecular Biology. It was on January 1st, 1969, that my family walked across the runway at Logan Airport with an outside temperature of 4°F and a massive wind blowing, to start a new life...

(i) Reproduced with permission.
(ii) From Les Prix Nobel. The Nobel Prizes 1992, Editor Tore Frängsmyr, [Nobel Foundation], Stockholm, 1993.

The Nobel Prize in Physiology or Medicine 1993

Phillip A. Sharp (0.5)

"for their discoveries of split genes"

Part of Sharp's biographical abstract reads [2]:

...a young professor at Union, Dr. Dan Foote, became a good friend and encouraged me to apply to the Department of Chemistry at the University of Illinois. This old and distinguished department must have recognized some hidden promise as I was offered a fellowship and soon began graduate studies under Victor Bloomfield in physical chemistry. Victor was an excellent mentor as he encouraged both my scientific as well as cultural growth. He provided funds for my participation in national scientific meetings and broadened my perspective on society and culture by being a long-haired liberal, well-read and artistic friend. Fortunately, I was deferred from the Vietnam draft for a number of years and was able to finish graduate school. My thesis dealt primarily with the description of DNA as a polymer using statistical and physical theories. My attempts at experimental science at this stage were juvenile. In spite of my youth on the farm, I was never very skilled in manual tasks; in fact, I soon lost interest in any complex "hands on" manipulations...

(i) Reproduced with permission.
(ii) From Les Prix Nobel. The Nobel Prizes 1992, Editor Tore Frängsmyr, [Nobel Foundation], Stockholm, 1993.

The Nobel Prize in Physiology or Medicine 1994

Martin Rodbell (0.5)

"for their discovery of G-proteins and the role of these proteins in signal transduction in cells"

Part of Rodbell's biographical abstract reads [2]:

...I received my Ph.D. in Biochemistry in 1954. We immediately left Seattle for Urbana, Illinois where I became a post-doctoral fellow under Dr. Herbert E. Carter, then chairman of the department of Chemistry at the U. of Illinois. It was a wonderful place to be at that time, not only because of the great chemists in the department but also because the department of Microbiology had such notables as Gunsalus, Luria and Spiegelman who enlivened seminars with their egocentric views and vivid arguments about everything. I took on the research problem of the biosynthesis of chloramphenicol, an antibiotic of note that interested Dr. Carter. The molecule contained a nitro group appended to its benzene ring and two chlorides in the aliphatic side chain. My interest was how inorganic chloride was taken up into the side chain. I had some good ideas toward the second and final year after spending a great deal of effort trying to crush the mycelia into cell-free extracts. Finally it came down to the understanding that chloride was taken up into an activated (radical?) carbon at the two position of acetylacetate derived from the metabolism of phenylalanine! That problem was ultimately solved. The challenge was exciting, it was time to move on. Dr. Carter asked me at what university would I wish to teach. I replied: none. I had experienced teaching his lecture courses for the first year students: few of the students passed my exams. Devastated I decided never to teach. I chose research as my metier. Dr. Anfinsen at the National Heart Institute accepted me for a position in his laboratory to work on "clearing factor"...

(i) Reproduced with permission.
(ii) From Les Prix Nobel. The Nobel Prizes 1994, Editor Tore Frängsmyr, [Nobel Foundation], Stockholm, 1995.

Section 2

Nobel Chemistry Laureates

1901

© Nobel Foundation [1]

Jacobus Henricus van't Hoff

"in recognition of the extraordinary services he has rendered by the discovery of the laws of chemical dynamics and osmotic pressure in solutions" [2]

In his presentation speech on 10th December 1901, CT Odhner, then President of the Royal Swedish Academy of Sciences and former Rector of the National Archives observed the following [2]:

> *...By applying these simple principles, which were originally borrowed from mechanics and thermodynamics, van't Hoff became one of the founders of chemical dynamics. His researches have been a substantial factor in bringing about the magnificent advances in physical chemistry, in which field his discoveries match the great contributions which other investigators, including those in our country, have made in the field of electrochemistry and in the theory of chemical reaction. This has opened up great prospects for scientific research. On the other hand, the investigations on the state of substances in solutions have had and will continue to have the greatest practical consequences-consequences whose benefit to mankind can best be appreciated if we remember that chemical reactions occur predominantly in solutions and that the vital functions of living organisms are maintained by metabolic processes which take place in solutions...*

Van't Hoff was proposed for the first chemistry Nobel Prize by E Bamberger, E Fischer, R Gnehm, P Klason, H Landolt, R Lorenz, G Lunge, W Ramsay, F Treadwell, PT Cleve and O Pettersson [3].

Van't Hoff was born in Rotterdam, The Netherlands, on 30th August 1852. He was the third child in a family of seven children of Jacobus Henricus van't Hoff, a physician, and Alida Jacoba Kolff. His school education was in Rotterdam. He married Johanna Francina Mees in 1878. They had two daughters, Johanna Francina (b. 1880) and Aleida Jacoba (b. 1882) and two sons, Jacobus Hendricus (b. 1883) and Govert Jacob (b. 1889).

Van't Hoff's suffered from lung infections during his later years and often spent periods of enforced rests in sanatoria. He died on 1st March 1911, at Steglitz near Berlin [2,4].

The Early Years

In 1869, van't Hoff entered the Polytechnic School at Delft and completed the three year course in technology in two years, obtaining his technology diploma in 1871. A year at Leiden, where he studied mainly mathematics, was followed by work with A.F. Kekulé (in Bonn) from autumn 1872 to spring 1873. This was followed by some time in Paris with A. Wurtz, when he attended a large part of the curriculum for 1873–1874. He returned to Utrecht, Holland in 1874 and obtained the degree of 'Doktor der Wis-en Natuurkunde' concerning research on cyanoacetic and malonic acids on 22nd December (under E. Mulder).

The Career

Several months prior to obtaining his doctorate, van't Hoff published a pamphlet consisting of twelve pages of text and one page of diagrams. It was titled 'Voorstel tot Uitbreiding der Tegenwoordige in de Scheikunde gebruikte Structuurformules in de Ruimte' (Proposal for the development of 3-dimensional chemical structural formulae). This small pamphlet paved the way for the development of stereochemistry. The concept of the "asymmetrical carbon atom", dealt with in this publication, supplied an explanation of the occurrence of numerous isomers, inexplicable by means of the then current structural formulae. At the same time he pointed out the existence of the relationship between optical activity and the presence of an asymmetrical carbon atom. Arrhenius was the first foreigner who came to work with him in Amsterdam (1888).

Year	*Post*	*Institute*
1876	Lecturer at the Veterinary College	Utrect
1878	Professor of Chemistry, Mineralogy and Geology	University of Amsterdam
1885	Appointed Member of the Royal Academy of Sciences	
1896–1905	Appointed Honorary Professor	Berlin University

The Amsterdam Period

Van't Hoff published several seminal publications including:

1875 *Chimie dans l'Espace*
1876 German translation of *Chimie dans l'Espace* with an introduction by J.Wislicenus
1884 *Études de Dynamique Chimique* (Studies in dynamic chemistry)
1885 *L'Équilibre chimique dans les Systèmes gazeux ou dissous à l'État dilué* (Chemical equilibria in gaseous systems or strongly diluted solutions)

This publication described aspects of osmotic pressure in sufficiently dilute solutions. He concluded that the osmotic pressure for dilute solutions is proportional to the concentration and to absolute temperatures. Thus his extended formula only deviates from that for gases by a pressure co-efficient 'i'. Van't Hoff went on to determine the value of 'i' by various methods. He concluded that the thermodynamic laws are not only valid for gases, but also for dilute solutions.

1891 Chemistry in Space (English Translation of Chimie dans l'Espace)

1892 Dix Années dans l'Histoire d'une Théorie (Ten years in the history of a theory) he drew attention to the fact that J.A. Le Bel had independently arrived at the same ideas, though in a more abstract form.

The Berlin Period

During his years in Berlin, van't Hoff studied the problem of the origin of oceanic deposits, with special reference to those formed at Stassfurt. In this extensive work, he was most notably assisted by W. Meyerhoffer – a former collaborator from van't Hoff's Amsterdam years. Van't Hoff was probably the first scientist to apply small-scale results, obtained in the laboratory, to phenomena occurring on a large scale in nature. His findings were mostly published in the Proceedings of the Prussian Academy of Sciences and were summarized by him in a two-volume work Zur Bildung ozeanischer Salzablagerungen, 1905–1909.

The thirty first volume of Zeitschrift für physikalische Chemie (of which journal he had been joint editor since its foundation in 1887) was dedicated to him by his pupils in honour of the twenty-fifth anniversary of his promotion to the degree of doctor.

Van't Hoff's Nobel acceptance speech was concerned with osmotic pressure from the historical perspective and his own findings. During the course of his presentation, van't Hoff considered a chemical equilibrium [2]:

This can be illustrated in the formula by introducing the sign for a reversible reaction instead of the sign of equality:

$$CH_4O + CH_2O_2 \rightleftarrows H_2O + C_2H_4O_2$$

Such transformations, then, take place in either direction: the final state being known as chemical equilibrium.

The laws governing this chemical equilibrium in the case of dissolved substances can now he deduced by using the basic relationship for osmotic pressure.

The first of these,

for non-electrolytes: $\Sigma n \log p = K$

for electrolytes: $\Sigma ni \log p = K$

determines the quantity of the product which will form in a given volume from given quantities with substances reacting with one another, provided that a single so-called equilibrium constant, K, is known.

Since the value i enters into this formula in the case of electrolytes, we see that it plays an important part in the relevant laws of equilibrium.

In the second law

$$\frac{dK}{dt} = \frac{W}{2T^2} EL$$

we find yet another factor affecting the equilibrium, namely the temperature T (so-called absolute temperature). W is the quantity of heat evolved during the reaction.

Finally, in the last formula

$$K = -\frac{A}{2T}$$

allowance is also made for the quantity of work done during a chemical reaction, e.g. a dynamite explosion, the quantity being linked with the equilibrium constants.

Van't Hoff used these equations of osmotic pressure to determine the molecular weight of dissolved substances. In addition, homogeneous mixtures of solid substances were demonstrated to be comparable to liquids extending these laws of osmotic pressure to solids.

Van't Hoff's work and equations continue to be used, and have been reported in most standard texts as well as in the more specialist texts relating to chemical thermodynamics [5–7].

Honours and Awards

Van't Hoff was awarded honorary degrees from the Universities of: Cambridge; Chicago; Greifswald; Harvard; Heidelberg (1908); Manchester (1903); Utrecht; Victoria; Yale (1901).

He was the recipient of numerous medals and awards including: the Davy Medal of the Royal Society (1893); the Chevalier de la Legion d'Honneur (1894); the Nobel Prize for Chemistry (1901); the Helmholtz Medal of the Prussian Academy of Sciences (1911); the Senator der Kaiser-Wilhelm-Gesellschaft (1911),

Van't Hoff received Membership, Fellowship or Foreign Fellow/Member of the following organizations: the Royal Netherlands Academy of Sciences (1885); the Chemical Society of London (1898); Royal Academy of Sciences Gottingen (1892); American Chemical Society (1898); Académie des Sciences Paris (1905); Chemical Society of London (1888); The Royal Society (1897); The Physical Society (1911); The German Chemical Society (1889); the German Bunsen Society (1895).

Van't Hoff's citation for election (date: 25/11/1897) as Foreign Member of the Royal Society reads [8]:

"Professor J.H. van' Hoff of Berlin distinguished for numerous papers and works relating to Physical Chemistry."

He was proposed by:
W T Thiselton Dyer; W H Gaskell; J A Ewing; W G Adams; R Meldola; T Clifford Allbutt

Bibliography

1. Reproduced with permission.
2. From Nobel Lectures, Chemistry 1901–1921, Elsevier Publishing Company, Amsterdam, 1966.
3. The Nobel Population 1901–1937, E. Crawford, J.L. Heilbron and R. Ullrich. Office of the History of Science and Technology, University of California, 1987, ISBN: 0-918102-15-4.
4. Proceedings 1911–12, 86, xxxix–xliii RS library.
5. Topics in Chemical Physics, Alfred Prock and Gladys McConkey, Elsevier, 1962, Library of Congress Catalog Card Number 61-18140.
6. Investigation of Rates and Mechanisms of Reactions, 4th Ed Part II, Editor Claude F. Bernasconi, John Wiley and Sons Inc, 1986, ISBN: 0-471-83096-8.
7. Chemical Thermodynamics, Irving M. Klotz and Robert M. Rosenberg, John Wiley and Sons Inc, 2000, ISBN: 0-471-33107-4.
8. Sackler database, Royal Society.

1902

© Nobel Foundation [1]

Hermann Emil Fischer

"in recognition of the extraordinary services he has rendered by his work on sugar and purine syntheses" [2]

During his presentation speech on 10th December 1902, Professor Hj Théel, then President of the Swedish Royal Academy of Sciences observed [2]:

> *...In the case of the compounds under discussion Fischer succeeded in determining fully not only the way in which the atoms are bound in the atomic complexes, the molecules, but also the position of the atoms relative to one another or, in other words, the stereometric configuration of the molecules. Since the degree of complexity here is great and the theory is verifiable down to the finest details, our conception of atomic grouping has attained such a degree of certainty through these studies that we can be convinced that the conception will never be substantially weakened, even were the conception of the nature of the atoms and particularly of valences to undergo radical change in the future...*

Fischer was nominated by G Bredig, PT Cleve, E Hjelt, K Mörner and O Widman. He was nominated for the years 1901, 1902, 1903*, 1905*, 1916 and 1919 (where * represents disallowed nominations – usually late submissions) [3].

Fischer was born on 9th October 1852, at Euskirchen near Bonn in the Cologne district of Germany the only son of Laurenz Fischer and Julie Poensgen [3,4]. His father ran the successful family lumber business. After three years with a private tutor, Fischer went to study at the local school. He then spent two years at school in Wetzlar, and two more at Bonn where he passed his final examination with great distinction in 1869. At the age of 18, before he went to the University of Bonn, Fischer suffered from gastritis which was to recur in later life.

Fischer married Agnes Gerlach (the daughter of J von Gerlach, Professor of Anatomy at Erlangen) in 1888, who died seven years after their marriage. They had three sons, of whom one was killed in the First World War; another took his own life at the age of 25 as a result of compulsory military training.

The third son, Hermann Otto Laurenz Fischer, who died in 1960, was Professor of Biochemistry in the University of California at Berkeley. Fischer died from cancer on 15th July (some report 14th July) 1919 in Berlin, Germany. Later that year the German Chemical Society instituted the Emil Fischer Memorial Medal [2,4].

The Early Years

Fischer's father sent him to the University of Bonn to study chemistry in 1871. There Emil attended the lectures of Kekulé, Engelbach and Zincke.

Fischer was persuaded by his cousin Otto to go with him to the newly established University of Strasbourg where Professor Rose was using Bunsen's analysis techniques. Here Fischer met, and studied under, Adolf von Baeyer.

Fischer worked on the phthalein dyes which Rose had discovered and took his Ph.D. at Strasbourg with a thesis on fluoresceine and orcin-phthalein in 1874. In the same year he was appointed assistant instructor at Strasbourg University. Fischer also isolated phenylhydrazine from the reduction of benzenediazonium chloride with potassium sulphite (1875). When von Baeyer was asked to succeed Liebig at the University of Munich (1875), Fischer went there with him to become an assistant in organic chemistry.

The Career

Fischer refused the Chair of Chemistry at Aix-la-Chapelle in 1879, and in 1883 he was asked by the Badische Anilin- und Soda-Fabrik to direct its scientific laboratory, which he also refused. He also refused an offer to follow Victor Meyer at the Federal Technical University at Zurich because of a re-occurrence of gastritis for which he had to take a year's leave of absence before he went to Würzburg in 1888.

The table below shows his appointments.

Year	*Post*	*Institute*
1878	Privatdozent	University of Munich
1879	Associate Professor of Analytical Chemistry	University of Munich
1881	Professor of Chemistry	University of Erlangen
1888	Professor of Chemistry	University of Würzburg
1892	Professor and Director of the Chemical Institute	University of Berlin

The Munich Period

Fisher continued to work on the hydrazines and, working there with his cousin Otto, who had followed him to Munich, worked out a new theory of the constitution of the dyes derived from triphenylmethane.

The Erlangen Period

At Erlangen Fischer studied caffeine and theobromine, the active principles of tea, coffee and cocoa and established the constitution of a series of compounds in this field, eventually synthesizing them.

In 1884, Fischer began his systematic and thorough work on the sugars including the oxidation of glucose into aldonic acid and reaction with phenylhydrazine to form phenylhydrazones and the osazones.

Fischer discovered phenylglucazone (1884) by reacting fructose and phenylhydrazine.

Investigations into uric acid and related compounds (1881–1899).

The Würzburg Period

Fischer continued his work on sugars and in 1888 established the relationship between glucose, fructose and mannose. He also studied the stereochemical nature and isomery of the sugars, and between 1891 and 1894 established the stereochemical configuration of all the known sugars.

The Berlin Period

Fischer's most memorable period was at Berlin where he further developed his studies of the purines and the sugars and between 1899 and 1908 he made contributions to knowledge of the proteins whilst undertaking studies concerning:

cyclic amino acids-proline and oxyproline
synthesis of proteins
the peptide bond
dipeptides and later the tripeptides and polypeptide
synthesis of the dipeptide, glycyl-glycine (in 1901 in collaboration with Fourneau)
hydrolysis of casein
synthesis of the oligopeptides culminated in an octodecapeptide
enzymes
chemical substances in lichens, used in tanning and fats
synthesis pf sigallic acid (1908)
synthesis of sambunigrin and prulaurasin (1917)
synthesis acetonecyanohydrin-glucoside and glycollonitrile-glucoside (1919)
studied the behaviour of synthetic carbohydrates with different families of yeast (1894)
studied the behaviour of carbohydrates and glucosides towards blood serum (1896)

During the course of his career Fischer also determined the structure of xanthine, hypoxanthine, adenine, guanine, caffeine, theophylline, theobromine and uric acid.

Fischer's acceptance speech concerned "syntheses in the purine and sugar group" [2]. Fischer commented:

> *...The application of the new knowledge to the superb natural process without which the living world could not exist, i.e. the assimilation of carbon dioxide from the atmosphere by plants, seems even more interesting. This leads as we know to the formation of sugar, Nature's first organochemical product, from which all other constituents of the plant and animal body are formed. As mentioned earlier, this transformation can also be accomplished with purely chemical resources, although only in a very roundabout way. But there still remains one distinction between the natural and the artificial synthesis. Above all, the latter invariably yields a mixture of dextro- and laevo- rotatory sugars which must first be separated by special operations. Nature, in contrast, produces exclusively the dextro sugar. Formerly*

this contrast seemed so wonderful that the direct preparation of optically active substances was regarded quite simply as the privilege of the living organism. The experimental knowledge gained with the sugar group has provided a simple explanation for that biological phenomenon, however, and in the light of the new conception it does not appear at all impossible to reproduce that asymmetrical synthesis artificially in the same way as it occurs in the natural formation of sugars.

And so, progressively, the veil behind which Nature has so carefully concealed her secrets is being lifted where the carbohydrates are concerned...

Many subsequent papers and texts on the use or synthesis of sugars exist in the literature, see for example [5–8].

Honours and Awards

Fischer was made a Prussian Geheimrat (Excellenz), and held honorary doctorates of the Universities of Christiania, Cambridge (England), Manchester and Brussels. He was also awarded the Prussian Order of Merit and the Maximilian Order for Arts and Sciences. He was the Faraday Lecturer to the Chemical Society (1892). In 1902, he was awarded the Nobel Prize in Chemistry for his work on sugar and purine syntheses. He was elected Foreign Member of the Royal Society (1899) and received the Davy Medal from the Royal Society (1890).

The Chemical Society of Great Britain, the forerunner of The Royal Society of Chemistry, published the following [9]...

"The Chemical Society considers that it is neither compatible nor consistent with its loyalty to the Crown, whence the Royal Charter under which it works was derived, to retain any alien enemies upon its List of Honorary and Foreign Members.

"It is therefore resolved that the names of A. von Baeyer, T. Curtius, E. Fischer, C. Graebe, P. H. von Groth, W. Nernst, W. Ostwald, 0. Wallach, and R. Willstätter, who were elected under happier conditions in recognition of their eminent services to Chemical Science—for which the Society still retains an undimindished appreciation and regard, are hereby removed from the list of Honorary and Foreign Members."

Fischer was subsequently awarded Honorary and Foreign Member of the Chemical Society. Fischer's citation for election (date 01/06/1899) as Foreign Member of the Royal Society reads [10]:

"Professor Emil Fischer of the University of Berlin distinguished as the author of numerous researches in organic chemistry."

He was proposed by:
Lister; M Foster; Arthur Schuster; N Story Maskelyne; James Dewar; T G Bonney.

Bibliography

1. Reproduced with permission.
2. From Nobel Lectures, Chemistry 1901–1921, Elsevier Publishing Company, Amsterdam, 1966.

3. The Nobel Population 1901–1937, E. Crawford, J. L. Heilbron and R. Ullrich. Office of the History of Science and Technology, University of California, 1987, ISBN: 0-918102-15-4.
4. The Royal Society Obituaries: Proc. Roy. Soc. Series A, 1920–1921, vol. 98, pp. l–lvii.
5. The Carbohydrates – Chemistry and Biochemistry 2nd ed. edited by Ward Pigman and Derek Horton, Academic Press, 1970, Library of Congress card Number 68-26647.
6. Carbohydrate Chemistry of Monosaccharides and their Oligomers, Hassan S. El Khadem, Academic Press Inc, 1988, ISBN: 0-12-236870-3.
7. Developments in Carbohydrate Chemistry ed Richard J. Alexander and Henry F. Zobel, 1992, American Accociation of Cereal Chemists, ISBN: 0-913250-76-7.
8. 22nd International Carbohydrate Symposium, SECC, Glasgow UK, 23-27th July 2004, published by the RSC.
9. J. Chem. Soc. Trans, 1917, **111**, 273–287.
10. Sackler database, Royal Society.

1903

Svante August Arrhenius

"in recognition of the extraordinary services he has rendered to the advancement of chemistry by his electrolytic theory of dissociation" [2]

During his presentation speech, delivered on 11th December 1903, HR Törnebladh then President of the Royal Swedish Academy of Sciences, observed [2]:

> *...The world of science already recognizes the importance and value of your theory, but its lustre will continue to increase in the days to come, as you yourself and others use it to advance the science of chemistry. Physical research has contributed to your discovery and this fact throws new light on the relationship-more sensed than proved-between the different natural sciences, the common objective of which is to solve the riddles of life...*

Arrhenius was nominated for the Chemisty Prize by CE Bergstrand, W Brögger, PT Cleve, E Fischer, H Goldschmidt, F von Hefner-Alteneck, TH Hiortdahl, P Klason, JO Rosenberg, K von Than, JH van't Hoff, and J Vogt. He was nominated in the years 1901, 1902 as well as 1903 for the Chemistry prize and 1901, 1902 and 1903 for the Physics Prize [3].

Svante August Arrhenius was born in Wijk near Uppsala on 19th February 1859, the son of Svante Gustaf Arrhenius and Carolina Christina Thunberg. His father was a land surveyor (for the von Essens) and was employed by the University of Uppsala and in charge of its estates at Vik, where Svante was born. The family moved to Uppsala in 1860.

He was twice married - in 1894 to Sofia Rudbeck, by whom he had one son (Olof Wilhelm), and in 1905 to Maria Johansson, by whom he had one son and two daughters. Arrhenius suffered a stroke in 1924 which he attributed to overwork. He retired from the Nobel Institute in February 1927 and died at Stockholm on 2nd October 1927, and was buried in the town cemetery at Uppsala on 8th October 1927 [2, 4, 5].

The Early Years

Educated at home until the age of eight, Arrhenius then attended the Cathedral school. From an early age Arrhenius showed an aptitude for arithmetical calculations, and at school he was greatly interested in mathematics and physics. He entered the University of Uppsala, studied mathematics, chemistry and physics, and then moved to Stockholm to work under Professor E Edlund at the Academy of Sciences in 1881.

Initially, Arrhenius assisted Edlund in his work on electromotive force measurements in spark discharges. He began independent work in the spring of 1882 concerning the decay of galvanic polarization with time, an account of which was published in the Bihung of the Swedish Academy in 1883. This was followed by the measurement of the conducting power of electrolytic solutions.

He measured the resistance of a considerable number of salts, acids and bases in dilute solutions and found for salts that the specific conductivity is nearly proportional to the concentration as Kohlrausch had discovered earlier. A discussion of the data led Arrhenius to the conclusion that:

> "*...if on dilution of a solution the conductivity does not change proportionally to the amount of electrolyte then a chemical change has occurred on the addition of the solvent...*"

He went on to introduce the concept of a coefficient of activity.

> *"the coefficient of activity of an electrolyte is the number expressing the ratio of the number of ions it actually contains to that which it would contain of the electrolyte were to be completely transformed into simple electrolytic molecules"*

From this and other work he deduced the following principle: "The heat of neutralization evolved by the transformation of a base and an acid, both perfectly active, into water and a simple salt is nothing but the heat of activation of water". He completed his experimental work in the spring of 1883.

Arrhenius presented these ideas to the University of Uppsala in 1884 as a dissertation for the doctorate of the university entitled 'Recherches sur la conductibilité galvanique des électrolytes' (Investigations on the galvanic conductivity of electrolytes). It was not received favourably and he was awarded a fourth class degree (Dr. phil, juris ct med). Whilst Bergman and Berzelius were less than impressed with the thesis, Ostwald and Lodge were impressed. Ostwald visited Arrhenius in August 1884 and in 1885 Arrhenius obtained a travelling scholarship from the Academy of Sciences which enabled Arrhenius to work with Ostwald in Riga, Kohlrausch in Würzburg, Boltzmann in Graz (1887) and van't Hoff in Amsterdam (1888).

The Career

Year	*Post*	*Institute*
1891	Declined chair	Giessen, Germany
1891	Lecturer in Physics	Högskola, Stockholm
1895	Professor of Physics	Högskola, Stockholm
1897–1905	Rector	Högskola, Stockholm
1905	Director of the Physico-Chemical Department	Nobel Institute for Physical Chemistry

Arrhenius returned to Stockholm following his time with van't Hoff and remained in Stockholm for the rest of his career. There were three distinct areas of his research:

- The theory of ionic dissociation;
- Climatology – where he developed a model of the magnitude of the greenhouse effect on global warming;
- Aspects of immunochemistry – particularly chemical reactions of serum therapy.

Arrhenius and van't Hoff published their joint ideas in the first volume of the *Zeitschrift für physikalisehe Chemie* in the latter half of 1887. Van't Hoff accepted Arrhenius's theory for electrolytes. Arrhenius gives the relationship between van't Hoff's constant *i* and the degree of dissociation *k* in the form

$$i = 1 + (k - 1)x,$$

where *k* is the number of ions into which the molecule of the electrolyte dissociates, *e.g.*, 2 for NaCI, 3 for Na_2S0_4. He compared the values of *i* calculated from Raoult's freezing-point data with molecular conductivity for some 80 different substances establishing good agreement. Arrhenius went on to discuss the properties of electrolytes in aqueous solutions, particularly their additive effect which he attributed to the independence of their ions.

Arrhenius used Fourrier's equations in 1894 to divide the surface of the planet into small squares and assessed the capacity of each segment to absorb and reflect solar radiation. He produced a series of temperature predictions for different latitudes and seasons determined by atmospheric concentrations of carbon dioxide that varied from two thirds to three times the then current value. His calculation that a doubling of carbon dioxide levels would cause a warming of 5 to 6 degrees Centigrade almost exactly mirrors the Intergovernmental Panel on Climate Change's assessment which puts 5.8 degrees Centigrade at the top of its likely warming range. Arrhenius presented a paper to the Stockholm Physical Society concerning climate effects [6]. He took up political causes after World War I, particularly campaigning for immunisation against small-pox.

From 1900 until the time of his death, Arrhenius was a member of the Nobel Committee for Physics [pxi of ref [6]], and an unofficial member of the Chemistry Committee. Such was his influence that he was instrumental in how the prizes were awarded. At his suggestion nominations were accepted from foreign members of the Academy of Sciences as well as previous prizewinners in physics or chemistry. The matter of peer evaluation was broadened to include national, international peers and members of foreign universities and individuals nominating candidates in their personal capacity [7].

Arrhenius published his *Lärobok i teoretisk elektrokemi* (Textbook of Theoretical Electrochemistry) in 1900, *Lehrbuch der kosmischen Physik* (Textbook of Cosmic Physics) in 1903, *Theorien der Chemie* (Theories of Chemistry) and *Immunochemistry* in 1906, as well as *Världarnas utveckling* (Worlds in the Making) in 1906, *Smittkopporna och deras bekämpande* (Smallpox and its Combating) in 1913, *Stjärnornas Öden* (Destiny of the Stars) in 1915, the Silliman lectures Theories of solutions in 1918 and *Kemien och det moderna livet* (Chemistry and Modern Life) in 1919.

Arrhenius' Nobel Prize acceptance speech is dated 11th December 1903 with the title 'Development of the theory of electrolytic dissociation' [4]. During the course of his presentation, Arrhenius commented [4]:

> *...The question now arose in what respect the active state of the electrolytes differs from the inactive state. I gave an answer to this question in 1887. At this time van't Hoff had formulated his*

far-reaching law that the molecules in greatly diluted solutions obey the laws which apply to the gaseous state if gas pressure in liquids is merely replaced by the osmotic pressure. As van't Hoff demonstrated, the osmotic pressure of a solute can be far more easily determined by measuring the freezing point of its solution than directly. However, both the direct measurements carried out by De Vries and the freezing points of electrolytic solutions showed a much higher osmotic pressure than would have been expected from the chemical formula. For instance, whilst the solution of one gram-molecule (mole) of ethyl alcohol, $C_2H_5OH = 46$ grams, in one litre of water gives a freezing point of $-1.85°C$, according to van't Hoff's calculations a solution of one mole of sodium chloride, $NaCl = 58.5$ grams in one litre of water gives a freezing point of $-3.26CC = -1.75 \times 1.85°C$. This characteristic could be explained in the same way as the abnormal density of gaseous ammonium chloride, i.e. by supposing a partial dissociation-up to 75%-of the sodium chloride molecules. On this assumption, therefore, the solution contains 0.25 moles of NaCI, 0.75 moles of Cl and 0.75 moles of Na, a total of 1.75 moles, and has a corresponding effect regarding the depression of the freezing point. We have seen in what way we can calculate from the electrical conductivity the number of active molecules in the same solution of sodium chloride, and from Kohlrausch's measurements we find exactly the same number, i.e. 0.75. This strengthens the conception that the active molecules in the salt NaCI have split up into the ions Na and Cl. These are absolutely free and they appear in the solutions exactly like other molecules. In the same way I calculated by both methods the degree of dissociation in all the electrolytes which had been determined at the time – there were about 80 of them – and I found that there was very good agreement between the two different methods of determination. In some cases the agreement was not so good and I therefore carried out new analyses of these bodies and some others. The new analyses were all completely in agreement with the theoretical assumptions...

Many subsequent papers and texts on the use of Arrhenius' concepts appear in the literature, see for example [5–8].

Honours and Awards

Arrhenius received honorary degrees from the Universities of Birmingham (1913), Cambridge (1908), Edinburgh, Greifswald, Groningen (1914), Heidelberg (1908), Leipzig (1909), Oslo (1904) and Oxford (1908).

Arrhenius was awarded the Royal Society's Davy medal (1902), the Nobel Prize in 1903 and the Faraday Medal of the Chemical Society (1914). He delivered the Tyndall Lectures to the Royal Institution in 1914 and the Silliman Lectures in the United States in 1911. He received the first Willard Gibbs Medal (1911).

He was elected an honorary Fellow of the Chemical Society (1898) and a Foreign Member of The Royal Society. He was a member of the Institute of Physics and the Royal Institution of Great Britain and of many European Institutes and Academies.

Arrhenius' citation for election (date: 30/06/1910) as Foreign Member of the Royal Society reads [11]:

"Professor Svante Arrhenius of the University of Stockholm distinguished for his researches on the Theory of Electrolytic Dissociation."

He was proposed by:

Arch Geikie; J Rose Bradford; A B Kempe; William Crookes; H B Baker; W H Perkin; D Prain; A E Shipley; Edward B Poulton; R A Sampson; H M Macdonald; Horace Lamb; P A MacMahon; A Strahan; Charles J Martin; Walter H Gaskell; J Larmor.

Bibliography

1. Reproduced with permission.
2. From Nobel Lectures, Chemistry 1901–1921, Elsevier Publishing Company, Amsterdam, 1966.
3. The Nobel Population 1901–1937, E. Crawford, J.L. Heilbron and R. Ullrich. Office of the History of Science and Technology, University of California, 1987, ISBN: 0-918102-15-4.
4. Proc Roy Soc Series A, 1928, vol. 119, pp. ix–xix.
5. Who was who: 1916–1928.
6. 'On the influence of carbonic acid in the air upon the temperature of the ground' 1895, Paper presented to the Stockholm Physical Society.
7. Arrhenius. From Ionic Theory to the Greenhouse Effect. Elisabeth Crawford, Watson Publishing International, ISBN: 0-88135-166-0 1996.
8. Reaction Kinetics by Michael J. Pilling and Paul W. Seakins, Oxford Science Publications, 1995, ISBN: 0-19-855527-X.
9. Fundamental Chemical Kinetics by Margaret Robson Wright, Horwood Publishing, 1999, ISBN: 1-898563-60-8.
10. Chemical Kinetics and Dynamics 2nd Ed by Jeffrey I. Steinfeld, Joseph S. Francisco and William L. Hase, Prentice Hall, 1999, ISBN: 0-13-737123-3.
11. Sackler database, Royal Society.

1904

© Nobel Foundation [1]

Sir William Ramsay

"in recognition of his services in the discovery of the inert gaseous elements in air, and his determination of their place in the periodic system" [2]

During his presentation speech on 10th December 1904, JE Cederblom, then President of the Royal Swedish Academy of Sciences, observed [2]:

> *...The discovery of an entirely new group of elements, of which no single representative had been known with any certainty, is something utterly unique in the history of chemistry, being intrinsically an advance in science of peculiar significance. The more remarkable is this advance when we recollect that all these elements are components of the atmosphere of the earth, and that, though apparently so accessible for scientific research, they have for so long a time baffled the acumen of eminent scientists...*

Ramsay was nominated in 1904 by A von Baeyer, M Bamberger, CA Bischoff, W Brögger, B Brauner, G Ciamician, PT Cleve, E Fischer, H Goldschmidt, PA Guye, TH Hiortdahl, K Hofmann, P Klason, W Koenigs, F Kohlrausch, A Lobry de Buuyun, J Lubbock, O Pettersson, O Pitoty, JO Rosenberg, H Söderbaum and J Vogt. Ramsay was also nominated for the Chemistry Prize in 1902 and 1903 and for the Physics Prize in 1904 and 1905, although the 1905 nomination was disallowed due to lateness [3].

Ramsay was born on 2nd October 1852 in Queen's Crescent, Glasgow, the only child of William Ramsay, a civil engineer (who died when Ramsay was 15 years old) and Catherine, née Robertson. He married Margaret, daughter of George Stevenson Buchanan (of Glasgow), in 1881. They had one son and one daughter [2,4,5].

Ramsay died on 23rd July 1916 in Hazlemere, near High Wycombe, Buckinghamshire.

The Early Years

Ramsay attended an elementary school from 4 to 10 years and then attended the Glasgow Academy, matriculating in November 1866. He read Graham's 'Chemistry' whilst at school, the only purpose for which was to learn how to make fireworks. In 1869, he went to the chemical laboratory of Robert Tatlock, where he began chemistry in earnest. He worked in Lord Kelvin's laboratory in Glasgow University for six months unrolling and taking out the kinks from a great mass of copper wire (used in Kelvin's electrometer), an operation that took nearly two weeks. A task that he judged to be worthwhile as he was able to spend time with Kelvin.

He went to Heidelberg at the end of the Franco-Prussian war, but in the spring of 1871 changed to the University of Tübingen, and entered the laboratory of Prof. Fittig. His Ph.D. was undertaken on orthotoluic acid and its derivatives and completed in two years. He returned to Glasgow and took the post of assistant in the Young Laboratory of Technical Chemistry, also and became assistant of Professor Ferguson at the University of Glasgow in 1874.

The Career

Year	*Post*	*Institute*
1874	Assistant to Professor Ferguson	University of Glasgow
1880	Principal and Professor of Chemistry	University College, Bristol
1887	Professor of Inorganic Chemistry	University College, London
1913	Retired	

During the course of his career, Ramsay worked in three academies

The Glasgow Period

Ramsay's early interests lay in organic chemistry. In addition to his doctorate thesis, he published work on picoline and, in conjunction with Dobbie, the decomposition products of the quinine alkaloids (1878–1879).

His investigations also concerned:

- a new bismuth mineral
- the action of heat on ethyl-sodium-thiosuiphate thereby proving the conformula for sodium dithionate
- the production of various pyridinic acids by the oxidation of the pyridine bases (previously prepared Prof. Anderson 30 years earlier)
- the silver salts of some of the acids obtained by the oxidation of the alkaloids quinine and cinchonine gave pyridine on heating, and were similar to the acids he had obtained by the oxidation of Anderson's pyridine bases; this was the first observation connecting that important class of organic compounds, the alkaloids, with pyridine

The Bristol Period

Almost as soon as he arrived in Bristol, Ramsay was elected Principal of the College. He started this period of his career working on some interests in physical chemistry namely stoichiometry, thermodynamics and evaporation, and dissociation (the latter in conjunction with Sidney Young). These investigations were completed in London and covered the years 1886–1889.

In addition to the work on physical chemistry, Ramsay published several notable papers concerning nitrogen oxides and reported on the discovery of helium, neon, krypton, and xenon (1885–1890). Many of Ramsay's Bristol papers were co-authored by his assistant Dr. S. Young. Together they published work concerning:

- Investigations into critical points
- Dissociation
- Evaporation
- thermal properties
- vapour densities
- the discovery of helium (1895)

Ramsay became an expert glass blower as many of the experiments that he devised required unique glassware. Some of his glassware designs are still in use today, principally those used for investigations concerning vapour pressure and boiling points.

The London Period

Ramsay was chosen to succeed Professor Williamson in the Chair of Chemistry at University College, London (1887), where he worked for over twenty-five years until retirement. Having deduced the need for the inert gases in the periodic table, Ramsay concluded that there must be others in the series and found neon, krypton, and xenon in 1898. In conjunction with Soddy, Ramsay later determined that helium is produced from radium (1903). Both Lord Rayleigh and Ramsay independently succeeded in proving that there must exist a previously unknown gas in the atmosphere. They subsequently worked in their separate laboratories on this problem but communicated the results of their work almost daily. At the meeting of the British Association in August 1894, they announced the discovery of argon.

In addition to his work on inert gases, Ramsay published work concerning:

- metallic solutions (1889)
- molecular surface-energy
- discovery of argon in conjunction with Lord Rayleigh. This was announced at the British Association Meeting at Oxford in 1894.

Ramsay's Nobel Prize acceptance speech was entitled 'The Rare Gases of the Atmosphere' [3]. He describes some of the work thus:

> *...That these are all monatomic gases was proved by determining the ratio of their specific heats by Kundt's method; and accurate determinations were made of their refractivities, their densities, their compressibilities at two temperatures, 11.2° and 237.3°; and the vapour pressures of argon, krypton, and xenon were determined, as well as their volumes at their boiling points, in the liquid*

state. The critical temperatures and pressures of the last three were also determined. It may be stated in general terms that these gases show a regular gradation of properties, from helium to xenon; and that they fill the gaps in the periodic table below and above argon, with the atomic weights: neon, 20; krypton, 82; and xenon, 128.

The amounts of neon and helium in air have since been measured; the former is contained in air in the proportion of I volume in 81,000; the latter, I volume in 245,000; the amounts of krypton and xenon are very much smaller-not more than 1 part of krypton by volume can be separated from 20,000,000, of air; and the amount of xenon in air by volume is not more than 1 part in 170,000,000...

The properties of the inert or noble gases were tabulated in 1988 [9].

Honours and Awards

Honorary doctorates (LL.D, D.Sc, M.D. and Ph.D.) were conferred on Ramsay by numerous universities in Great Britain and abroad.

Ramsay received many awards/medals including: Davy Medal 1895; Nobel Prize (Chemistry) 1904; Longstaff Medal 1897; the Barnardo Medal and a prize of $5,000 from the Smithsonian Institution; a prize of Fr. 25,000 from France (together with Moissan); the A.W. Hoffmann Medal in Gold (Berlin, 1903). He was appointed K.C.B. (Knight Commander of the Order of Bath) in 1902, and was also a Knight of the Prussian Order "Pour le mérite", Commander of the Crown of Italy, and Officer of the Legion d'Honneur of France. He was a Commander of the Crown of Italy.

Many institutions elected Ramsay to Fellowships or Memberships including: The Institut de France; the Royal Academies of Ireland, Berlin, Bohemia; the Netherlands, Rome, Petrograd, Turin, Romania, Vienna, Norway and Sweden; the Academies of Geneva, Frankfurt and Mexico; the German Chemical Society; the Royal Medical and Chirurgical Society of London; the Académie de Médecine de Paris; the Pharmaceutical Society; the Philosophical Societies of Manchester, Philadelphia and Rotterdam. Ramsay was a Fellow of the Chemical Society and The Royal Society (elected 1888). He was made a Foreign Member of scientific societies in every country where they existed; Russia, Germany, Hungary, Austria, Italy, Switzerland, France, Holland, Belgium, Spain, Bohemia, Sweden, Norway, Romania, Bengal, The United States, Mexico, and Brazil.

In 1898, he was made a Member of the Royal Commission on Sewage Pollution. For many years he was the chairman. Endless subjects engaged his attention, from improvements in the fixation of nitrogen from the atmosphere to the composition of wax candles used in Roman Catholic churches, or to the perforation of the pipes used to convey water to the Coolgardie Goldfields. Ramsay was appointed adviser to the Indian Government on the question of university and technical education (1900), President of the International Congress (1909), Foreign Secretary to the Chemical Society (for five years) and President in 1907.

Many books have been written concerning Sir William Ramsay. References to a few of them are provided below [6–8].

Ramsay was proposed for election (date 07/06/1888) as Fellow of the Royal Society by [10]:

H Perkin; H E Roscoe; G D Liveing; Lyon Playfair; J H Gladstone; Henry E Armstrong; William A Tilden; Geo Fras Fitzgerald; John G McKendrick; Frederick Guthrie; Hugo Muller; J H Gilbert; Herbert McLeod; W N Hartley; John Beddoe; Isaac Bayley Balfour; G Carey Foster; A Liversidge; R Meldola; Harold B Dixon

Bibliography

1. Reproduced with permission.
2. From Nobel Lectures, Chemistry 1922–1941, Elsevier Publishing Company, Amsterdam, 1966.
3. The Nobel Population 1901–1937, E. Crawford, J. L. Heilbron and R. Ullrich. Office of the History of Science and Technology, University of California, 1987, ISBN: 0-918102-15-4.
4. The Royal Society Proceedings A 1916–1917, 93, xlii–liv.
5. Who was who: 1916–1928.
6. A Life of Sir William Ramsay by Morris W. Travers FRS, Edward Arnold Publishers Ltd 1956.
7. Sir William Ramsay by Sir William A. Tilden FRS, MacMillan and Co. 1918.
8. A life of Sir William Ramsay by Morris W. Travers FRS, pub by Edward Arnold Ltd 1956.
9. Thermophysical Properties of Neon, Argon, Krypton and Xenon (National Standard Reference Data Service of the USSR, Vol 10) by V. A. Rabinovich, A. A. Vassermna, V. I. Nedostup and L. S. Veksler Hemisphere Pub, 1988, ISBN: 0891166750.
10. Sackler database, Royal Society.

1905

© Nobel Foundation [1]

Johann Friedrich Wilhelm Adolf von Baeyer

"in recognition of his services in the advancement of organic chemistry and the chemical industry, through his work on organic dyes and hydroaromatic compounds" [2]

Adolf von Baeyer is the only Chemistry Nobel Laureate to date, to be prevented from being present in Stockholm to receive the prize in person through ill health, and so it was conveyed to him through His Excellency the German Ambassador. Even though von Baeyer was unable to attend, Professor A. Lindstedt (then President of the Royal Swedish Academy of Sciences) read his statement on 10th December 1905 during the Nobel Prize ceremony. His comments regarding the impact of Baeyer's work concerning the structure elucidation and first synthesis of indigo (which led others to develop a commercial synthesis of indigo) were [2]:

> *...the result is striking. Already the price of indigo has fallen to a third of its former price, and Germany's export of synthetic indigo in 1904 could be valued at over 25 million Marks...*

Although awarded the Nobel Prize in 1905 (having been proposed by R. Behrend, E. Erlenmeyer, R. Nietzki, H. Ost, H. Rupe, K. Seubert, J. Thiele, T.W. Richards, E. Schaer and J. Volhard [3]) Baeyer was also nominated in the years 1902, 1903 and 1904.

Baeyer was born in Berlin (Prussia) on 31st October 1835 to Johann Jakob Baeyer (a captain in the Prussian Army) and Eugenie (daughter of Julius Eduard Hitzig, Director of the Criminal Court in Berlin). Baeyer and his wife Lida (daughter of Geheimrat Bendemann, Berlin) married in 1868. They had one daughter, Eugenie, (married Oskar Piloty who became Professor of Chemistry in Munich and fell in the war in 1915) and three sons. Sadly one son died in infancy. The other two Hans (born 1875) and Otto (born 1877) went on to chairs in their own right. Hans to the Chair of Orthopaedic surgery in Heidelberg and Otto to the Chair of Physics in Berlin [2].

The Early Years

Bayer's interest in chemistry developed at a young age. Indeed his father gave him a copy of Stockhardt's "School of Chemistry" at the age of 9. Baeyer built a laboratory in a small room in his parents house and worked through every experiment. He made his first new substance aged 12, which was later investigated by Struve and Gentele and found to be $CuCO_3$,Na_2CO_3,$3H_2O$, a double salt of copper and sodium carbonates prepared by mixing copper sulphate with sodium carbonate and allowing to stand the solution for several days.

Baeyer was given a two thaler piece on his 13th birthday with which he bought a sample of indigo, an unusual purchase for a 13 year old. The fascination with indigo remained until he had solved both the structure and the synthesis, albeit many years later. However, Baeyer's interest in the sciences was much broader than chemistry. He seriously considered reading mathematics and physics at university through the influence of K. Schellbach whom Baeyer had met when 13 years old.

At age 15, however, his spare time was devoted to the natural sciences of botany and geology. Two years later (aged 17) he undertook a six month walk through Dalmatia and Montenegro with Ferdinand von Triest, the main object for which was the collection of material for his studies. He returned to his first scientific endeavours and in 1856 found himself a pupil of Robert Bunsen. His contemporaries at the time included Roscoe, Lothar Meyer, Lieben and Beilstein.

The Career

Baeyer's professional career started in 1860 in Berlin where he started to assemble a group containing some of the most notable scientists of the day. The ability to attract the best available students continued during his next two chairs, most notably upon the arrival of Emil Fischer whilst Baeyer was at Strassburg.

Year	*Post*	*Institute*
1856	Student	Worked under Bunsen in Heidelberg
1857	Student	Worked under Kekule in Heidelberg
1858/9 (winter) followed Kekule to Ghent	Student	Worked under Kekule in Ghent
1858	Student	Doctorate (University of Berlin) for work on cacodyl compounds under Kekule
1860	Privatdozent followed by teacher	Gewerbe-Akademie in the Klosterstrasse, Berlin, Graebe; Liebermann and Victor Meyer joined von Baeyer at this time
1866	Senior lecturer (unpaid)	Also in Berlin
1870		Turned down the Chair of Chemistry in Konigsberg as successor to Werther
1872	Professor of Chemistry and Director	Strassburg. Emil and Otto Fischer joined the group at this time
1875	Professor	Munich – as successor to Liebig

Baeyer's first task during the Bunsen-Heidelberg years was to help Bunsen and Roscoe with their well known experiments on the combination of chlorine and hydrogen under the influence of light [4], and the second which was suggested by Bunsen, was a short account on methyl chloride [5]. On completion of these studies, Baeyer felt that he had moved away from his interests of theoretical chemistry, especially organic chemistry mainly because Bunsen's interests lay elsewhere.

Somewhat disillusioned with the technical direction of his work, Baeyer asked Kekule if he could join him in a small laboratory in Heidelberg as Kekule's first pupil. His doctorate thesis, submitted in Berlin in 1858, was written and presented in Latin, entitled "*De arsenici cum methylo conjunctionibus*".

The Ghent Period

Shortly after starting to work for Kekule in Heidelberg, Kekule decided to move to Ghent. So it was then that Baeyer decided to follow Kekule. It proved to be an interesting journey for the young Baeyer as he met Adolf Schlieper on the way (in the winter of 1858/9). Schlieper gave Baeyer a box containing some compounds which Schlieper had used in his studies of uric acid. Whilst thoroughly examining the box left to him by Schlieper, Baeyer also completed some work on the esters of allophonic acid (begun in Heidelberg). Baeyer subsequently published some work on pseudo-uric acid in conjunction with Schlieper. It was to be the start of Baeyer's interest in uric acid and its derivatives. The Ghent period was perhaps the least prolific period for publications partly because of the work that he had undertaken and partly interrupted by military service.

The Berlin Period

Whilst continuing to work on uric acid and its derivatives, the Baeyer group were active in a number of areas including:

- work on indigo in 1865 which led to the discovery of indole and to the partial synthesis of indigotin
- the reduction of the colouring matter of the madder plant to anthracene (Graebe and Liebermann)
- the synthesis of alizarin from coal tar anthracene (Graebe and Liebermann)
- the use of the zinc-dust distillation technique developed by von Baeyer, to clarify the structure of alizarin (once again Lieberman and Graebe)
- studies initiated on condensation reactions which led to the discovery of the phthaleins [6];
- carbon dioxide assimilation in formaldehyde
- and proposed a formula for grape sugar.

The Strassburg Period

W. H. Perkin Jr., a former student of Baeyer, recalled [7]...

> *...At that time the instruction of inorganic chemistry was in the able hands of Rose, who had recently come from Heidelberg, and he at once introduced the system of teaching, and especially analytical methods, of Bunsen...Emil Fischer commented ... how attractive these simple and rapid methods of analysis were in comparison with the long and tedious processes being taught elsewhere...*

Perhaps the most significant event during the early Strassburg period was indeed the arrival of Otto and Emil Fischer. Baeyer asked Emil Fischer to join in on the investigations of the condensation products of aldehydes with aromatic hydrocarbons and phenols. Such was Fischer's success that he soon published his first paper on the subject [8]. Baeyer's fame at the time attracted so many students that it soon became clear that the available laboratory space was insufficient for the group's needs. A new building was commissioned with the laboratories opened the following year. Unfortunately for the authorities, Baeyer was invited to the Chair in Munich as Liebig's successor, which after due deliberation he accepted. Perhaps the most significant original work to originate from the Strassburg period was the first preparation of nitroso-dimethylaniline and its conversion to nitrosophenol, which was published by Baeyer and Caro [9,10].

The Munich Period

The Munich period was to be Baeyer's golden era. During this period work was brought to successful conclusions in so many areas. Amongst them were:

- Studies concerning acetylenes and polyacetylenes
- the Baeyer strain theory
- studies on the constitution of benzene
- the Baeyer-Villiger oxidation of ketones by means of per acids
- the mechanism of the condensation of phthalic anhydride with phenols was clearly established
- studies on the connection between constitution and colour were undertaken
- finally, the total synthesis of indigo was achieved by four methods
    - the conversion of isatin chloride into indigotin by reduction with hydriodic acid and phosphorus
    - the use of o-nitrophenylpropiolic acid by the action of alkaline reducing agents
    - the use of o-nitrobenzaldehyde and acetone (the Baeyer-Drewson synthesis published in 1882), a schematic of which is given below
    - the use of di[-o-nitrophenylacetylene] also published in 1882

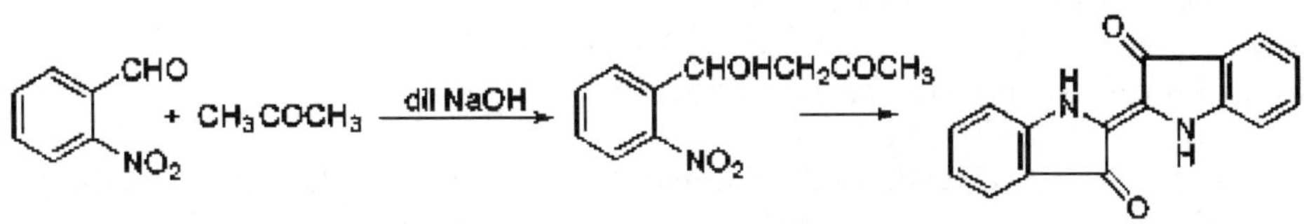

The Baeyer Drewson synthesis of Indigo

Writing to Chakravarti in February 1973, Sir Robert Robinson remarked [[11], page 184].

> *...I wonder whether chemistry would have been richer or poorer if Baeyer had been able to put indigo into a machine and get the structure right away; we should probably know nothing about the isatins and numerous important synthetic reactions, and so on. It would have been a serious loss, and that is what will happen in the future if the disciples of the physical methods of end-all and be-all in organic chemistry get their way...*

As if the above accomplishments were insufficient, Baeyer published work (between 1884 and 1893) conerning:

- the synthesis of hydrindenecarboxylic acid
- the synthesis of tetrahydronaphthalenedicarboxylic acid
- the synthesis of naphthalene
- a discussion of the synthesis of acetoacetic ester
- the synthesis of phloroglucinol
- the conversion of phloroglucinol into trioxome
- a study of the properties of succinylsuccinic ester
- the centric formula for benzene (which had also been put forward by Armstrong)
- the synthesis of p-diketohexamethylene and its conversion into p-dihydroxyhexamethylene (chinitol)
- a systematic investigation into the reduction of benzene derivatives in particular the reduction of the phthalic acids
- a study of succinylsuccinic ester led to the first synthesis of a terpene
- studies into cyclic terpenes
- an investigation concerning organic peroxides and oxonium compounds
- and a series of papers investigating Caro's acid

Baeyer published his last paper in 1915 when he was 80 years old, which was a discusssion concerning the reduction of pyrones [12].

Von Baeyer will forever be remembered for his work on indigo, which is still a cornerstone of dye production, see for example [15–17].

Honours and Awards

As one might imagine, many awards; honours and milestones were bestowed upon von Baeyer during the course of his long and illustrious career, including:

- Founder of the German Chemical Society (1867)
- Davy Medal of the Royal Society (1881)
- Hereditary nobility by King Ludwig II of Bavaria (1885)
- Elected Foreign Member of the Royal Society (10/12/1885)
- Nobel Prize for Chemistry (1905)
- Carl Duisberg donated funds to the German Chemical Society thereby establishing the Adolf Baeyer Denkmuenze – a prize awarded from time to time ever since

A more comprehensive list of these milestones appears in 'Erinnerungen aus meinem Leben' which Baeyer wrote as an introduction to the two volumes of his collected papers which were published in 1905 on the occasion of his 70th birthday.

There were, however, some less pleasant milestones. The Chemical Society of Great Britain, the for-runner of The Royal Society of Chemistry, published the following...

> *"The Chemical Society considers that it is neither compatible nor consistent with its loyalty to the Crown, whence the Royal Charter under which it works was derived, to retain any alien enemies upon its List of Honorary and Foreign Members.*

"It is therefore resolved that the names of A. von Baeyer, T. Curtius, E. Fischer, C. Graebe, P. H. von Groth, W. Nernst, W. Ostwald, 0. Wallach, and R. Willstätter, who were elected under happier conditions in recognition of their eminent services to Chemical Science—for which the Society still retains an undimindished appreciation and regard, are hereby removed from the list of Honorary and Foreign Members."

This extract was taken from J. Chem. Soc. Trans. [13] and was written in 1917 during the First World War and would have been a significant blow to Baeyer. Baeyer died a few months after he was removed as a Fellow of the Chemical Society and so was never re-instated.

Von Baeyer's citation for election (date: 10/12/1885) as Foreign Member of the Royal Society reads [18]:

"Professor Adolf Baeyer of Munich, distinguished for his researches in Organic Chemistry, is recommended by us as a proper person to be placed on the List of Foreign members of the Royal Society"

He was proposed by:

Thomas H. Huxley; John Evans; M. Foster; G. G. Stokes; Robt Etheridge; Hugo Muller; W. H. M. Christie.

Bibliography

1. Reproduced with permission.
2. From Nobel Lectures, Chemistry 1922–1941, Elsevier Publishing Company, Amsterdam, 1966.
3. The Nobel Population 1901–1937, E. Crawford, J.L. Heilbron and R. Ullrich. Office of the History of Science and Technology, University of California, 1987, ISBN: 0-918102-15-4.
4. A. Baeyer, *Annalen*, 1857, **103**, 178.
5. A. Baeyer, *Annalen*, 1857, **103**, 181.
6. A. Baeyer, *Annalen*, 1866, **140**, 297.
7. W. H. Perkin Jr., *J. Chem. Soc. Trans*, 1923, **123**, 1520–1546.
8. E. Fischer, *Chem. Ber*, 1874, **7**, 1211.
9. A. Baeyer and H. Caro, *Chem. Ber*, 1874, **7**, 809.
10. A. Baeyer and H. Caro, *Chem. Ber*, 1874, **7**, 963.
11. Robert Robinson Chemist Extraordinary by Trevor Williams, Clarendon Press, 1990, ISBN: 0-19-858180-7.
12. A. Baeyer and J. Piccard, *Annalen*, 1915, **407**, 332.
13. *J. Chem. Soc. Trans.*, 1917, **111**, 273–287.
14. A Life of Magic Chemistry, G. Olah, Wiley Interscience, 2001, ISBN: 0-471-15743-0.
15. Indigo from Seed to Dye by Dorothy Miller, Indigo Pr., 1982, ISBN: 0960406018.
16. A Handbook of Indigo Dyeing by Vivien Prideaux, Search Press, 2004, ISBN: 0855329769.
17. Indigo by Jenny Balfour-Paul (Editor), Fitzroy Dearborn Publishers, 2000, ISBN: 1579583067.
18. Sackler database, Royal Society.

1906

Henri Moissan

"in recognition of the great services rendered by him in his investigation and isolation of the element fluorine, and for the adoption in the service of science of the electric furnace called after him" [2]

Professor P Klason, President of the Royal Swedish Academy of Sciences, observed the following during his presentation speech on 10th December 1906 [2]:

> *...The whole world has admired the great experimental skill with which you have isolated and studied fluorine-that savage beast among the elements. With the aid of your electric furnace you have solved the riddle of how diamonds are formed in nature. You have unleashed a mighty wave into the world of technology, a wave which has not yet attained its full height...*

Moissan was nominated for the 1906 Chemistry Prize by A von Baeyer, V Czerny, G Darboux, L Henry, P Klason, P Lenard, W Ramsay and E Rosenberg. He was also nominated for the years 1901, 1902, 1903, 1904, 1905 and 1907. The latter nomination was disallowed [3].

Moissan was born in Paris on the 28th September 1852. His father was from Toulouse and his mother (nee Mitelle) was from Orleans. He died suddenly in Paris on 20th February 1907, shortly after his return from the prize-giving ceremony in Stockholm [2,4].

The Early Years

His advanced education began in the Collège de Meaux (1872) and later in Edmond Frémy's laboratory at the Musée d'Histoire Naturelle, where he attended lectures by E H Sainte-Claire Deville and Henri Debray. A year later, he moved to Dehérain's laboratory in the École Pratique des Hautes Études and subsequently he directed a small laboratory of his own, before joining Debray and Troost in the laboratories of the Sorbonne. He graduated with the degree of Bachelier in 1874 and

Licencié in 1877. Moissan became "Pharmacien de Première Classe" (1879) and "Docteur ès Sciences Physiques" (1880) with a thesis on the cyanogen series.

The Career

Moissan's career appointments are listed in the table.

Year	*Post*	*Institute*
1879	Répétiteur de Physique	Agronomic Institute
1880	Assistant lecturer	School of Pharmacy
1883	Senior demonstrator	School of Pharmacy
1886	Professor of Toxicology	School of Pharmacy
1899	Professor of Inorganic Chemistry	School of Pharmacy
1900	Assessor to the Director and Professor of Inorganic Chemistry	University of Paris
1900	Honorary Professor	School of Pharmacy

Moissan's initial experiments concerned some aspects of biology (the interchange of oxygen and carbon dioxide in plant leaves) and iron group oxides. From 1884 onwards his two principal research interests were the chemistry of fluorine and the development and use of the Moissan furnace, both areas of work being undertaken in Paris. It is, therefore, appropriate to group Moissan's work by subject matter rather than by location.

Fluorine and its Derivatives

Moissan's initial experiment was the preparation of phosphorous fluoride by heating copper phosphide with lead fluoride. He prepared pure arsenic by first preparing arsenic fluoride by reacting arsenious oxide, sulphuric acid and calcium fluoride followed by electrolysis. He obtained some organic fluorides (C_2H_5F) by reacting ethyl fluoride with silver fluoride. Similar work followed creating methyl and isobutyl fluorides.

He first discovered fluorine by discovering that KF.3HF remains liquid at –23°C and conducts electricity electrolytically. Further experiments continued including:

The electrolysis of arsenious fluoride;
The preparation of SF_6, PF_3, PF_5, BrF_3, SiF_4 and BF_3;
Attempts to react fluorine with argon and helium were unsuccessful.

The Furnace

Moissan's furnace consisted of a rectangular block of lime which had a hole scooped out of the middle. Carbon electrodes were aligned to the centre of the furnace and a block was used as a cover. The early experiments used a current of 100 to 125 amps at 50 to 60 volts applied to generate the arc. During its operation white hot lime dust was generated, which came out of the furnace through the electrode holes. Later versions of the furnace were constructed out of limestone as a cost saving measure. For some experiments Moissan used an inclined tube feeding in materials such as chromium oxide. This produced a continuous flow of the pure metal.

Moissan produced the following materials using these and other techniques:

- Crystallized lime
- Strontium
- Distilled copper, silver, platinum, tin, gold, iron and uranium
- Volatized carbon and silicon
- Many different carbides
- Small artificial diamonds
- Carborundum
- Metallic chromium, manganese, molybdenum, tungsten, vanadium, zirconium and titanium
- Carbides of the following metals lithium, calcium, barium, strontium, cerium, lanthanum, yttrium, thorium, aluminium, manganese and uranium
- The preparation on calcium by heating sodium with calcium iodide
- Metallic niobium and tantalum
- Silicon borides
- A lithium silicide Li_6Si_2

Combining his research interests of fluorine with the electric furnace, Moissan was able to prepare:

Thionyl fluoride, SOF_2 and sulphuryl chloride SO_2F_2.

Moissan is credited with over three hundred publications, his greatest works being *Le Four Électrique* (The electric-arc furnace) (1897), *Le Fluor et ses Composés* (Fluorine and its compounds) (1900) and *Traité de Chimie Minerale* (Treatise on Inorganic Chemistry) (five volumes 1904–1906).

No acceptance speech or lecture was given by Moissan at the award ceremony [3].

More details of the chemistry of fluorine can be found, for example in references [5–8].

Honours and Awards

Moissan received many awards, honours and memberships/fellowships in the period leading up to the award of the Nobel Prize, but he sadly died within a few months of receiving it. The following were all awarded prior to his visit to Stockholm.

Awards and medals:

The Institute awarded him the Prix Lacaze (1887), Davy medallist (1896), Hofmann medallist (1903). He obtained honours from the Franklin Institute of Philadelphia, the Société d'Encouragement pour l'Industrie Nationale, and the Société Industrielle dii Nord de la France, and Nobel Prize for Chemistry (1906).

Memberships/Fellowships:

Moissan was elected to Commandeur de la Légion d'Honneur, was member of the Académie de Médecine (1888), Académie des Sciences (1891), Conseil d'Hygiène de la Seine (1895), and the Comité Consultatif des Arts et Manufactures (1898). He was honoured by the Franklin Institute of Philadelphia and awarded Fellowships of the Royal Society of London and the Chemical Society (London), Honorary Member of the Royal Institution and of the Academies of Denmark, Vienna, Belgium, Uppsala, Haarleni, Amsterdam, New York, and Turin, besides numerous others.

Moissan's citation for election (date: 11/05/1905) as Foreign Member of the Royal Society reads [9]:

"Professor Henri Moissan, of Paris, distinguished as having isolated Fluorine, and for his application of the Electric Furnace to Chemistry"

He was proposed by:
William Huggins; William A Tilden; Percy F Frankland.

Bibliography

1. Reproduced with permission.
2. From Nobel Lectures, Chemistry 1922–1941, Elsevier Publishing Company, Amsterdam, 1966.
3. The Nobel Population 1901–1937, E. Crawford, J.L. Heilbron and R. Ullrich. Office of the History of Science and Technology, University of California, 1987, ISBN: 0-918102-15-4.
4. Proc. Roy. Soc. Series A, 1907–1908, vol. 80, pp. xxx–xxxvii.
5. Halogen chemistry volume 1,2 and 3, ed V. Gutmann, Academic Press, 1967, Library of Congress Card Number 66-30147.
6. Chemistry of the Elements 2nd ed, N.N. Greenwood and A. Earnshaw, Reed International and Professional Publishing Ltd., 1997, ISBN: 0-7506-3365-4.
7. Inorganic Fluorine Chemistry: Toward the 21st Century (Acs Symposium, No 555) Joseph S. Thrasher, Steven H. Strauss, An American Chemical Society Publication, 1994, ISBN: 0841228698.
8. Fluorine in Organic Chemistry by R. D. Chambers, OTHER, 2004, ISBN: 0849317908.
9. Sackler database, Royal Society.

1907

© Nobel Foundation [1]

Eduard Buchner

"for his biochemical researches and his discovery of cell-free fermentation" [2]

The abstract of the presentation speech by Professor the Count KAH Mörner [2], President of the Royal Swedish Academy of Sciences, dated 10th December 1907, given below, was not delivered orally. The presentation ceremony was cancelled owing to the decease of King Oscar II two days earlier,

...By Buchner's work, the fermentation mentioned and various other processes analogous to it have been freed from the shackles which previously held them and which prevented any progress in research. Now, no special difficulty is encountered in obtaining from yeast cells and various other cells an ample amount of powerfully active cell substance which is free from live cells. Numerous clarifying investigations into its properties have also been made, partly by Buchner himself and partly by others. Hitherto inaccessible territories have now been brought into the field of chemical research, and vast new prospects have now been opened up to chemical science...

Such was the importance of Buchner's work concerning fermentation that Professor Tiselius commented on his work during his address at the award ceremony of 1946. At that time (1946) Professor A tiselius was a member of the Nobel Committee for Chemistry of the Royal Swedish Academy of Sciences:

...In 1897 Eduard Buchner, the German research worker, discovered that sugar can be made to ferment, not only with ordinary yeast, but also with the help of the expressed juices of yeast which contain none of the cells of the Saccharomyces. The discovery was considered so important that in 1907 Buchner was awarded the Nobel Prize for Chemistry...

Buchner was nominated for the Chemistry Prize in 1907 by H von Euler-Chelpin and C Harries, having previously been nominated in 1905 [2].

Buchner was born in Munich on 20th May 1860, the son of Dr. Ernst Buchner, Professor Extraordinary of Forensic Medicine and Physician at the University, and Friederike *née* Martin [3].

He married Lotte Stahl in 1900. He was wounded on 3rd August 1917, whilst serving as a Major in a field hospital at Folkschani in Roumania. He died of the wounds that he received in action in Munich on the 12th of the same month.

The Early Years

Buchner matriculated at the Grammar School in his birth place and after a short period of study at the Munich Polytechnic in the chemical laboratory of E Erlenmeyer senior, he started work in a preserve and canning factory, with which he later moved to Mombach on Mainz.

He was originally destined for a commercial career but, after the early death of his father in 1872, his older brother Hans, ten years his senior, made it possible for him to take a more general education. Buchner's two principal interests were chemistry and botany. He studied the former with Adolf von Baeyer and the latter with Professor C von Naegeli at the Botanic Institute, Munich. Whilst with Professor C von Naegeli, he studied under the special supervision of his brother Hans (who later became well-known as a bacteriologist). Indeed his first publication arose from that work *Der Einfluss des Sauerstoffs auf Gärungen* (The influence of Oxygen on Fermentations) in 1885. Buchner was awarded a three year Lamont Scholarship by the Philosophical Faculty which made it possible for him to continue his studies.

The Career

After one term in Erlangen in the laboratory of Otto Fischer, Buchner took his Doctor's degree in the University of Munich (1888). His subsequent academic posts are detailed in the table.

Year	*Post*	*Institute*
1889	Assistant lecturer	University of Munich
1891	Lecturer	University of Munich
1893	Analytical department supervisor	University of Kiel
1895	Professor	University of Kiel
1896	Professor Extraordinary for Analytical and Pharmaceutical Chemistry	University of Tübingen.
1898	Chair of General Chemistry, he also held lectureships on agricultural chemistry and agricultural chemical experiments as well as on the fermentation questions of the sugar industry	Agricultural College in Berlin
1909	Professor	University of Breslau
1911	Professor	University of Würzburg

Buchner published his discoveries on the alcoholic fermentation of sugar in collaboration with his brother Professor Hans Buchner, and Martin Hahn *Die Zymasegärung* (Zymosis), in 1903.

Buchner's Nobel Prize acceptance speech was entitled "Cell-Free Fermentation" [3]. During his presentation, Buchner observed...

...If we now summarise the results of all these experiments, we establish that a separation of the fermentation effect from the live yeast cells can be carried out. To start off a fermentation process, no such complicated apparatus is needed, as the yeast cell is; rather is there a 'cell-free fermentation'. But even the assumption that the action of the expressed juice is attributable to the presence of still-living splinters of protoplasm must be regarded as refuted, since these hypothetical structures would need to be so small that they would pass through a Chamberland biscuit-candle and, although deprived of the protection of a cell membrane, survive attacks which kill all micro-organisms. The active agent in the expressed yeast juice appears rather to be a chemical substance, an enzyme, which I have called 'zymase'. From now on one can experiment with this just as with other chemicals...

...There are now two directions in which my colleagues and I have occupied ourselves particularly in recent years. On the one hand it was obvious that we should strive for a preparation of zymase, and a separation from the proteins and the rest of the enzymes of the expressed juice. Experiments in this direction did not meet with any success worth mentioning, however, since all the separation processes used up to now had principally led merely to the destruction of a large part of the fermentation power. Even an enrichment of the live yeast, and hence the expressed juice, with zymase succeeded only within narrow limits.

On the other hand it appeared urgently necessary to follow up most care the chemical processes in cell–free fermentation.

Fermentation technology and the underlying principles are better understood now than during the early part of the 20th century. Reference [4] outlines some of the industrial applications in current use.

Honours and Awards

He was awarded the Nobel Prize in 1907 for his biochemical investigations and his discovery of non-cellular fermentation.

Bibliography

1. Reproduced with permission.
2. The Nobel Population 1901–1937, E. Crawford, J.L. Heilbron and R. Ullrich. Office of the History of Science and Technology, University of California, 1987, ISBN: 0-918102-15-4.
3. From Nobel Lectures, Chemistry 1901–1921, Elsevier Publishing Company, Amsterdam, 1966.
4. Principles of Fermentation Technology by P. F. Stanbury, A. Whitaker and S. Hall Butterworth-Heinemann, 2nd ed. 1999, ISBN: 0750645016.

1908

Ernest Rutherford

"for his investigations into the disintegration of the elements, and the chemistry of radioactive substances" [2]

Professor K. B. Hasselberg, President of the Royal Academy of Sciences, observed the following during his presentation speech on 10th December 1908 [2]:

> *. . .The above-mentioned disintegration theory and the experimental results upon which it is based, are synonymous with a new departure in chemistry, involving a fresh and decidedly extended comprehension of the very basis of that science. To the chemists of the 19th century the atom and the element represented each in its sphere the uttermost limit of chemical subdivision or disintegration, and at the same time the point beyond which it was impossible for experimental investigation to proceed. If it were queried what there was beyond, nothing but more or less vague and fruitless speculations were forthcoming. . .*

Rutherford was nominated for the Chemistry Prize in 1908 by S. Arrhenius, R Wegscheider and O Widman. He was also nominated for the chemistry prize in 1907 and 1923 and for the physics prize for 1907, 1908, 1909 (disallowed), 1922, 1923, 1924, 1931, 1932, 1933, 1935, and 1937 [3].

Ernest Rutherford was born on 30th August 1871, in Spring Grove, Brightwater, Nelson, New Zealand, the fourth child and second son, in a family of seven sons and five daughters. Two of his brothers died in a swimming accident. His father, James Rutherford, a Scottish wheelwright, engineer and flax miller, emigrated to New Zealand with Ernest's grandfather and the whole family in 1842. His mother, née Martha Thompson, was an English schoolteacher, who, with her widowed mother, also went to live there in 1855. Rutherford married Mary Newton, only daughter of Arthur and Mary de Renzy Newton, in 1900. Their only child, Eileen was born in 1901, and marry the physicist R.H. Fowler FRS in 1925. She died of an embolism at the age of 29, nine days after the birth of her fourth child and just two days before Christmas of 1930, shortly before Rutherford's peerage was announced.

Rutherford died in Cambridge on 19th October 1937, died aged 66, as a result of delays in operating on his partially strangulated umbilical hernia. His ashes were buried in the Nave of Westminster Abbey, just west of Sir Isaac Newton's tomb and beside that of Lord Kelvin [2, 4].

The Early Years

Rutherford obtained his first science book at the age of 10. It suggested an experiment that involved using the speed of sound to determine the distance that could be reached by firing a cannon. Rutherford attempted to make a cannon from a hat peg, a marble and blasting powder. Fortunately, the exploding cannon caused no injury.

Rutherford boarded at Nelson College and was head boy in 1889 (the Dux of the school, hence his nick-name 'quacks'). He obtained a scholarship (on the second attempt) to assist attendance at Canterbury College of the University of New Zealand (1890–1894).

He obtained B.A. in 1892 (Pure Mathematics and Latin (both compulsory), Applied Mathematics, English, French and Physics), M.A. in 1893 (double First Class Honours, in Mathematics and Mathematical Physics and in Physical Science (Electricity and Magnetism)) and B.Sc. in 1894 [6] and a D.Sc. degree of the University of New Zealand (1901).

Rutherford's initial researches in New Zealand, were concerned with the magnetic properties of iron exposed to high-frequency oscillations. This work resulted in his thesis which was entitled *Magnetization of Iron by High-Frequency Discharges*. His second paper, *Magnetic Viscosity*, was published in the *Transactions of the New Zealand Institute* (1895) and contains a description of a time-apparatus capable of measuring time intervals of a hundred-thousandth of a second.

He was awarded the 1851 Scholarship in 1895 which enabled him to go to Trinity College, Cambridge, as a research student at the Cavendish Laboratory under J. J. Thomson.

Rutherford invented a detector for electromagnetic waves, an essential feature being a magnetizing coil containing tiny bundles of magnetized iron wire. He worked with Thomson on the behaviour of the ions observed in gases which had been treated with X-rays (1897) and on the mobility of ions in relation to the strength of the electric field (also 1897), and on related topics such as the photoelectric effect. Rutherford reported the existence of alpha and beta rays in uranium radiation (1898).

The Career

Rutherford departed the Cavendish laboratory to take his first chair, in Canada, the details for which appear in the table below.

Year	*Post*	*Institute*
1898	Macdonald Chair of Physics	McGill University, Montreal, Canada
1907	Langworthy Professor of Physics	University of Manchester, England
1919	Cavendish Professor of Experimental Physics	Cambridge University, England

In addition to the above posts, Rutherford became Chairman of the Advisory Council, H.M. Government, Department of Scientific and Industrial Research, Professor of Natural Philosophy, Royal Institution, London, and Director of the Royal Society Mond Laboratory, Cambridge (1936).

The Montreal Period

During his time in Canada, Rutherford worked on radioactive bodies, particularly:

- the emission of alpha rays
- the "emanation" of thorium (with R. B. Owens)
- discovered a new noble gas, an isotope of radon, which was later to be known as thoron.
- "disintegration theory" of radioactivity (with Frederick Soddy)

The Manchester Period

Rutherford continued to work on various aspects of radiation and its effects including work concerning:

- the properties of the radium emanation and of the alpha particles
- a method of detecting a single alpha particle and counting the number emitted from radium (with H Geiger in 1910)
- investigations into the scattering of alpha rays and the nature of the atomic nucleus
- the use of cathode rays to bombard atoms of various elements (H. G. Moseley in 1913)
- discovered that the nuclei of certain light elements, such as nitrogen, could be "disintegrated" by the impact of energetic alpha particles coming from some radioactive source, and that during this process fast protons were emitted (Blackett later proved, with the cloud chamber, that the nitrogen in this process was actually transformed into an oxygen isotope)
- He first split the atom in 1917, but the work was not published until 1919

Niels Bohr joined him at Manchester (in 1912) and adapted Rutherford's nuclear structure to Max Planck's quantum theory and so obtained a theory of atomic structure which, with later improvements, mainly as a result of Heisenberg's concepts.

Rutherford worked on acoustic detection of submarines during World War I as part of the war effort.

The Cambridge Period

In many ways, Rutherford's second period in the Cavendish laboratories were, for Rutherford, a period of consolidation, but one of the better periods for the Cavendish laboratories. Rutherford was Director of the Cavendish Laboratories whilst:

- James Chadwick discovered the neutron (1932)
- John Cockcroft and Ernest Walton split the atom using protons (1932)
- Mark Oliphant (at Rutherford's suggestion) built a lower voltage accelerator but with a much improved particle flux. Following the gift of dueterium (from Gilbert Lewis of Berkeley), they bombarded dueterium with deuterium and discovered tritium (H^3 the third isotope of hydrogen) and the light isotope of helium (He^3)

Rutherford published several books: *Radioactivity* (1904); *Radioactive Transformations* (1906), *Radiation from Radioactive Substances*, with James Chadwick and C. D. Ellis (1919, 1930), *The Electrical*

Structure of Matter (1926), *The Artificial Transmutation of the Elements* (1933), *The Newer Alchemy* (1937).

Many books have been written about Rutherford see for example [7–18].

For his Nobel Prize acceptance speech, Rutherford commented upon "The Chemical Nature of the Alpha Particles from Radioactive Substances".

Rutherford was one of the pioneers of the study of radioactive materials, a subject which is still ongoing, see for example, references [19–22].

Honours and Awards

Rutherford was awarded about 32 honorary doctorates including L.L.D. those from the Universities of Pennsylvania, Wisconsin, McGill, Birmingham, Edinburgh, Melbourne, Yale, Glasgow; an Honorary Ph.D. from Giessen University; Honorary D.Sc. degrees from Copenhagen, Cambridge, Dublin, Durham, Oxford, Liverpool, Toronto, Bristol, Cape Town, London and Leeds.

Amongst the 34 awards and medals that were presented to Rutherford were: the Rumford Medal (1905 some say 1904); Nobel Prize (Chemistry) 1908; the Bressa Prize (1910) of the Turin Academy of Science; the Copley Medal (1922); the Franklin Medal (1924) from the Franklin Institute; the Albert Medal (1928) of the Royal Society of Arts; the Faraday Medal (1930) of the Institution of Electrical Engineers; the Barnard Medal from Columbia University.

Many Fellowships were bestowed on Rutherford including: Fellow of the Royal Society in 1903 (Vice President 1922–1923; President from 1925 to 1930); Hon RSNZ (1904); elected Fellow of Trinity College, Cambridge (1919); FInstP (president 1931–1933). He was President of the British Association for the Advancement of Science (1922), Professor of Natural Philosophy at the Royal Institution, President of the Institute of Physics and an 1851 Exhibition Commissioner. President of the Academic Assistance Council to aid displaced academics (1933). In 1937 he was elected President of the twenty-fifth (silver jubilee) Meeting of the Indian Science Congress. His presidential address, prepared about two months before his untimely death, was read (January, 1938) by his successor, Sir James Jeans.

Rutherford was knighted (1914), appointed to the Order of Merit in 1925, and was created First Baron Rutherford of Nelson, New Zealand, and Cambridge (1931).

Rutherford has appeared on the stamps of four countries: Sweden (1968); Canada (1971); Russia (1971); New Zealand (1971 and 2000). In November of 1992 he featured on the new NZ$100 banknote. Because it is the highest denomination banknote, his image regularly appears as background on TV news items, and TV and newspaper advertisments, involved with finance.

He was Bakerian lecturer in 1904 and 1920.

Rutherford's citation for election (date: 11/06/1903) as Fellow of the Royal Society reads [23]:

> **'Professor of Experimental Physics, McGill University, Montreal. As one who has made discoveries in the science of Electricity, and as the author of the following papers:- 'A Magnetic Detector of Electromagnetic Waves and some of its Applications' (Phil Trans, 1897); 'On the Passage of Electricity through Gases exposed to Rontgen Rays' (with Prof J J Thomson) (Phil Mag, vol xlii, p 392); 'On the Electrification of Gases exposed to Rontgen Rays and the Absorption of Rontgen Rays by Gases and Vapours' (ibid, vol xliii, p 241); 'On a Radio-active Substance emitted by Thorium compounds' (ibid, vol xlix, p 1); 'On Radio-activity produced in a Substance by the action of Thorium compounds' (ibid, vol xlix, p 161); 'The Discharge of Electrification by Ultra-Violet Light' (Proc Camb Phil Soc, vol ix, p 401); 'Dependence of the Current through Conducting**

Gases on the Direction of the Electric Field' (Phil Mag, series VI, vol ii, p 210); 'Discharge of Electricity from Glowing Platium and the Velocity of the Ions' (Physical Review, vol xiii, p 321); 'On the Effect of Temperature on Radio-active Substances' (Physikal Zeits, 1901); 'On the Energy of Rontgen and Becquerel Rays' (with McClung) (Phil Trans, 1901)'.

He was proposed by:
J J Thomson; Rayleigh; O J Lodge; J H Poynting; [George Gabriel Stokes]; C T R Wilson.

Bibliography

1. Reproduced with permission.
2. From Nobel Lectures, Chemistry 1901–1921, Elsevier Publishing Company, Amsterdam, 1966.
3. The Nobel Population 1901–1937, E. Crawford, J. L. Heilbron and R. Ullrich. Office of the History of Science and Technology, University of California, 1987, ISBN: 0-918102-15-4.
4. Obituary Notices Fellows of the Royal Society 1936–1938, 2, 395–423.
5. Who was who: 1929–1940.
6. http://www.rutherford.org.nz/biography.htm.
7. Rutherford: Scientist supreme by John Campbell, AAS Publications, 1999, ISBN: 0-473-05700-X.
8. The life of Baron Rutherford of Nelson by Ivor B. N. Evans, pub Stanley Paul & Co. Ltd. 1939.
9. Rutherford being the life and letters of Rt. Hon. Lord Rutherford OM by A SA Eve FRS, pub Cambridge University Press 1939.
10. Rutherford: Recollections of the Cambridge days by Mark Oliphant, Elsevier Publishing Co, 1972, ISBN: 0-444-40968-8.
11. Rutherford and physics at the turn of the century ed Mario Bunge and William R. Shea, pub by Dawson and Science History Publications, 1979, ISBN: 0-7129-0918-4.
12. Ernest Rutherford: Atom Pioneer by John Rowland pub by T. Werner Laurie Limited 1955.
13. Ernest Rutherford by C. L. Boltz distributed by Heron Books, 1970.
14. Rutherford: Simple genius by David Wilson, Hodder and Stoughton, 1983, ISBN: 0-340-23805-4.
15. Ernest Rutherford and the atom by P. B. Moon, Priory Press Ltd, 1974, ISBN no: 85078 187 6.
16. Rutherford and the nature of the atom by E. N. da C. Andrade, Anchor Books, 1964.
17. Lord Rutherford by Norman Feather, Blackie and son Ltd, 1940.
18. Rutherford at Manchester ed J. B. Birks, pub Heywood and Co. Ltd., 1962.
19. Invisible Rays: A History of Radioactivity by G. I. Brown, Sutton Publishing, 2002, ISBN: 0750926678.
20. Practical Applications of Radioactivity and Nuclear Radiations by Gerhart Lowenthal and Peter Airey, Cambridge University Press, New ed. 2005, ISBN: 052101980X.
21. Radiochemistry and Nuclear Chemistry by Gregory Choppin, Jan Rydberg and Jan-Olov Liljenzin Butterworth-Heinemann, 3rd ed. 2001, ISBN: 0750674636.
22. Nuclear and Radiochemistry: Fundamentals and Applications by Karl Heinrich Lieser Wiley-VCH, 2nd ed. 2001, ISBN: 3527303170.
23. Sackler database, Royal Society.

1909

© Nobel Foundation [1]

Wilhelm Ostwald

"in recognition of his work on catalysis and for his investigations into the fundamental principles governing chemical equilibria and rates of reaction" [2]

Dr. H. Hildebrand, President of the Royal Academy of Sciences and former Rector General of National Antiquities, made the following observations during the presentation ceremony on 10th December 1909 [2]:

> *...Although the Nobel Prize for Chemistry is now being awarded to Professor Ostwald in recognition of his work on catalysis, he is a man to whom the chemical world is indebted also in other ways. By the spoken and the written word he, perhaps more than any other, has carried modern theories to a rapid victory and for several decennia he played a leading part in the field of general chemistry. In other ways too he has furthered chemistry by his versatile activity with numerous discoveries and refinements in both the experimental and the theoretical spheres...*

Ostwald was nominated for the 1909 Chemistry Nobel Prize by S Arrhenius, J H van't Hoff and G Bredig. Ostwald was also nominated for the Chemistry Prize in 1904, 1905, 1906, 1907, 1908, and 1910 (disallowed) and for the Physics Prize for 1920, 1921, 1923 and 1924 – all of the physics nominations were disallowed [3].

Wilhelm Ostwald was born on 2nd September 1853, in Riga, Latvia, the son of master-cooper Gottfried Wilhelm Ostwald and Elisabith Leuckel. He married Helene von Reyher in 1880. They had two daughters and three sons. Ostwald died at his country home near Leipzig on 3rd April 1932 [2, 4].

The Early Years

He was educated at the "Real gymnasium" and was admitted to Dorpat University in 1872 to study chemistry. Upon graduation, Ostwald obtained the post of Assistant at the Physics Institute under

Professor Arthur von Oettingen, and subsequently took a similar position in the Chemistry Laboratory under Carl Schmidt.

The Career

To many chemists, Ostwald is regarded as the founding father of physical chemistry as van't Hoff (Nobel Prize 1901), Arrhenius (Nobel Prize 1903), Nernst (Nobel Prize 1920), Tammann and Wislicenus were some of his pupils. His career began in 1877 and ended on his retirement in 1906 – see table.

Year	*Post*	*Institute*
1877	Lecturer (unpaid)	Dorpat University
1881	Professor of Chemistry	Polytechnicum in Riga
1887	Professor of Physical Chemistry	Leipzig University

His earliest experiments (1875) were concerned with problems of chemical affinity as they relate to the law of mass action of water. Of particular concern was the emphasis on electrochemistry and chemical dynamics.

With the exception of a brief period of one term (in 1904–1905) when he was first "Exchange Professor" at Harvard University, Cambridge (Mass.), Ostwald remained at Leipzig until his retirement. The two chairs ccupied by Ostwald resulted in involvement in different activities.

The Riga Period

Law of dilution named after him.

The Leipzig Period

Founded the "Deutsche Elektrochemische Gesellschaft" in 1894 (German Electrochemical Society) which in 1902 expanded to become the "Deutsche Bunsen-Gesellschaft für angewandte physikalische Chemie" (German Bunsen-Society for Applied Physical Chemistry).

Retirement

Ostwald studied and published work on philosophy, such as:

Der energetische Imperativ (*The Energetic Imperative*),
Moderne Naturphilosophie (*Modern Natural Philosophy*),
Die Pyramide der Wissenschaften (*The Pyramid of the Sciences*). He supported the middle-class pacifist movement, and was interested in educational reforms and in monism – see, for example, *Monistische Sonntagspredigten (Monistic Sunday Sermons)* and *Arbeiten zum Monismus (Works on Monism)*.

Ostwald studied colours and shapes, the goal for which was a scientific standardization for colours – see for example:

Die Farbenfibel (*The Colour Primer*)
Die Farbenlehre (*Colour Theory*)
Die Harmonie der Farben (*Harmony of the Colours*)
He also published a periodical *Die Farbe* (*Colour*)

Ostwald used most of his Nobel Prize money to finance an organization, Die Brücke ("The Bridge"), an "international institute for the organizing of intellectual work," which he founded in Munich with Karl Wilhelm Bührer and Adolf Saager (in June 1911). The aims of the The Bridge, (called "*The Organising of Intellectual Work*") was published in Esperanto and German (in 1911).

It advocated "the monographic principle" (hypertext), technical standards, the use of the Universal Decimal Classification, and the concept of a World Brain. The Bridge ended in 1913. One lasting legacy of his work, however, is the international standard for paper sizes (A4 *etc.*) [4].

Ostwald published numerous textbooks, including *Lehrbuch der allgemeinen Chemie (Textbook of General Chemistry)* in 1884. This was followed by *Grundriss der allgemeinen Chemie (Outline of General Chemistry)* in 1889 and *Hand- und Hilfsbuch zur Ausführung Physikalisch-Chemischer Messungen (Handbook and Manual for Physicochemical Measurements)* in 1893. Numerous other scientific works on analytical chemistry, electrochemistry, inorganic chemistry followed.

In 1902 Ostwald founded *Annalen der Naturphilosophie* and edited 14 volumes up to 1921, *Klassiker der exakten Wissenschaften* in 1889, of which some 250 volumes have been published. He also founded and edited the *Zeitschrift für physikalische Chemie* in 1887; Ostwald edited 100 volumes, up to 1922.

Ostwald's Nobel Prize acceptance speech was titled "On Catalysis". He comments [2]...

> *...It is not quite a year ago that I wrote in a different context: "I myself have no cause for complaint whatsoever about lack of recognition of the scientific work which I have been so happy to perform in the field of general chemistry; yet the purely intellectual achievement of conceiving catalytic phenomena as accelerations of possible processes which are in progress, whereby the whole vast field of catalysis was first opened to exact study, lay at the time so far outside general scientific thought that where the broader masses of scientists are concerned it is today still at the incubation stage, notwithstanding the passing of some twenty years. I presume that subsequently this advance will become so naturally part and parcel of the overall context of scientific thought that, compared with before, the gap will cease to be apparent at all and I shall be deprived of that measure of personal glory to which I am justly entitled by virtue of this advance in theory. Yet this will not upset me any further since in the meantime I have come to regard such matters as object studies in the reactions of the collective psyche."*
>
> *That I have fortunately been proved wrong in that prediction demonstrates how far I underestimated that as science progressively developed and as its nature and attributes became more and more familiar, mankind's appreciation and acceptance of scientific progress has steadily accelerated...*

Ostwald's work still forms a part of the modern literature – see for example [5, 6].

Honours and Awards

Ostwald received honorary doctorates from several universities in Germany, Great Britain and the USA.

He was made an honorary member of learned societies in Germany, Sweden, Norway, the Netherlands, Russia, Great Britain and the USA. In 1899, he was made a "Geheimrat " by the King of Saxony. Unfortunately, there were some side effects of the war in Europe.

> *"The Chemical Society considers that it is neither compatible nor consistent with its loyalty to the Crown, whence the Royal Charter under which it works was derived, to retain any alien enemies upon its List of Honorary and Foreign Members.*
>
> *"It is therefore resolved that the names of A. von Baeyer, T. Curtius, E. Fischer, C. Graebe, P. H. von Groth, W. Nernst, W. Ostwald, 0. Wallach, and R. Willstätter, who were elected under happier conditions in recognition of their eminent services to Chemical Science—for which the Society still retains an undimindished appreciation and regard, are hereby removed from the list of Honorary and Foreign Members."*

This extract was taken from J. Chem. Soc. Trans. [7] and was written in 1917 during the First World War.

Bibliography

1. Reproduced with permission.
2. From Nobel Lectures, Chemistry 1901–1921, Elsevier Publishing Company, Amsterdam, 1966.
3. The Nobel Population 1901–1937, E. Crawford, J. L. Heilbron and R. Ullrich. Office of the History of Science and Technology, University of California, 1987, ISBN: 0-918102-15-4.
4. http://www.tu-harburg.de/b/hapke/ostwklas.html.
5. Chemical Kinetics and Catalysis by Richard I. Masel, John Wiley and Sons Inc., 2001, ISBN: 0-471-24197-0.
6. Chemical Kinetics and Catalysis by R. A. van Santen and J. W. Niemantsverdriet, Plenum Press, 1995, ISBN: 0-306-45027-5.
7. J. Chem. Soc. Trans, 1917, 111, 273–287.

1910

© Nobel Foundation [1]

Otto Wallach

"in recognition of his services to organic chemistry and the chemical industry by his pioneer work in the field of alicyclic compounds" [2]

Professor O. Montelius, President of the Royal Swedish Academy of Sciences and the Rector General of National Antiquities, observed the following during his presentation speech on 10th December 1910 [2]:

> *...The Royal Academy of Sciences has awarded to you this year's Nobel Prize for Chemistry in recognition of the momentous services you rendered in the development of organic chemistry and the chemical industry by your pioneering work in the field of alicyclic compounds. Once again it has been proved that results obtained by scientific research, which at first seem to be solely of theoretical interest may actually be of great practical importance...*

Wallach was nominated by O. Aschan and B. Tollens for the 1910 Chemistry Nobel Prize and also in the years 1907 and 1908 [3].

Wallach was born on 27th March 1847, in Königsberg, Germany, the son of Gerhard Wallach (a high-ranking civil servant, who later became Auditor General at Potsdam) and his wife, Otillie *née* Thoma. Wallach remained a bachelor throughout his life and died on 26th February 1931 [2].

The Early Years

Wallach's early academic interests lay in history and art, as chemistry was not taught at the "Gymnasium" at Potsdam. He went to Göttingen to study chemistry with Wöhler, Fittig and Hübner (in 1867) but soon left for Berlin to study under A.W. Hofmann and G. Magnus (for one semester). After his return to Göttingen, he obtained his doctor's degree under Hübner (in 1869) after only five semesters of study. (At that time, working hours at the Wöhler laboratory were from 7 a.m. till 5 p.m., after which gas was turned off and some work had to be rounded-off under the light of privately bought candles.) His thesis dealt with the position isomers in the toluene series.

Wallach's academic career had some false starts. He was Assistant to H. Wichelhaus in Berlin (in 1869 and 1870), with whom he worked on the nitration of b-naphthol. He traveled to Bonn to work with Kekulé (Easter 1870) at Kekulé's suggestion. Unfortunately, Wallach had to leave Bonn for military service in the Franco-Prussian war later that year.

The Career

After the war, Wallach moved to Berlin and worked for the newly founded firm "Aktien-Gesellschaft für Anilin-Fabrikation" (later "Agfa"), but his fragile health could not cope with the noxious fumes of the factory and in 1872, he returned to Bonn.

Year	*Post*	*Institute*
1870	Privatdozent	University of Bonn
1876	Professor Extraordinary	University of Bonn
1879	Chair of Pharmacology	University of Bonn
1889	Professor of Chemistry and Director of the Chemical Institute	Göttingen
1915	Retirement	

The Bonn Period

During his period as Chair of Pharmacology, he discovered the iminochlorides by the action of phosphorus pentachloride on the acid amides. Kekulé drew his attention to the existence of an old forgotten cupboard full of bottles containing essential oils, and invited Wallach to make a study of the contents.

Wallach undertook a study of the $C_{10}H_{16}$ family which led to the correct classification of the various isomers. Until his study, this group was referred to by a multitude of names ranging from terpene to camphene, citrene, carvene, cinene, cajuputene, eucalyptine, hesperidine, *etc*. By using simple reagents such as hydrogen chloride and hydrogen bromide, he successfully characterized the differences between the isomers. A year later, he could establish that many of them were indeed identical. He published the results of his work in his book *Terpene und Campher* (in 1909), a volume of 600 pages.

Wallach's other work during this period included studies on:

- the conversion of chloral into dichloroacetic acid
- a series of studies on the amide chlorides, imide chlorides
- amidines, glyoxalines, *etc*

- azo dyes and diazo compounds
- and many others

The Göttingen Period

In 1889, he was appointed Victor Meyer's successor in Wöhler's Chair and Director of the Chemical Institute at Göttingen. As with many other workers during the war years, his group suffered losses. He retired in 1915 from his posts when six of his assistants were killed in action.

Wallach's Nobel acceptance speech was on alicyclic compounds. During the talk he observed...

...By suitably converting hydrocarbons into oxygenous compounds and, reciprocally, oxygenated substances into hydrocarbons, and by effectuating the transitions, it was practicable to find the connection between a very large number of substances occurring in essential oils and appertaining to the terpene group and to trace the genetic relationship between the various substances. It would be hopeless to try and attempt to discuss within the short time available the long series of experiments that were successfully carried out with this purpose in mind. I have given a short tabular of my experiments on pages 38–51 of my book Terpene und Campher" (Terpenes and Camphors).

Listing only a few examples here, I want to mention that I succeeded in converting the main component of the ordinary turpentine oil, namely pinene, $C_{10}H_{16}$*, into the odiferous constituent of caraway oil, namely car vone* $C_{10}H_{14}0$*, and then reconverting this into the isomers of pinene, i.e. limonene and terpinene. It was possible to convert this same pinene via the lilac-scented alcohol terpineol to eucalyptole,* $C_{10}H_{18}0$*, the main constituent of worm-seed oil (chenopodium oil) and of the oil from Eucalyptus glohulus, and to convert this in turn to l-limonene, and so on...*

Nowadays, however, artificial flower oils - such as rose-blossom oil, jasmine-flower oil, orange-blossom oil, are produced in excellent qualities on a scientific basis; these are not at all, or only slightly, inferior to the natural oils, but are not so exorbitantly expensive; an idea of the price of the natural oils may be given by the calculation that a single kilogramme of essential violet-blossom oil would cost 80,000 Mark and would require 33,000 kilogrammes of blooms.

The clever utilization for practical purposes of scientific results is the only cause of a highly flourishing essential oil industry in Germany, with the result that this northern country is now an important competitor in the production of perfumery to the climatically much more favoured Romanic countries (especially the South of France).

The present output of the German essential oil and artificial scent industry is estimated at 40-50 million Marks annually. According to their own information, the firm of Schimmel in Miltitz alone have a stock of material of an average value of 2 million Marks...

Wallach makes the point elsewhere that the total chemical industry in Germany doubled its production value in the period 1550–1895, whereas the essential oil industry quadrupled during the same period.

Some terpenes form the basis of many fragrances used in modern products. Their fragrances and chemistry are still widely reported, for example [4–6].

Honours and Awards

Wallach received honorary doctorates from Universities of Manchester, Leipzig and the Technological Institute of Braunschweig.

In addition to the Chemistry Nobel Prize (1910), Wallach received the Kaiserlicher Adlerorden III. Klasse (Imperial Order of the Eagle) in 1911, the Davy Medal in Gold and Silver in 1912, and the Königlicher Kronorden II. Klasse (Royal Order of the Crown) in 1915.

He was elected Honorary Fellowships of the Chemical Society (1908) and Honorary Member of the Verein Deutscher Chemiker (1912). Unfortunately, the Chemical Society later pronounced:

"The Chemical Society considers that it is neither compatible nor consistent with its loyalty to the Crown, whence the Royal Charter under which it works was derived, to retain any alien enemies upon its List of Honorary and Foreign Members.

"It is therefore resolved that the names of A. von Baeyer, T. Curtius, E. Fischer, C. Graebe, P. H. von Groth, W. Nernst, W. Ostwald, 0. Wallach, and R. Willstätter, who were elected under happier conditions in recognition of their eminent services to Chemical Science—for which the Society still retains an undimindished appreciation and regard, are hereby removed from the list of Honorary and Foreign Members."

This extract was taken from J. Chem. Soc. Trans. [7] and was written in 1917 during the First World War.

Bibliography

1. Reproduced with permission.
2. From Nobel Lectures, Chemistry 1901–1921, Elsevier Publishing Company, Amsterdam, 1966.
3. The Nobel Population 1901–1937, E. Crawford, J. L. Heilbron and R. Ullrich. Office of the History of Science and Technology, University of California, 1987, ISBN: 0-918102-15-4.
4. A fragrant introduction to terpenoid chemistry by Charles S. Sell, RSC, 2003, ISBN: 0-85404-681-X.
5. Progress in Terpene Chemistry ed D. Joulain, 1986, Editions Frontières, ISBN: 2-86332-045-9.
6. Chemistry of Terpenes and Terpenoids ed A.A. Newman, 1972, Academic Press Inc, ISBN: 0-12-517950-2.
7. J. Chem. Soc. Trans., 1917, 111, 273–287.

1911

© Nobel Foundation [1]

Marie Curie, née Sklodowska

"in recognition of her services to the advancement of chemistry by the discovery of the elements radium and polonium, by the isolation of radium and the study of the nature and compounds of this remarkable element" [2]

Dr. E.W. Dahlgren, Head Librarian of the National Library and President of the Royal Swedish Academy of Sciences, commented (on 10th December 1911) [2]:

...In 1903, the Swedish Academy of Sciences had the honour of conferring upon you the Nobel Prize for Physics for the part which you, together with your late husband, took in the momentous discovery of spontaneous radioactivity.

This year, the Academy has decided to award you the prize for Chemistry in recognition of the eminent services you have rendered to this science by your discovery of radium and polonium, by your description of the characteristics of radium and its isolation in the metallic state, and by your research into the compounds of this remarkable element.

During the eleven years in which Nobel Prizes have been awarded, this is the first time that the distinction has been conferred upon a previous prizewinner. I beg you, Madam, to see in this circumstance a proof of the importance which our Academy attaches to your most recent discoveries...

Curie was nominated for the 1911 Chemistry Nobel Prize by S. Arrhenius and G. Darboux, and for the Physics Prize in 1902 and 1903 (winner in 1903) [3].

Marie Curie, *née* Maria Sklodowska, was born in Warsaw on 7th November 1867. Both her parents were teachers. Curie had one elder sister, Bronya. She received a general education in local schools and some scientific training (physics and chemistry) from her father. She met Pierre Curie, Professor in the School of Physics, in 1894 and in July of the following year, they were married at the town hall at Sceaux, where Pierre's parents lived. The Curies had two daughters, Iréne, born in September 1897 (married Frédéric Joliot in 1926, they were joint recipients of the Nobel Prize for Chemistry in 1935) and Eve (married the American diplomat H.R. Labouisse. As Director of the United Nations' Children's Fund, he received on its behalf the Nobel Peace Prize in Oslo in 1965).

Pierre Curie was run over and killed by a horse-drawn wagon near the Pont Neuf in Paris (19th April 1906). Mme. Curie died in Savoy, France, after a short illness (leukemia), on 4th July 1934.

Marie and Pierre Curie's bodies were exhumed from their place of burial at Sceaux on 20th April 1995. Their bodies were laid to rest under the dome of the Panthéon. Marie Curie thus became the first woman to be accorded this mark of honour on her own merit.

The papers they left behind (stored at the Bibliothèque Nationale) give off pronounced radioactivity. Researchers wishing to study the three black notebooks, in which their work from December 1897 and the three following years is recorded, have to sign a certificate saying that you do so at your own risk [2, 4].

The Early Years

Marie was active in a students' revolutionary organization whilst in Warsaw, which at the time was in the part of Poland dominated by Russia. She left for Cracow, which at that time was under Austrian rule. From there she moved to the Sorbonne in Paris (aged 24) to continue her studies, obtaining Licenciateships in Physics and the Mathematical Sciences (1891). Curie attended lectures given by Marcel Brillouin, Paul Painlevé, Gabriel Lippmann, and Paul Appell. Two years later she took her degree in physics (1893) and the following year a degree in mathematics. Her goal was to take a teacher's diploma, which she passed coming first in her group (1896), and then to return to Poland. She never returned to work in Warsaw.

Following her marriage to Pierre Curie and the birth of her daughter Iréne, Curie decided to undertake a systematic investigation of 'uranium rays' for her doctorate. Curie's first breakthrough was the discovery that thorium produced the same rays as uranium, furthermore, the strength of the radiation was independent of the element under study but was dependent upon the quantity of the element being studied. Curie drew the conclusion that the ability to radiate was linked to the interior of the atom and not on the arrangement of the atoms in a molecule. Following a comprehensive study of the rest of the periodic table, Curie determined that only uranium and thorium gave off this radiation. Curie then obtained geological samples from museums and found that pitchblende was four to five times more active than uranium and postulated that it contained an element not previously discovered. Her isolation of one decigram of almost pure radium chloride from several tons of the original material allowed for the determination of the atomic weight of radium (225), the results of which appear in her doctoral thesis which she presented on 25th June 1903.

Whilst studying for her doctorate, Curie also had a part-time teaching post at the École Normale Supérieure de Sèvres for girls (from 1900).

The Career

Curie's illustrious career was concerned with the isolation, purification and study of radioactivity. Her career details are given in the table below.

Year	Post	Institute
Pre-1903	Head of the Physics Laboratory	Sorbonne, Paris
1906	Professor of General Physics	Faculty of Sciences, Sorbonne
1918	Director of the Curie Laboratory in the Radium Institute – completed in 1914	University of Paris

Amongst the highlights of Curie's work were:

- The isolation of polonium
- The development of methods for the separation of radium from radioactive residues
- Actively promoted the use of radium to alleviate suffering during World War I (assisted by her daughter, Iréne). Indeed, President Hoover of the United States presented her with a gift of $50,000, donated by American friends of science, to purchase radium for use in the laboratory in Warsaw (1929). She coined the term radioactivity.
- The award of half the Nobel Prize in Physics (with Pierre Curie) in 1903. Pierre explained that neither of them were able to come to Stockholm to receive the prize (in 1903). They could not get away because of their teaching obligations. He added, "Mme Curie has been ill this summer and is not yet completely recovered." They went to Stockholm in June 1905 when Pierre gave a Nobel lecture.
- The isolation of radium in metallic form (in collaboration with André Debierne)
- The international acclaim afforded the Radium Institute

No other family has achieved the honour of so many Nobel awards. Not only did Pierre and Marie share half of the Physics Prize in 1903 and Marie, the sole recipient of the Chemistry 1911, their daughter and son-in-law were awarded the Chemistry Prize in 1935.

The full effects of radiation were at the time unknown. The health of both Marie and Pierre Curie gave rise to concern. Their friends tried to make them work less. All their symptoms were ascribed to the drafty shed and to over-exertion. We now have a better understanding of the longer term effects of radiation exposure.

Curie's work is recorded in numerous papers in scientific journals and she is the author of *Recherches sur les Substances Radioactives (Investigations on Radioactive Substances)* (1904), *L'Isotopie et les Eléments Isotopes (Isotopy and Isotopic Elements)* and the classic *Traité de Radioactivité (Treatise on Radioactivity)* (1910).

Curie's Nobel Lecture dated December 11, 1911 and titled 'Radium and the New Concepts in Chemistry' provides insights concerning both the extraction and studies of radium and polonium, and develops her concept of radioactivity. Furthermore, Curie postulated the existence of other elements which at the time were unknown. She states [5]...

> *...I have so far considered radium and polonium only as chemical substances. I have shown how the fundamental hypothesis which states that radioactivity is an atomic property of the substance has led to the discovery of new chemical elements. I shall now describe how the scope of this hypothesis has been greatly enlarged by the considerations and experimental facts which resulted in establishing the theory of atomic radioactive transformations...*

Many books have been written about Curie's life and work, see for example [6–13].

Honours and Awards

Curie received many honorary science, medicine and law degrees.

Amongst her many awards and medals were the Nobel Prize for Physics (1903), the Davy Medal of the Royal Society in 1903 and, in 1921, President Harding of the United States, on behalf of the women of America, presented her with one gram of radium in recognition of her service to science, and the Nobel Prize for Chemistry in 1911.

She was a member of the Conseil du Physique Solvay from 1911 until her death, and since 1922, she had been a member of the Committee of Intellectual Co-operation of the League of Nations.

Bibliography

1. Reproduced with permission.
2. From Nobel Lectures, Chemistry 1901–1921, Elsevier Publishing Company, Amsterdam, 1966.
3. The Nobel Population 1901–1937, E. Crawford, J. L. Heilbron and R. Ullrich. Office of the History of Science and Technology, University of California, 1987, ISBN: 0-918102-15-4.
4. Who was who: 1929–1940.
5. Marie and Pierre Curie and the Discovery of Polonium and Radium by Nanny Fröman, Originally delivered as a lecture at the Royal Swedish Academy of Sciences in Stockholm, Sweden, on February 28, 1996 (Translation from Swedish to English by Nancy Marshall-Lundén).
6. Marie Curie by Eve Curie (her daughter), Windmill Press Surrey, 1930.
7. Marie Curie A Life by Susan Quinn, Mandarin, 1996.
8. Curie, Marie, *Pierre Curie* and *Autobiographical Notes*, The Macmillan Company, New York, 1923. Subsequently Marie Curie refused to authorize publication of her *Autobiographical Notes* in any other country.
9. Pflaum, Rosalynd, *Grand Obsession: Madame Curie and Her World*, Doubleday, New York, 1989.
10. Reid, Robert, Marie Curie, William Collins Sons & Co Ltd, London, 1974.
11. Nuclear and Radiochemistry: Fundamentals and Applications by Karl Heinrich Lieser Wiley-VCH, 2nd ed. 2001, ISBN: 3527303170.
12. Aqueous Chemistry Of Polonium by Sue A. Brown Albion/Horwood Pub Jan 1, 2006, ISBN: 1898563926.
13. Practical Applications of Radioactivity and Nuclear Radiations by Gerhart Lowenthal and Peter Airey, Cambridge University Press, New ed. 2005, ISBN: 052101980X.

1912

© Nobel Foundation [1]

Victor Grignard (0.5)

'for the discovery of the so-called Grignard reagent, which in recent years has greatly advanced the progress of organic chemistry" [2]

Professor H.G. Söderbaum, then President of the Royal Swedish Academy of Sciences, observed the following during the Nobel award ceremony on 10th December 1912 [2]:

...Professor Grignard's method is the culmination of a series of investigations, carried out with remarkable technical skill, which revealed that metallic magnesium in the presence of ether reacts on organic derivatives of chlorine, bromine and iodine by forming ether-soluble organic compounds of magnesium. These latter compounds in turn react extremely readily with a large number of other organic substances, so that carbon combines with carbon, i.e. forming a true organic synthesis.

The importance of Grignard's method to organic chemistry lies in the number of essential virtues that it possesses, namely simplicity of procedure, and therefore saving of time; low cost of materials employed; satisfactory yield and, most important of all, a very wide application. From this last point of view no method of organic synthesis superior to that of Grignard's is known, and indeed there is scarcely any sphere of organic chemistry outside its scope.

The services which this method has rendered to Science are great, but no less great are those which it can be expected to render in future. Shall we, for instance, ever learn how to produce artificially alkaloids or vegetable organic bases, which are so important in medicine but which are still so difficult to prepare and therefore so costly? If we do there is little doubt that this will only be with the aid of Grignard's reagent...

...Anyone who has been concerned with organic chemistry during recent years knows how greatly you have contributed to the advancement of this branch of science. The number of remarkable and important investigations which have been made possible by the reagent invented by and named after you is already most imposing, and it continues to increase every day. Your method has thus pushed

back appreciably the frontiers of our knowledge, of our ability to observe; and not only that, it is also opening up prospects of new conquests for our science...

Grignard was nominated for the 1912 Chemistry Prize by A. von Baeyer and O. Widman, and also for the 1906, 1909, 1910, and 1928 Chemistry Prize [3].

François Auguste Victor Grignard was born in Cherbourg on 6th May 1871. He married Augustine Marie Boulant (1910). They had one son, Roger and one daughter. Grignard died on 13th December 1935 [2,4].

The Early Years

Grignard attended schools in Cherbourg (1883–87), winning a scholarship to the École Normale Spécial at Cluny (1889). Two years after joining, the school was closed because of a dispute between supporters of the "classic" and "modern" methods of secondary education. Grignard joined the University of Lyons, where he was attached to the Faculté des Sciences. He left for military service in 1892. Returning to Lyon following the demobilization (towards the end of 1893), he gained the degree Licencié ès Sciences Mathématiques (1894).

Grignard accepted a junior post in the Faculté des Sciences (December, 1894), working with Louis Bouveault followed by promotion to Préparateur where he began his long association with Philippe Barbier. Grignard obtained the degree Licencié ès Sciences Physiques and became Chef des Travaux Pratiques (1898). He submitted his thesis on organic magnesium compounds *Sur les Combinaisons organomagnésiennes mixtes* and was awarded the degree Docteur ès Sciences de Lyons (1901).

The Career

As with most if not all scientists of this period, Grignard worked on the war effort during 1914–18. He was drafted into the army at his former rank of corporal, but was soon commissioned. The table below lists his career posts.

Date	*Post*	*Institute*
1905	Maître de Conférences	University of Besançon
1906	Maître de Conférences	University of Lyon
1908	Professeur-adjoint de Chimie Générale	University of Lyon
1909	Head, Department of Organic Chemistry	University of Nancy
1910	Professor of Organic Chemistry	University of Nancy
1914–18	War work	University of Nancy
1917–18	Chemical representative on the Tardieu Committee	visited the United States
1919	Professor of General Chemistry	University of Lyon
1921	Professor of General Chemistry and Director of l'École de Chimie Industrielle de Lyons and member of the University Council	University of Lyon
1929	Dean of the Faculty of Sciences	University of Lyon

Grignard studied:

- ethyl-isopropylacetobutyrate and the stereoisomeric diisopropylbutenedicarboxylic acids
- studies of branched unsaturated hydrocarbons
- studied organomagnesium compounds (1899)
- preparation of magnesium alkyl halides (first communicated by Henri Moissan to the Académie des Sciences on 11th May 1900)

He used the agents to prepare and study:

- alcohols, ketones, keto-esters, nitriles and terpene compounds
- developed a method for the synthesis of fulvenes
- constitution of unsaturated compounds by quantitative ozonization
- condensation of aldehydes and ketones
- ketone splitting of tertiary alcohols
- the cracking of hydrocarbons in presence of aluminium chloride
- catalytic hydrogenation and dehydrogenation processes under reduced pressures

Grignard was the author of some 170 publications on his research work. Two volumes of his *Traité de Chimie Organique* (Treatise on Organic Chemistry) had already been published, two more were ready for the press and the editorial work for another two was well advanced: it was later to be finished by his collaborators. In 1937, two of his students, Jean Cologne and Roger Grignard, published *Précis de Chimie Organique* (Survey of Organic Chemistry) which is based on Grignard's lecture course in organic chemistry.

During his acceptance speech [2], Grignard spoke on the use of organomagnesium compounds in preparative organic chemistry thus...

...The reactions of organic magnesium compounds are of two kinds - reactions of substitution and reactions of addition.

It is in this way that these compounds react with substances containing one or more atoms of reactive hydrogen, irrespective of whether this hydrogen is joined to the carbon, as we have already seen, or to the nitrogen, as in ammonia, the primary and secondary amines, the amides, etc., or again to the oxygen as in alcohols, phenols, acids and pseudoacids, etc. The organic radical is then released as hydrocarbon, the volume of which, if one is careful to take one of the first members, can be used for making a very accurate determination of the reactive hydrogen (Tschugaeff, B. Oddo, Zerewitinoff).

The action of water, in particular, can be used for preparing hydrocarbons (Tissier and Grignard). The organic halogenated derivatives and, more specifically, the allyl halides (Tiffeneau), the chloromethyl esters (Hamonet), readily react, accompanied by joining of the organic radicals, to give hydrocarbons or ethers.

The reactions of addition take place with a divalent element such as oxygen, which gives alcohols or phenols, and sulphur which results in mercaptans, thiophenols, etc; but they are of more special importance with the functional groups in which two atoms are united by a multiple bond. Such are:

$$(C = O)'' \; (C = S)'' \; (C \sim N)' \; (N = O)' \text{ or } ``(S = O)''$$

In all these cases one of the bonds in the functional group is broken, whilst the radical R becomes fixed to the element of maximum valence and MgX becomes fixed to the other. However, the organic magnesium compounds do not become fixed to an ethylenic or acetylenic bond...

The use of Grignard reagents has been taught on undergraduate chemistry courses for many years, and is used in both the research and industrial context. References [5, 6] provide the interested reader with a starting point for further reading.

Honours and Awards

Grignard was presented with honorary doctorates from the Universities of Brussels and Louvain.

He shared the Cahours Prize (Institut de France) in 1901 and 1902, he won the Berthelot Medal, was awarded the Prix Jecker (1905), the Lavoisier Medal (1912) and the Nobel Prize for Chemistry (1912).

He was appointed Chevalier (1912), Officier (1920) and Commandeur (1933) in the order of the Légion d'Honneur, he was also Honorary Professor, University of Nancy (1931), Honorary Fellow of the Chemical Society (London) and Foreign Member of the Royal Swedish Academy of Sciences.

Bibliography

1. Reproduced with permission.
2. From Nobel Lectures, Chemistry 1901–1921, Elsevier Publishing Company, Amsterdam.
3. The Nobel Population 1901–1937, E. Crawford, J. L. Heilbron and R. Ullrich. Office of the History of Science and Technology, University of California, 1987, ISBN: 0-918102-15-4.
4. Who was who: 1929–1940.
5. Grignard Reagents: New Developments edited by H. G. Richey John Wiley & Sons, 1999, ISBN: 0471999083.
6. Handbook of Grignard Reagents (Chemical Industries) by Gary S. Silverman, CRC, 1996, ISBN: 0824795458.

1912

© Nobel Foundation [1]

Paul Sabatier (0.5)

"for his method of hydrogenating organic compounds in the presence of finely disintegrated metals whereby the progress of organic chemistry has been greatly advanced in recent years" [2]

Professor H.G. Söderbaum, then President of the Royal Swedish Academy of Sciences, commented on 10th December 1912 [2]:

> *...in particular your general method of hydrogenating organic compounds in the presence of reduced metals and catalysts, a method which in the sphere of organic chemistry has during the past few years made possible much progress and the opening-up of so many new fields to science.*
>
> *It is for the discovery of this brilliant method that the Academy is pleased to bestow on you its highest reward...*

Sabatier was nominated for the Chemistry Prize by G. Darboux and O. Widman and was also nominated in the years 1907, 1909, 1911, 1928 and 1929 [3].

Sabatier was born at Carcassonne in Southern France on 5th November 1854 (some say 1859). He married Mlle Herail and they had four daughters. He died in Toulouse, France on 14th August 1941 [2, 4].

The Early Years

Sabatier was educated at the local Lycée and the École Normale Supérieure which he entered in 1874, graduating three years later the first in his class. He taught physics for a year at the Lycée of Vîmes at Nîmes before going to the Collège de France (1878) as Assistant to Berthelot. He received the degree of Doctor of Science in 1880 having presented a thesis on metallic sulphides. Sabatier's early work lay in the fields of physical and inorganic chemistry. In the later years he published a series of papers on the sulphides, including one on hydrogen disulphide (1886).

During work concerning inorganic chemistry, he:

- discovered a number of new metallic nitrides
- worked on the deep blue nitroso-disulphonic acid
- worked on a series of complex cupric salts

His early investigations in physical chemistry concerned:

- a series of thermochemical measurements
- a study of the velocity of transformation of metaphosphoric acid
- the partition of a base between two acids

Sabatier worked in the Faculty of Sciences at Bordeaux teaching physics until January 1882.

The Career

Sabatier entered the University of Toulouse (in January 1882) where he spent the rest of his career. During his time at Toulouse, however, he occupied a variety of posts, which are presented in the table.

Year	*Post*	*Institute*
1883	Responsible for courses in chemistry	University of Toulouse
1884	Professor of Chemistry	University of Toulouse
1905	Dean of the Faculty of Science	University of Toulouse
1908	Turned down Chair as successor to Moissan	University of Sorbonne
1930	Retired	University of Toulouse

Sabatier's influence in all aspects of catalysis in organic synthesis was profound. Amongst his many activities he:

- demonstrated the selectivity of catalytic action
- demonstrated selectivity of catalysts to poisons
- introduced the use of supports and showing the resultant enhanced activity
- made a close study of catalytic hydration and dehydration
- observed that zinc oxide decomposes formic acid into hydrogen and carbon dioxide exclusively but at the same temperature titanium oxide gives carbon monoxide and water exclusively
- formulated a chemical theory of catalysis in which the temporary formation of unstable chemical compounds as intermediary steps is postulated
- developed (in a systematic manner) his concept of a temporary unstable combination as the explanation of catalysis in general
- invented a nickel catalyst which is effective in vapour phase hydrogenation
- was the first to examine the catalytic behaviour of various oxides, not only as oxidation catalysts, but also in dehydrogenation and dehydration
- showed that most oxides are capable of bringing about two different reactions when alcohol is passed over them at suitable temperatures

Sabatier was not only interested in practical applications of catalysts, he also investigated the intermediates formed during the reaction. Indeed he isolated intermediates for example, the copper induced catalytic oxidation of organic compounds, or the decomposition of carbon monoxide on nickel produced the intermediate copper oxide and nickel carbonyl respectively. He postulated the possibility of organo-metallic intermediates during heterogeneous catalysis with acetylene for example.

Sabatier concluded:

...'As far as I am concerned this idea of temporary unstable intermediate compounds has been the beacon light that has guided all my work on catalysis; its light may perhaps be dimmed by the glare of lights, as yet unsuspected, which will arise in the better explored fields of chemical knowledge. Actually, such as it is, in spite of its imperfections and gaps, the theory appears to us good because it is fertile and permits in a useful way to foresee reactions.'...

He observed that in some cases the type of catalyst can determine the reaction products. For example ethyl alcohol is converted into ethyl ether, ethylene and water in the presence of alumina. Whereas the use of molybdenum and zinc oxide the reaction is predominately a dehydrogenation for example ethyl alcohol is converted into acetaldehyde and alcohol.

Furthermore, Sabatier observed the reversibility of the reaction

$$\text{Alcohol} \rightleftharpoons \text{aldehyde} + \text{hydrogen}$$

It would be thirty years before the thermodynamic and physico-chemical conditions of this equilibrium were fully established.

As with some of the early German dye chemists, Sabatier has left his mark on industry. Indeed the margarine, oil hydrogenation, and synthetic methanol industries owe their existence to the use of finely-divided hydrogenation catalysts. In keeping with the other chemists, the work of commercialization was largely left to others, although Sabatier did write many patents.

Sabatier's work is published in the journals of learned societies. His most important book, *La Catalyse en Chimie Orgarnique* (Catalysis in Organic Chemistry), was first published in 1913, with a second edition in 1920, of which an English translation by E. E. Reid was published in 1923.

Sabatier's Nobel Lecture, dated 11th December 1912 was titled 'The Method of Direct Hydrogenation by Catalysis', Sabatier explained [2]...

How can we explain hydrogenation by catalysis? I assume that hydrogen acts upon the metal by very rapidly producing a compound on its surface. The hydride thus produced is readily and rapidly dissociated, and if it is placed in the presence of substances capable of using hydrogen it gives it up to them, at the same time regenerating the metal, which again produces the same effect, and so on.

The distinction which I have made between several kinds of activity of nickel would suggest that there are several stages of combination. For instance, well-prepared and pure nickel might produce the perhydride NiH_2 which is capable of hydrogenating benzene. On the other hand, nickel prepared at too high a temperature or containing some impurities, would give only a poorer hydride, such as

$$\begin{array}{l} Ni-H \\ | \\ Ni-H \end{array}$$

which would be incapable of reacting with benzene, though active in respect of ethylenic hydrocarbons or nitro derivatives.

If this theory is correct, it can be deduced that nickel and the other active metals (copper, cobalt, iron, platinum) must be capable not only of fixing free hydrogen, but must also take hydrogen from such substances as are capable of giving it up, and that consequently they must be dehydrogenation catalysts.

A general account of his life appeared in Chemistry World in 2004 [5]. Catalysts used in hydrogenation have continued to be developed since Sabatier's Nobel Prize award. Some of these classic metals and their derivatives are described in [6]. The more recent and selective catalysts are the subject of [7].

Honours and Awards

He was elected Doctor Honoris Causa of the Universities of Louvain, Saragossa and Philadelphia.

Sabatier was awarded the Lacaze prize of the Academy of Science of Paris (1897), the Prix Lacate (1897), the Jecker prize in 1905, the Nobel Prize in Chemistry (1912), he received the Davy Medal from the Royal Society (1915) and the Franklin Medal of the Franklin Institute (1933).

He was elected as Foreign Member of the Royal Society (28th February 1918). He was a Foreign Member of several other scientific societies, notably of the Royal Institution of the Academies of Sciences of Amsterdam and Madrid, the Royal Society of Bohemia and the American Chemical Society. Sabatier was a Member of the French Academy of Sciences and Commander of the Légion d'Honneur.

Sabatier's citation for election (date: 28/02/1918) as Foreign Member of the Royal Society reads [8]:

'M. le Prof. Paul Sabatier of the University, Toulouse distinguished for his researches in Chemistry'

He was proposed by:

J. J. Thomson; A. B. Kempe; Arthur Schuster; W. B. Hardy; W. A. Herdman; Gilbert C. Bourne; W. H. R. Rivers; C. S. Sherrington; A. R. Cushny; Hugh K. Anderson; J. Swinburne; J. H. Jeans; J. W. L. Glaisher; R. J. Strutt; H. G. Lyons; W. W. Watts; W. H. Lang; B. Hopkinson; George Beilby.

Bibliography

1. Reproduced with permission.
2. From Nobel Lectures, Chemistry 1901–1921, Elsevier Publishing Company, Amsterdam, 1966.
3. The Nobel Population 1901–1937, E. Crawford, J. L. Heilbron and R. Ullrich. Office of the History of Science and Technology, University of California, 1987, ISBN: 0-918102-15-4.
4. Obituary Notices of Fellows of the Royal Society 1942–1944, 4, 63–66.
5. Chemistry World December 2004, No. 12, 46.
6. Handbook of Heterogeneous Catalytic Hydrogenation for Organic Synthesis by Shigeo Nishimura, Wiley-Interscience, 1st ed. 2001, ISBN: 0471396982.
7. Heterogeneous Enantioselective Hydrogenation: Theory and Practice (Catalysis by Metal Complexes) by Evgenii Klabunovskii, Gerard V. Smith and Ágnes Zsigmond Springer, 1st ed. April 1, 2006, ISBN: 1402042949.
8. Sackler database, Royal Society.

1913

© Nobel Foundation [1]

Alfred Werner

"in recognition of his work on the linkage of atoms in molecules by which he has thrown new light on earlier investigations and opened up new fields of research especially in inorganic chemistry" [2]

On 10th December 1913, former Councillor T. Nordström, then President of the Royal Swedish Academy of Sciences, observed [2]:

> *...we can with good reason agree with the remark of an eminent research worker, that Werner's theoretical and experimental work in inorganic chemistry has opened up new paths of chemical research and is of positively revolutionary significance. It is substantially his researches which have during the last few decades set the trend of development in inorganic chemistry and have newly inspired this branch of science which had been somewhat neglected during the last quarter of the 19th century,...*

Werner was nominated for the 1913 Chemistry Prize by E. Fischer, A. R. Hantzsch, P. Sabatier and S. Arrhenius. Werner was also nominated in 1907, 1908, 1911 and 1912. Nominations were disallowed for the years 1912 and 1913 [3].

Alfred Werner was born on 12th December 1866, at Mülhausen in Alsace, the son of factory foreman J. A. Werner and his wife Jeanne, née Tesche. Werner married Emma Giesker of Zurich in 1894. They had one son, Alfred, and one daughter, Charlotte.

He was already suffering from arteriosclerosis when he was awarded the Nobel Prize for Chemistry. By 1915 this illness forced him to give up presenting his general lectures on chemistry and in 1919 ill health forced him to resign from his Professorship. He died at the early age of 53 on 15th November 1919 [2,4].

The Early Years

Werner attended school in Mülhausen, Alsace, where he showed an interest in chemistry and undertook his first independent chemical research at the age of 18.

Military service in Karlsruhe (1885 to 1886) was followed by attending the lectures of Engler at the Technical High School in Karlsruhe. He went on to attend lectures at the Federal Technical, where he was appointed Assistant in Professor Lunge's laboratory (1889). Werner took his degree in the University of Zurich (1890) with a thesis on the spatial arrangements of the atoms in molecules containing nitrogen. He carried on this work in Paris under Professor Berthelot at the Collège de France (1890 to 1891). Werner's doctorate thesis, submitted in 1892, covered some aspects of the spatial relationships of atoms in the molecule, particularly the many compounds of tervalent nitrogen.

The Career

Werner's interests lay in the field of inorganic coordination chemistry. His career posts were:

Year	*Post*	*Institute*
1892	Lecturer	Technical High School, Zurich
1893	Associate Professor	University of Zurich
1895	Professor of Chemistry	University of Zurich
1895	Acquired Swiss nationality	

Werner declined posts in Vienna, Basle and Wurzburg.

Werner continued some of the concepts that he formulated for his pc, amongst which were:

- the theory of affinity and valence (1891), i.e. that affinity is an attractive force exerted from the centre of the atom which acts uniformly towards all parts of the surface of the atom
- his theory of variable valence – which included the concept of coordination number (the most important being 3, 4, 6 and 8, the number 6 occurring more often)
- the preparation and study of a new series of compounds (over a 20 year period). This study resulted many papers, over 150 of which he was the sole author
- the discovery of optical active isomers of the complexes (more than 40 optically active octahedral complexes were separated with the coordination number of 6)
- the coordination theory extended throughout the whole of inorganic chemistry

Werner wrote two books *Neuere Anschauungen auf dem Gebiete der anorganischen Chemie* (New ideas in Inorganic Chemistry) and *Lehrbuch der Stereochemie* (Textbook of Stereochemistry), both published in 1904.

The title of Werner's Nobel acceptance speech was 'On the constitution and configuration of higher order compounds'. During his talk Werner commented [2]...

> *We are now concerned with the question as to the manner in which the six groups which in the complex radicals MeA_6 are linked with the centre atom, are arranged about this atom in space. This question can be answered by experimental examination of the consequences arising from the various possibilities of arrangement with regard to the occurrence of isomerism phenomena. That the arrangement of the six groups about the centre atom must be symmetrical, is rendered experimentally probable by the fact that in spite of many experiments we have never been able to observe*

isomerism phenomena in compounds with complex radicals: [$MeAB_5$] Of symmetrical arrangements, the following may be involved: (1) plane, (2) prismatic, (3) octahedron (Fig. 1).

Plane and prismatic arrangements require, for compounds with complex radicals [MeA_4B_2], the occurrence of three isomeric forms, whereas the octahedron arrangement only envisages the existence of two isomeric compounds of this nature (Fig. 2). The latter isomers are characterised in that one form contains the two B groups in two adjacent positions of the octahedron arrangement, whereas the other form contains it in more remote (diagonal) positions.

The experimental investigations have shown that compounds with complex radicals [MeA_4B_2] do in fact occur in two isomeric forms; in no case were three isomeric forms found to exist (see Fig. 2). The elucidation of these isophenomena which were discovered first in the platinum series and then in the cobalt series, occupied my mind for many years, and a large number of compounds had to be prepared to provide the material by means of which we were able to test the theoretical conclusions. Without this extensive material of compounds it would have been quite impossible to obtain information about the arrangement of the atoms in these molecules. We now know, in the case of cobalt, more than 20 different series of compounds which show this isomerism.

There are many aspects of coördination chemistry. References [5–7] attempt to show the breadth of the literature from complexes of biological importance, those references of an acknowledged expert, and classic inorganic complexes.

Honours and Awards

Werner held an Honorary Doctorate of the University of Geneva.

He was awarded the Leblanc Medal of its Societe Chimique and the distinction of Officier de l'Instruction Publique.

Werner was corresponding member of the Royal Society of Sciences (Königliche Gesellschaft der Wissenschaften) at Göttingen and of the Physico-Medical Society (Physikalisch-medizinische Sozietät) of Erlangen. He was an Honorary Member of the Society of Physics and Natural History, the Physical Association (Physikalischer Verein) of Frankfurt/Main, the German Bunsengesellschaft, the Société Vaudoise des Sciences Naturelles at Lausanne and of the Chemical Society of London. He was

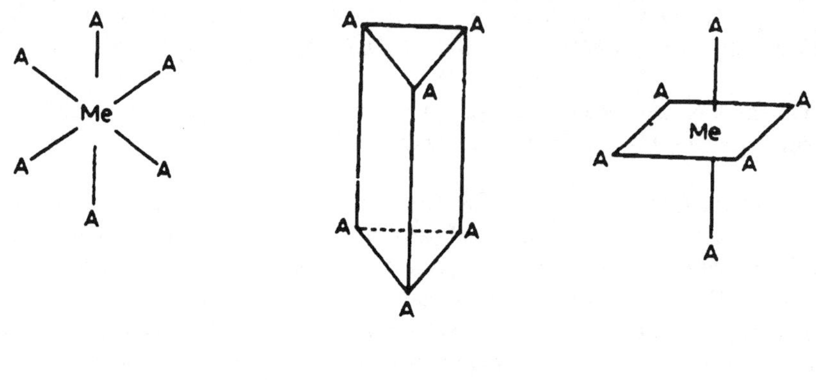

Figure 1

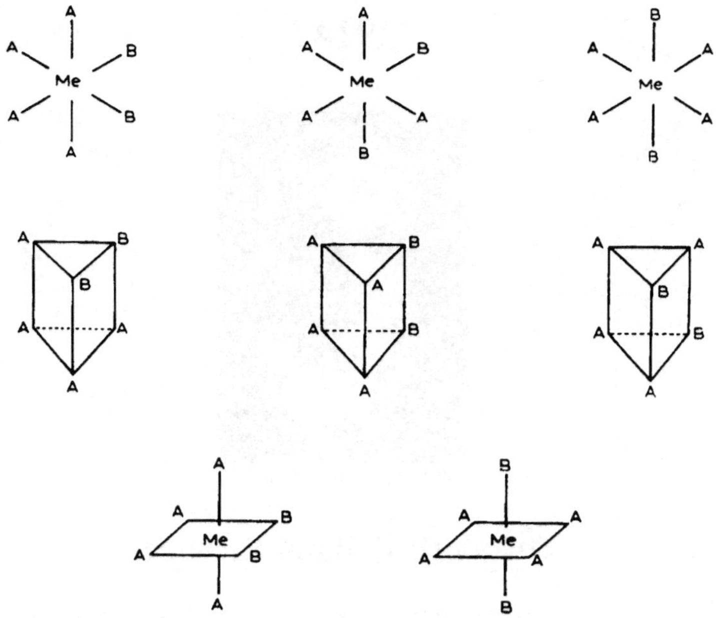

Figure 2

also a permanent Member of the Imperial Society of Friends of Natural History, Anthropology and Ethnography of Moscow.

Bibliography

1. Reproduced with permission.
2. From Nobel Lectures, Chemistry 1901–1921, Elsevier Publishing Company, Amsterdam, 1966.
3. The Nobel Population 1901–1937, E. Crawford, J. L. Heilbron and R. Ullrich. Office of the History of Science and Technology, University of California, 1987, ISBN: 0-918102-15-4.
4. Who was who: 1916–1928.
5. The Biological Chemistry of the Elements: The Inorganic Chemistry of Life by J. J. R. Frausto da Silva and R. J. P. Williams, Oxford University Press, 2nd ed. 2001, ISBN: 0198508484.
6. On Being Well-Coordinated: A Half-Century of Research on Transition Metal Complexes: Selected Papers Fred Basolo (World Scientific Series in 20th Century Chemistry, V. 13) edited by Fred Basolo and John L. Burmeister, World Scientific Pub Co Inc., 2003, ISBN: 9812380876.
7. Metal Complexes and Metals in Macromolecules: Synthesis, Structure and Properties by Dieter Wöhrle and Anatolii D. Pomogailo, Wiley-VCH, 2003, ISBN: 3527304991.

1914

© Nobel Foundation [1]

Theodore William Richards

"in recognition of his accurate determinations of the atomic weight of a large number of chemical elements" [2]

The Nobel Prize in Chemistry for 1914 was announced on 11th November 1915. In regard to Richards' work Pofessor H.G. Söderbaum, then a member of the Nobel Committee for Chemistry of the Royal Swedish Academy of Sciences, made the following statement [2]:

...that no less than thirty atomic weights have been redetermined with a degree of accuracy undoubtedly never before attained, and by the employment moreover of methods that, by comparison with those in earlier use, mark a very appreciable advance...

...Twenty-one of the atomic weights referred to have been determined by Richards himself or under his immediate guidance, while the determination of the others has been carried out in accordance with his methods by pupils trained under him.

It should be pointed out that the determinations in question embrace all those elements (viz. oxygen, silver, chlorine, bromine, iodine, potassium, sodium, nitrogen and sulphur), whose atomic weights are customarily termed fundamental, on account of their forming the basis for the determination and calculation of the atomic weights of all the other elements.

From a purely quantitative point of view alone, the work involved in these determinations is very considerable indeed. Apart from Berzelius, the great pioneer in the department of atomic weight determination, there is assuredly no one investigator who has contributed anything like so much as Richards, either in bulk or scope, towards the development of experimental stoichiometry...

...Most of Richards' atomic-weight determinations are so exact, that any variation there may be is measured in thousandths, i.e. is of the same degree of magnitude as the apparent volumes of the electrons. Their correctness has been controlled in several ways, both by Richards' own

> *counter-experiments and by the work of other investigators, and in almost every case the agreement has been found to be entirely satisfactory...*

Richards was nominated for the 1914 Chemistry Prize by S. Arrhenius, M. T. Bogert, B. Brauner, Carlheim-Gyllensköld, C. R. van Hise and W. Ramsay. Richards was also nominated for the years 1902, 1905, 1908, 1909, 1910, 1911, 1912, 1913 and 1915. There was at least one nomination disallowed for the years 1913 and 1914 [3].

Theodore William Richards was born in Germantown, Pennsylvania, USA on 31st January 1868. His father, William T. Richards (1833–1905) was a landscape and seascape painter, his mother, Anna, *née* Matlack, a poet. Richards married Miriam Stuart Thayer (daughter of Professor Joseph H. Thayer) in 1896; they had one daughter (Grace) and two sons. Richards died in Cambridge, Massachusetts, USA (02nd April 1928) [2, 4, 5].

The Early Years

During his childhood, Richards travelled to England and France and, up to the age of 14 he was educated by his mother. When he was 13, he received a small box of chemicals. This inspired him to set up a small laboratory in his home. In 1883, he entered Haverford College, Pennsylvania, to graduate in science, and topped in his class prior to entering Harvard University (1885). He received the degrees, B.A. in 1886, M.A. and Ph.D. in 1888. The following twelve months were spent in Germany where he studied under Victor Meyer (University of Göttingen), G. Kruss (University of Munich) and W. Hempel (University of Dresden).

The Career

Richards spent the majority of his working life at Harvard University.

Year	*Post*	*Institute*
1889	Assistant	Harvard University
1891	Instructor	Harvard University
1894	Assistant Professor	Harvard University
1901–1928	Professor	Harvard University
1901	Declined full professorship	University of Göttingen
1903	Chairman of the Department of Chemistry	Harvard University
1912–1928	Erving Professor of Chemistry and Director of the Wolcott Gibbs Memorial Laboratory	Harvard University

He spent part of 1895 in Europe where he visited and worked in the laboratories of Ostwald and Nernst. Richards worked in the two fields of atomic weight determination and molecular volume studies, each area of study occupying about half of his time.

Atomic Weights

Richards' work was exacting given the available technology. He nevertheless worked on:

- the atomic weight determination of oxygen and copper (1886)
- developed a new technique for the determination of halide ratios
- invented the nephelometer and demonstrated the effect of occluded moisture in gases and solids
- redetermined (with the highest accuracy) the atomic weights of over thirty important elements (by 1912)
- determined the atomic weight of some isotopes. This was accomplished for lead isotopes by studying the lead derived from thorium decay compared with that from uranium

Molecular Volume

Richards also studied atomic and molecular volume during which he:

- formulated a hypothesis of compressible atoms
- carried out a series of measurements of compressibilities of many elements and compounds
- introduced the use of transition temperatures of pure hydrated salts as fixed points in the standardization of thermometers
- developed the fundamentals of adiabatic calorimetry
- established, to a first approximation, the internal pressure of many solids and liquids

His researches are recorded in some three hundred technical papers published mainly in the *Proceedings of the American Academy of Arts and Sciences*, the *Journal of the American Chemical Society* and the publications of the Carnegie Institution of Washington.

For his Nobel prize acceptance speech, Richards presented his work concerning atomic weights (dated 6th December 1919) [2]. His rationale for his work and the impact of radioactive isotopes are provided below...

> *...The investigations now to be brought to your attention were first inspired by an intense philosophic desire to know more about the fundamental nature of matter, and its relations to energy. Later I perceived more and more clearly that a better understanding of the behavior of matter–the carrier of those complex chemical changes called Life – must give mankind more power over the circumstances of living. In short then, scientific curiosity on the one hand, and the hope of providing a basis for bettering man's condition on the other, were the leading impulses of the work...*

> *...The discovery of radioactivity with its amazing concomitants has thus put quite a new aspect upon the study of atomic weights. Because of this recent knowledge, some investigators seem to believe that these so-called "constants of nature" are of less significance than of old, since some atoms may be transitory, and since a given substance may have different atomic weights according to the circumstances of its life history.....the question as to whether or not the supposed constants of physical chemistry are really not constants, but are variable within small limits, is of profound interest and of vital importance to the science of chemistry and to natural philosophy in general. If this latter alternative is true, the circumstances accompanying each possible variation must be determined with the utmost precision in order to detect the ultimate reason for its existence. As*

Democritus said long ago, "The word chance is only an expression of human ignorance." Every variation must have a cause, and that cause must be one of profound effect throughout the physical universe. Thus the idea that the supposed constants may possibly be variable, adds to the interest which one may reasonably take in their accurate determination, and enlarges the possible field of investigation instead of contracting it."

Most, if not all, inorganic chemistry text books will contain atomic weights of all of the elements, usually within a periodic table at the front or back of the book. There are recent texts that have been written concerning the periodic table, however, that contain much more information concerning atomic weights in some cases providing atomic weights of common isotopes, see for example [6,7].

Honours and Awards

Richards received Honorary Doctorate degrees in Science from Yale (1905), Harvard (1910), Cambridge, Oxford, Manchester (1911) and Princeton (1923); in Philosophy from Prague (1909) and Christiania (1911); in Law from Haverton (1908), Pittsburgh (1915) and Pennsylvania (1920); in Chemistry from Clark (1909); and in Medicine from Berlin (1910).

He received: the Davy Medal of Royal Society (1910); the Faraday Medal (1911); the Willard Gibbs Medal of the American Chemical Society (1912); the Nobel Prize for Chemistry (1914); the Franklin Medal (1916); the Le Blanc and Lavoisier Medals in 1922.

He was President of the American Chemical Society (1914), the American Association for the Advancement of Science (1917) one of the early Members of the Advisory Board of the Carnigie Institutions and the President of the American Academy of Arts and Sciences (1919–21). He was appointed Officier de la Lègion d'Honneur in 1925 and he held Fellowships or Memberships of academies and learned societies in the United States, the British Isles, France, Germany and Scandinavia. He was elected Foreign Member of the Royal Society (26/06/1919); Honorary Fellow of the Chemical Society and Honorary Foreign Fellow of The Royal Society of Edinburgh (elected 2nd July 1923) [5].

Richards' citation for election (date: 26/06/1919) as Foreign Member of the Royal Society reads [8]:

'Prof. T. W. Richards of Cambridge, Mass distinguished for his researches in Chemistry'

He was proposed by:

W. B. Hardy; W. A. Herdman; W. H. R. Rivers; J. Swinburne; F. W. Dyson; J. W. L. Glaisher; F. W. Gamble; James J. Dobbie; Arthur Schuster.

Bibliography

1. Reproduced with permission.
2. From Nobel Lectures, Chemistry 1901–1921, Elsevier Publishing Company, Amsterdam, 1966.
3. The Nobel Population 1901–1937, E. Crawford, J.L. Heilbron and R. Ullrich. Office of the History of Science and Technology, University of California, 1987, ISBN: 0-918102-15-4.
4. Proc. Roy. Soc. Series A 1928, vol. 121, pp. xxix–xxxiv.
5. Proc. Roy. Soc. Edinb. 1927–28, 48.
6. Periodic Table of the Elements by Yasmin Patell, Wiley, 2003, ISBN: 0471474126.
7. The Periodic Table of Elements (Oxford Chemistry Series) by R. J. Puddephatt and P. K. Monaghan, Oxford University Press, USA, 2nd ed. 2005, ISBN: 0198555164.
8. Sackler database, Royal Society.

1915

© Nobel Foundation [1]

Richard Martin Willstätter

"for his researches on plant pigments, especially chlorophyll" [2]

Professor O. Hammarsten, Chairman of the Nobel Committee for Chemistry of the Royal Swedish Academy of Sciences, made the following comments upon the occasion of the presentation of the 1915 Nobel Chemistry Prize to Richard Willstätter [2]:

> *...in these investigations, which concerned the pigment nucleus both in chlorophyll and in haemoglobin, he has made several new and important observations regarding pyrroles and their position in this nucleus; in particular, however, he has shown that from these two pigments the same parent porphyrin, "aetioporphyrin", can be prepared, whose molecule has retained the essential characteristics of the pigment nucleus. By doing this he has produced the most interesting and decisive proof of the relationship between the two most biologically important pigments in Nature – haemoglobin and chlorophyll...*

Willstätter was nominated for the 1915 Nobel Chemistry Prize by A. Baeyer, E. Bosshard, A.G. Ekstrand, H. von Euler-Chelpin, P. Klason, T.W. Richards, H. Staudinger, H. Thoms, and E. Beckmann. There were also five nominations that were disallowed as they were received after the deadline. He was also nominated in 1908, 1909, 1911, 1912, 1913 and 1914 [3].

Willstätter was born in Karlsruhe in Baden on 13th August 1872. His parents were Max and Sophie of Jewish race which was destined to have a profound effect on his life because of the growing anti-semitism in Germany at that time. He married Sopie Leser (daughter of a Heidelberg professor). Sopie died in 1908 after five years of marriage, the cause of her death being a delayed operation for appendicitis. Willstätter also lost his only son Ludwig, who died at the age of 10 years, apparently in a diabetic coma. His daughter, Ida Margarete studied mathematics, married Dr. Ernst Bruch and settled in the United States. Ida and Ernst had one son, Ludwig Willstätter Bruch, born in January 1940 in Winnebago, Illinois. Willstätter's dedicated his autobiography to his grandson. Such were his experiences and feelings concerning the growing anti-Semitism that he resigned from his Chair in Munich on 24th June 1924. He nominated Heinrich Wieland (who was later awarded the 1927 Nobel Prize) as his successor.

On 10th November 1938 his house was searched by the Gestapo, and in 1939 he fled Germany to Switzerland, (with the help of a pupil A. Stoll) and lived at the Villa Eremitaggio at Muralto-Locarno. He died on 3rd August 1942 after a heart attack. A memorial was unveiled in Muralto in 1956 [2,4].

The Early Years

At the age of six he made a collection of minerals, and at twelve declared that he wanted to study chemistry. He was desperately ill with diphtheria whilst of school age which attacked his central nervous system and produced widespread paralysis. After a six month convalescence he re-joined school only to fall victim to scarlet fever followed by inflammation of the kidneys. He attended the Gymnasium for two years in Karlsruhe, until 1883, then moved to the Technical School in Nuremberg when his family moved to Nuremberg. In 1890 (aged 18), he travelled to the University of Munich and studied at both the University and Technische Hochschule. In 1893, he took the pre-doctorate examination and was assigned to Alfred Einhorn as a research student where he started working on some aspects of cocaine. At that time there was no room for Willstätter in von Baeyer's group. Whilst he was climbing on a snow-slope of the Sojen mountains he suffered serious injuries. Professor von Zittel visited him and reported the incident to Adolf von Baeyer. Willstätter was thus allowed to study chemistry whilst convalescing. The accident disqualified him from military service, although he was declared fit some 25 years later.

The Career

His first paper, written with Einhorn, was published in 1893 on hydrogenised paratoluic acids [5]. Buchner, Rupe and Bamberger amongst others were some of his early tutors. Unfortunately, his relationship with Einhorn deteriorated to the extent that the two scientists rarely spoke in later life. Willstätter started independent work in 1894 upon completion of his doctorate.

Year	*Post*	*Institute*
1896	Lecturer	in Munich
1905	Professor	Federal Technical College in Zurich
1912	Honorary Professor and Director	Kaiser Wilhelm Gesellschaft at Dahlem in Berlin
1915	succeeded Adolf von Baeyer	in Munich
24th June 1924	resigned from his Chair	continued to direct a cell of research in the laboratory

The First Munich Period

Willstätter's first Munich period involved investigations into:

- Ketones of the Tropine group – some 14 papers
- The products of decomposition of atropine and cocaine
- The action of potassium permanganate on bases
- The synthesis of hygric acid
- The synthesis of tropidine
- The synthesis of ecgonic acid
- The synthesis of tropilidene
- Studies of betaines
- The separation of gold and platinum
- Synthesis of the tropine group – a series of papers
- The theory of dyeing
- Quinonoid compounds – a series of papers

The Zurich Period

Many scientists of world renown were working in Zurich at this time including J. Wislicenus, Victor Meyer, Lunge, Bamberger, Bredig, Werner, Debye and Einstein. Willstätter's own work involved investigations into:

Continuing work on quinonoid compounds
Chlorophyll – 18 papers were written in this period
A series of papers on cyclo-octane – 5 papers were written during this period
Reduction of platinum
Linking of the iron to the colour of blood
The colouring matter of tomatoes

The Berlin Period

Investigations into:

Hydrogenation of aromatic compounds using platinum and hydrogen
Continuation of the studies of chlorophyll
Continuation of the cyclo-octane series
Hydrolysis oc cellulose
Blood pigments
Anthrocyanins

The Second Munich Period

Richard Kuhn (Nobel winner 1938) was a student at this time. Willstätter's group:

- Continued work on anthocyanins
- Studied the assimilation of carbon dioxide

- Completed the synthesis of tropione
- Produced a series of papers concerning maltase
- Studied the nature and activity of enzymes in general
- Undertook some studies of fermentation
- Studied hydrates and hydrogels
- Published a synthesis of natural cocaine

Post Munich

In his memorial lecture presented on 6th November 1952 (p 1025 of [6]), Sir Robert Robinson remarks of Willstätter's resignation...

> *...His senior colleagues also sent him a letter expressing their great appreciation of his services to science and the University and begging him to reconsider his decision. But irrevocable that decision remained, and he only stayed long enough to aid his students to complete their researches and to make the installation of Heinrich Wieland, his successor, as smooth as possible. Then came a spate of attractive offers, a new Research Institute to be built to his own specification, Head of the National Research Foundation for Physics and Chemistry in Spain, chairs at Heidelberg, Leipzig and three American universities. He was also offered the Directorship of Research in the Siemens Company, and in 1926 his old position in the Institute in Berlin. All were declined...*

Willstätter continued to help his students complete their respective degrees during which he published work concerning:

Enzymes
Maltase
Hydrates and hydrogels-this series ran to 11 papers
Plant proteases enzymes of leucoctes-this series ran to 10 papers
Willstätter left the university in 1925 and did not return.

Writing in Willstätter's obituary notice, Sir Robert Robinson [4] remarked ...

> *...Willstätter was a great technician and was always alert for observations that might lead to new methods of research. It is significant that the techniques he developed involve catalysts for reduction and oxidation, differential absorption on surfaces, and partition between partially immiscible solvents...He popularized the process of catalytic reduction in the presence of noble metals and gave careful directions for the preparation of platinum black, far more active than any previously examined...*

Willstatter will forever be remembered for his contributions to the study of chlorophyll. That the complex structure is a mixture of two compounds named chlorophyll-a and chlorophyll-b is just one of his discoveries.

The methyl group with an asterisk is replaced by an aldehyde in chlorophyll-b. The obituary notice, written by Robert Robinson [4] includes a more comprehensive list of papers.

See, for example, the following references for more contempory accounts of the various aspects of Willstätter's contributions to the organic chemistry of natural products [9–11].

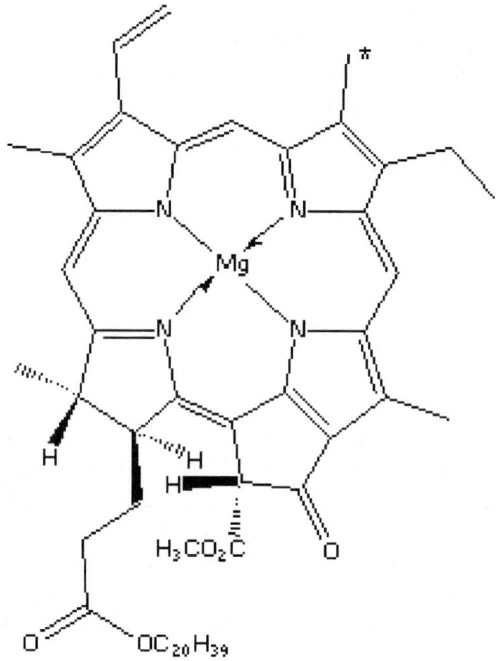

Honours and Awards

A. Stoll wrote an epilogue to Willstatter's autobiography where the list of honours and awards occupied three pages [7]. With so much world wide recognition it is difficult to present all of the awards and honours. A selection are included to demonstrate the depth and breadth of his recognition.

He was awarded the Iron Cross second class for civilians in recognition of the pioneering work, which started just before the First World War, on the fillings for gas masks intended as protection for chemical warfare. Over 30 million containers were said to have been put into service.

Willstätter:

- gave the Dunham lectures in Boston and to receive the Willard Gibbs Medal in 1933
- was awarded the Davy Medal from the Royal Society in 1932
- delivered the Faraday Lecture of the Chemistry Society in 1927
- received various Honorary degrees including the University of Manchester, and Honorary Doctorate from Oxford University

There were, however, some less pleasant milestones. The Chemical Society of Great Britain, the forerunner of The Royal Society of Chemistry, published the following [8]...

> *"The Chemical Society considers that it is neither compatible nor consistent with its loyalty to the Crown, whence the Royal Charter under which it works was derived, to retain any alien enemies upon its List of Honorary and Foreign Members.*

"It is therefore resolved that the names of A. von Baeyer, T. Curtius, E. Fischer, C. Graebe, P. H. von Groth, W. Nernst, W. Ostwald, 0. Wallach, and R. Willstätter, who were elected under happier conditions in recognition of their eminent services to Chemical Science—for which the Society still retains an undimindished appreciation and regard, are hereby removed from the list of Honorary and Foreign Members."

His post resignation honours included:

- An Honorary Fellowship of The Chemical Society (1927)
- Foreign Member of the Royal Society (21/06/28)

Willstatter was also a corresponding or Foreign Member of many National Academies of Science, the first to recognize him having been the Academy at Turin.

Willstätter's citation for election (date: 21/06/1928) as Foreign Member of the Royal Society reads [12]:

'Richard Willstätter of Munich distinguished as the foremost of German organic chemists is recommended by us as a proper person to be placed on the List of Foreign Members of the Royal Society'

He was proposed by:

E. Rutherford; J. H. Jeans; H. H. Dale; R. T. Glazebrook; J. E. Littlewood; Gilbert T. Walker; T. L. Heath; F. W. Aston; F. A. Lindemann; Arthur Lapworth; James Walker; F. A. Bather; C. Tate Regan; G. Elliot Smith; Hugh K. Anderson; E. D. Adrian; J. C. G. Ledingham; D. Prain.

Bibliography

1. Reproduced with permission.
2. From Nobel Lectures, Chemistry 1901–1921, Elsevier Publishing Company, Amsterdam, 1966.
3. The Nobel Population 1901–1937, E. Crawford, J. L. Heilbron and R. Ullrich. Office of the History of Science and Technology, University of California, 1987, ISBN: 0-918102-15-4.
4. Robert Robinson Obituary Notices of Fellows of the Royal Society Vol.8, No.22 (Nov.,1953), 609–634.
5. R. Willstätter and A. Einhorn, *Chem Ber*, 1893, 26, 2009.
6. Willstätter memorial lecture delivered by Sir Robert Robinson on 6th November 1952 (J. Chem. Soc. 1953, 999–1026).
7. Aus Meinem Leben ed A. Stoll, Verlag Chemie, Weinheim 1949, English edition From My Life, Benjamin, New York, 1965.
8. J. Chem. Soc. Trans. 1917, 111, 273–287.
9. Heme, Chlorophyll, and Bilins: Methods and Protocols edited by Alison G. Smith and Michael Witty, Humana Press, 1st ed. 2001, ISBN: 1588291111.
10. Plant Pigments and Their Manipulation: Annual Plant Reviews, Volume Twelve by Kevin Davies, 2004, ISBN: 0849323509.
11. Plant Pigments And Their Manipulation edited by Kevin M. Davies, Blackwell Pub, 2005, ISBN: 1405117370.
12. Sackler database, Royal Society.

1916

The prize money was allocated to the Special Fund of this prize section

1917

The prize money was allocated to the Special Fund of this prize section

1918

© Nobel Foundation [1]

Fritz Haber

"for the synthesis of ammonia from its elements" [2]

The 1918 Prize ceremony was delayed until early June 1920 having been announced on 13th November 1919. Owing to the death of the Crown-Princess Margaret, no member of the Royal family was present at the Prize ceremony. Dr. A.G. Ekstrand, then President of the Royal Swedish Academy of Sciences, observed the following on 1st June 1920 [2]:

> *...This country's Academy of Sciences has awarded you the 1918 Nobel Prize for Chemistry in recognition of your great services in the solution of the problem of directly combining atmospheric nitrogen with hydrogen. A solution to this problem has been repeatedly attempted before, but you were the first to provide the industrial solution and thus to create an exceedingly important means of improving the standards of agriculture and the well-being of mankind. We congratulate you on this triumph in the service of your country and the whole of humanity...*

Haber was nominated for the 1918 Chemistry Prize by W. Prandlt. He was also nominated in the years 1911*, 1912, 1913, 1915*, 1916, 1917*, 1919, 1920*, where * signifies that at least one nomination was disallowed in that year [3].

Fritz Haber was born on 9th December 1868 in Breslau, Germany, the son of Siegfried Haber, a merchant. Haber suffered from heart disease and died on 29th January 1934, at Basle, on his way from England to Switzerland [2].

The Early Years

Harber attended the St. Elizabeth school at Breslau. He was interested in chemistry whilst at school and undertook many chemical experiments.

He studied chemistry under Bunsen (1886 until 1891) at the University of Heidelberg, under A.W. Hoffmann at the University of Berlin, and under Liebermann at the Technical School at

Charlottenberg. An early interest in chemical technology resulted in him volunteering to work in his father's chemical business whilst continuing to study under Professor Georg Lunge at the Institute of Technology at Zurich. Having decided to take up a scientific career he spent one and a half years working under Ludwig Knorr at Jena, publishing a joint paper with Knorr on diacetosuccinic ester. Harber accepted an Assistantship working with Hans Bunte the Professor of Chemical Technology at Karlsruhe (1894), remaining there until 1911. Harber studied combustion chemistry and the study of petroleum at Karlsruhe which formed the basis of his subsequent work. He subsequently qualified as a Privatdozent with a thesis on his experimental studies of the decomposition and combustion of hydrocarbons (1896).

The Career

Most of Harber's career was spent in Germany:

Year	*Post*	*Institute*
1906	Professor of Physical Chemistry and Electrochemistry and Director of the Institute	Karlsruhe
1911	Director of the Institute for Physical and Electrochemistry	Berlin-Dahlem
1933	Resigned	
1933		Cambridge

As with all leading scientists of the period, Harber worked for the war effort during 1914–1918. He was appointed a consultant to the German War Office and organised gas attacks and defences against gas attacks. He helped to create the German Relief Organisation and served on the League of Nations Committee on Chemical Warfare. Unfortunately, his work undermined his health, particularly his work on gas attacks which forced him to spend sometime on administrative work.

Harber's resignation in 1933 concerned the political situation at the time. Most of his staff felt compelled to resign against the Nazi laws, Harber chose to resign rather than agree with the newly introduced laws. Sir William Pope invited Harber to visit him at Cambridge which he did for a while, but chose to return to Switzerland to avoid the English winter.

Harber concentrated his efforts into providing solutions to industrial processes resulting in investigations concerning:

- electrolytic oxidation and reduction (1898)
- the reduction of nitrobenzene in stages at the cathode
- the electrolysis of solid salts (1904)
- the establishment of the quinone-hydroquinone equilibrium at the cathode, which laid the foundations for Biilmann's quinhydrone electrode for determining the acidity of a liquid
- the glass electrode which he invented in collaboration with Cremer
- the first experimental investigations of the potential differences that occur between solid electrolytes and their aqueous solutions
- the loss of energy by steam engincs, turbines and motors driven by fuels

- seeking methods of limiting steam energy loss by electrochemical means
- a fundamental solution for the laboratory combustion of carbon monoxide and hydrogen
- a study of flames and did fundamental researches on the Bunsen flame
- work leading to a chemical method of determining flame temperatures
- the fixation of nitrogen from the air (for which he was given the Nobel Prize in Chemistry)
- first synthesis of ammonia from hydrogen and nitrogen by circulating nitrogen and hydrogen over the catalyst at a pressure of 150–200 atmospheres at a temperature of about 500°C.
- a modification to the Haber process provided ammonium sulphate for use as a soil fertilizer the synthesis of methyl alcohol (by Alwin Mittasch)
- the hydrogenation of coal by the method
- the production of nitric acid

The post First World War saw many inventions, particularly:

- his firedamp whistle for the protection of miners
- his quartz thread manometer for low gas pressures
- his observation that adsorption powers can be due to unsaturated valence forces of a solid body, which was used by Langmuir for his theory of adsorption
- he experimented on the recovery of gold from sea water (1920 until 1926) which was largely unsuccessful

Harber also devoted some time and effort to the re-organisation of his Institute, to which he appointed sectional directors with complete freedom in their work, the principal work related to colloid chemistry and atomic physics. Among these colleagues were James Franck, Herbert Freundlich, Michael Polanyi and Rudolf Ladenburg.

Haber spent his own research time on mechanisms of oxidation and on hydrogen peroxide catalysis. Harber published two books. They were:

- his textbook on *Electrochemistry* (1898)
- his book on the thermodynamics of technical gas reactions (1905), in which he recorded the production of small amounts of ammonia from N_2 and H_2 at a temperature of 1000°C with the help of iron as a catalyst

Harber's Nobel lecture, dated 2nd June 1920, was titled 'The synthesis of ammonia from its elements' [2]. Harber commented...

> *At that point it seemed to me, in 1905, useless to pursue the problem further. A combination of the elements had certainly been achieved, and the requirements for large scale synthesis had been outlined; but these requirements appeared so unfavourable that they deterred one from a deeper study of the problem. The discovery of catalysts which would provide a rapid adjustment of the point of equilibrium in the vicinity of 300°C and at normal pressure seemed to me quite unlikely: and they have not been found anywhere in the 15 years that have since elapsed.*

The synthesis of ammonia which had been demonstrated at normal pressure could be carried out at high pressure on a laboratory scale without any great difficulties. It needed only a slight modification of the pressure oven, such as that used by Hempel 15 years earlier to carry out nitrogen absorption in the case of indirect ammonia synthesis under pressures of up to 66 atmospheres. But I did not think it worth the trouble; at that time I supported the widely-held opinion that a technical realization of a gas reaction at the beginning of red heat under high pressure was impossible. Here the matter rested for the next three years.

Already in 1906, on the other hand, a new determination of the ammonia equilibrium proved necessary. In the course of his investigations into the heat theorem which has been named after him, Nernst succeeded in finding an approximate formula which permitted a prediction of the equilibria based on the values of the heat effect and the so-called chemical constants. In the case of ammonia this gave a deviation from the values obtained at my first measurements which, as later became apparent, was caused by the original value of the conventional chemical constant of hydrogen then used. This deviation led to fresh determinations of the equilibrium which Nernst had carried out at his Institute in a pressure oven indicated by him while I, in collaboration with Robert le Rossignol, repeated the determinations at normal pressure with greater care than before... During the course of these investigations, together with my young friend and co-worker Robert le Rossignol, whose work I would like to mention here with particular sincerity and gratitude, I took up once again, in 1908... The cycle of ammonia production and removal must clearly be achieved by the simplest possible means at a constant high pressure. It seemed essential that the heat produced during the synthesis of ammonia should he removed from exhaust gases, where it had only a deleterious effect, and be led to the fresh gas so that the process itself yielded the heat required for its operation. The construction and operation (carried out in collaboration with Robert le Rossignol) of a small-scale plant which suited these requirements, together with the performance of the new catalysts mentioned, was indeed sufficient to persuade the "Badische Anilin- und Sodafabrik" which thus far had devoted its attention to the indirect method of producing ammonia by means of the nitrides of aluminium, silicium and titanium, to undertake high-pressure synthesis from the elements.

The company then studied the catalysts on a large scale, using far more substantial means, and discovered ways, in the temperature employed in their production plant and particularly in the deliberate use of inert materials, of improving the performance of poor catalysts to the level of osmium and uranium. Their results were, indeed, important in the case of the classic ammonia catalyst employed by Ramsay and Young, namely iron. They also discovered an improvement in the design of the oven which overcame the effect of hydrogen on the carbon content of steel which they had observed over a long period of operation.

The main work of the company however, was in substituting electrolytic hydrogen, with which we conducted our experiments, for water-gas hydrogen which introduced impurities. The difficulties encountered by the Technical Director Dr. Bosch resembled those which his predecessor Knietsch had overcome with equal success in the course of his technical application of the sulphuric acid contact process. Dr. Bosch has made a large-scale industry of ammonia synthesis.

Present-day industrial working pressures in the vicinity of 200 atmospheres, a working temperature of about 500–600°C, circulation under constant high pressure, and the method of heat exchange from exhaust to inlet gas are all main features of laboratory work which have been retained.

Persistence was the key. Ammonia is still produced on the commercial scale, one of the latest plants for its production being in the Caribbean [4].

Honours and Awards

Apart from the Nobel Prize, Haber received many honours during his life. At Max von Laue's instigation, the Institute for Physical and Electrochemistry at Berlin-Dahlem was renamed the Fritz Haber Institute after his death.

Bibliography

1. Reproduced with permission.
2. From Nobel Lectures, Chemistry 1901–1921, Elsevier Publishing Company, Amsterdam, 1966.
3. The Nobel Population 1901–1937, E. Crawford, J.L. Heilbron and R. Ullrich. Office of the History of Science and Technology, University of California, 1987, ISBN: 0-918102-15-4.
4. TRINIDAD & TOBAGO: Construction plans for proposed 132,000 metric ton per year ammonia production capacity expansion project, POTASH CORP. OF SASKATCHEWAN ... in Latin America & the Caribbean [HTML] Worldwide Projects, Inc.; ISBN: B00084ABXE, 2004.

1919

The prize money was allocated to the Special Fund of this prize section

1920

© Nobel Foundation [1]

Walther Hermann Nernst

"in recognition of his work in thermochemistry" [2]

The 1920 Nobel Prize in Chemistry was announced on 10th November 1921, and the ceremony took place on 10th December 1921. During his presentation speech Professor Gerard de Geer, then President of the Royal Swedish Academy of Sciences, commented [2]:

> *...For long years chemists eagerly sought the suspected connection between the evolution of heat and the chemical affinity during the combustion of coal and in other chemical reactions.*
>
> *Your work has now brought this connection to light.*
>
> *You have used brilliant acuteness during your masterly experimental researches on specific heat and chemical equilibria.*
>
> *Using the heat theorem discovered by you it has now become possible on the one hand to calculate from the heat evolution during chemical reactions and the specific heats, the chemical affinity and the maximum possible output of energy during chemical reactions, and on the other hand to calculate the equilibrium in reactions not yet studied...*

Nernst was nominated for the 1920 Chemistry Prize by M. Le Blanc, C. Drucker, F. Harber, A. R. Hantzsch, B. Naunyn, W. Palmaer, W. von Walseyer-Hartz and F. Weigert. Nernst was also nominated for the Chemistry Prize for the years 1906, 1907, 1908, 1909, 1910, 1911, 1912, 1913, 1913*, 1914, 1914*, 1915, 1915*, 1916, 1916*, 1917, 1918, 1918*, 1919*, 1921, 1921* and 1922* and for the Physics Prize for the years 1911, 1912, 1913, 1914, 1915, 1916, 1917 and 1921, where * signifies disallowed nominations. Although friends when young, Nernst and Arrhenius developed a rift in later life so much so that Arrhenius blocked Nernst's Nobel Prize for some 15 years. In total, Nernst received 93 nominations across the Chemistry and Physics Prizes[3].

Walther Hermann Nernst was born in Briesen (Wabrzezno), West Prussia, on 25th June 1864. His father, Gustav Nernst, was a district judge, his mother was Ottilie née Nerger. He married Emma

Lohmeyer, the daughter of a distinguished surgeon, in 1892. They had two sons, who were both killed in the First World War, and three daughters. Most sources, including the Year Book of Austrian Academy, give his date of death as 18th November 1941 (a few quote 19th November 1941) in Berlin [2, 4].

The Early Years

Nernst spent his early school years at Graudentz Gymnasium where he was interested in classics and literature. He subsequently went to the Universities of Zurich (Ludwig Boltzmann), Berlin and Graz (Albert von Ettinghausen), studying physics and mathematics, before moving to Wurzburg (Friedrich Kohlrausch), where he graduated in 1886 with a thesis on electromotive forces produced by magnetism in heated metal plates. Nernst's first two papers were written with von Ettinghausen (1886) [5,6]. Their joint work studied the effects of a magnetic field applied perpendicular to a temperature gradient. This gave rise to corresponding thermo-magnetic and galvo-magnetic effects. He then joined Wilhelm Ostwald at Leipzig University working alongside van 't Hoff and Arrhenius.

The Career

The whole of Nernst's career was in various posts within Germany.

Year	*Post*	*Institution*
1894	Refused two Chairs	Munich and in Berlin
1894	Professor of Physics	University of Göttingen
	Founded the Institute for Physical Chemistry and Electrochemistry becoming its Director	University of Göttingen
1905	Professor of Chemistry, later of Physics	University of Berlin
1924	Director	"Physikalisch-Chemisches Institut"
1933	Retirement	

Nernst was taken on as an army driver during the First World War participating in the German push through Belgium into France. He played a part in the introduction of gas warfare and was placed on the list of those to be extradited (1919).

The Göttingen Period

Amongst Nernst's achievements whilst at Göttingen were:

- that he elucidated the theory of galvanic cells (1889)
- that he wrote his book '*Theoretical Chemistry*' (first published in 1893 but underwent translations and further editions)
- that he performed some calculations on ammonia synthesis

- that he derived equations which defined the conditions by which solids precipitate from saturated solutions
- that he proposed Third Law of Thermodynamics (published in 1905)

The Berlin Period

During the Berlin period, Nernst continued to test the Third Law of Thermodynamics. He spent some time with Max Planck following on from his realization that his specific heat curves approximated to the form predicted by Einstein on the basis of quantum theory. Einstein's prediction thus linked the specific heat curves to Planck quanta. He also worked on:

- determinations of specific heats of solids at very low temperatures
- vapour densities at high temperatures
- atom chain reaction theory (1918)
- aspects of photochemistry
- aspects of cosmology
- improvements to the electric light called the Nernst Lamp
- an electrical piano
- astrophysical theories using the heat theorem

A full list of Nernst's publications can be found in the obituary notice of the Royal Society [4], as can a full list of his published books which cover areas such as electrochemistry, various aspects of thermodynamics, the theory of solutions and photochemistry.

Nernst was influential throughout the world for many years, no more so than in Berlin. Indeed it was Nernst who was responsible for Einstein being offered a post at the Berlin Academy of Sciences.

Nernst's Nobel acceptance speech [2] was titled 'Studies in Chemical Thermodynamics' and was dated 12th December 1921...

> *In very lengthy, tedious calculations I then made a closer check on these points which I believe to have clarified in principle soon afterwards by the publication of my work Űber die Berechmung chemischer Gleichgewicht aus thermischen Messungen (The Calculation of Chemical Equilibria from Thermal Measurements) (1906). The simple principle at which I arrived at that time may be stated as follows: that in all cases chemical affinity and evolution of heat become identical at low temperatures. Not, and this is the essential point, in the sense that they intersect at absolute zero, but rather in the sense that they invariably become practically identical some distance before absolute zero is reached; in other words the two curves become mutually tangential in the vicinity of absolute zero.*
>
> *If we frame this principle in quite general terms, i.e. if we apply it not only to chemical but to all processes, then we have the new heat theorem which gives rise to a series of very far-reaching consequences:*
>
> (i) *The best working capacity attainable in the most favourable case during a process can be computed from the heat evolution which must, however, be known down to the lowest temperatures.*
>
> (ii) *This can also be expressed by stating that the evolution of heat at an arbitrary temperature, and the specific heats of the substances concerned must be known to very low temperatures.*

(iii) *'The specific heats must be strictly additive at very low temperatures, or in other words the evolution of heat at very low temperatures must always be independent of the temperature.*

Of course, the above principles, although applicable to all processes ocin Nature, are particularly useful in connection with chemical ones.

Several books have been written concerning Nernst and his work, the references for three of which are provided below [6, 7, 8]. Many books have also been written concerning thermodynamics, of which Nernst's work is a fundamental part, references for three of which are also provided below [6, 7, 8].

Honours and Awards

Nernst received many honours culminating in the Nobel Prize for Chemistry in 1920. He was awarded the order 'Pour le Mérite' (corresponding to the Order of Merit), the highest honour of his country.

He was elected an honorary Member of the Royal Institution (1899) and Foreign Membership of the Royal Society (1932). One of the last honours he received was Doctor of Science (hc) of the University of Oxford (1937). This was the last occasion on which he visited England.

Unfortunately, Nernst was not immune to the effects of the First World War and suffered, along with other distinguished Foreign Members and Fellows of the Chemical Society [12].

"The Chemical Society considers that it is neither compatible nor consistent with its loyalty to the Crown, whence the Royal Charter under which it works was derived, to retain any alien enemies upon its List of Honorary and Foreign Members.

"It is therefore resolved that the names of A. von Baeyer, T. Curtius, E. Fischer, C. Graebe, P. H. von Groth, W. Nernst, W. Ostwald, 0. Wallach, and R. Willstätter, who were elected under happier conditions in recognition of their eminent services to Chemical Science—for which the Society still retains an undimindished appreciation and regard, are hereby removed from the list of Honorary and Foreign Members."

Nernst's citation for election (date: 02/06/1932) as Foreign Member of the Royal Society reads [13]:

'Walter Nernst of Berlin distinguished for researches on the theory of electrical potential of heat'

He was proposed by:

Rayleigh; R. H. Fowler; G. I. Taylor; W. L. Bragg; C. E. Inglis; A. C. Egerton; N. V. Sidgwick; W. N. Haworth; G. Barger; O. T. Jones; F. E. Smith.

Bibliography

1. Reproduced with permission.
2. From Nobel Lectures, Chemistry 1901–1921, Elsevier Publishing Company, Amsterdam, 1966.
3. The Nobel Population 1901–1937, E Crawford, J. L Heilbron and R Ullrich. Office of the History of Science and Technology, University of California, 1987, ISBN: 0-918102-15-4.
4. Obituary Notices of Fellows of the Royal Society 1942–1944 vol. 4 pp. 101–112.
5. W. Nernst and A.v. Ettinghausen, Ueber das Hall'sche Phaenomen. S. B. Akad. Wiss. Wien. 1886, 94, 560–610.

6. W. Nernst and A. v. Ettinghausen, Ueber das Auftreten Elektromotorischer Kraefte in Metaliplatten, weiche von einem Waermestrome durchflossen werden und sich im magnetischen Felde befinden, Ann. Phys., 29, 343–347.
7. Walter Nernst and the Transition to modern Physical Chemistry by Diana Kormos Barkan, Cambridge University Press, 1999, ISBN: 0 521 44456 X.
8. J. S. Rowlinson, 'The Third Law', review of D. K. Barkan, Walther Nernst and the Transition to Modern Physical Science in NR, 2000, vol. 54, pp. 123–124.
9. Thermodynamics 2nd ed by J.M. Honig, Academic Press, 1995, ISBN: 0-12-355045-9.
10. Concise Chemical Thermodynamics 2nd ed by J.R.W. Warn and A.P.H. Peters, Chapman and Hall, 1996, ISBN: 0-412-46650-3.
11. Principles of Thermodynamics by Myron Kaufman, Marcel Dekker Inc, 2002, ISBN: 0-8247-0692-7.
12. J. Chem. Soc. Trans, 1917, 111, 273–287.
13. Sackler database, Royal Society.

1921

© Nobel Foundation [1]

Frederick Soddy

"for his contributions to our knowledge of the chemistry of radioactive substances, and his investigations into the origin and nature of isotopes" [2]

The Nobel Prize in Chemistry for 1921 was announced on 9th November 1922 and presented on 10th December 1922. In his presentation speech by Professor H.G. Söderbaum, then a member of the Nobel Committee for Chemistry of the Royal Swedish Academy of Sciences, observed [2]:

> *...The Royal Swedish Academy of Sciences, of which institution you have for several years been a highly valued member, is sure that it is acting in complete accordance with the opinion of the scientific world in awarding to you the Nobel Prize in Chemistry for 1921, on account of your important contributions to our knowledge of the radioactive bodies and of your pioneer works on the existence and nature of isotopes...*

Soddy was nominated for the 1921 Nobel Chemistry Prize by E. Rutherford. He was also nominated for the Chemistry Prize for the years 1908, 1918, 1919 and 1922 and for the Physics Prize for 1923 (disallowed) [3].

Frederick Soddy, the son of Benjamin Soddy (a London merchant), was born at Eastbourne, Sussex, England, on 2nd September 1877 and he married Winifred Beilby (1908), the daughter of Sir George Thomas Beilby F.R.S. Winifred died in 1936 of a thrombosis and he died on 22nd September 1956 at Brighton, Sussex [3,4,5]. He established the Frederick Soddy Trust by his will, giving grants to groups studying the whole life of a community [2, 4].

The Early Years

Soddy was educated at Eastbourne College, the University College of Wales, Aberystwyth and Oxford University. Whilst still at school, and aged 17, he published his first chemical paper [6] with Hughes.

His year in Aberystwyth, which was essentially to mark time whilst he applied for a Merton College Open Science Postmastership (which he won in 1895), was extended over a difficulty with his Latin studies. Nevertheless, in 1896, he was in Merton College, Oxford University, from where he graduated with a first class honours degree in chemistry (1898). Sir William Ramsay was his external examiner and with whom he later collaborated.

After two years of research at Oxford, he worked as a University Demonstrator in the Chemistry Department of McGill University, Montreal (1900 to 1902). Whilst in McGill, he worked with Professor Sir Ernest Rutherford on problems of radioactivity. Together they published a series of papers on radioactivity, introducing the concept of the disintegration theory. This work was written up in a series of eight papers, the references for which appear in the Obituary Notice for Soddy which appears in the biographical memoirs of Fellows of the Royal Society [4]. They also investigated the gaseous emanation of radium.

Soddy went on to work with Sir William Ramsay at University College, London, where he continued the study of radium emanation. Using spectroscopic means, Soddy and Ramsay demonstrated that the element helium was produced in the radioactive decay of radium bromide and that helium was evolved in the decay of emanation.

The Career

Soddy's career interests changed during his later years where his work took on more of a social and economic outlook. This is reflected in his books – see later.

Year	*Post*	*Institute*
1904	Lecturer in Physical Chemistry and Radioactivity	University of Glasgow
1914	Professor of Chemistry	University of Aberdeen
1919	Dr. Lees Professor of Chemistry	Oxford University
1937	Retired on the death of his wife.	

The Aberdeen Period

During the war years, Soddy undertook a variety of tasks, some of which relate to the extraction of ethylene from coal gas (carried out in the Aberdeen Gas Works). He also produced reports for the Marine Sub-Committee of the Board of Inventions and Research which concerned steam raising by the addition of water to strong or anhydrous soda. These activities were a diversion from his well established work concerning the isotopes.

The Glasgow Period

In Glasgow, Soddy returned to his studies concerning radiation and the isotopes, producing work concerning:

- the disintegration theory
- the "Displacement Law" namely that the emission of an alpha-particle from an element causes that element to move back two places in the Periodic Table
- the formulation of the concept of isotopes (1913) which stated that certain elements exist in two or more forms which have different atomic weights but which are indistinguishable chemically

The Oxford Period

This final phase of Soddy's career was more concerned with social injustice than experimental chemistry, for he wrote about and was concerned with:

- Social outlook
- Virtual wealth theory
- Application of the Laws of Conservation of Economics
- Some aspects of mathematics

His books include *Radioactivity (1904), The Interpretation of Radium (1909), The Chemistry of the Radioactive Elements (1912–1914), Matter and Energy (1912), Science and Life (1920), Cartesian Economics (1922), Inversion of Science (1924), Wealth, Virtual Wealth and Debt (1926), The Wrecking of a Scientific Age (1927), Money versus Man (1931), The Interpretation of the Atom (1932), The Role of Money (1934), The Arch-Enemy of Economic Freedom (1943), The Story of Atomic Energy (1949),* and *Atomic Transmutation (1953).*

Soddy's Nobel acceptance speech was titled 'The origins of the conceptions of isotopes' and is dated 12th December 1922. He summarises the story of isotope discovery thus [2]...

(i) *Experimental methods are available, uniquely for the radio-elements, which enable isotopes to be severally recognised, by a suitable combination of chemical analyses at appropriate intervals of time, whereby, owing to the successive changes of the constituents, they may be separated, although chemical analysis alone is quite unable to effect this separation. This dates from 1905.*

(ii) *The complete chemical identity of isotopes, as distinct from close chemical similarity, came gradually to be recognized. McCoy and Ross were the first to express a definite opinion in this sense (1907).*

(iii) *The existence of chemical identities among the radio-elements led to the deduction that they might exist among the common elements and be responsible for the exceptions in the Periodic classification, and for the fact that the atomic weights in some cases depart widely from integral values. Strömholm and Svedherg first made this deduction (1909).*

(iv) *The recognition of the effect of the expulsion of, first, the* α*- and then, the* β*-particle (1911 and 1913) led to the correct placing of all three disintegration series from end to end in the Periodic Table. On the experimental side the names of A. Fleck, and on the theoretical side that of G. von Hevesy and A. S. Russell, but pre-eminently that of Kasimir Fajans, are associated with this advance.*

(v) *The identity of isotopes was extended to include their electrochemistry (Paneth and Hevesy) and their spectra (Russell and Rossi), though here infinitesimal differences were subsequently found (Harkins and Aronberg, Merton).*

(vi) *Isotopes, on Rutherford's theory of atomic structure, are elements with identical external electronic systems, with identical nett positive charge on the nucleus, but with nuclei in which the total number of positive and negative charges and therefore the mass is different. The originator of the view that the places in the Periodic Table correspond with unit difference of intra-atomic charge is Van den Broek.*

(vii) *Moseley extended this view to the non-radioactive elements, and ultimately for the whole Periodic Table, and the definite determination of the number and sequence of the places in it became possible.*

(viii) *The chemistry of the radioactive elements and the lacunae previously existing in the radioactive series, especially in connection with the origin of actinium, have been cleared up, and this led to the discovery of a new element, proto-actinium (eka-tantalum) in uranium minerals. Occupying the place between uranium and thorium and existing in sufficient quantity for the compounds of the element to be prepared in a pure state, and its spectrum and atomic weight ultimately to be determined, as Mme. Curie did for radium. Cranston and I share with Hahn and Meitner the original discovery, but the subsequent developments are due to the latter.*

(ix) *The preparation from radioactive minerals of different isotopes of lead followed and the determination of their atomic weight, spectrum, density and other properties, established that the same chemical character and atomic volume can coexist with differences of atomic weight. The work on the iomum-thorium mixture from pitchhlende is a second example.*

(x) *The last result, and perhaps the most important of all, is the subject of the award of the Nobel Prize for Chemistry for 1922.*

More information concerning radioactivity, including radioactive isotopes, can be found, for example in [7–10].

Honours and Awards

Soddy was awarded honorary degrees from Glasgow University (L.L.D. 1934) and Oxford University.

He received the Cannizzaro Prize (Rome 1913), the Nobel Prize in Chemistry (1921) and the Albert Medal in 1951.

Soddy was elected a Fellow of the Royal Society (election date 05/05/1910). He was also a Foreign Member of the Swedish, Italian and Russian Academies of Science.

He was President of the Röntgen Society (1905–6), Science Masters' Association (1943) and President of the Institute of Atomic Information for the Layman.

Soddy's citation for election (date: 05/05/1910) as Fellow of the Royal Society reads [11]:

'Lecturer in Physical Chemistry and Radioactivity, University of Glasgow. Distinguished for his investigations in Radioactivity. Author of numerous papers, including:-'The Radioactivity of Uranium' (Journ. Chem. Soc, 1902); 'Production of Radium from Uranium' (Phil. Mag., 1905); 'Calcium as an Absorbent of Gases' (Proc. Roy. Soc., 1907); 'Relation between Uranium and Radium, Part III' (Phil. Mag., 1908); 'Attempts to detect the Production of Helium from the Primary Radioactive Elements' (ibid, 1908); 'Radioactivity' (Elect. Publ. Co., 1904); 'The Evolution of Matter as Revealed by the Radioactive Elements' (Wilde Lecture, Manch. Lit. and Phil. Soc., 1904); 'Present Position of Radioactivity' (Pres Address, Journ Rontgen Soc); 'Internal Energy of Elements' (Proc. Inst. Elect. Eng., 1906); 'Annual Reports on Radioactivity, 1904–1907' (Journ Chem Soc). With Prof Rutherford the following:-'Radioactivity of Thorium Compounds, Parts I and II' (Journ. Chem. Soc, 1902); 'Cause and Nature of Radioactivity, Parts I and II' (Phil. Mag., 1902); 'Radioactivity of Uranium' (ibid, 1903); 'A Comparative Study of the Radioactivity of

Radium and Thorium' (ibid, 1903); 'The Condensation of the Radioactive Emanations' (ibid, 1903); 'Radioactive Change' (ibid, 1903). With Sir William Ramsay:-'The Production of Helium from Radium' (Proc Roy Soc, 1903); 'Further Experiments on the Production of Helium from Radium' (ibid, 1904). With T D Mackenzie:-'The Relation between Uranium and Radium' (Phil. Mag., 1907); 'The Electric Discharge in Monatomic Gases' (Proc. Roy. Soc., 1908). With Baron R von Hirsch:-'A Gas Generated from Aluminium Electrodes' (Phil. Mag., 1907).)'

He was proposed by:

E. Rutherford; Joseph John Thomson; H. B. Dixon; William Ramsay; J. N. Collie; J. Joly; S. Young; A. V. Harcourt; Henry E. Roscoe; James Dewar; William A. Tilden.

Bibliography

1. Reproduced with permission.
2. From Nobel Lectures, Chemistry 1901–1921, Elsevier Publishing Company, Amsterdam, 1966.
3. The Nobel Population 1901–1937, E Crawford, J. L Heilbron and R Ullrich. Office of the History of Science and Technology, University of California, 1987, ISBN: 0-918102-15-4.
4. Biographical Memoirs of Fellows of the Royal Society 1957, vol. 3, pp. 203–216.
5. Papers deposited in the Bodleian Library, Oxford.
6. F. Soddy and R. E. Hughes. The action of dried ammonia on dried carbon dioxide gas. Chem. News, 22, 111.
7. Invisible Rays: A History of Radioactivity by G. I. Brown, Sutton Publishing, 2002, ISBN: 0750926678.
8. Practical Applications of Radioactivity and Nuclear Radiations by Gerhart Lowenthal and Peter Airey, Cambridge University Press, New ed. 2005, ISBN: 052101980X.
9. Radiochemistry and Nuclear Chemistry by Gregory Choppin, Jan Rydberg and Jan-Olov Liljenzin, Butterworth-Heinemann, 3rd ed. 2001, ISBN: 0750674636.
10. Nuclear and Radiochemistry: Fundamentals and Applications by Karl Heinrich Lieser, Wiley-VCH, 2nd ed. 2001, ISBN: 3527303170.
11. Sackler database, Royal Society.

1922

Francis William Aston

"for his discovery, by means of his mass spectrograph, of isotopes, in a large number of non-radioactive elements, and for his enunciation of the whole-number rule" [2]

Professor H. G. Söderbaum, member of the Nobel Committee for Chemistry of the Royal Swedish Academy of Sciences, made the following remarks during his presentation speech on 10th December 1922 [2]:

> *...for your discovery, made with the help of the mass spectrograph, of the isotopes of a great number of inactive elements and of the whole-number rule, discoveries which are of fundamental importance for the study of nature in general and for chemical science in particular. While expressing a cordial hope that it may be vouchsafed to you in the future to add further scientific successes to those which you have already obtained,...*

Aston was nominated for the 1922 Chemistry Nobel Prize by E. Baur. He was also nominated for the 1923 Chemistry Prize which was disallowed, and for the Physics Prize for 1922 which was also disallowed [3].

Francis William Aston was born on 1st September 1877 at Camomile Green, Harborne, Birmingham, England, the third child and only surviving son, of a family of seven children. His father (William) was a metal merchant. His mother was Fanny Charlotte Hollis. He was a life-long bachelor. He died on 20th November 1945 in Cambridge. After payment of death duties and bequests to members of his family, Aston directed an appreciable sum to various institutions. Amongst those that benefited were the Cavendish Laboratory, the Royal Society, Birmingham University, the British Association, the Cambridge Preservation Society, the Cambridge Philosophical Society, the Fitzwilliam Museum and Malvern College. All his scientific apparatus was left to the Cavendish Laboratory [2,4].

The Early Years

His earliest scientific memory concerned the nature and formation of soap bubbles and the action of sulphuric acid on zinc (which was conducted in a disused pigsty). He also made picric acid bombs cased in empty sparklet bulbs which he used as fireworks. Aston attended Harborne Vicarage School (September 1891). He entered Malvern College and later Mason College (later Birmingham University, September 1893) where he studied chemistry under Frankland and Tilden. He won the Forster Scholarship (1898) enabling him to work on the optical properties of tartaric acid derivatives, the results of which were published in 1901. Aston left academia for a while, taking a job as a chemist in a brewery. At this time his interests lay in physics and mechanical engineering, designing and constructing a new type of pump for evacuating vessels in his spare time. Aston was one of the early pioneers of motor-cycling, building his own machine (1902) and racing on it in the Gordon Bennett Race in Ireland (1903).

Aston obtained a scholarship to Birmingham University, as it had become (1903), and worked on the properties of the Crookes Dark Space in discharge tubes. Within a short time he had discovered the phenomenon which is now known as the Aston Dark Space. Towards the end of 1909, he moved to the Cavendish Laboratory, Cambridge, accepting the invitation of Sir J.J. Thomson to work as his assistant on studies of positive rays. During this period he obtained definitive proof of the existence of two isotopes of neon. Aston refined his high vacuum techniques and designed a camera which he used in his analyses. This camera became the standard apparatus for this work at that time. He was awarded the D.Sc. from Birmingham University (1914).

The Career

Aston's career started at the comparatively late age of 37 – see table.

Year	*Post*	*Institute*
1914–1918	Technical Assistant	Royal Aircraft Establishment, Farnborough
1919	Staff Scientist	Cavendish Laboratory, Cambridge University
1920	Research Fellow of Trinity College	Cambridge University

Aston studied the effect of atmospheric conditions on aeroplane fabrics and dopes (i.e. synthetic coatings) during the war years and returned to the Cavendish Laboratories after the war to concentrate his efforts on the separation of neon isotopes.

The Cambridge Period

Whilst at Trinity College, Aston worked on:

- the continuation of the work on the separation of neon isotopes
- the construction of the mass-spectrograph (1919) for which he was elected Fellow at Trinity College

- the proposal of the whole number rule formulated
- the concept of percentage diversions from whole numbers (using oxygen 16 as the standard) that led to the term 'packing fractions' (the results were given as the Bakerian lecture in 1927)
- 'Eclipse expeditions' (organized by the Joint Permanent Eclipse Committee of the Royal Society and the Royal Astronomical Society)
- the discovery of 212 naturally occurring isotopes
- the first table of stable isotopes (1936)

The results of his work were published in the *Proceedings of the Royal Society* and in the *Philosophical Magazine*. He was also the author of the books *Isotopes* (1922; revised edition 1941) and of *Structural Units of the Material Universe* (1923).

Aston's Nobel acceptance speech was titled 'Mass spectra and isotopes' and is dated 12th December 1922. During his lecture, Aston observed [2]:

A statistical study of the results given above shows that the natural restrictions can be stated in the form of rules as follows:

> ***In the nucleus of an atom there is never less than one electron to every two protons.*** *There is no known exception to this law. It is the expresssion of the fact that if an element has an atomic number N the atomic weight of its lightest isotope cannot be less than 2N. Worded as above, the ambiguity in the case of hydrogen is avoided. True atomic weights corresponding exactly to 2N are known in the majority of the lighter elements up to* ^{36}A*. Among the heavier elements the difference between the weight of the lightest isotope and the value 2N tends to increase with the atomic weight; in the cases of mercury it amounts to 37 units. The corresponding divergence of the mean atomic weights from the value 2N has of course been noticed from the beginning of the idea of atomic number.*
>
> ***The number of isotopes of an element and their range of atomic weight appear to have definite limits*** *— since the atomic number only depends on the net positive charge in the nucleus, there is no arithmetical reason why an element should not have any number of isotopes.*
>
> *So far the largest number appears to be 9 in the case of xenon, which also shows the maximum difference between its lightest and heaviest isotopes, 12 units. The greatest proportional difference calculated on the lighter weight is recorded in the case of lithium, where it amounts to one-sixth. It is about one-tenth in the cases of boron, neon, argon, selenium, krypton, and xenon. No element of odd atomic number has more than two isotopes.*
>
> ***The number of electrons in the nucleus tends to be even.*** *-This rule expresses the fact that in the majority of cases even atomic number is associated with even atomic weight, and odd with odd. If we consider the three groups of elements, the halogens, the inert gases and the alkali metals, this tendency is very strongly marked. Of the halogens - odd atomic numbers - all 6 (+1?) atomic weights are odd. Of the inert gases-even atomic numbers - 13 (+2?) are even and 3 odd. Of the alkali metals - odd atomic numbers - 7 are odd and I even. In the cases of elements of other groups the preponderance, though not so large, is still very marked. Beryllium and nitrogen are the only elements yet discovered to consist entirely of atoms whose nuclei contain an odd number of electrons.*
>
> *If we take the natural numbers 1 to 40, we find that those not represented by known atomic weights are 2, 3, 5, 8, 13, (17), (18), 21, (33), 34, (38). It is rather remarkable that these gaps, with the exception of the four in parentheses, are represented by a simple mathematical series of which any*

term is the sum of the two previous terms. In consequence of the whole-number rule there is now no logical difficulty in regarding protons and electrons as the bricks out of which atoms have been constructed. An atom of atomic weight m is turned into one of atomic weight m+1 by the addition of a proton plus an electron. If both enter the nucleus, the new element will be an isotope of the old one, for the nuclear charge has not been altered. On the other hand, if the proton alone enters the nucleus and the electron remains outside, an element of next higher atomic number will be formed. If both these new configurations are possible, they will represent elements of the same atomic weight but with different chemical properties. Such elements are called 'isobases''.

The study and uses of isotopes have continued to occupy the time of many chemists and physicists since Aston's award of the Nobel Prize. Some of the current isotope data is available in [6–8].

Honours and Awards

Aston was awarded Honorary Doctorates from the Universities of Birmingham (1923) and Dublin (1929, degree of Ph.D. at the Senate House(1924), Sc.D. (hc) Cambridge University (1924), the degree of Hon.Ph.D. of the University of Freiburg (1930), D.Sc. Benares (hc) (1937) and D.L. (hc) Calcutta (1938).

He was awarded the following awards/medals: Mackenzie Davidson Medal of the Röntgen Society(1920); Royal Society's Hughes Medal (1922); Nobel Chemistry Prize (1922); John Scott Medal (1923); the Paterno Medal (1923); the Royal Medal (1938), the Joykissen Mookherjee Medal (Calcutta 1938); and the Duddell Medal of the Physical Society (1941).

He was elected to the following Fellowships/distinguished posts: Fellowship at Trinity College (1920); Fellow of the Royal Society (elected 12/05/1921); honorary Member of the Russian Academy of Sciences and of the Accademia dei Lincei; Fellowship of the Institute of Chemistry (1923); elected to the Council of the Royal Society (1924); Member of the Athenaeum Club (1925); Member of the Russian Academy of Science (1925); Foreign Member of the Accademia Nazionale dei Lincei (1926); Council Member of the Royal Society (1927); Member of the Board of Governors of Birmingham University (1927); Chairman of the International Committee on Atoms of the Union Internationale de Chimie (1935); President of the Cambridge Philosophical Society (1935); President of Section A at the Norwich Meeting of the British Association (1935); Hon Silver Jubilee Member Indian Science Congress Association (1939).

Aston's portrait was painted for the National Gallery in 1944.

Aston's citation for election (date: 12/05/1921) as Fellow of the Royal Society reads [9]:

'Distinguished for contributions to our knowledge of the passage of electrical discharge through gases, and particularly for his proof of the complexity of the atoms of many of the ordinary elements of the positive ray method. In a series of five papers published in Proc. Roy. Soc. between 1907 and 1913 an account is given of experiments to determine the length of the cathode dark space in different gases under varied conditions. 'A Simple Form of Microbalance' (Proc. Roy. Soc., 1914) was designed to test the separation of the isotopes of neon by diffusion. In 'A Positive Ray Spectrograph' (Phil. Mag, 1919) a new method is described for determining the mass spectra of the chemical elements. In the 'Constitution of Atmospheric Neon' and 'The Mass Spectra of the Chemical Elements' (Phil. Mag, 1920, and Nature, 1919 and 1920), a number of the elements are shown to consists of two or more isotopes whose masses have been measured. Author of numerous other papers on electric discharge in gases.'

He was proposed by:

Ernest Rutherford; J. J. Thomson; William Pope; J. W. Nicholson; O. W. Richardson; F. Soddy; Joseph Larmor; J. S. Townsend; F. A. Lindemann; C. T. R. Wilson; Frank Dyson.

Bibliography

1. Reproduced with permission.
2. From Nobel Lectures, Chemistry 1922–1941, Elsevier Publishing Company, Amsterdam, 1966.
3. The Nobel Population 1901–1937, E. Crawford, J. L. Heilbron and R. Ullrich. Office of the History of Science and Technology, University of California, 1987, ISBN: 0-918102-15-4.
4. Obituary Notices of Fellows of the Royal Society 1945–1948, vol. 5, pp. 635–650.
5. Papers deposited Cambridge University Library (covering the period 1911–1945).
6. Isotopes: Principles and Applications by Gunter Faure and Teresa M. Mensing, Wiley, 3rd ed. 2004, ISBN: 0471384372.
7. Handbook of Stable Isotope Analytical Techniques, Volume I by P.A. de Groot, Elsevier Science, 2004, ISBN: 0444511148.
8. Principles of Stable Isotope Geochemistry, 1/e by Zachary Sharp, Prentice Hall, 2006, ISBN: 0130091391.
9. Sackler database, Royal Society.

1923

Fritz Pregl

"for his invention of the method of micro-analysis of organic substances" [2]

During his presentation speech on the 10th December 1923, Professor O. Hammarsten then Chairman of the Nobel Committee for Chemistry of the Royal Swedish Academy of Sciences, commented [2]:

> *...Pregl succeeded, by introducing new apparatus and techniques, in reducing to the almost incredibly small amount of 5.3 milligrams and even less the quantity normally required for the quantitative determination of various elements in organic compounds. The same amount of substances which up to now would have been required for a single carbon-hydrogen determination, is now with Pregl's methods sufficient for up to 50 different analyses. In this way this quantity suffices for several determinations not only of carbon and hydrogen, but also of other elements forming the compound; it also makes possible certain other investigations important for the study of the chemical structure of the compound, its constitution...*

Pregl was nominated for the 1923 prize by J. Sjöqvist, K. A. Vesterberg and by H. von Euler-Chelpin. He was also nominated for 1917 and 1922 [3].

Fritz Pregl was born in Laibach on 3rd September 1869. He never married, and died after a short illness at Graz on 13th December 1930. Shortly before his death, he donated a large sum of money to the Vienna Academy of Sciences. This was to be used for micro-chemical research and the interest from this fund to be used each year to award a prize for outstanding work to Austrian micro-chemists. The "Fritz Pregl Prize" has been awarded since that time by the Vienna Academy of Sciences [2].

The Early Years

Pregl attended the local Gymnasium (grammar school) proceeding to the University of Graz to study medicine. He received his M.D. in 1894. Prior to graduation he became Assistant Lecturer for physiology and histology under Alexander Rollett, taking over the Chair (1903) upon the death of Rollett.

The Career

Pregl went to Germany (1904) studying for short periods under Gustav v. Hüfner (Tübingen), W. Ostwald (Leipzig) and Emil Fischer (Berlin), returning to Graz (1905) to work under K.B. Hofmann (at the Medico-Chemical Institute). His career started in earnest in 1907 – see table.

Year	*Post*	*Institute*
1907	Forensic Chemist for the Graz circuit	Medico-Chemical Institute, Graz
1910	Professor	Innsbruck University
1913	Professor	Graz University
1916–1917	Dean of the Medical Faculty	Graz University
1920–1921	Vice-Chancellor	Graz University

Pregl's initial research interests (1907) were the investigation of the components of albuminous bodies and the analysis of bile acids. He was hampered in this work by the lack of large quantities of starting materials which forced him to look at using smaller quantities of materials for his studies.

His Innsbruck years (1910–1913), were almost entirely devoted to developing micro-analysis techniques for quantitative organic analysis, which continued when he was recalled to Graz University. Pregl's techniques of quantitative micro-analysis enabled him to make measurements of carbon, hydrogen, nitrogen, sulphur and halogen, using only 5–13 mg. His results were just as accurate as those obtained by macro-analysis undertaken 10 years previously. Pregl's interest in micro-techniques and methods were extended allowing these methods to be used for measuring atomic groups. Of necessity, he was instrumental in developing a sensitive microbalance for this work.

Pregl published a monograph entitled *Die Quantitative Microanalyse* (published by J. Springer, Berlin) (1917) with the 2nd edition in 1923 and a 3rd revised edition (enlarged) appeared in 1930. It has been translated into French and English.

During Pregl's Nobel acceptance speech, titled 'Quantitative Micro-Analysis of Organic Substances' and dated 11th December 1923 [2] he commented...

> *Finally it might not be superfluous to raise the question what positive results micro-analysis has achieved so far.*
>
> *In this connection we may mention that one of the first results was the discovery that the yellow colouring matter in corpora lutea of cattle ovaries was by micro-analytical carbon hydrogen determination found to be carotin, the substance which is in carrots widely found in the vegetable kingdom. This was extracted from approximately 10,000 ovaries in the Willstätter Laboratory in Zurich and would have been sufficient for only one macro-analysis. Thus, in this case already micro-analysis has made possible important progress in our knowledge, by proving that the same substance appears as colouring matter in both natural kingdoms, the vegetable as well as the animal kingdom.*
>
> *Also, in view of the minute quantity of material available, I want to draw your attention to the valuable information on molecular structure obtained by means of micro-analysis, through the work*

of Adolf Windaus on cymarin and cholesterol, and Wieland's work on the decomposition of bile acids, where micro-analysis proved its value.

Last but not least I want to mention that quite recently a Swede, C.A. Yllner, applied micro-analytical methods successfully at my Institute in Graz for the identification of alkaloids in forensic cases, that is, when these had to be extracted from dissected corpses. During the last few years my methods have also been applied in technology for electrolytic metal analysis; and for the analysis of wine very clever micro-analytical methods have been developed, apart from those discussed here.

Pregl's methods of micro-analylsis are now standard techniques in all laboratories. Perhaps a worthy example of their use is in the conservation, restoration and study of materials and artefacts of culturo-historical value, see [4] for further details.

Honours and Awards

Pregl was awarded an Honorary Doctorate in Philosophy from the University of Göttingen (1920).

Pregl received the Lieben Prize for Chemistry from the Imperial Academy of Sciences in Vienna (1914), Nobel Prize for Chemistry by the Swedish Academy of Sciences (1923).

He was a Corresponding Member by the Academy of Sciences in Vienna.

Bibliography

1. Reproduced with permission.
2. From Nobel Lectures, Chemistry 1922–1941, Elsevier Publishing Company, Amsterdam, 1966.
3. The Nobel Population 1901–1937, E. Crawford, J. L. Heilbron and R. Ullrich. Office of the History of Science and Technology, University of California, 1987, ISBN: 0-918102-15-4.
4. Non-destructive Micro Analysis of Cultural Heritage Materials, Volume XLII, First Edition (Comprehensive Analytical Chemistry) by K. Janssens and R. van Grieken, Elsevier Science, 2005, ISBN: 0444507388.

1924

The prize money was allocated to the Special Fund of this prize section

1925

© Nobel Foundation [1]

Richard Adolf Zsigmondy

"for his demonstration of the heterogenous nature of colloid solutions and for the methods he used, which have since become fundamental in modern colloid chemistry" [2]

The Nobel Prize in Chemistry for 1925 was announced on 11th November 1926, the presentation ceremony for which was held on 10th December 1926. Professor H.G. Söderbaum, then Secretary of the Royal Swedish Academy of Sciences observed [2]:

> *...but should surely suffice to show how it pioneered the way and opened up new regions in a field of research which had so far been difficult of access, a field which must be recognized as having the very greatest importance for human knowledge. Let us only remember in this connection that all manifestations of organic life are finally bound to the colloidal media of the protoplasm... for proving the heterogeneous nature of colloidal solutions and for the methods used which have laid the foundation of modern colloid chemistry...*

Zsigmondy was nominated for the 1925 prize by G. Meyer, V. Rothmund and T. Svedberg. He was also nominated for the years 1913, 1915, 1919*, 1921*, 1922, 1923, 1924, and 1926, where * signifies one or more disallowed nominations for that year [3].

Richard Adolf Zsigmondy was born in Vienna on 1st April 1865, one of four children of Adolf Zsigmondy and Irma, née von Szakmary. His father, a doctor, invented several surgical instruments and devices for dentistry, and had published several scientific and medical works. His four children were all encouraged to develop an interest in the natural sciences. He died when Zsigmondy was only 15 years old.

Zsigmondy married Laura Luise, née Müller (the daughter of Professor Wilhelm Müller) in Jena (1903), they had two daughters, Annemarie and Käthe. Annemarie married Dr. Erich Huckel of Göttingen (1925), who was Assistant to Professor Debye (whilst in Zurich). Zsigmondy died in Göttingen on 24th September 1929 [2].

The Early Years

Whilst still very young, Zsigmondy studied Stoeckhardt's textbook *Schule der Chemie* and carried out many of the experiments mentioned there in his own small home laboratory. He learned the basic facts about quantitative analysis from Professor E. Ludwig of the Medical Faculty in Vienna before more formal study at the Technische Hochschule also in Vienna. He then studied organic chemistry under Professor W. von Miller in Munich (1887). After taking his doctorate, he remained as Professor von Miller's Assistant, later accepting a similar position as Assistant to the physicist Professor Kundt (Berlin).

The Career

Year	*Post*	*Institute*
1893	Lecturer	Technische Hochschule (Graz)
1897	Scientist	Schott und Genossen glass works (in Jena)
1900	Private Scholar	Jena
1907	Professor and Director of the Institute of Inorganic Chemistry	University of Göttingen
February 1929.	Retired	

Zsigmondy's early work concerned lustre colours for glass and china. This work helped him further his studies on colloids in more detail. He discovered how to reproducibly prepare gold hydrosols and developed the slit-ultramicroscope (jointly with Siedentopf).

Zsigmondy published a book *Über das kolloide Gold* (with P.A. Thiessen) and his collected papers were published as *Kolloidforschung in Einzeldarstellungen.*

Zsigmondy's Nobel acceptance speech, titled 'Properties of Colloids' on 11th December 1926, in which he observed [2]...

> *Had I conceded that my gold solutions were simply to be regarded as suspensions, it would not only have given a completely wrong picture of their character, but if I had classified my gold divisions under such well known systems as suspensions, there would have been no reason for me to concern myself with the following questions which have become extremely important for colloid chemistry:*
>
> (i) *What particle size have colloidal particles? Are they molecules or aggregates of molecules? How many atoms approximately does a single gold particle contain?*
> (ii) *Why is the emission of osmotic energy so low with gold solutions? Do colloidal solutions perhaps really exert no osmotic pressure, as was frequently believed at the time?*
> (iii) *Is the Faraday-Tyndall light cone, which by the way is observed in all colloidal solutions, an essential characteristic of them, or is this light cone rather incidental?*
>
> *Question 1 regarding the particle size was decided in 1902 and later by ultramicroscopy. The correct answer to Question 2 is given by the masterly work of The Svedberg and Perrin. Question 3 I hoped to be able to decide already in 1900 by viewing the light cone under the microscope.*

Further in his acceptance speech, Zsigmondy provided a discourse concerning:

- the nature of electrical charge and coagulation
- colloids in gel structures
- the transport of electricity by colloidal particles and
- the structure of the particles

The application of colloid science in the development of commercial products is now commonplace in many industries. This is reflected in the number of texts that have been written since Zsigmondy was awarded the Nobel Prize, see for example [4–6].

Honours and Awards

Zsigmondy was awarded the Nobel Prize for Chemistry for the year 1925.

Bibliography

1. Reproduced with permission.
2. From Nobel Lectures, Chemistry 1922–1941, Elsevier Publishing Company, Amsterdam, 1966.
3. The Nobel Population 1901–1937, E. Crawford, J.L. Heilbron and R. Ullrich. Office of the History of Science and Technology, University of California, 1987, ISBN: 0-918102-15-4.
4. Handbook of Surface and Colloid Chemistry, Second Edition by K.S. Birdi, CRC, 2nd ed. 2002, ISBN: 0849310792.
5. Colloid Chemistry I (Topics in Current Chemistry) edited by Markus Antonietti Springer, 1st ed. 2003, ISBN: 3540004157.
6. Principles of Colloid and Surface Chemistry (Undergraduate Chemistry Series) by Paul Hiemenz, CRC, 3rd ed. 1997, ISBN: 0824793978.

1926

© Nobel Foundation [1]

The (Theodor) Svedberg

"for his work on disperse systems – (the ultracentrifuge)" [2]

During his presentation speech dated 10th December 1926, Professor H.G. Söderbaum, then Secretary of the Royal Swedish Academy of Sciences, commented [2]:

> *...During the last few years Svedberg has completed an extremely ingenious invention, the so-called ultracentrifuge, which enables highly interesting investigations to be made also on such molecular-disperse systems. We know that when a slurry, an emulsion, is put into a rapidly rotating motion, its heavier constituents are thrown outwards in the direction of the periphery of the motion. This happens in the most used of all centrifuges, the milk separator, where the skimmed milk is pressed outwards whilst the lighter fat particles, the cream, accumulate inwards and can therefore be separated. Similarly in a solution, when centrifuging is sufficiently rapid, the molecules of the dissolved substance must accumulate outwards if they are considerably heavier than the molecules of the solvent...*

Svedberg was nominated for the 1926 Chemistry Nobel Prize by W. Nernst. Svedberg was also nominated for the Chemistry Prize for the years 1910*, 1911*, 1915*, 1916, 1918, 1919, 1920, 1923, 1924 and 1925*. He was also nominated for the Nobel Physics Prize for the years 1913 and 1919, where * represents one or more disallowed nominations for that year [3].

Theodor (The) Svedberg was born at Fleräng, Valbo, Gävleborg, Sweden on 30th August 1884 (some sources record 13th), the son of Elias Svedberg (a works manager) and his wife, Augusta, née Alstermark. Svedberg married four times: Andrea Andreen (in 1909, two children); Jane Frodi (in 1916, four children); Ingrid Blomquist (in 1938, three children); and Margit Hallén (in 1948, three children). There were six sons and six daughters. Svedberg died in Orebro, Sweden on 25th February (some sources record 26th February) 1971 and is buried at Ljusnarsberg cemetery, Kopparberg [2,4].

The Early Years

Svedberg attended Köping School, Örebro High School and Gothenburg Modern School from which he passed his student examination (December 1903). He attended Uppsala University (January 1904) gaining his Bachelor of Arts degree (1905), his Master's degree (1907) and his Doctor of Philosophy (1908). His doctoral thesis was titled *Studien zur Lehre von den kolloiden Lösungen* and is now regarded as a classic. He described a new method of producing colloid particles and provided supportive evidence for Brownian motion (Einstein and von Smoluchowski's theory on the Brownian motion was novel at the time). His first paper, published in 1905, concerned the preparation metal organosols [5].

The Career

Svedberg had a lifelong association with Uppsala Univeristy. His appointments were:

Year	*Post*	*Institute*
1905	Assistant	Chemical Institute at Uppsala
1907	Additional position of Lecturer in Chemistry	University of Uppsala
1909	Lecturer and Demonstrator of physical chemistry	University of Uppsala
1913[a]	Professor of Physical Chemistry	University of Uppsala
1949	Emeritus Professor	University of Uppsala
1949	Director of the Gustaf Werner Institute for Nuclear Chemistry	University of Uppsala

[a] permanent personal Chairs were rarely given in Sweden at that time. Svedberg's required a Royal Appointment which was granted on 29th June 1912, effective from January 1913. His inaugural lecture (on 1st February 1913) was titled 'Atomic research in modern physics and chemistry'

Although Svedberg was associated with the University of Uppsala for the whole of his professional life he travelled widely, visiting numerous laboratories in foreign countries and cities: Germany (1908); Holland and France (1912); Berlin (1913); Vienna (1916); London and Paris (1920); England and Denmark (1922); the United States and Canada (1922–1923). His sojourn at the University of Wisconsin (USA) lasted eight months during which he gave lectures, organized research in centrifugal fields, electrophoresis, diffusion, electrical colloid synthesis and some work on the photographic process.

Svedberg's contribution to colloids included:

- the physical properties of colloids(diffusion and light absorption)
- a study of sedimentation (for which he constructed his well-known ultracentrifuge)
- showing that the molecules of certain pure proteins are all of one size
- demonstrated the use of the ultracentrifuge to show the presence of contaminants
- further studies concerning proteins
- studies on ovalbumin, phycocyanin, phycoerythrin and haemocyanin
- publishing a large monograph concerning the different methods of preparing colloidal solutions of inorganic substances (1909)

His other work covered:

- problems of nuclear chemistry
- radiation biology

- photographic processing
- assisting in photographing the famous *Codex Argenteus* in colour

He also worked out a method for making synthetic rubber during the Second World War.

Svedberg has contributed a large number of scientific papers to Swedish and foreign journals on colloid and high-molecular solutions, nuclear chemistry and radiation biology. A full list of his papers appears in the Biographical Memoirs of Fellows of the Royal Society [4].

Among his publications the following books should be mentioned: *Die Methoden zur Herstellung kolloider Lösungen anorganischer Stoffe* (1909), *Die Existenz der Moleküle* (1912), *Colloid Chemistry* (1914; 2nd Enl. Ed. with A. Tiselius 1928), *The Ultracentrifuge* (with K.O. Pedersen 1940; also in German). He has written several excellent popular books in Swedish, namely *Materien* (1912; 2nd Ed. 1919; also German Ed. 1914), *Arbetets dekadens* (1915, 1920; German Ed. 1923) and *Forskning och Industri* (1918).

Svedberg's Nobel acceptance speech, titled 'The Ultracentrifuge' dated 19th May 1927, in which he observed [2]...

> *The values obtained by sedimentation equilibrium and sedimentation speed correspond to one another within the test error limits. It is interesting to note how the molecular weight of phycoerythrin is approximately twice the molecular weight of phycocyanin. These two proteins occur together in red algae (the material studied originates from a Swedish species of the genus Ceramium). According to analyses by Kylin (1910) and Kitasato (1925), their elementary composition is identical within the analysis error limits. According to the centrifugation examination carried out by Lewis, it is therefore probable that the main difference between them is that phycoerythrin has twice the molecular weight of phycocyanin. Phycoerythrin is red (strong absorption band in green), while phycocyanin is blue (strong absorption band in yellow). Since these two proteins are relatively difficultly soluble and they also contain no metallic constituents, it was impossible to determine or estimate their molecular weight in any way other than by centrifugation.*
>
> *Examination of highly disperse coiloids and other solutions of similar properties by centrifugation is still at its very beginning. The results obtained clearly show that in this way it will be possible gradually to collect a valuable fund of knowledge concerning such medicinally and technically important groups as the proteins, polymeric hydrocarbons (for example rubber), polymeric carbohydrates (for example starch, cellulose), etc.*

The centrifuge is now regarded as a commonplace instrument in many laboratories, each generation of which producing greater spin speeds or other features. Perhaps for this reason there is not an extensive collection of texts concerning the centrifuge. The reader is directed to [6] as an example of one of the later texts.

Honours and Awards

Svedberg was awarded Honorary Degrees from the following Universities: Groningen (1914); Madison, Wisconsin (1923); medicine at Uppsala (1932); Harvard (1936); Oxford (1937); Delaware (1937); Paris (1950).

He was awarded the following awards/medals: Scheele (1907, and 1916), Edlund (1909), Wallmark (1910) and the Berzelius Medal (1941) (all from Royal Swedish Academy of Sciences); John Ericsson

Medal (1942); the Medal of the Franklin Institute (1949); Nobel Prize in Chemistry (1926); Bergstedt Reward (1910); the Oscar II Reward (University of Uppsala 1908); the Björkén Prize (1913, 1923 and 1926); and the Gustaf Adolf Medal (1948). He also received the Medal of the City of Uppsala (1948).

Svedberg was elected Member or Honorary Member of more than 30 learned societies all over the world. He was a member of the Swedish Academy of Sciences (elected when 28 years old), the Halle Academy, The Chemical Society (London), the Indian Academy of Science, the American Philosophical Society (Philadelphia), the New York Academy of Sciences, the Royal Society (London, elected 15/06/1944) and the National Academy of Science (Washington), Honorary Fellow, U.S.S.R. Academy of Sciences, Moscow (1966).

Svedberg's citation for election (date: 15/06/1944) as Foreign Member of the Royal Society reads [7]:

> **'Theodor Svedberg of Uppsala University distinguished for his work in physical and colloid chemistry and the development of the ultracentrifuge.'**

He was proposed by:

H. H. Dale; A. V. Hill; A. C. Egerton; H. T. Tizard; J. R. Merton; H. W. Turnbull; D. R. Hartree; E. N. da C Andrade; H. R. Robinson; E. J. Salisbury; E. J. Bowen; J. L. Simonsen; Percival Hartley; V. B. Wigglesworth; W. E. Le Gros Clark; F. A. E. Crew; E. B. Bailey; H. J. Fleure; Paul Fildes; Thomas Lewis.

Bibliography

1. Reproduced with permission.
2. From Nobel Lectures, Chemistry 1922–1941, Elsevier Publishing Company, Amsterdam, 1966.
3. The Nobel Population 1901–1937, E. Crawford, J.L. Heilbron and R. Ullrich. Office of the History of Science and Technology, University of California, 1987, ISBN 0-918102-15-4.
4. Biographical Memoirs of Fellows of the Royal Society 1972, vol 18, pp 595–627.
5. Ueber die Elektrische Darstellung Einiger Neuen Colloidalen Metalle. Ber. Dt. Chem. Ges., 1905, 38, 3616–3620.
6. Decanter Centrifuge Handbook by A. Records and K. Sutherland, Elsevier Science, 1st Ed 2001, ISBN: 1856173690.
7. Sackler database, Royal Society.

1927

© Nobel Foundation [1]

Heinrich Otto Wieland

"for his investigations of the constitution of the bile acids and related substances" [2]

The Nobel Prize in Chemistry 1927 was announced on 13th November 1928, and the presentation took place on 10th December 1928. During his presentation speech, Professor H.G. Söderbaum, then Secretary of the Royal Swedish Academy of Sciences, remarked of Wieland [2]:

> *..Wieland succeeded in producing from bile a saturated acid which can be regarded as the mother substance or parent acid of the bile acids, and which he studied and characterised in detail. When Windaus then produced this same parent acid, cholanic acid, from cholesterol by means of a complicated and very ingenious series of experiments, this indicated very clearly the close relationship between cholesterol and the bile acids. It should be pointed out in this connection that Wieland's investigations into bile acids themselves gave a deeper insight of the mechanism of the action of the bile in the resorption of food in the intestines...*

Both Wieland and Windaus (Chemistry Nobel prizewinners for the years 1927 and 1928) were presented on the same day in 1928.

Wieland was nominated for the 1927 Prize by S. Arrhenius and E. Vahlen. He was also nominated for the years 1924, 1925*, 1926, 1928, 1933 and 1934, where * represents one or more disallowed nomination for that year [3].

Heinrich Otto Wieland the son of Theodor Wieland and Elise, née Blom, was born on 4th June 1877, in Pforzheim, Germany, where his father was a pharmaceutical chemist. He married Josephine Bartmann (1908) of Munich. They had three sons, Wolfgang (a doctor of pharmaceutical chemistry), Theodor (Professor of Chemistry in the University of Frankfurt), and Otto (Professor of Medicine in the University of Munich), and one daughter Eva. Eva married Feodor Lynen, Professor of Biochemistry in the University of Munich, Nobel Laureate in Physiology or Medicine (1964).

Wieland died on 5th August 1957 in Starnberg, Germany [2,4].

The Early Years

After leaving his local grammar school (1896) Wieland went on to study for two semesters at the University of Munich followed by the winter semester (1897) at the University of Berlin. This was followed by a year's study at the Technical High School in Stuttgart. He then returned to the Baeyer Laboratory in Munich where he received his doctorate under Johannes Thiele (1901). For the following three years Wieland remained in Munich, after which he qualified as a lecturer.

The Career

Wieland's career details are listed in the table below. During his career Wieland undertook pioneering work in both organic chemistry and biochemistry. His return to the University of Munich at the specific request of Willstätter who asked Wieland to succeed him in the Chair in Munich, whose previous occupants had been von Baeyer and Liebig.

Year	*Post*	*Institute*
1909	Extraordinary Professor	Koniglich Bayern Technische Hochschule
1913	Senior Lectureship in the University Chemical Laboratory	Koniglich Bayern Technische Hochschule
1917	Professor	Koniglich Bayern Technische Hochschule
1917–1918	Defence work	Kaiser Wilhelm Institute in Berlin-Dahlem
1921	Ordinary Professor of Chemistry	University of Freiburg
1925	Ordinary Professor of Chemistry	University of Munich
1952	Retired	

Organic chemistry contributions included:

- studies of the reaction of nitrogen oxides with orefins and aromatics
- the clarification of furoxanes
- classical experiments with fulminic acid
- succeeded in the first production of stable organic nitrogen radicals (diphenyl nitrogen and its N-oxide)

Biochemistry contributions include:

- the clarification of the structure of morphine
- the clarification of the structure of strychnine
- the constitution and synthesis of the lobelia alkaloid
- research into the curare alkaloid

- work on the poisonous agent in the "death cap" mushroom
- the isolation of the crystalline cyclopeptides phalloidine and amanitine
- work on the pigment of butterflies
- the discovery of the biologically important class of pterin compounds
- the publications on the bile acids (publications from 1912–1932)
- the clarification of the carbon framework of the steroids
- investigations into the oxidation processes in living cells
- theory of primary activation of hydrogen
- the dehydrogenation theory
- a study of the toad poisons

In his biographical memoir, Karrer observed [4]:

> *...With characteristic perseverance, Wieland also solved the problem of the constitutions of the most important of the then known pterins, a task that was more difficult than would be expected from the relatively simple structures of these compounds. For a long time, unusual analytical difficulties and ambiguity in the interpretation of the experimental results prevented him from putting forward well-founded formulae. Eventually, however, all these difficulties were overcome and the structures of leucopterin, xanthopterin and isoxanthopterin (8-desoxyleucopterin) were confirmed by synthesis...*

Wieland's scientific work was recorded in some four hundred publications. An extensive list was compiled by Karrer in his biography of Wieland [4].

Wieland's Nobel acceptance speech was titled 'The Chemistry of the Bile Acids' dated 12th December 1928. In describing the path towards determining the structure of the bile acids, Wieldand commented [2]...

> *...The problem is not very attractive from the experimental viewpoint. There is no nitrogen, which adds interest and variety to the treatment of alkaloids. Only carbon, hydrogen and a little oxygen, all in the traditional combination, which does not lead us to expect any surprising results...*

> *In order to get a comprehensive picture of the biological relationships in the wide field of chemically related natural substances which, in addition to the bile acids, includes the sterols, and very likely also the vegetable cardiac poisons from the group of saponins, the poisons from the skin secretions of toads, and probably also other important substances such as certain vitamins, we must above all be clear as to the fundamentals of their chemical structure. My object, of briefly describing the extent to which the chemical nature of ihe bile-acid molecule has been determined, can best be achieved by basing my account on the formulae which so far appear to us to be the most likely ones...*

> *When the last puzzles of our constitution problems have been solved, we can expect as usual the synthesis of the compounds. In cholanic acid there are seven asymmetric carbon atoms, and in cholic acid there are ten; as yet no procedure has been developed for the experimental production of the peculiar bonding of the numerous rings. While I acknowledge that it is the duty of the organic analyst to carry out syntheses such as have been carried out so excellently in the investigation of the coloured components of the colouring matter of blood, I must decidedly repudiate this in the field being discussed. I am fully conscious, however, of the duty to follow the course that has been embarked upon right to the end...*

The structure of cholanic acid is:

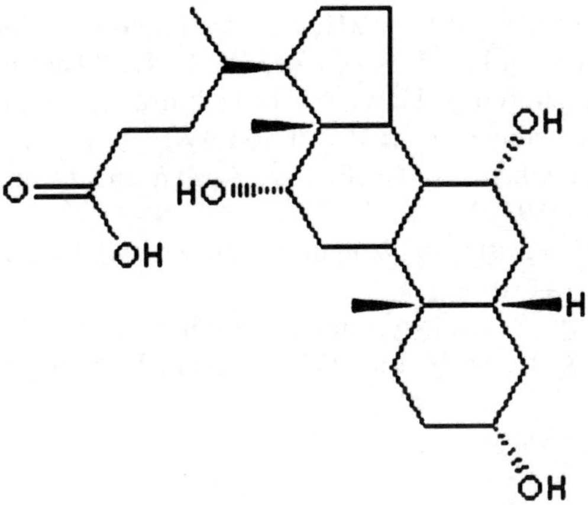

Some of the more important bile acids are now covered in modern general text books – see for example [5, 6]. A more specialist text concerning the bile acids has not been written for nearly twenty years – see for example [7].

Honours and Awards

Honorary degrees were conferred on Wieland by the Technical High School, Darmstadt (Dr.Ing.), the University of Freiburg (Dr.Med.), and the University of Athens (Dr.Phil.).

He received the Order of Merit from Germany, the Nobel Chemistry Prize (1927) and was the first recipient of the Otto Hahn Prize (1955).

Wieland was a member of the great learned societies of the world including: Foreign Member of the Royal Society (Election Date: 25/06/1931): the National Academy of Sciences (Washington); the Academy of Arts and Sciences (Boston); an Honorary Member of the Chemical Society (London); the Romanian Chemical Society; the Japanese Academy; the Indian Academy; a Member of the Academies of Munich, Gottingen, Heidelberg, and Berlin; a Corresponding Member of the Academy of Sciences of Moscow.

For 20 years Wieland was Editor of *Justus Liebigs Annalen der Chemie*.

Wieland's citation for election (date: 25/06/1931) as Foreign Member of the Royal Society reads [8]:

'Heinrich Wieland of Munich distinguished for his work in organic chemistry'

He was proposed by:

F. G. Hopkins; H. H. Dale; Rayleigh; G. H. Hardy; G. I. Taylor; E. A. Milne; J. A. Ewing; R. V. Southwell; E. V. Appleton; H. Hartley; W. W. Watts; A. B. Rendle; D. M. S. Watson; E. S. Goodrich; George Barger; P. P. Cathcart; Thomas Lewis; A. C. Boycott; H. G. Lyons; F. E. Smith.

Bibliography

1. Reproduced with permission.
2. From Nobel Lectures, Chemistry 1922–1941, Elsevier Publishing Company, Amsterdam, 1966.
3. The Nobel Population 1901–1937, E. Crawford, J. L. Heilbron and R. Ullrich. Office of the History of Science and Technology, University of California, 1987, ISBN: 0-918102-15-4.
4. Biographical Memoirs of Fellows of the Royal Society 1958, vol. 4, pp. 341–352.
5. Modern Physical Organic Chemistry by Eric V. Anslyn and Dennis A. Dougherty, University Science, 2005, ISBN: 1891389319.
6. Introduction to Organic Chemistry by William H. Brown and Thomas Poon, John Wiley & Sons, 3rd ed. 2004, ISBN: 0471444510.
7. The Bile Acids: Chemistry, Physiology, and Metabolism: Methods and Applications edited by K.D.R. Setchell, David Kritchevsky and Padmanabhan P. Nair, Springer, 1st ed. 1988, ISBN: 0306426226.
8. Sackler database, Royal Society.

1928

© Nobel Foundation [1]

Adolf Otto Reinhold Windaus

" for the services rendered through his research into the constitution of the sterols and their connection with the vitamins" [2]

During his presentation speech dated, 10th December 1928, Professor H.G. Söderbaum, then Secretary of the Royal Swedish Academy of Sciences, observed [2]:

> *...As a result of patient and skilful work, Windaus succeeded in producing several of the digitalis glucosides and their components in the pure state. In this way it was shown that these vegetable cardiac poisons are directly related on the one hand to cholesterol and the bile acids, and on the other hand to the animal cardiac poison bufotoxin, which Wieland studied with great success...*

> *...Another sterol which Windaus has studied in detail, is ergosterol, which occurs partly in ergot and partly in yeast. The research carried out in recent years, in which Windaus himself has also played a leading part, has revealed the very important fact that, on being irradiated with ultraviolet light, this ergosterol assumes exactly the same properties as the antirachitic vitamin, "vitamin D", i.e. it will cure rachitis (rickets)...*

Windaus was nominated for the 1928 Nobel Chemistry Prize by F. Harber, O. Munn, S. Suida, W. Wien and R. Willstätter. He was also nominated for the years 1925*, 1926, 1927 and 1929*, where * represents one or more disallowed nominations [3].

Adolf Windaus was born in Berlin on 25th December 1876, the son of Adolf Windaus and Margarete Elster. He married Elisabeth Resau (1915) and had three children, Günter (1916), Gustav (1918), and Margarete (1921) [2,4].

Windaus died on 9th June 1959.

The Early Years

After attending the Französisches Gymnasium (French grammar school) in Berlin, Windaus started to study medicine in Freiburg and Berlin (1895), passing his preliminary medical examinations ("Physikum") in 1897. Influenced by Emil Fischer's lectures during his stay in Berlin, Windaus began to study chemistry under Kiliani in Freiburg whilst also continuing his medical studies. Windaus obtained his Dr. Phil. degree (in the winter of 1899–1900), the subject of his thesis being the cardiac poisons of the Digitalis plant.

Windaus moved to Berlin to work with Emil Fischer after his graduation. Whilst there, he was to form a lifelong friendship with Otto Diels (see entry for 1950). Returning to Freiburg (1901), Windaus began work on cholesterol and related sterols (on Kiliani's suggestion).

The Career

For the majority of his career, Windaus worked at the University of Göttingen.

Year	*Post*	*Institute*
1906	Assistant Professor	Innsbruck University
1913	Professor of Applied Medical Chemistry	Innsbruck University
1915	Professor of Chemistry and Director of the Laboratory for General Chemistry	University of Göttingen
1944	Retired	

His principal research interests were all concerned with biochemical investigations including:

- the sterols (which occur in every cell) which he considered as the parent substance of other groups of natural substances
- successfully transforming cholesterol into cholanic acid (1919), thereby demonstrating that the bile acids are closely related to the sterols
- an investigation of imidazole derivatives (in collaboration with Knoop). Windaus attempted to prepare natural amino acids by reacting sugar with ammonia and thus to establish conversion of sugar into proteins. Through the production of the resultant imidazole derivatives, Windaus demonstrated the presence of a prototype of the imidazole ring in proteins
- the confirmation that histidine (a building-block of proteins) is an imidazole alanine
- the discovery of histamine (imidazole-ethylalanine). Many groups were interested in his work. The German chemical industry saw the potential of the histamines as did Alfred Hess (the American physiologist). Thus Windaus travelled to New York to work on a joint study of the antirachitic vitamin (1925).
- the structure of the antineuritic vitamin. Windaus was able to prove that it is not an imidazole-ring (as suggested by Jansen and Donath) and furthermore that a thiazole- and a pyrimidine-ring are present in vitamin B_1
- the stereochemical problems of the *cis*- and *trans*-linking of hydrogenated ring systems
- his investigations of the possible use of chemotherapy in cancer research

Windaus' Nobel Prize acceptance speech was titled 'Constitution of Sterols and their connection with other substances occurring in nature' and is dated 12th December 1928 [2].

> *...Although the animal thus is in a position to absorb cholesterol from its food with the aid of bile salts, nevertheless not even the carnivore is* ***dependant*** *on this cholesterol introduced from outside. As Beumer has demonstrated, puppies fed on a diet deficient in cholesterol exhibit after four weeks an increase of cholesterol which is twenty times greater than the amount given in the food. Particularly painstaking have been the investigations of Randles and Knudson into the question of a cholesterol synthesis. They have succeeded in maintaining white rats on a diet completely free of sterols, on the basis of the discovery that pulverized alfalfa, on extraction with cold ether, relinquishes its sterols completely, but retains enough vitamins for the maintenance of rats. The scientists mentioned reached the incontestable conclusion that the cholesterol content of fully-grown rats fed on a diet free of cholesterol is many times greater than the cholesterol content of new-born rats. The only possible explanation for this is that the organism of the rat is capable of forming cholesterol from substances which are different from the sterols and other substances soluble in ether.*
>
> *There is thus in the animal organism a cholesterol synthesis, and the synthetic capabilities of the animal have been gravely underestimated here, as in many other cases...*
>
> *...Using coprosterol as a starting-point, an important conversion from the sterols to bile acids has also been discovered. The familiar bile acids derive from an acid $C_{24}H_{40}0_2$, Wieland's cholanic acid; that is, they contain 3 carbon atoms less than cholesterol and coprosterol. These three carbon atoms may be separated out of coprosterol as acetone via the corresponding hydrocarbon, coprostane, by using an oxidizing agent; and in this way Wieland's cholanic acid is arrived at. Moreover, some time ago Wieland carried out the appropriate synthesis, and succeeded in returning from cholanic acid to coprostane.*
>
> *The chemical connection between cholesterol and bile acids is completely clarified by this, and it is interesting that bile acids also derive not from normal cholesterol, or rather dihydrocholesterol, but allocholesterol, or coprosterol...*

The structure of cholesterol is:

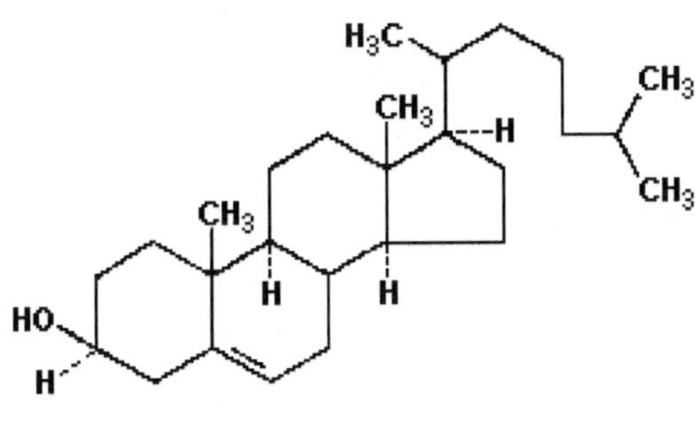

Further details concerning sterols are available in [5] and on the biosynthesis of vitamins [6].

Honours and Awards

Windaus was presented with honorary doctorates from the Universities of Göttingen, Munich, Freiburg, and Hanover.

Windaus was awarded: the Nobel Prize in Chemistry (1928); the Pasteur Medal (1938); the Goethe Medal (1941); the "Groszes Verdienstireuz" (Grand Order of Merit) (1951); the Order pour le Merite, Peace Class (1952); and the "Groszes Verdienstkreuz mit Stern" (Grand Order of Merit with Star) (1956).

He was a Member of the Berlin Academy of Science and the Göttingen Academy of Sciences, a Honorary Member of the Chemical Society, a Foreign Member of the Prussian Academy of Sciences, a Corresponding Member of the Bavarian Academy of Sciences and of many more learned societies and academies.

Bibliography

1. Reproduced with permission.
2. From Nobel Lectures, Chemistry 1922–1941, Elsevier Publishing Company, Amsterdam, 1966.
3. The Nobel Population 1901–1937, E. Crawford, J. L. Heilbron and R. Ullrich. Office of the History of Science and Technology, University of California, 1987, ISBN: 0-918102-15-4.
4. Who was who: 1951–1960.
5. Biochemistry and Function of Sterols by Edward J. Parish and W. David Nes, CRC-Press, 1st ed. 1997, ISBN: 0849376742.
6. Biosynthesis: Polyketides and Vitamins (Springer Desktop Editions in Chemistry) by F.J. Leeper (Editor) and J.C. Vederas, Springer, 1st ed. 2000, ISBN: 3540669698.

1929

© Nobel Foundation [1]

Hans Karl August Simon von Euler-Chelpin (0.5)

"for their investigations on the fermentation of sugar and fermentative enzymes" [2]

During his presentation speech dated 10th December 1929, Professor H.G. Söderbaum, then Chairman of the Nobel Committee for Chemistry of the Royal Swedish Academy of Sciences, observed of von Euler-Chelpin [2]:

...It is a great pleasure to the Swedish Academy of Sciences to be able to award this time the Alfred Nobel's Prize also to one of her members, and so much more since during a long series of years we have been in the position to follow from nearby your energetic, persevering, and systematic investigations. The Academy is also firmly convinced that the distinction which has fallen upon you today, will not contain for you the temptation to rest on laurels already obtained, but that on the contrary it will mean a stimulus to continued and, as we all hope, successful work in the service of biochemistry...

Von Euler-Chelpin was the joint recipient for the 1929 Nobel Chemistry Prize (shared with Harden). He was nominated for the 1929 prize by O. Aschan, M. Bodenstein, R. Willstätter and C. Neuberg. He was also nominated for the years 1924, 1925*, 1926, 1927, and 1928, where * represents one or more disallowed nominations for that year [3].

Hans von Euler-Chelpin was born on 15th February 1873 in Augsburg. His father was a captain in the Royal Bavarian Regiment. Von Euler-Chelpin was married twice. He and his first wife (Astrid Cleve) had five children. He had a further four children with his second wife, Elisabeth Baroness af Ugglas (married in 1913), who was one of his work collaborators.

Von Euler-Chelpin died in Stockholm on 6th November 1964, at the age of 91. One of his sons won the Nobel Prize for Medicine (1970) [2].

The Early Years

Von Euler-Chelpin spent most of his childhood at Wasserburg with his grandmother. At that time his father was stationed in Munich (in the Royal Bavarian Regiment). He attended schools in Munich, Wurzburg and Ulm, and studied art at the Munich Academy of Painting (1891 until 1893), first under Schmid-Reutte and later under Lenbach. During his years studying art, von Euler-Chelpin developed an interest in colour, particularly the colours of the spectrum. This interest led von Euler-Chelpin to take up the study of science (in 1893), moving to Berlin where he studied chemistry at the University under Emil Fischer and A. Rosenheim, and physics under E. Warburg and Max Planck. He presented his doctorate at the University of Berlin (1895). Further studies followed, firstly a short course in physical chemistry in Berlin, followed by a period with Nernst (1896 until 1897) at University of Göttingen, and then with Arrhenius in Stockholm where he was appointed Assistant.

The Career

Von Euler-Chelpin's career was concerned with unraveling physico-chemical and biochemical problems – see the table for career details.

Year	*Post*	*Institute*
1899	Assistant	Royal University at Stockholm
1899–1900	visited the laboratory of van't Hoff	
1906	Professor of General and Organic Chemistry	Royal University, Stockholm
1929	Director	Vitamin Institute and Institute of Biochemistry, Stockholm
1941	Retired from teaching, but carried on his researches	

One of von Euler-Chelpin's early studies concerned the catalytic hydrolysis of substrates by salt formation with the catalyst. This was developed further in some later work (in collaboration with E. Rudberg and A. Ölander).

His first biochemical work was to investigate the action of enzymes considered in relation to his earlier work on catalysis (published in 1904).

Many biochemical investigations followed (from 1906) and include:

- the chemistry of plants (1908–1909)
- the chemistry of the fungi (in collaboration with P. Lindner, work published in 1915)
- the chemistry of enzymes (1925–1930)

Of specific interest was a study of the enzymes saccharase and catalase (in collaboration with K. Josephson) and in the preparation of triose-reductone (with C. Martius).

His studies of fermentation were begun in 1905. Later investigations concentrated on phosphorylation, and the first phases of fermentation and their catalysis. Of particular interest was the action of co-zymase and also related activators (in collaboration with K. Myrback). Clarification of the structure of co-zymase was achieved by von Euler-Chelpin's group and his collaborators. Some of the other issues, however, were not confirmed until Lord Todd's involvement (1956).

The various studies of vitamins were begun in 1924 and had several collaborators, for example B. von Euler, Paul Karrer and Margareta Rydbom. A study of the marked vitamin A activity of carotene (1928) was published from this collaboration.

Von Euler-Chelpin investigation of the biochemistry of tumours (1935), especially the nucleic acids in tumours resulted in the use of his invention, a technique of labelled compounds (in collaboration with G. de Hevesy).

In addition to the numerous papers, von Euler-Chelpin wrote several books and monographs:

He published a book on the chemistry of yeast and alcoholic fermentation (1914); a monograph entitled *Chemie der Enzyme* (which covered his work in enzymology for the period 1925 to 1934); The *Biochemistry of Tumours* (in collaboration with Boleslaw Skarzynski 1942); a book on the chemistry and biochemistry of reductones (published with B. Eistert, 1957); a monograph on the reductones (in collaboration with Professor K. Yamafuji and Drs. Namura and Adachi, 1958); and *The Chemotherapy and Prophylaxis of Cancer* (1962).

Von Euler-Chelpin's Nobel acceptance speech was titled 'Fermentation of sugars and fermentative enzymes' and is dated 23rd May 1930. During his speech von Euler-Chelpin commented [2]...

> *...We also know that the decomposition of sugar during fermentation is identical with the reactions by which sugar begins to burn during respiration, and there is therefore no organism whose life is not dependent on the sugar reaction under discussion.*
>
> *What I shall deal with mainly in this lecture relates to the means by which the living cell decomposes its sugar and thus produces energy for its life and growth. By these means the reactions of the living cell differ from other chemical processes.*
>
> *Within the living organism, the majority of reactions are brought about by special substances already active in minimum quantities, such substances being known as enzymes or ferments. Every group of substances, and in fact practically every substance, requires its specific enzyme for its reaction. Only a few enzyme types were known in the early days, such as pepsin in the gastric juice, which splits proteins, or amylase in saliva and in malt, which converts starch into sugar, but in more recent times the number of enzymes whose existence has been proved or substantiated has risen to over 100...*
>
> *...Another way adopted by us to discover something of the chemical nature of the enzyme is as follows:*
>
> *It was known that certain chemical substances even in very small doses are able to inactivate dissolved enzymes. Phenols, amines, mercury salts and hyacid are among them, i.e. substances which even the layman knows as poisons. Their fatal effect on the organism consists precisely in that they inactivate the enzymes necessary to life. Systematic examination has shown that such poisoning of isolated enzymes is of a purely chemical nature, i.e. the poison binds the enzyme, which is thus inactivated. By studying this effect quantitatively, we determined enzyme equivalents of poison. Such poisoning can readily be shown with various yeast enzymes...*

Many texts concerning alcoholic fermentation have been written over the intervening period – see for example some of the later texts [4–6].

Honours and Awards

Von Euler-Chelpin held honorary doctorates of the Universities of Stockholm, Zurich, Athens, Kiel, Berne, Turin, Rutgers and New Brunswick.

He was awarded the Lindblom Prize of the German Academy of Sciences (1898), Nobel Prize for Chemistry (jointly with Arthur Harden, 1929) and Grand Cross for Federal Services with Star, Germany (1959).

Von Euler-Chelpin was a Fellow of the Academies of Sciences of Bangalore, Berlin, Munich, Rome, Paris, Vienna, Copenhagen, Helsinki, Moscow, Leningrad, Halle, Göttingen, Tokio and New Delhi. He was a Member of: the Royal Swedish Academy of Sciences; the Royal Swedish Academy of Engineering Sciences; the Royal Institution (London); the Finnish Academy of Sciences; Academy of Sciences of New Delhi; a Foreign Member of the Max Planck Society; a Corresponding Member of the Academy of Sciences (Paris). He was a Honorary Member of the Academies of Sciences of Helsinki and Japan, the Indian Academy of Sciences, the Chemical Societies of France, Italy and Berlin, and of the Japanese Cancer Association, Tokyo.

Bibliography

1. Reproduced with permission.
2. From Nobel Lectures, Chemistry 1922–1941, Elsevier Publishing Company, Amsterdam, 1966.
3. The Nobel Population 1901–1937, E. Crawford, J.L. Heilbron and R. Ullrich. Office of the History of Science and Technology, University of California, 1987, ISBN: 0-918102-15-4.
4. Batch Fermentation: Modeling, Monitoring, and Control (Chemical Industries, Vol. 93) (Chemical Industries) by Ali Cinar, CRC, 2003, ISBN: 0824740343.
5. Fermentation Microbiology and Biotechnology by Mansi El-Mansi, CRC, 1st ed. 1999, ISBN: 0748407332.
6. Principles of Fermentation Technology by P. F. Stanbury, A. Whitaker and S. Hall, Butterworth-Heinemann, 2nd ed. 1999, ISBN: 0750645016.

1929

© Nobel Foundation [1]

Arthur Harden (0.5)

"for their investigations on the fermentation of sugar and fermentative enzymes" [2]

During his presentation speech dated 10th December 1929, Professor H.G. Söderbaum, then Chairman of the Nobel Committee for Chemistry of the Royal Swedish Academy of Sciences, observed [2]:

> *...When the Royal Swedish Academy of Sciences resolved to adjudge to you, together with Professor von Euler, this year's Nobel Prize in Chemistry on account of your important contributions to our knowledge of alcoholic fermentation, the Academy had let herself be guided by a firm conviction that these contributions had opened indeed a new chapter in the investigation of that very complicated matter...*

Harden was nominated by C. Neuberg for the 1929 Chemistry Prize. He was not nominated for any other year [3].

Arthur Harden was born in Manchester (England) on 12th October 1865, the only son of Albert Tyas Harden, a Manchester businessman, and Eliza, née Macalister, originally from Paisley. Although the only son, there were several daughters. Harden married Georgina Bridge (daughter of C. Wynyard Bridge of Christchurch, New Zealand) in 1900 (some say 1890). They had no children. Georgina died in January 1928 and Harden on 17th June 1940, at his home in Bourne End, Buckinghamshire. Towards the end of his life Harden suffered from a progressive nervous disease [2, 4, 5].

The Early Years

Harden was educated (from the age of seven years) at a private school in Victoria Park (1873–1877) run by Dr. Ernest Adam and at Tettenhall College, Staffordshire (1877–1881). He entered the Owens College in the University of Manchester (in 1882), studied under Sir H.E. Roscoe, and graduated with first-class honours in chemistry (1885). He was awarded the Dalton Scholarship in Chemistry (in 1886) and spent 12 months (1887–1888) working with Otto Fischer at Erlangen. (Otto Fischer was the cousin of Emil Fischer (Chemistry 1902). Both Otto and Emil worked for von Baeyer (Chemistry, 1905). Harden prepared β-nitroso-naphthylamine and investigated its properties for his Ph.D. (under the direction of Otto Fischer).

The Career

Year	*Post*	*Institute*
1888	Lecturer and Demonstrator	University of Manchester
1897	Chemist	Jenner Institute of Preventive Medicine in London, later renamed the Lister Institute
1907	Head of the Biochemical Department	Jenner Institute of Preventive Medicine in London, later renamed the Lister Institute
1912	Emeritus Professor of Chemistry	University of London
1930	Retired	From the Lister Institute

Harden's early career interests lay in lecturing and publishing. His research work developed during his time in London.

The Manchester Period

Whilst Harden's years in Manchester were principally concerned with writing and lecturing, he did undertake some research, for example:

- the action of light on mixtures of carbon dioxide and chlorine
- a new form of stereometer (in collaboration with W. Haldane Gee [6])
- the composition of some iron and bronze implements (found by Professor Flinders Petrie [7])
- experiments on the combination of carbon monoxide and chlorine

During his time in Manchester, Harden unsuccessfully applied for two posts. One was the post of Principal at the Wandsworth Technical Institute and the other that of Inspector under the Science and Arts Department at South Kensington.

The London Period

Harden was appointed to succeed Luff as Chemist at the Jenner Institute of Preventive Medicine in London (1897). He enjoyed a considerable reputation for his wide knowledge of chemistry and his teaching abilities, which are thought to be the reasons behind his appointment. Harden was responsible for the chemical course at the Institute, which was mostly concerned with the analysis of water and foods. Some years after joining the Jenner Institute, the London medical schools started to teach chemistry themselves, leaving Harden free to undertake research. Additionally the two chemistry departments at the Institute, Chemical (under Harden), Biochemical (under Hedin) were merged. Harden was invited to preside over the newly formed department (1907).

Harden's principal research interests at the Lister Institute concerned:

- bacterial chemistry
- alcoholic fermentation
- the breakdown products of glucose
- the chemistry of the yeast cell
- the chemistry of the fermentation of sugar by yeast juice
- the vitamins.

Studies which led to a series of papers on the antiscorbutic and antineuritic vitamins and their occurrence in food and drinks.

Harden received the title of Professor of Chemistry, University of London, in recognition of his work concerning bacterial chemistry and alcoholic fermentation.

Although Harden retired from his professorship at the Lister Institute in 1930, he continued to visit his laboratory everyday carrying on experimental work and correcting proofs for the *Biochemical Journal* until a year or two before his death.

Harden was Joint Editor (with W.M. Bayliss) of *The Biochemical Journal* from 1913 to 1938, and, in addition wrote numerous papers in scientific journals. He is the author of *Alcoholic Fermentation*, *A New View of the Origin of Dalton's Atomic Theory*, and *Chemistry for Advanced Studies* (with Sir H.E. Roscoe), *Inorganic Chemistry for Advanced Students*, and *An Elementary Course of Practical Organic Chemistry* (with F.C. Garrett).

Harden's Nobel Prize acceptance speech, dated 12th December 1929, was titled 'The function of phosphate in alcoholic fermentation', during the course of which he shared some observations [2] concerning...

Fermentation by living yeast

...I have spoken hitherto on the assumption that the processes in the living cell are essentially of the same kind as those which occur in the various preparations made from the dead cell, but differ from these mainly in the relative intensity of some of the reactions, and I know no valid argument against this assumption.

The cycle undergone by the phosphate in the series of changes which constitutes ordinary fermentation clearly consists in the alternate formation of a phosphoric ester and the hydrolysis of this to free phosphoric acid. A simple calculation based on the phosphorus content of living yeast shows that the whole of this phosphate must pass through the stage of phosphoric ester every five or six minutes in order to maintain the normal rate of fermentation, whereas in an average sample of yeast-juice the cycle, calculated in the same way, would last nearly two hours...

Nature and function of the phosphoric esters produced

...If we next consider the exact nature of these phosphoric esters and the relation of their formation and hydrolysis to the decomposition of the sugar molecule, we are met with a singularly complex condition of affairs, which cannot yet be interpreted satisfactorily... The mechanism of the fermentation of the monoester has not yet been worked out in sufficient detail to afford valid

evidence either for or against the theory, but Dr. Robison and I have made experiments (about to he published) which show that the monophosphate itself reacts with a further quantity of phosphate and that this reaction is accompanied by an enhanced production of carbon dioxide and alcohol...

Sugar metabolism in vegetable and animal organisms

...After the publication of Meyerhof's preliminary papers in which these observations were recorded I wrote the following passage in concluding a short review of the work (Nature, 118 (1926) Dec. 18th) which I may perhaps be allowed to quote: 'The striking similarity established by Meyerhof between the changes of carbohydrates in muscle and in the yeast cell is seen to be much closer than has been believed. The remarkable phenomena accompanying alcoholic fermentation are now duplicated in the case of lactic acid production, and it may reasonably be expected that most of the fermentative decompositions of the sugars will he found to be initiated in a similar manner.'

Direct proof is still wanting in many cases, but some instances are known among bacteria (Virtanen), moulds (von Euler and Kullberg) and higher plants (Ivanoff, Bodnar). It is not too much to say that the fundamental biological mode of attack on carbohydrates is that revealed by the study of alcoholic fermentation...

Many texts concerning alcoholic fermentation have been written over the intervening period – see for example some of the later texts [8–10].

Honours and Awards

Harden was Honorary Doctor of Science, University of Athens and Honorary Doctor of Laws, Universities of Manchester and Liverpool.

Harden received many awards and prizes including Nobel Prize (Chemistry, 1929), the Davy Medal (1935).

He was elected to many learned societies, including Fellow of the Royal Society (serving on its council from 1921 to 1923) and the Kaiserlich Leopold Deutsche Akademie der Naturforscher of Halle. Harden was knighted in 1936.

Harden's citation for election (date: 06/05/1909) as Fellow of the Royal Society reads [11]:

'Head of the Chemical Department of the Lister Institute (1897). Ex-Assistant Lecturer and Demonstrator of Chemistry, Owens College (1888–1897). Distinguished for his knowledge of Chemistry, especially with reference to its applications to Biology. Author of numerous original researches in Chemistry, amongst which are the following: 'Action of Silicon Tetrachloride on Amido-compounds' (Journ. Chem. Soc. Trans, 1887, Dalton Scholarship); 'Beta-nitroso-alpha-naphthylamine' (Annalen, 1889); 'Combination of Chlorine with Carbon Monoxide' (with G. Dyson, Journ. Chem. Soc. Trans, 1903); The following on Bio-chemical subjects: 'Chemical Action of B 'coli communis' on Carbohydrates' (ibid, 1901); 'Biological Resolution of Inactive Acids' (with G. S. Walpole, Proc. Roy. Soc, 1906); 'Voges and Proskauer's Reaction' (ibid, 1906); 'Autofermentation of Pressed Yeast' (with S. Rowland, Journ. Chem. Soc. Trans, 1901); 'Glycogen from Yeast' (with W. J. Young, ibid, 1902); 'Buchner's Zymase mit Blutserum' (Ber Chem Ges, 1903); 'Ueber Pressaft aus Obergahriger Hefe' (with W. J. Young, ibid, 1904); 'Alcoholic Ferment of Yeast-Juice' (with W. J. Young, Proc Roy Soc, 1906); ditto, Part II, 'The Co-ferment of Yeast Juice' (ibid, 1906). Joint Author of the following: 'A New View of Dalton's Atomic Theory' (with

H. E. Roscoe, Macmillan, 1896); 'Inorganic Chemistry for Advanced Students' (with H. E. Roscoe, Macmillan, 1899); 'An Elementary Course of Practical Organic Chemistry' (with F. C. Garrett, Longmans). Joint Editor of the Treatise on Chemistry by Roscoe and Schorlemmer, vol i, 1894, 1905, vol ii, 1897, 1906'

He was proposed by:

Henry E. Roscoe; C. J. Martin; H. B. Dixon; W. H. Perkin; H. B. Baker; R. Meldola; H. T. Brown; J. R. Bradford; [Michael Foster]; H. E. Armstrong; H. McLeod; William Ramsay; A. Smithells; P. F. Frankland; F. W. Pavy; Joseph Wilson Swan.

Bibliography

1. Reproduced with permission.
2. From Nobel Lectures, Chemistry 1922–1941, Elsevier Publishing Company, Amsterdam, 1966.
3. The Nobel Population 1901–1937, E. Crawford, J.L. Heilbron and R. Ullrich. Office of the History of Science and Technology, University of California, 1987, ISBN: 0-918102-15-4.
4. Obituary Notices of Fellows of the Royal Society 1942–1944, vol. 4, pp. 3–14.
5. Who was who: 1929–1940.
6. A. Harden and W. Haldane Gee, Transactions of the Manchester Literary and Philosophical Society, 1891.
7. A. Harden, Transactions of the Manchester Literary and Philosophical Society, 1897.
8. Batch Fermentation: Modeling, Monitoring, and Control (Chemical Industries, Vol. 93) (Chemical Industries) by Ali Cinar, CRC, 2003, ISBN: 0824740343.
9. Fermentation Microbiology and Biotechnology by Mansi El-Mansi, CRC, 1st ed. 1999, ISBN: 0748407332.
10. Principles of Fermentation Technology by P. F. Stanbury, A. Whitaker and S. Hall, Butterworth-Heinemann, 2nd ed. 1999, ISBN: 0750645016.
11. Sackler database, Royal Society.

1930

© Nobel Foundation [1]

Hans Fischer

"for his researches into the constitution of haemin and chlorophyll and especially for his synthesis of haemin" [2]

During his presentation speech dated 10th December 1930, Professor H.G. Söderbaum then Chairman of the Nobel Committee for Chemistry of the Royal Swedish Academy of Sciences, observed [2]:

...It was discovered that the haemin molecule can be converted by certain chemical procedures into iron-free substances, called porphyrins, and that the porphyrins can be broken down, by other procedures, into pyrrole derivatives, i.e. into compounds containing four carbon atoms and one nitrogen atom in a closed ring. It was clear that the road to an exact knowledge of the structure of the blood pigments would involve detailed examination of these pyrroles and porphyrins.

By your analytical and synthetic work on the porphyrins, on the blood pigment, on leaf pigments, and other related substances, you have accomplished an achievement which can only be described a great feat in chemistry which will undoubtedly have a beneficial effect on the most diverse branches of natural sciences...

Fischer was nominated for the 1930 chemistry Noble Prize by A. Fovorskij, K. Freudenberg, C. G. Schwalbe, M. Trautz, H. Wieland and E. Wilke. Fischer was also nominated for the years 1928 and 1929. [3].

Hans Fischer was born in 27th July 1881 at Hoechst (on the river Main) in Germany. His father (Dr. Eugen Fischer) was a Director of the firm of Kalle & Co, Wiesbaden, and Privatdozent at the Technical High School, Stuttgart. His mother was Anna, née Herdegen. Fischer married Wiltrud Haufe in 1935. He died in Munich on 31st March 1945 [3].

The Early Years

Fischer attended primary school in Stuttgart, and later the Humanistische Gymnasium (grammar school with emphasis on the classics) in Wiesbaden (1899). He read both chemistry and medicine, initially at the University of Lausanne and then at Marburg. He received his chemistry degree under T. Zincke at Marburg (1904). His licence for medicine was conferred on him in Munich (1906) and his M.D. under F. von Müller (1908), also at Munich.

Fischer's early career was spent at the Second Medical Clinic in Munich and at the First Berlin Chemical Institute (under Emil Fischer), returning to Munich in 1911. He qualified as Lecturer on internal medicine in 1912.

The Career

Fischer's career centred upon various investigations of biochemistry.

Year	*Post*	*Institute*
1913	Lecturer (physiology)	Physiological Institute, Munich
1916	Professor of Medical Chemistry (succeeded Windaus)	University of Innsbruck
1918	Professor	University of Vienna
1921	Professor of Organic Chemistry (succeeded to Wieland)	Technische Hochschule (Technical University) in Munich
1945	Died in post	

His investigations included:

- investigations of the constitutive properties of the pigments in blood and bile
- investigations concerning leave pigments
- the chemistry of pyrrole particularly the synthesis of natural pyrrole pigments
- the synthesis of bilirubin

His many papers were mostly published in *Justus Liebigs Annalen der Chemie* and *Hoppe-Seylers Zeitschrift für Physiologische Chemie.*

Fischer's Nobel Prize acceptance speech, dated 11th December 1930, was titled 'On haemin and the relationships between haemin and chlorophyll'. During his speech he made the following observations [2]...

> *...Haemin has the formula $C_34H_{32}O_4N_4FeCl$. Countless variations of this atomic arrangement are possible. By analytical methods of Nencki, Küster, Piloty, Wilistätter, H. Fischer and their pupils, the pyrrole nature of the haemm was demonstrated, and much information was obtained about its constitution. When haernin is deprived of its iron, porphyrins are formed, which are substances having a sensitizing effect, as Hausmann has shown.*

These porphyrins are widely distributed in Nature, and it was possible to assume as a working hypothesis from the start that there existed a relationship with the porphyrins of blood pigment; it was moreover to be expected that the determination of their constitution would also provide information about the constitution of haemin itself, just as the results of disintegration of bilirubin which is formed from blood pigment by biological means had stimulated determination of the constitution of blood pigment itself. This was the reason why I tackled systematically the investigation of natural porphyrins, and the biological degradation of blood pigment itself...

For further information concerning the porphyrins, see for example [4–6].

Honours and Awards

Fischer was bestowed the title of Honorary Doctor by Harvard University (1936).

He was awarded the Liebig Memorial Medal (1929), the Nobel Prize for chemistry (1930) and the Davy Medal (1937).

Fischer was conferred the title of a "Geheimer Regierungsrat" (Privy Councillor) in 1925.

Bibliography

1. Reproduced with permission.
2. From Nobel Lectures, Chemistry 1922–1941, Elsevier Publishing Company, Amsterdam, 1966.
3. The Nobel Population 1901–1937, E. Crawford, J. L. Heilbron and R. Ullrich. Office of the History of Science and Technology, University of California, 1987, ISBN: 0-918102-15-4.
4. The Porphyrin Handbook (10-Volume Set) by Karl Kadish, Kevin M. Smith and Roger Guilard, Academic Press, 1999, ISBN: 0123932009.
5. Photosensitization of Porphyrins and Phthalocyanines by Ichiro Okura Taylor & Francis, 1st ed. 2001, ISBN: 9056993542.
6. Heme, Chlorophyll, and Bilins: Methods and Protocols edited by Alison G. Smith and Michael Witty, Humana Press, 1st ed. 2001, ISBN: 1588291111.

1931

© Nobel Foundation [1]

Friedrich Bergius (0.5)

"in recognition of their contributions to the invention and development of chemical high pressure methods" [2]

Professor W. Palmær, then a Member of the Nobel Committee for Chemistry of the Royal Swedish Academy of Sciences, made the following observation during his presentation speech (dated 10th December 1931) [2]:

> *...General Director Bergius. You undertook to tackle a problem which, in its importance for humanity, can be compared with the solution of the nitrogen question. You have shown how, by the injection of hydrogen under pressure, pit coal, brown coal, and other carbon-bearing materials can be processed to liquid fuels which are considered indispensable in modern life for the propulsion of ships and vehicles. You have thereby obviated the danger which threatened of exhaustion of petroleum deposits, an event which must have happened sooner or later. In your work, you arrived at the high-pressure method quite independently. On the basis of your work, a powerful industry has already been formed...*

Bergius was nominated for the 1931 Chemistry Nobel Prize by E. Hägglund. He was also nominated for the year 1929 [3].

Friedrich Bergius was born on 11th October 1884, in Goldschmieden near Breslau, Silesia, son of Heinrich and Marie Bergius. His father owned a chemical factory in Goldschmieden and his grandfather was Professor of Economics in Breslau. Bergius married Ottilie Kratzert and they had two sons and one daughter. Bergius emigrated to Argentina after the Second World War and died in Buenos Aires on 30th March 1949 [2, 4].

The Early Years

Whilst undertaking his early education in Breslau, Bergius found time to visit his father's chemical plant. Bergius, therefore, became acquainted with chemico-technical processes from an early age. The time spent in the laboratories and the works there later stood him in good stead, as he was able to obtain practical knowledge of both industrial technology and practical uses of chemistry. Indeed Bergius senior sent his son to the Ruhr for six months, where Bergius junior studied the practical aspects of a large metallurgical plant, before he entered university. Thus Bergius entered Breslau University to read chemistry (1903) under Ladenburg, Abegg and Herz. His studies were interrupted by a year of military service after which he entered Leipzig University (1905) to work under Hantzsch on his thesis *Über absolute Schwefelsäure als Lösungsmittel* (On Absolute Sulphuric Acid as a Solvent). Bergius completed this work in Breslau under Abegg receiving his degree at Leipzig (1907). Bergius went on to study for a further two terms under Nernst, at the latter's Institute in Berlin, followed by one term under Haber (1909). Bergius' work with Harber set the tone for his early independent work which concerned an investigation on the dissociation of calcium peroxide. He also developed a practical method for laboratory work at pressures up to 300 atmospheres.

The Career

Although Bergius collaborated with Universities, and indeed lectured part time for some of his career, his interests and accomplishments were conducted from industrial posts.

Year	*Post*	*Institute*
1910	established his own private laboratory	in Hanover
1911-1914	lectured on technical gas reactions, equilibrium theory, and metallurgy	Technische Hochschule, Hanover
1914	transferred his laboratory	the Essen works of the firm Th. Goldschmidt A.G, Berlin
1921	Industrial chemist and director	Mannheim-Rheinau plant

His main achievements include:

- The hydrogenating effect of hydrogen on coal (1912)
- The hydrogenating effect of hydrogen on heavy oils under pressure (1913)
- Setting up a hydrogenating process in Rheinau near Mannheim
- Establishing a group of German companies, the Shell Trust and a number of British enterprises (particularly from the coal industry)
- Investigations into the origins of coal
- Concluding his own work on the liquefaction of coal
- I.G. Farbenindustrie and Imperial Chemical Industries taking up the work on an industrial scale
- Inventing a process of obtaining sugar from cellulose in wood
- The establishment of an industrial plant to extract sugar from wood (after 15 years research and development a plant was built in the Rheinau works)

Even though the mid war years were financially challenging, Bergius was able to have the hydrogenation process implemented in major chemical companies throughout Europe.

Bergius wrote a monograph '*The Application of High Pressures to Chemical Processes and Simulation of the Genesis of Bituminous Coal*' (1912), Verlag Willhelm Knapp, Halle, Germany

During his Nobel Prize acceptance speech, dated 21st May 1932, and titled 'Chemical reactions under pressure', Bergius commented on several areas of his work, a few of which are mentioned below [2]...

Study of the Dissociation of CaO

My main reason for mentioning the first series of experiments in 1910, the measurement of the dissociation equilibrium of calcium peroxide, is that it played a significant part in the development of our high-pressure work. The economical and technical starting-point of these researches was the interest which the production of peroxides and perborates had gained at the time. Production methods of hydrogen peroxide and its starting materials were anything but perfect...

Production of Hydrogen

Work on catalytic hydrogenation and the development of technical hydrogenation reactions of hardening of fats and of the ammonia synthesis directed the attention of expert circles at that time to methods for the production of pure and cheap hydrogen. At that time, the process of the Badische Anilin- & Soda-Fabrik for the catalytic conversion of carbon monoxide of water gas into hydrogen and carbon dioxide was still unknown. The best available method for producing hydrogen seemed to be that according to Linde in which the carbon monoxide of the water gas was separated from the hydrogen by liquefaction at low temperatures.

In 1911 I conceived the notion that it should be possible to carry out the water gas reaction, which occurs at a technically satisfactory velocity and then according to the formula

$$C + H_2O = CO + H_2$$

only at temperatures of the order of 900°, at considerably lower temperatures, at which the position of the equilibrium is practically such that the reaction corresponds to the formula

$$C + 2H_2O = CO_2 + 2H_2$$

Hardening of Fats

As early as 1912 we worked on the problem of the hydrogenation of organic substances with the aid of highly compressed hydrogen. By virtue of the increased concentration of hydrogen at high pressure we hoped to achieve effects similar to those made possible at normal pressure by the action of catalysts. Special interest attached at that time to the hydrogenation of unsaturated vegetable and animal oils and fats. It was an obvious idea to study the hardening of fats with high pressure hydrogen, in particular with strongly unsaturated fats, such as the train oils. The results of treating such substances with high pressure hydrogen at about 300°, a temperature at which these oils in a reducing atmosphere can still be treated undecomposed, was, as the iodine numbers showed, a considerable decrease of the unsaturated character. It was however discovered that this phenomenon also occurred if treatment was carried out under the same conditions not with compressed hydrogen, but with compressed nitrogen. What was involved was, as could be established, a polymerization process. No addition of hydrogen at high pressure without catalysts to the unsaturated fractions of the fatty substances could be observed.

Hydrogenation of Coal

The investigations into the conversion of cellulose into coal by coalification in conjunction with our studies on the hydrogenation of heavy oils opened the way to the discovery of the process for liquefying coal. We had observed that a product forms from cellulose with splitting of certain quantities of carbon dioxide and water which we considered to be similar to natural coal, and we had further observed that the hydrogen in this so-called synthetic coal is unstable. The attempt to formulate this process of cellulose decomposition led me to a hypothetical concept of the constitution of this coal-like substance formed from cellulose. I assumed that our artificial coal is a compound of carbon, hydrogen, and oxygen-that coal is not carbon, was not obvious at the time-whose structure must resemble that of certain unsaturated compounds of terpene-like character, and that this compound should be capable of absorbing not inconsiderable quantities of hydrogen. By hydrogen absorption, such a compound would approach the class of heavy petroleum hydrocarbons. As a consequence of these trains of thought of whose hypothetical character I have always been aware, it was an obvious step to attempt hydrogen addition to coal and to endeavour at the same time to discover whether the hydrogen-charged coal can be split into medium and light hydrocarbons...

High pressure reactions are a routine part of industrial process, some of the technology for which is included in [5–7].

Honours and Awards

Bergius received the degree of Dr. Phil. from the University of Heidelberg and a Honorary Doctorate from the University of Hanover and a Honorary D.Sc. from Harvard University.

He was awarded the Liebig Medal and shared the Nobel Prize (1931) with Carl Bosch.

Bergius was elected to the Board of Directors of many associations and companies interested in coal and oil.

Bibliography

1. Reproduced with permission.
2. From Nobel Lectures, Chemistry 1922–1941, Elsevier Publishing Company, Amsterdam, 1966.
3. The Nobel Population 1901–1937, E. Crawford, J. L. Heilbron and R. Ullrich. Office of the History of Science and Technology, University of California, 1987, ISBN: 0-918102-15-4.
4. Who was who: 1941–1950.
5. High Pressure Process Technology: fundamentals and applications (Industrial Chemistry Library) by A. Bertucco and G. Vetter, Elsevier Science, 1st ed. 2001, ISBN: 0444504982.
6. High Pressure Vessels by Donald M. Fryer and John F. Harvey, Springer, 1st ed. 1998, ISBN: 0412074516.
7. High Pressure Chemistry: Synthetic, Mechanistic, and Supercritical Applications edited by Rudi van Eldik and Frank-Gerrit Klärner, Wiley-VCH, 2002, ISBN: 3527304045.

1931

© Nobel Foundation [1]

Carl Bosch (0.5)

"in recognition of their contributions to the invention and development of chemical high pressure methods" [2]

Professor W. Palmær, then a Member of the Nobel Committee for Chemistry of the Royal Swedish Academy of Sciences commented of Bosch during his presentation speech (dated 10th December 1931)[2]:

> *...You, Professor Bosch, were the first to enrich the chemical industry with this powerful tool, for the production of ammonia from the elements. Between the presentation of the scientific basis of this synthesis and its industrial exploitation lay a mighty chasm, which you bridged, inter alia, by the brilliant invention and construction of the high-pressure apparatus. By doing so, you made it possible for nitrogen to be made available to mankind in inexhaustible quantities, in a form suitable for agriculture, and even at lower prices than hitherto. Furthermore, you developed high-pressure methods for the production of other important substances...*

Bosch was nominated for the 1931 Chemistry Nobel Prize by C. Benedicks, N. G. Dalén, A. Enström, E. Hubendick, T. Lindmark, G. Richert, K. Sondén and E. Hägglund. Bosch was also nominated for the years 1915*, 1916, 1920, 1926, 1929 and 1930, where the * indicates one or more disallowed nomination for that year [3].

Carl Bosch was born in Cologne on 27th August 1874. He died after a prolonged illness on 26th April 1940 [2,4].

The Early Years

Bosch studied metallurgy and mechanical engineering at the Technische Hochschule in Charlottenburg (1894 to 1896). He started reading chemistry at Leipzig University (1896) under Professor Wislicenus, graduating (in 1898) with a paper in organic chemistry.

The Career

Bosch spent his entire career working in various industrial posts within the German chemical industry.

Year	*Post*	*Institute*
April 1899	Chemist	Badische Anilin- und Sodafabrik, Ludwigshafen, Rhine
1919	Managing Director	Badische Anilin- und Sodafabrik
1925	Principal	I.G. Farbenindustrie Aktiengesellschaft
1935	Chairman of the Board of Directors	I.G. Farbenindustrie AG

During his career, Bosch was involved with:

- the development of the then new industry of synthetic indigo (under the guidance of Dr. Rudolf Knietsch)
- the problem of fixing of nitrogen. His first experiments in this field were done with metal cyanides and nitrides at the turn of the century
- starting a pilot plant for the production of barium cyanide (1907)

The Badische Anilin-und Sodafabrik acquired the process of high-pressure synthesis of ammonia, which had been developed by Fritz Haber at the Technische Hochschule in Karlsruhe (1908). This was to be a turning point in Bosch's career, as he was given the task of developing the Harber process on a large industrial scale. Several challenges presented themselves to him including:

- the need to change the catalyst to pure iron with certain additives (for cost and availability) Bosch worked with collaborators in this activity
- solving the construction of safe high-pressurized blast furnaces
- the need of a cheap way of producing and cleaning the gases necessary for the synthesis of ammonia.

Bosch successfully went on to using increasingly larger manufacturing units, creating an industry dealing with the production of synthetic ammonia using the high-pressure Harber process.

Bosch also succeeded developing methods for the industrial production of nitrogen fertilizers. This work resulted in the Stickstoffwerke (Nitrogen works) in Oppau opening (1913), followed by the even larger Leunawerke, near Merseburg (1917). Methods for the manufacture of methanol and the hydrogenation of oil were developed and produced from these factories.

During his Nobel Prize acceptance speech, dated 21st May 1932, and titled 'The Development of the Chemical High Pressure Method during the establishment of the new ammonia industry', Bosch made the following comments [2]:

> *...The technical development of these processes (high pressure) within the I.G. Farhenindustrie concern, a task with which I have been entrusted for about 25 years, posed a profusion of scientific, technical and economic problems the solution of which occupied hundreds of scientists and which affords a living to many thousands of people.*

I can only report here on a small section of our sphere of work and with an eye to the motives underlying the award of the prize I have chosen to discuss the solution to the question of apparatus...

...When the management at the time entrusted me with the task of expanding the conversion of this high pressure synthesis to an industrial scale with a view to its economic application, it was obvious that there were three main problems which had necessarily to be settled before the construction of a plant could be undertaken. These were:

- *supply of raw materials, i.e. of the gases hydrogen and nitrogen, at a lower price than hitherto possible*
- *manufacture of effective and stable catalysts*
- *the construction of the apparatus.*

These problems were dealt with simultaneously.

...To begin with, sufficient pure hydrogen for the tests was available from chlorine electrolysis and yielded the required nitrogen-hydrogen mixture by combustion with air. None of the then known hydrogen processes was suitable for large-scale manufacture. We examined all of them but without exception they were too expensive or yielded insufficiently pure gases. It must not be forgotten that the procurement of hydrogen is the largest item in the prime cost, at least nowadays since because of the efficiency of the high pressure synthesis the conversion of the ready-made gas mixture into ammonia is only a minor cost factor. Since we were dependent on coal as a starting material we at once decided on the sole eligible source in our situation, water gas from which nowadays, after a short transition period-separation of the hydrogen by Linde's low temperature liquefaction method-hydrogen is manufactured in enormous quantities by a catalytic process developed by us.

The accomplishment of the second task was no less significant. Osmium, a very good catalyst, was difficult to maintain since when in its active, i.e. finely disperse, form it comes into contact with air (never wholly avoidable under industrial conditions) it readily volatilises as osmium tetroxide, but in particular because the whole world's supply amounted to only a few kilograms, hence at best we could have based only a very moderate manufacture on osmium. Uranium was expensive, of course, yet nevertheless obtainable in some quantity but proved extremely sensitive to oxygen and water and it could not, and still cannot, be converted by any means into a form that can be used for a mass-production process. However, as a result of very broadly based series of experiments using our rapidly expanded experimental technique, we succeeded in preparing relatively fast-acting, technically perfect, easily manipulable, stable and inexpensive catalysts, chiefly those with iron as the active substance (molybdenum, particularly in mixtures, was also found to be an excellent catalyst) which are currently in universal use presumably exclusively for the synthesis of ammonia. They are a completely new type of catalyst, their characteristic being that they no longer consist of pure elements in varyingly fine form, but are mixtures. It was only after studies extending over many years and also only in the light of recent theoretical knowledge.

The successful conclusion of the third task, which is the only one I intend to discuss today in more detail, related to the apparatus. There were no examples in industry. The only technique, and that of a physical nature, involving high pressures, was Linde's liquefaction of air. The conventional apparatus for this was made from soft soldered copper and was hence unsuitable for temperatures up to 600°C and locally still higher. In true awareness that the manufacture of the apparatus would be one of the most difficult tasks, quite apart from surprises yet to come and soon forthcoming, an immediate start (a typical feature of our work) was made on fitting up a workshop specially adapted to our needs which first had to make the apparatus necessary for the laboratory work. Right from the start this test apparatus was of very servicable design since the individual tests all took a very long time as it was particularly important to ascertain the stability of the new catalysts in experiments of long duration...

High pressure reactions are a routine part of industrial process, some of the technology for which is included in [5–7].

Honours and Awards

Bosch received the honorary doctorates from: Technische Hochschule in Karlsruhe (1918); Landwirtschaftliche Hochschule (Agriculture College), Berlin (1921); Technische Hochschule in Munich (1922); Halle University (1927); Technische Hochschule in Darmstadt (1928).

He received: the Liebig Memorial Medal (the Association of German Chemists); the Bunsen Medal (German Bunsen Society); the Siemens Ring; the Golden Grashof Memorial Medal of the VDI (Association of German Engineers); the Exner Medal from the Austrian Trade Association; the Carl Lueg Memorial Medal from the Association of German Metallurgists. He was awarded the Nobel Prize for Chemistry (1931), jointly with Friedrich Bergius.

Bosch held memberships/fellowships with various German and foreign scientific academies. He was chairman of the Kaiser Wilhelm Society, of which he became its President (1937). He was also conferred with Honorary Senator of the Universities of Heidelberg (1922) and Leipzig (1939), and Honorary Citizen of Frankfurt (1939).

Bibliography

1. Reproduced with permission.
2. From Nobel Lectures, Chemistry 1922–1941, Elsevier Publishing Company, Amsterdam, 1966.
3. The Nobel Population 1901–1937, E. Crawford, J. L. Heilbron and R. Ullrich. Office of the History of Science and Technology, University of California, 1987, ISBN: 0-918102-15-4.
4. Who was who: 1929–1940.
5. High Pressure Process Technology: fundamentals and applications (Industrial Chemistry Library) by A. Bertucco and G. Vetter, Elsevier Science, 1st ed. 2001, ISBN: 0444504982.
6. High Pressure Vessels by Donald M. Fryer and John F. Harvey, Springer, 1st ed. 1998, ISBN: 0412074516.
7. High Pressure Chemistry: Synthetic, Mechanistic, and Supercritical Applications edited by Rudi van Eldik and Frank-Gerrit Klärner, Wiley-VCH, 2002, ISBN: 3527304045.

1932

© Nobel Foundation [1]

Irving Langmuir

"for his discoveries and investigations in surface chemistry" [2]

Professor H. G. Söderbaum, then Chairman of the Nobel Committee for Chemistry of the Royal Swedish Academy of Sciences, observed of Langmuir during his presentation speech dated 10th December 1932 [2]:

> *...The work which has been awarded the prize does not, however, comprise only investigations concerning the adsorption in the boundary surface between gases and solid bodies, but also investigations concerning the thin films in the boundary surface between gases and liquids and concerning the adsorption on the surfaces of liquids. The well-known phenomenon of the spreading of an oil film has served as a starting-point for these investigations and the determination of the lateral pressure with which such a film opposes attempts to diminish its surface...*

Langmuir was nominated for the 1932 Chemistry Nobel Prize by J. Kendall and H. von Euler-Chelpin, having also been nominated for the Chemistry Prize for the years 1927, 1928, 1929 and 1931 and for the Physics Prize for the years 1916, 1928, 1929 and 1933 [3].

Irving Langmuir was born in Brooklyn, New York, on 31st January 1881, the third of four sons of Charles Langmuir and Sadie, *neé* Comings. Langmuir married Marion Mersereau in April 1912. They had a son, Kenneth, and a daughter, Barbara. Langmuir died after a short illness on 16th August 1957 in Falmouth, Massachusetts [2, 4, 5].

The Early Years

Langmuir attended the local public schools in Brooklyn, with the exception of a three year interval when the family moved to Paris (1892–1895), where his father was in charge of the New York Life Insurance agency. Whilst at boarding school in Paris, one of the teachers encouraged Langmuir to learn the use of logarithms and solve problems in trigonometry. These subjects were not part of the curriculum. In addition, his brother Arthur encouraged Langmuir to have a workshop at the age of nine. He graduated as a metallurgical engineer from the School of Mines at Columbia University in 1903. For his postgraduate work in physical chemistry under Nernst (in Göttingen), Langmuir was awarded the degrees of M.A. and Ph.D. (1906).

The Career

Year	*Post*	*Institute*
1906	Instructor in Chemistry	Stevens Institute of Technology
1909	Scientist, rising to Associate Director	Research Laboratory of the General Electric Company

With the exception of a brief period in academia, Langmuir spent most of his working life in the industrial laboratories of the General Electric Company. His work covered a broad range of chemistry, physics, and engineering. Some of the specific areas of interest were:

- studies of vacuum phenomena
- investigations into the properties of adsorbed films
- the nature of electric discharges in high vacuum and in certain gases at low pressures
- thermal conduction and convection in gases (the subject of his first paper [6])
- filaments in gases
- the invention of the gas filled incandescent lamp
- the discovery of atomic hydrogen
- the development of the atomic hydrogen welding process
- that he was the first to observe the very stable adsorbed monatomic films on tungsten and platinum filaments
- investigations of oil films on water
- the formulation of a general theory of adsorbed films
- the study of the catalytic properties of adsorbed films
- space charge effects and related phenomena
- the study of structure and valence

Langmuir also contributed to the development of the Lewis theory of shared electrons.

Many books have been written about Langmuir, references to two of which are presented in the bibliography [7,8].

During the early pages of Langmuir's Nobel Prize acceptance speech, dated 14th December 1932 and titled 'Surface Chemistry' [2], he commented:

> ...*When I first began to work in 1909 in an industrial research laboratory, I found that the high-vacuum technique which had been developed in inlamp factories, especially after the introduction of the tungsten filament lamp, was far more advanced than that which had been used in university laboratories. This new technique seemed to open up wonderful opportunities for the study of chemical reactions on surfaces and of the physproperties of surfaces under well-defined conditions. I*

resolved to try the effect of introducing very small quantities of various gases into a highly evacuated bulb containing a tungsten filament...

He went on to briefly cover the following topics and provides the interested reader with 37 references for further reading.

- Clean-up of hydrogen
- Oxygen films on tungsten
- Interaction of oxygen with hydrogen in contact with tungsten filament
- Thorium on tungsten
- The interaction of carbon dioxide with carbon filaments
- Surface energies of pure liquids
- Oil films on water
- Absorbed films on solutions
- Condensation-evaporation theory of adsorption on solids
- Adsorption isotherms
- Effect of forces acting between adatoms
- Caesium films on tungsten
- Types of adsorption
- Activated adsorptions
- Catalytic action of surfaces

The study of surface forces and thin films is still ongoing, perhaps more so today, with the use of surfactants as thin film coating aids amongst other technologies. Langmuir's concepts are still considered to be of importance and are reported in many texts including [9, 10, 11].

Honours and Awards

Langmuir received honorary degrees from the following Colleges and Universities: Northwestern, Union, Edinburgh, Columbia, Kenyon, Princeton, Lehigh, Harvard, Oxford, Johns Hopkins, Rutgers, Queens, and Stevens Institute of Technology.

Some of the awards and prizes presented to Langmuir were: the Nichols Medal (1915 and 1920); the Hughes Medal (1918); the Rumford Medal (1921); the Cannizzaro Prize (1925); the Perkin Medal (1928); the School of Mines Medal (Columbia University, 1929); the Chardler Medal (1929); the Willard Gibbs Medal (1930); the Popular Science Monthly Award (1932); the Nobel Prize in Chemistry (1932); the Franklin Medal (1934); the Holly Medal (1934); the John Scott Award (I937); the Egleston Medal (1939); the "Modern Pioneer of Industry" (1940); the Faraday Medal (1944); the Medal of Merit (1948); the John Carty Medal (1950) and the Mascart Medal (1950).

He was elected a Foreign Member of the Royal Society of London; Fellow of the American Physical Society; Honorary Member of the British Institute of Metals; Honorary of the Chemical Society (London) and academies in France, Italy, Sweden, Brazil, and India.

He had served as President of the American Chemical Society and as President of the American Association for the Advancement of Science. He gave the Pilgrim Trust Lecture of The Royal Society in 1938.

Langmuir was a consultant to the Research Laboratories of the General Electric Company.

Langmuir's citation for election (date: 27/06/1935) as Foreign Member of the Royal Society reads [12]:

"Irving Langmuir of Schenectady distinguished for his contributions to our knowledge of moloecular and atomic phenomena"

He was proposed by:

F. Gowland Hopkins; H. H. Dale; A. C. Seward; J. C. McLennan; E. T. Whittaker; A. W. Conway; G. P. Lenox-Conyngham; W. H. Eccles; Robert Robinson; G. T. Morgan; D. L. Chapman; E. J. Butler; Charles S. Martin; T. R. Elliott; P. P. Laidlaw; E. D. Adrian; F. H. A. Marshall; W. T. Calman; H. G. Lyons; F. E. Smith.

Bibliography

1. Reproduced with permission.
2. From Nobel Lectures, Chemistry 1922–1941, Elsevier Publishing Company, Amsterdam, 1966.
3. The Nobel Population 1901–1937, E. Crawford, J. L. Heilbron and R. Ullrich. Office of the History of Science and Technology, University of California, 1987, ISBN: 0-918102-15-4.
4. Biographical Memoirs of Fellows of the Royal Society, 1958, vol. 4, pp. 167–184.
5. Who was who: 1951–60.
6. Thermal conduction and convection in gases at extremely high temperatures, Trans. Amer. Electrochem. Soc. 1911, 20, 225.
7. The Quintessence of Irving Langmuir by Albert Rosenfeld, Pergamon Press, 1966, Library of Congress catalog card no 65-19839.
8. Langmuir: The man and the scientist Pergamon Press 1962 - also contains the Quintessence of Irving Langmuir, Library of Congress card no 62-12992.
9. Langmuir-Blogett Films Ed Gareth Roberts, Plenum Press, 1990, ISBN: 0-306-43316-8.
10. Surface Chemistry by Elaine M. McCash, Oxford University Press, 2001, ISBN: 0-19-850328-8.
11. Surface Forces by B.V. Dejaguin, N.V. Churaev and V.M. Muller, Consultants Bureau, 1987, ISBN: 0-306-11011-3.
12. Sackler database, Royal Society.

1933

The prize money was with 1/3 allocated to the Main Fund and with 2/3 to the Special Fund of this prize section

1934

Harold Clayton Urey

"for his discovery of heavy hydrogen" [2]

In his presentation speech dated 10th December 1934, Professor W. Palmær, then Chairman of the Nobel Committee for Chemistry of the Royal Swedish Academy of Sciences, observed [2]:

> *...When Urey set about his attack upon the question (that of isotopes of hydrogen), it was not altogether untrodden ground, for there had been suppositions expressed as to the existence of a hydrogen isotope, a number of experiments having even been made to prove it, without however having led to any definite result. Urey proved to be the man to tackle the question in a rational way and to solve it...The merit of having made the actual fundamental discovery appears, however, to belong mainly to Urey and hence the Academy of Sciences has resolved to award this year's Nobel Prize for Chemistry to him, in recognition of his discovery of heavy water...*

Urey did not attend the presentation ceremony as his wife gave birth on Commemoration Day – the normal day for the Nobel Prize presentation ceremony.

Urey was nominated for the 1934 Chemistry Nobel Prize by T. Svedberg. He was also nominated for the 1934 Physics Nobel Prize [3].

Harold Clayton Urey was born in Walkerton, Indiana, on 29th April 1893, as the son of the Rev. Samuel Clayton Urey and Cora Rebecca née Reinoehl. He married Frieda Daum (1926) and together they had four children, Elizabeth, Frieda Rebecca, Mary Alice and John Clayton, and nine grandchildren. Urey died on 5th January 1981 in La Jolla, California [2,4,5].

The Early Years

His early education in rural schools was uneventful, although he only narrowly passed the entrance examination for high school. His graduation from high school (1911) and subsequent three months educational training led to a three year period of teaching in country schools. During his last teaching post (at a school in a mining camp in the Gallatian Mountains), he lodged with a family whose son, Brian 0. Wilson, decided to go to college. Urey decided that he also wanted a college education and enrolled at the University of Montana (1914) with the intention of studying psychology. He changed his mind and studied biology with a minor in chemistry, graduating with a Bachelor of Science degree in Zoology (1917).

The next two years were spent as an industrial research chemist after which he returned to Montana as an instructor in Chemistry. His Ph.D. studies were undertaken at the University of California (1921–1923) under Professor Lewis. Urey's doctoral work dealt with the rotational contributions to heat capacities and entropies of gases, during which he postulated that given the internal partition functions, the thermodynamic quantities of a molecular gas at any temperature can be predicted. He graduated in 1923 and travelled to the University of Copenhagen as an American-Scandinavian Foundation Fellow. He spent the following year working at Niels Bohr Institute for Theoretical Physics, returning to the United States in 1924. His first paper [6], published in 1923, concerned heat capacities and entropies of diatomic and polyatomic gases. A full list of his other papers also appears in [4].

The Career

Year	*Post*	*Institute*
1924	Associate in Chemistry	Johns Hopkins University
1929	Associate Professor of Chemistry	Columbia University
1934	Professor of Chemistry	Columbia University
1945	Distinguished Service Professor of Chemistry, Institute for Nuclear Studies	University of Chicago
1952	Martin A. Ryerson Distinguished Service Professor of Chemistry	University of Chicago
1956–1957	George Eastman Visiting Professor	University of Oxford
1958	Emeritus Professor of Chemistry-at-large	University of California
1970	Retirement	

Urey's interest in isotopes, initially of hydrogen but later uranium, led him to spend the period 1940–1945 as Director of War Research, Atomic Bomb Project, Columbia University in addition to his other duties. A detailed account of Urey's war work, particularly his work concerning the separation of uranium isotopes and the use of the isotopes in the development of the atomic bomb appears in his biographical memoir [4]:

> *...In early 1940 it was definitely established that ^{235}U was the isotope fissionable by thermal neutrons. Urey, together with a group of Columbia faculty members headed by Dean Pegram, promptly made a proposal to the Naval Research Laboratory and to Lyman Briggs, chairman of the President Roosevelt's Committee for Uranium, for research on the separation of the uranium isotopes to be carried out under Urey's direction. The methods to be investigated were thermal diffusion, chemical separation methods, and the centrifugal fractionation column method. A contract was executed on 9 August 1940, although work had begun in May.*

Practically simultaneously Urey was appointed chairman of an Advisory Committee on Nuclear Research, to give technical advice to the Committee on Uranium. In this capacity he coordinated experimental centrifuge studies by Professor Beams at the University of Virginia; an investigation of the separation of uranium isotopes by gaseous diffusion through a membrane, using pressures near atmospheric, which was being investigated by a group at Harvard University led by Professor G. B. Kistiakowsky and the Columbia studies...

Urey's principal academic interests concerned:

- the entropy of diatomic gases
- problems of atomic structure, absorption spectra and the structure of molecules
- the concentration of any possible heavy hydrogen isotopes by the fractional distillation of liquid hydrogen (1931)
- the discovery of deuterium
- the electrolytic method for the separation of hydrogen isotopes (in collaboration with E.W. Washburn)
- investigations of the properties of hydrogen isotopes, particularly the vapour pressure of hydrogen and deuterium
- the equilibrium constants of exchange reactions
- the separation of uranium isotopes
- the measurement of paleotemperatures
- investigations into the origin of the planets
- the chemical problems of the origin of the earth

Urey is the author of the books *Atoms, Molecules, and Quanta* (1930, with A.E. Ruark), and *The Planets* (1952). He was editor of the *Journal of Chemical Physics* (1933–1940).

Urey's Nobel Prize acceptance speech is dated 14th February 1935 and is titled 'Some thermodynamic properties of hydrogen and deuterium', during which he observed [2]:

...In the twenty-odd years which have elapsed since the existence of isotopes was first well established, isotopes of most of the chemical elements have been discovered and their individual masses determined. Of the several hundred now known, by far the majority have been revealed by Aston's mass spectrograph'. In recent years rare isotopes of exceptional interest have been found by the molecular spectrum method, the first being the isotopes 17 and 18 of oxygen, discovered by Giauque and Johnston. Then the isotopes 15 of nitrogen and 13 of carbon, detected in studies of molecular spectra by Naudé and by Birge and King respectively, were added to the list of known isotopes.

Due to these discoveries the abundant isotopes of the light elements were known when the existence of the hydrogen isotope of atomic weight two, now called deuterium, was first demonstrated in 1931. Previous to the experimental detection of deuterium an empirical regularity among the known isotopes which indicated the possible existence of $^{2}H(D)$, $^{3}H(T)$ *and* ^{5}He *had been recognized...*

Urey went on to discuss:

- the theory of hydrogen and deuterium vapour pressures
- the equilibria of exchange reactions involving hydrogen and deuterium, and
- the separation of isotopes

Urey concluded:

> *...The discovery of deuterium and the marked differences in the physical and chemical properties of hydrogen and deuterium, together with an efficient method for the separation of these isotopes, have opened an interesting field of research in several of the major branches of science. It is my expectation that the next few years will witness the separation of the isotopes of the lighter elements in sufficient quantities for effective research in chemistry, physics and biology. If this can be effected, the work on deuterium is only the beginning of a very interesting scientific development...*

Deuterium is still used in the separation of radioactive isotopes and therefore the isolation of deuterium is still of fundamental importance. Perhaps well known is the use of deuterium in nmr studies [7,8,9].

Honours and Awards

Urey received honorary doctor of science degrees from: Montana; Princeton; Newark; Columbia; Oxford; Washington and Lee; McMaster; Yale; Indiana; Birmingham; Athens; Durham and Saskatchewan; also Honorary Doctor of Law degree from Wayne University and the University of California.

He received: the Willard Gibbs Medal (1934); Nobel Prize, Chemistry (1934); Davy Medal (1940); Franklin Medal (1943); Medal of Merit (1946); Cordoza Award (1954); Honor Scroll Award (1954); Joseph Priestley Award (1955); Alexander Hamilton Award (1961); the J. Lawrence Smith Award (National Academy of Sciences) (1962); Remsen Memorial Award (1963); University of Paris Medal (1964); National Science Medal (1964); the Gold Medal RAS (1966); the Chemical Pioneer Award (1969); the Leonard Medal (1969); the Arthur L. Day Award (1969); the Linus Pauling Award (1970); the Johann Kepler Medal (1971); the Gold Medal Award (1972); the Priestley Award (1973); the NASA Exceptional Scientific Achievement Award (1973); the Headliner Award (1974).

He was a member of many of the more important scientific societies of the world including Foreign Member of the Royal Society, Honorary Fellow of: the Chemical Society (London), the National Institute of Sciences of India and the Weizmann Institute of Science (Israel).

Urey's citation for election (date: 1/05/1947) as Foreign Member of the Royal Society reads [10]:

> **"distinguished for his discovery of deuterium, for his investigations into the separation and properties of isotopes and for the application of these discoveries in the development of methods for producing atomic energy"**

He was proposed by:

Thomas R. Merton; A. V. Hill; E. J. Salisbury; E. D. Adrian; M. L. Oliphant; W. T. Asbury; E. L. Hirst; M. H. A. Newman; S. Goldstein; R. E. Stradling; A. C. Chibnall; C. H. Andrewes; N. Hamilton Fairley; R. A. Fisher; H. H. Read; W. Brown; E. C. Bullard.

Bibliography

1. Reproduced with permission.
2. From Nobel Lectures, Chemistry 1922–1941, Elsevier Publishing Company, Amsterdam, 1966.
3. The Nobel Population 1901–1937, E. Crawford, J. L. Heilbron and R. Ullrich. Office of the History of Science and Technology, University of California, 1987, ISBN: 0-918102-15-4.
4. Biographical Memoirs of Fellows of the Royal Society 1983, vol. 29, pp. 623–659.

5. Who was who: 1981–90.
6. J. Am. Chem. Soc. 1923, **45**, 1445–1455.
7. The design of a continuously operated 1-keV deuterium-ion extractor (UCID) by J. H. Fink, The National Technical Information Service, 1978, ASIN: B0006X69VK.
8. Deuterium concentration by chemically-refluxed ammonia-hydrogen exchange: Final report (M.I.T. Dept. of Nuclear Engineering) by Edward A. Mason, M.I.T. Dept. of Nuclear Engineering, 1969, ASIN: B0007FNYCE.
9. Deuterium and Shift Calculation (Nmr-Basic Principles and Progress, Vol. 23) by P. Diehl, Springer, 1991, ISBN: 0387529497.
10. Sackler database, Royal Society.

1935

© Nobel Foundation [1]

Frédéric Joliot (0.5)

"in recognition of their synthesis of new radioactive elements" [2]

On the 10th December 1935 Professor W. Palmær, then Chairman of the Nobel Committee for Chemistry of the Royal Swedish Academy of Sciences, addressed the husband and wife team of Frédéric and Irène who jointly shared the Nobel Chemistry Prize for 1935. This in itself was unique for the chemistry prize, made even more so as Irène's mother was Marie Curie, Laureate of the 1911 chemistry prize and joint winner of half of the 1903 Physics Prize with her husband. Professor W. Palmær commented [2]:

> *...Madame, Sir. In awarding to both of you in equal shares the Nobel Prize in Chemistry for this year, the Royal Academy of Sciences has been pleased to be able to reward, in a brilliant way, the synthesis, achieved by your united efforts, of new radioactive elements.*
>
> *Thanks to your discoveries, it has become possible, for the first time, to transform artificially one element into another hitherto unknown. At last the old dream of the alchemists has become reality. Their main object, the production of gold, has been attained, though by a path, it is true, less direct than they thought would lead them there. The results of your researches are of capital importance for pure science, but in addition, physiologists, doctors, and the whole of suffering humanity hope to gain from your discoveries, remedies of inestimable value...*
>
> *...Madame and Sir, have quickly attracted the attention of the scientific world, and its importance has been amply and universally recognized...*

Frédéric Joliot was nominated for the 1935 Chemistry Prize by H. von Euler-Chelpin, E. Rutherford and T. Svedberg. Joliot was also nominated for the Chemistry Prize for 1936* and for the Physics Prize for the years 1934, 1935 and 1936*, where * represents one or more disallowed nominations for that year [3].

Jean Frédéric Joliot was born in Paris on 19th March 1900, and he was the youngest of the six children. His father, Henri Joliot, was 57 when Frédéric was born, and had fought in the 1870 war and afterwards, in the Commune. As a result of which he had to flee to Belgium. Joliot senior returned after the amnesty and became a businessman, devoting himself to the education of his family His mother was Emilie, née Roederer. Jean Frederic married Irène Curie (1926). They had one daughter, Helene, and one son, Pierre. Joliot died in Paris on 14th August 1958 [2,4,5].

The Early Years

Joliot spent seven years at the Lycée Lakanal, two at the Ecole Lavoisier, then entering the Ecole de Physique et de Chimie Industrielles de la Ville de Paris in 1919. Langevin, a teacher at the Ecole de Physique et de Chimie Industrielles, was to have a profound influence on Joliot's life. Joliot spent the next two years working in the steel industry in Luxembourg (1922–24), followed by a year of military service. This was followed by his appointment, on Langevin's recommendation, to the post of assistant to Marie Curie at l'Institut du Radium de Paris (1925). His thesis for the Doctor of Science degree, submitted in 1930, concerned the electrochemistry of radio-elements.

The Career

During the first ten years of their married life, Frédéric and Irène collaborated on most of their scientific work. They published 26 joint papers during this period.

Year	*Post*	*Institute*
1935	Lecturer	Faculty of Science, Paris
1937	Professor	Collège de France
1956	Director of the Curie Laboratory of the Radium Institute (In addition to his Chair at the Collège de France Professor of Nuclear Physics (following his wife's death))	Sorbonne, Paris

Scientific Interests

Joliot's first research work involved a new method of studying the electrolytic deposition of radioactive elements. This involved creating a very thin film of mica and gold on the bottom of a vessel. Polonium was electrolyzed out of solution on to this surface. The emitted alpha particles passed through the mica sheet into an ionization chamber below. The rate of deposition of polonium could be followed quantitatively and continuously as a function of the electrode potential. His later works concerned:

- investigations on the structure of the atom (1935 onwards)
- the projection of nuclei, which was an essential step in the discovery of the neutron (Chadwick, 1932) the positron (Anderson, 1932)
- the discovery of artificial radioactivity (1934), arguably his greatest work. This was achieved by bombarding boron, aluminium, and magnesium with alpha particles, producing ^{13}N, ^{30}P, ^{27}Si

and ^{28}Al. These isotopes are not found naturally and decompose spontaneously, producing either positive or negative electrons. It was for this very important discovery that Frédéric and Irène received in 1935 the Nobel Prize for Chemistry
- building the first cyclotron in Western Europe
- working on chain reactions

Joliot also worked on the requirements for the successful construction of an atomic pile using uranium and heavy water (five patents were taken out in 1939 and 1940). Mindful of the advance of the German forces (1940), Joliot managed to get the documents and materials relating to this work transported to England.

In addition to his academic posts, Joliot was also appointed:

- Director of the Centre National de la Recherche Scientifique (1945)
- The first High Commissioner for Atomic Energy (1946)

He also directed the construction of the first French atomic pile (1948).

He was relieved of these duties in 1950 for political reasons. However, he retained control of his laboratories. Joliot devoted the last two years of his life to the inauguration and development of a large centre for nuclear physics at Orsay.

Political Interests

Throughout his life, Joliot took a considerable interest in politics:

- he joined the Socialist Party
- he joined the S.F.I.O. (1934)
- he also joined the League for the Rights of Man (1936)
- he took an active part in the Resistance during the French occupation
- he was President of the National Front
- he formed the French Communist Party
- he was relieved of some of his duties in 1950 for political reasons although retained his research laboratory
- he was elected President of the World Peace Council

At the conclusion of his Joliot's Nobel Prize acceptance speech, dated 12th December 1935 and titled 'Chemical evidence for the transmutation of elements', Joliot observed [2]:

> *...If, turning towards the past, we cast a glance at the progress achieved by science at an ever-increasing pace, we are entitled to think that scientists, building up or shattering elements at will, will be able to bring about transmutations of an explosive type, true chemical chain reactions.*
>
> *If such transmutations do succeed in spreading in matter, the enormous liberation of usable energy can be imagined. But, unfortunately, if the contagion spreads to all the elements of our planet, the consequences of unloosing such a cataclysm can only be viewed with apprehension. Astronomers sometimes observe that a star of medium magnitude increases suddenly in size; a star invisible to the naked eye may become very brilliant and visible without any telescope - the appearance of a Nova.*

This sudden flaring up of the star is perhaps due to transmutations of an explosive character like those which our wandering imagination is perceiving now - a process that the investigators will no doubt attempt to realize while taking, we hope, the necessary precautions...

Many texts are available concerning all aspects of radioactivity, perhaps the interexted reader would use [6,7] as entry points into the subject.

Honours and Awards

Joliot was holder of a honorary doctor degrees from several universities.

Joliot was presented with many awards including: Commander of the Legion of Honour; the Croix de Guerre; the Nobel Prize for Chemistry jointly with his wife (1935); the Hughes Medal (1947).

He was a Member of the French Academy of Sciences and of the Academy of Medicine and also a member of numerous foreign scientific academies and societies.

Joliot's citation for election (date: 09/05/1946) as Foreign Member of the Royal Society reads [8]:

"distinguished for his researches into the nuclear constitution of matter, and in particular for the discovery of induced radio-activity"

He was proposed by:

Robert Robinson; E. J. Salisbury; A. V. Hill; W. V. D. Hodge; W. T. Astbury; H. W. Melville; S. Goldstein; D. R. Pye; R. A. Peters; I de Burgh Daly; Gordon Holmes; R. A. Fisher; S. Zuckerman; C. Forster Cooper; F. E. Fritsch; E. C. Bullard; Thomas R. Merton; A. C. Egerton

Bibliography

1. Reproduced with permission.
2. From Nobel Lectures, Chemistry 1922–1941, Elsevier Publishing Company, Amsterdam, 1966.
3. The Nobel Population 1901–1937, E. Crawford, J. L. Heilbron and R. Ullrich. Office of the History of Science and Technology, University of California, 1987, ISBN: 0-918102-15-4.
4. Biographical Memoirs of Fellows of the Royal Society, 1960, vol. 6, pp. 87–105.
5. Who was who: 1951–60.
6. Invisible Rays: A History of Radioactivity by G. I. Brown, Sutton Publishing, 2002, ISBN: 0750926678.
7. Practical Applications of Radioactivity and Nuclear Radiations by Gerhart Lowenthal and Peter Airey, Cambridge University Press, New ed. 2005, ISBN: 052101980X.
8. Sackler database, Royal Society.

1935

© Nobel Foundation [1]

Irène Joliot-Curie (0.5)

"in recognition of their synthesis of new radioactive elements" [2]

On the 10th December 1935, Professor W. Palmær, then Chairman of the Nobel Committee for Chemistry of the Royal Swedish Academy of Sciences, addressed the husband and wife team of Frédéric and Irène who jointly shared the Nobel Chemistry Prize for 1935. This in itself was unique for the Chemistry Prize, made even more so as Irène's mother was Marie Curie, Laureate of the 1911 Chemistry Prize and joint winner of half of the 1903 Physics Prize with her husband. Professor W. Palmær commented [2]:

> *...Thanks to your discoveries, it has become possible, for the first time, to transform artificially one element into another hitherto unknown. At last the old dream of the alchemists has become reality. Their main object, the production of gold, has been attained, though by a path, it is true, less direct than they thought would lead them there. The results of your researches are of capital importance for pure science, but in addition, physiologists, doctors, and the whole of suffering humanity hope to gain from your discoveries, remedies of inestimable value.*
>
> *Madame, 24 years ago, your mother, Madame Marie Sklodowska-Curie, was present at the Nobel festival in order to receive the Chemistry Prize, as a reward for her discovery of radium in the first place; and you, Madame, were also present on that occasion, as a little girl.*
>
> *Her husband, your father Pierre Curie, was already dead; but earlier on, in 1903, he had shared with her, half the Physics Prize, an award well deserved for their work on the phenomena of radiation. In collaboration with your husband, you have worthily maintained those brilliant traditions. The work of both of you, Madame and Sir, have quickly attracted the attention of the scientific world, and its importance has been amply and universally recognized...*

Joliot-Curie was nominated for the 1935 Chemistry Nobel Prize by H. von Euler-Chelpin, E. Rutherford and T. Svedberg. She was also nominated for the Chemistry Prize for the 1936* and for the Physics Prize for the years 1934, 1935 and 1936*, where * represents one or more disallowed nominations for that year [3].

Irène Curie was born in Paris on 12th September 1897, the daughter of Pierre and Marie Curie. Irène married Jean Frederic Joliot (1926). They had one daughter, Hélène, and one son, Pierre. Joliot-Curie died in Paris on 17th March 1956 [2].

The Early Years

Joliot-Curie studied at the Faculty of Science in Paris, serving as a nurse radiographer during the First World War. She presented her Doctor of Science thesis on the alpha rays of polonium in 1925.

The Career

Much of the early works of Joliot-Curie were undertaken with her husband. The list of accomplishments presented here for Joliot-Curie may well be shown under Joliot's career details in other texts.

Year	*Post*	*Institute*
1932	Lecturer	Faculty of Science, Paris
1937	Professor	Faculty of Science, Paris
1946	Director of the Radium Institute	Paris

Either alone or in collaboration with her husband, Joliot-Curie undertook research concerning:

- important work on natural and artificial radioactivity
- transmutation of elements
- nuclear physics
- research on the action of neutrons on the heavy elements (1938), this was an important step in the discovery of uranium fission

Joliot-Curie was Commissioner for Atomic Energy for six years, and in that capacity took part in the creation and construction of the first French atomic pile (1948).

She was also involved with the creation of the large centre for nuclear physics at Orsay, for which she worked out the plans. This centre was equipped with a synchro-cyclotron of 160 MeV. Its construction was continued after her death, by her husband.

The work for which Joliot-Curie and Joliot were awarded the Chemistry Nobel Prize was reported in their joint paper *Production artificielle d'éléments radioactifs. Preuve chimique de la transmutation des éléments* (1934).

Joliot-Curie's Nobel Lecture dated 12th December 1935 was titled 'Artificial Production of Radioactive Elements' during which she commented:

...We have shown that it is possible to create a radioactivity characterized by the emission of positive or negative electrons in boron and magnesium, by bombardment with alpha rays. These artificial radio-elements behave in all respects like the natural radio-elements

Returning to our hypothesis concerning the transformation of the aluminium nucleus into a silicon nucleus, we have supposed that the phenomenon takes place in two stages: first there is the capture of the alpha particle and the instantaneous expulsion of the neutron, with the formation of a radioactive atom which is an isotope of phosphorus of atomic weight 30, while the stable phosphorus atom has an atomic weight of 31. Next, this unstable atom, this new radio-element which we have called "radio-phosphorus" decomposes exponentially with a half-life of three minutes.

We have interpreted in the same way the production of radioactive elements in boron and magnesium; in the first an unstable nitrogen with a half-life of 11 minutes is produced, and in the second, unstable isotopes of silicon and aluminium...

Many texts are available concerning all aspects of radioactivity, perhaps the interested reader would use [4,5] as entry points into the subject.

Honours and Awards

Joliot-Curie was presented with honorary doctor degrees from several universities.

She was an Officer of the Legion of Honour and shared the 1935 Chemistry Nobel Prize with her husband (1935).

She was a Member of several foreign academies and of numerous scientific societies, a Member of the Comité National de l'Union des Femmes Françaises and of the World Peace Council.

Joliot-Curie was appointed Undersecretary of State for Scientific Research (1936).

Bibliography

1. Reproduced with permission.
2. From Nobel Lectures, Chemistry 1922–1941, Elsevier Publishing Company, Amsterdam, 1966.
3. The Nobel Population 1901–1937, E. Crawford, J. L. Heilbron and R. Ullrich. Office of the History of Science and Technology, University of California, 1987, ISBN: 0-918102-15-4.
4. Invisible Rays: A History of Radioactivity by G. I. Brown, Sutton Publishing, 2002, ISBN: 0750926678.
5. Practical Applications of Radioactivity and Nuclear Radiations by Gerhart Lowenthal and Peter Airey, Cambridge University Press, New ed. 2005, ISBN: 052101980X.

1936

© Nobel Foundation [1]

Petrus (Peter) Josephus Wilhelmus Debye

"for his contributions to our knowledge of molecular structure through his investigations on dipole moments and on the diffraction of X-rays and electrons in gases" [2]

During his presentation speech, dated 10th December 1936, Professor A. Westgren then Secretary of the Nobel Committee for Chemistry of the Royal Swedish Academy of Sciences, commented of Debye [2]:

...Debye has elaborated a theory of the effect of electric fields on molecules and has worked out methods for the determination of their dipole moments. These can be determined by measurements of the variation of the insulating power and of the density with the temperature. Debye's theory applies strictly only for diluted gases, in which there is no need to reckon with a reciprocal effect between the molecules. It is difficult, however, to prepare the experimental material for the gases necessary for the calculation of the dipole moments. It is therefore of value that the theory, as has been established experimentally, can also be applied without noteworthy errors to diluted solutions of substances in non-polar solvents...Your rich scientific activity has been aimed in particular to research into the structure of matter. Your wealth of ideas, your penetration and your secure mastery of mathematical methods have yielded great success to your endeavours, and your results have enriched chemistry to an extraordinary degree in all kinds of ways. By your investigations on dipole moments and also on X-ray and electron interferences in gases you have widened and deepened our knowledge of molecular structure to such an extent that the Royal Academy of Sciences has awarded you the Nobel Prize for Chemistry...

Debye was nominated for the 1936 Chemistry Nobel Prize by C. Ingold and H.C. Urey. He was also nominated for the years 1927, 1928, 1929, 1931, 1931*, 1932, 1933, 1934 and 1935. He was nominated for the Physics Prize for the years 1916, 1917, 1921, 1922, 1924, 1925, 1926, 1927, 1928, 1929, 1930, 1931, 1932, 1933 and 1936, where * represents those years where one or more nominations were disallowed [3].

Petrus (Peter) Josephus Wilhelmus Debye was born on 24th March 1884, at Maastricht, the Netherlands. His parents, Joannes Wilhelmus Debije (1859–1937) and Maria Anna Barbara Ruemkens (1859–1940) were married on 23rd May 1883. Debye had one younger sister (by 4 years) and no brothers. He married Mathilde Alberer and had a son, Peter Paul Rupprecht (1916) and a daughter, Mathilde Maria (1921). Debye died on 2nd November 1966 in Ithaca, New York, USA [2,4].

The Early Years

Debye spoke the Maastricht patois when young and knew little Dutch. In later life he chose to write to his old friends in the Netherlands in the Maastricht patois. He received his early education at the elementary school in Maastricht, attending the Hoogere Burger School at Maastricht for five years, from the age of twelve. Debye studied a range of subjects at the Hoogere Burger School, but crucially, did not study Greek and Latin, a requirement for entry into a Dutch university. Family finances dictated that Debye enrol at the Technische Hochschule at Aachen. Thus for some years, Debye left his home soon after five to catch the early morning train to Aachen, thirty kilometres away, and across the border in Germany. He gained a degree in electrical technology from Aachen in 1905.

His first published work was undertaken as part of his diploma and concerned a theoretical treatment of Foucault currents in a rectangular conductor [5]. Debye's doctoral thesis was completed at Munich (July 1908), dealing with the radiation pressure experienced by spherical particles of varied refractive properties.

The Career

Year	*Post*	*Institute*
1904	Assistant in Technical Mechanics	Aachen Technological Institute
1906	Assistant in Theoretical Physics	Munich University
1910	qualified as a University Lecturer	Munich University
1911	Professor of Theoretical Physics – successor to Einstein	Zurich University
1912	Professor of Theoretical Physics	Utrecht University
1914	Head of the Theoretical Department, then Director of the entire Institute	Physical Institute, University of Göttingen
1920	Professor of Physics, and Principal of the Eidgenössische Technische Hochschule	Zurich University
1927	Professor of Physics	Leipzig University
1934	Director of the Max Planck Institute	Kaiser Wilhelm Institute for Physics in Berlin-Dahlem
1935	Professor of Physics	University of Berlin
1940	Professor of Chemistry and Principal of the Chemistry Department	Cornell University, Ithaca, New York
1952	Resigned as Head of the Chemistry Department and Emeritus Professor of Chemistry	Cornell University, Ithaca, New York

Life as a Dutch citizen in 1939 was not easy, however, and so Debye chose to move to the USA in 1940. Some of the work undertaken by Debye included:

The First German Period

- Propagation problems of significance to radar and waveguide systems (with Hondros, 1910)
- The interaction of radiation with atomic and molecular systems
- Stationary and quasi-stationary electric fields
- Specific heats
- Scattering of X-rays by a crystal lattice

The Utrecht and Göttingen Period

- New representation of dielectric dispersion
- Treatment of the influence of lattice vibrations upon X-ray diffraction intensities
- The theory of anomalous dispersion in the region of long-wave electromagnetic radiation
- Rationalized and more general version of the quantization condition
- An account of the Zeeman effect in terms of the quantum hypothesis
- The evaluation of the van der Waals cohesive forces

The Second German Period

- X-ray scattering and quantumn theory
- freezing point depression and related phenomena
- the evaluation of the concentration dependence of the ion mobilities in solution (with Hückel)
- the influence of other electrolytes upon the solubility of sparingly soluble salts
- electric permittivity factor in the electrolyte (with Linus Pauling)
- some remarks on magnetization at low temperatures
- Achieved clear X-ray diffraction by isolated molecules (for the first time)
- Molecular structural studies by X-ray diffraction in gases
- the study of the scattering and diffraction of light by ultrasonic waves

No new work was started during the Berlin period. Debye commented:

...'At the time I accepted to go to Berlin I was still a Dutch citizen. At the same time the German government conceded in a letter signed by the Minister of Education, Dr Rust, that in accepting the positions offered, I did not become and would not be asked to become a German citizen. The positions, as usual, were lifetime positions. During the time the laboratory was still under construction, I received an offer from Harvard University, which I declined because I did not feel free to quit before having finished what I had undertaken to do... The war broke out and one day without previous warning I was informed by Dr Telschow from the Kaiser-Wilhelmthat I could no longer enter the laboratory except by becoming a German citizen. I refused. I was advised by the Minister of Education to stay at home and occupy myself by writing a book. Instead, I was able to overcome the difficulties put in my way by different German authorities and to leave for the U.S.A.,

by way of Italy, in order to give the Baker Lectures at Cornell University, to which I had been invited.'...

The Cornell University Period

On completion of the George Fischer Baker lectures (1940), Debye accepted the chairmanship of the Department at Cornell University, becoming a US citizen in 1946. His work at Cornell included:

- Polymeric substitutes for Malayan rubber (part of the war effort)
- Developed an empirical model for polymer molecular weight determination from the viscosity
- Various studies of light scattering
- Thermal diffusion
- Various polymer studies

His last paper, published posthumously in 1968 was titled '*Direct visual observation of concentration fluctuations in a critical mixture* '[6]. A comprehensive list of papers and books appears in reference [4].

Debye was Editor of *Physikalische Zeitschrift.*

Debye's Nobel Prize acceptance speech was dated 12th December 1936 and titled 'Methods to determine the electrical and geometrical structure of molecules'. He concluded thus [2]:

..."In summary I would like to emphasize that the physical methods discussed here besides furnishing a precise definition of atomic positions have given brilliant confirmation of structures obtained by purely chemical means"...

These principles still receive recognition by inclusion in modern texts, see for example [7–9].

Honours and Awards

Debye had honorary doctorates conferred upon him by the following Universities and Institutes: Brussels and Liège; Oxford; Sofia; Mainz; Technische Hochschule, Aachen; Eidgenösissche Technische Hochschule, Switzerland; Harvard; St. Lawrence; Colgate; Notre Dame; Holy Cross; Brooklyn Polytechnic; Boston College; Providence College.

He held: the Rumford Medal (1930); the Franklin Medal; the Nobel Prize, Chemistry (1936); the Faraday Medal; the Lorentz Medal; the Max Planck Medal (1950); the Willard Gibbs Medal (1949); the Nichols Medal (1961); the Kendall Award (1957); the Priestley Medal (1963).

Debye was associated with scientific academies in many countries including: Brussels and Liège (Belgium); Denmark; Royal Irish Academy, Dublin; Berlin, Göttingen, and Munich, Germany; the Royal Institute of Great Britain and the Royal Society, London; the Indian Academy, Bangalore and the National Institute of Science, India; The Netherlands; the Papal Academy, Rome; the Real Sociedad Española de Fisica y Quimica and Academia de Ciencias, Madrid, Spain; Washington, New York, Boston, and Philadelphia in the United States of America; the Academies of Sciences of the U.S.S.R., Hungary and Argentina.

He was appointed Kommandeur des Ordens Leopold II (1956).

Debye was a Visiting Lecturer at many universities - Columbia, California, Paris, Liège, Oxford, Cambridge, Harvard, Michigan and South California.

Debye's citation for election (date: 25/05/1933) as Foreign Member of the Royal Society reads [8]:

"Peter Debye of Leipzig distinguished for his researches on specific heats, X-ray spectroscopy, the theory of solutions and dipole properties of molecules.."

He was proposed by:

F. Gowland Hopkins; J. A. Arkwright; A. V. Hill; Wilfred Trotter; E. Mellanby; D'arcy W. Thompson; A. G. Tansley; A. Hutchinson; G. Udny Yule; R. Robinson; N. V. Sidgwick; A. C. Egerton; C. H. Desch; G. M. B. Dobson; W. L. Bragg; J. G. Littlewood; Rayleigh; H. H. Dale; F. E. Smith; H. G. Lyons.

Bibliography

1. Reproduced with permission.
2. From Nobel Lectures, Chemistry 1922–1941, Elsevier Publishing Company, Amsterdam, 1966.
3. The Nobel Population 1901–1937, E. Crawford, J. L. Heilbron and R. Ullrich. Office of the History of Science and Technology, University of California, 1987, ISBN: 0-918102-15-4.
4. Biographical Memoirs of Fellows of the Royal Society, 1970, vol. 16, pp. 175–232.
5. Wirbelströme in Stäben von techteckigem Querschnitt. Z. Math. Phys. 1907, **54**, 418.
6. P. Debye and R. T. Jacobsen, J. Chem. Phys, 1968, **48**, 203.
7. Topics in Chemical Physics – based on the Havard Lectures of Peter Debye by Alfred Prock and Gladys McConkey, Elsevier, 1962, Library of Congress Catalog Card Number 61-18140.
8. Chemical Thermodynamics 6th Ed by Irving M. Klotz and Robert M. Rosenberg, John Wiley and Sons Inc., 2000, ISBN: 0-471-33107-4.
9. Principles of Thermodynamics by Myron Kaufman, Marcel Dekker Inc, 2002, ISBN: 0-8247-0692-7.
10. Sackler database, Royal Society.

1937

© Nobel Foundation [1]

Walter Norman Haworth (0.5)

"for his investigations on carbohydrates and vitamin C" [2]

Professor W. Palmær, then Chairman of the Nobel Committee for Chemistry of the Royal Swedish Academy of Sciences, commented of Haworth during his presentation speech dated 10th December 1937 [2]:

> *...What has Haworth then accomplished within this domain? The answer may be thus formulated that he has, above all, made clear the chemical structure of vitamin C...The knowledge regarding the constitution of a vitamin does not only possess a theoretical interest but is also of very great practical importance. On the one hand it may be found possible, by minor changes in the known composition which may be brought about in an artificial way, to produce compounds which in some cases may prove to be more suitable as medicine. And above all, it opens the way to the artificial production of the compound, a thing of very great importance in the case of vitamins which do occur in nature only in a state of very great dilution. Thus vitamin C is already produced on a technical scale and at a price very much lower than that of the natural product...*

Haworth was nominated for the 1937 Chemistry Nobel Prize by A. Virtanen and H. von Euler-Chelpin. He was nominated for the years 1929, 1930, 1931, 1934, 1934*, 1935, and 1936, where * represents one or more disallowed nomination for that year [3].

Walter Norman Haworth was born at Chorley, Lancashire, on 19th March 1883, the second son and fourth child of Thomas and Hannah Haworth. Thomas was the manager of the Chorley Rylands factory. Haworth married Violet Chilton (1922), second daughter of Sir James Dobbie, LLD., FRS. They had two sons. Haworth died suddenly on 19th March 1950, at their home in Barnt Green, Birmingham, Worcestershire [2,4,5].

The Early Years

Haworth attended the local school in Chorley until the age of 14, after which he joined his father, where he learned linoleum design and manufacture. Part of his work involved the use of dyestuffs which led Haworth to develop an interest in chemistry. Wanting to further his education, he sought private tuition in Preston, which enabled Haworth to pass the entrance examination of the University of Manchester (1903). He entered the Chemistry Department as a pupil of W.H. Perkin, Junior, graduating with a first class honours degree in 1906. Three years of research with Perkin during which his first paper [6] was published (1908), led to the award of an 1851 Exhibition Scholarship which he chose to undertake at Wallach's laboratory at Göttingen (1909). He gained his Ph.D. (Gottingen), and after only one year of study with Wallach, returned to Manchester for the second year of his Scholarship, where he continued work on the terpene group as a Research Fellow. He was awarded the D.Sc. degree of the University of Manchester (1911) in the minimum time permissible.

The Career

Year	*Post*	*Institute*
1911	Demonstrator	Imperial College, London
1912	Lecturer and later Reader in Chemistry	St. Andrews, Scotland
1920	Professor of Chemistry	University of Durham
1921	Director of the Department and Professor of Chemistry	University of Durham
1925	Professor and Director of the Department of Chemistry	University of Birmingham
1947–1948	Dean of the Faculty of Science and acting as Vice-Principal	University of Birmingham
1948	Retired	

Although all of Haworth's professional career was spent in academia, he worked for the war effort during the two World Wars.

Academic Achievements

- investigations on the constitution of terpenes (initially with Perkin)
- the synthesis of sylvestrene (1912)
- investigations concerning carbohydrates where he extended Emil Fischer's method of reacting sugars with methanol producing methylated derivatives which were used to characterize the constitution of sugars in (association with T. Purdie and J.C. Irvine)
- the structures of maltose, cellobiose, lactose, gentiobiose, melibiose, gentianose, raffinose and the glucoside ring structure of normal sugars
- the study of lactones from sugars and co-related structure with optical rotatory powers
- devised a method for the determination of chain length in methylated polysaccharides
- helped with the basic features of the starch, cellulose, glycogen, inulin and xylan molecules

- the synthesis of ascorbic acid
- investigations of bacterial polysaccharides

War effort

During the First World War, Haworth organized the laboratories at St. Andrews for the production of fine chemicals and drugs;

During the Second World War:

- Haworth was active in the preparation of highly purified uranium metal
- He became Chairman of the British Chemical Panel concerned with the atomic energy project (1941)
- He worked on the preparation and properties of organic fluorine compounds – an extensive effort
- He undertook research work on fundamental aspects of the chemical problems
- Various aspects of the organization of the Colonial Products Research Council for the study of cane-sugar and starches and their transformation products (from 1944)
- Continued his duties of Chair and Dean of the Faculty of Science (1943–1946)
- He was war-time President of the Chemical Society (1944–1946)

Haworth was active in retirement, serving on many Boards and Committees; he represented the Royal Society at the Seventh Pactfic Science Congress in New Zealand during February, 1949.

Haworth wrote numerous scientific papers and contributed to *Advances in Carbobydrate Chemistry*. His book *The Constitution of Sugars* was published in 1929.

Full list of papers appears in his obituary [4].

Haworth's Chemistry Nobel Prize acceptance speech, dated 11th December 1937, was titled 'The structure of carbohydrates and vitamin C'. He observed of the structure elucidation of vitamin C [2]:

> *...An important observation was that, when newly formed, the primary reversible oxidation product from ascorbic acid does not possess acidic properties but behaves in all respects as a lactone, which develops acidity when kept in aqueous solution. It followed that the acidic character of ascorbic acid is due to an enolic hydroxyl group and not to a free carhoxyl group and, in order to determine the structure of ascorbic acid, it remained only to discover the nature of the lactone ring in the primary oxidation product. The main features of the constitution of ascorbic acid were now established and its formulation as a lactone of 2-keto-1-gulonic acid, capable of reacting in various tautomeric modifications, was first announced from the University of Birmingham, early in 1933.*
>
> *The elucidation of the nature of the lactone ring followed from a study of the oxidation product obtained when tetramethylascorbic acid reacts with ozone. It had been shown by Micheel that the dimethyl derivative of ascorbic acid, obtained by Karrer by the action of diazomethane, gives a di-pnitrobenzoyl derivative and that the latter reacts with ozone giving a neutral ester containing the same number of carbon atoms as the unoxidized material. It followed that a ring system must he present in ascorbic acid, but the products obtained on hydrolysis of the neutral ester (oxalic acid and 1-threonic acid) did not permit of deductions being made concerning the nature of the ring...*

Further information concerning modern syntheses of carbohydrates and vitamin C may be found in [7–10].

Honours and Awards

Haworth received honorary science degrees from the Universities of Belfast, Oslo and Zurich, a honorary D.Sc. from Cambridge and a honorary Doctor of Law from the University of Manchester.

He was awarded: the Longstaff Medal (1933); the Davy Medal (1934); the Nobel Prize, Chemistry (1937); the Royal Medal (1942).

Haworth was elected Fellow, and later, Vice-President (1947–1948) of the Royal Society (1928). He was also elected Fellow of the Chemical Society and later President (1944–1946). He also held memberships of nine foreign scientific academies.

He delivered the Bakerian lectures (1944) and was knighted in 1947.

Haworth's citation for election (date: 10/05/1928) as Fellow of the Royal Society reads [11]:

> **"Distinguished as an Organic Chemist. Author or joint author of some 35 memoirs, published chiefly in the Journal of the Chemical Society, on Organic Synthesis and on the Constitution of some Terpenes and Carbohydrates. He has determined the structure of many of the di- and tri-saccharides, including sucrose, maltose, lactose, melibiose, cellobiose, gentiobiose, raffinose, and gentiacrose, and has synthesised amygdalin".**

He was proposed by:

Robert Robinson; N. V. Sidgwick; W. H. Mills; H. B. Baker; W. H. Perkin; J. C. Irvine; George Barger; James Walker; Arthur Lapworth, H. B. Dixon

Bibliography

1. Reproduced with permission.
2. From Nobel Lectures, Chemistry 1922–1941, Elsevier Publishing Company, Amsterdam, 1966.
3. The Nobel Population 1901–1937, E. Crawford, J. L. Heilbron and R. Ullrich. Office of the History of Science and Technology, University of California, 1987, ISBN: 0-918102-15-4.
4. Obituary Notices of Fellows of the Royal Society, 1950–1951, vol. 7, pp. 373–404.
5. Who was who: 1941–1950.
6. W. N. Haworth and W. H. Perkin Jr. J. Chem. Soc, 1908, p. 573.
7. The Molecular and Supramolecular Chemistry of Carbohydrates: Chemical Introduction to the Glycosciences by Serge David and Rosemary Green Beau, Oxford University Press, 2005, ISBN: 0198500475.
8. Carbohydrate Chemistry (Oxford Chemistry Primers, 99) by B. G. Davis and A. J. Fairbanks, Oxford University Press, USA, 2002, ISBN: 0198558333.
9. Essentials of Carbohydrate Chemistry and Biochemistry by Thisbe K. Lindhorst, John Wiley & Sons, 2nd ed. 2003, ISBN: 3527306641.
10. Classics in Total Synthesis II: More Targets, Strategies, Methods by K. C. Nicolaou and S. A. Snyder, John Wiley & Sons, 2003, ISBN: 3527306846.
11. Sackler database, Royal Society.

1937

© Nobel Foundation [1]

Paul Karrer (0.5)

"for his investigations on carotenoids, flavins and vitamins A and B2" [2]

Professor W. Palmær, then Chairman of the Nobel Committee for Chemistry of the Royal Swedish Academy of Sciences, commented of Karrer during his presentation speech dated 10th December 1937 [2]:

> *...The Royal Academy of Sciences has decided also to award to Professor Paul Karrer in Zurich one half of the Nobel Prize in Chemistry this year in recognition of his researches concerning carotenoids and flavins, and the vitamins A and B_2...The splendid research concerning the carotenoids, made by Karrer, received its coronation, when it led to the isolation, the production in a pure form and the determination of the chemical structure of vitamin A. This vitamin, which had been known to exist from its biological effects already since 1906 and the synthesis of which in a pure form had been tried in vain in many laboratories all over the world, was successfully isolated by Karrer in 1931 from cod-liver oil, and it was the first of the vitamins of which the chemical structure was clarified. It forms a growth factor, i.e. a substance necessary for the growth of the body...*

Karrer was nominated for the 1937 Chemistry Nobel Prize by M. Duboux, P. Dutoit, A. von Eiselsberg, H. von Euler-Chelpin, H. Goldstein, R. Mellet and H. Fischer having already been nominated for years 1932, 1933, 1934, 1935, and 1936*, where * represents one or more disallowed nominations in that year [3].

Paul Karrer was born in Moscow on 21st April 1889. His parents, Paul Karrer and Julie, née Lerch, were Swiss nationals (the family originated from Teufenthal and Oberin the canton of Aargau), although his father was a practicing Moscow dentist at the time of Karrer's birth. Karrer married Helena Froelich (1914) who was the daughter of the director of the Psychiatric Clinic, Konigsfelden. They had three sons, one died in infancy. Karrer died on 18th June 1971 [2,4,5].

The Early Years

The family returned to Switzerland (1892), living for the first three years in Erlenbach near Zurich, where Karrer received his early education at the primary school in the nearby Moriken. He moved on to the district school in Lenzburg and then the gymnasium in Aarau. He matriculated from Aarau in 1908 and went on to study chemistry under Professor Alfred Werner at University of Zurich. He gained his Ph.D. (1911) in six semesters under Werner (studying cobalt complexes). Karrer spent a further year as Assistant (to Werner) in the Chemical Institute (1911–1912). He wrote a comedy '*Drehen und Spaltern*' in 1911 in which he performed, along with some of his fellow students at a Christmas party (held in the Casino Hottingen). Karrer's first publication, the first of over a thousand, concerned organic arsenic compounds [6]. Karrer was invited to be Scientific Assistant to P. Ehrlich in Frankfurt-am-Main (autumn of 1912) because of his work on arsenic compounds. He later became a member of the George Speyer House. For a short time, Karrer served on the Swiss border in his capacity of an artillery officer in the Swiss army.

The Career

Year	*Post*	*Institute*
1918	Reader then Professor	University of Zurich
1919	Director	Chemical Institute, University of Zurich
1959	Retired	

Whilst Karrer's early research concerned the preparation and investigation of the properties of complex metal compounds, his most important work concerns plant pigments. Some of his major contributions were:

- an investigation of the yellow carotenoids
- the structure elucidation of the carotenoids and the conversion of some of them into vitamin A in the body
- the constitution of some of the anthocyanins (1927–28)
- an investigation of the structure of crocin, the yellow colour of saffron (1928)
- the structure of β-carotene (1930)
- the clarification of the structure of vitamin A (1931)
- that he confirmed the structure of ascorbic acid (vitamin C)
- some investigations into vitamin B_2 and E fields
- the chemistry of the flavins
- the identification of lactoflavin
- some investigations into co-enzymes
- some work on curare and other alkaloids
- the preparation of pure vitamin K_1 (1939)
- some work on amino acids and carbohydrates
- some investigations into organo-arsenic
- the structure and partial synthesis of toxiferine and related alkaloids (1958)
- the synthesis of canthaxanthin from β-carotene
- the stereochemistry of the carotenoids

Karrer published over 1,000 scientific papers. His *Lehrbuch der Organischen Chemie* (1930) has been translated in full into English, Italian, Spanish, French, Polish and Japanese. His monograph on carotenoids (1948) has also been translated into English. He also published 78 patents.

He founded the 'Fritz Hoffman-La Roche foundation for the advancement of interdisciplinary groups in Switzerland' and also the 'Foundation for scholarships in chemistry'.

Karrer's Nobel acceptance speech, dated 11th December 1937 and titled 'Carotenoids, flavins and vitamin A and B_2'. Karrer commented [2]:

> *...It has often been said that it is only the first discovery which is difficult and that the ensuing discoveries are usually only the continuations of the first. Perhaps vitamins and carotenoids offer examples of this.*
>
> *In a lecture which I gave in 1932 I remarked: °The chemistry of the vitamins has made great progress in the last few years; it has overtaken the chemistry of the hormones and left that of the enzymes far behind. Soon we shall be so far advanced that all kinds of positive statements can be made about the chemical nature of the different vitamins. Then the chemist will be able to hand back the vitamin problem to the physiologist, so that the latter may establish in what way these substances develop their effect in the organism. Once physiology has solved this problem, it will have become enriched by one of its greatest findings.*
>
> *This prophecy has been fulfilled to a large extent within the short space of five years. The chemical side of the vitamin problem is solved in its essential points; relationships to enzymes were found. It is the task of physioltoday to explain the intervention of these agents in the cell processes. As this is, however, a matter of chemical processes, these will in the end probably have to be elucidated by the chemist once again...*

Further biographical data concerning Karrer can be obtained from [7, 8] and from his many research interests [9–12].

Honours and Awards

Karrer received honorary degrees from a number of Universities and Institutes including: M.D. from Basel, Breslau, Lausanne, Zurich; D.Sc. (or Ph.D.) Bruxelles, London, Lyon, Paris, Rio de Janeiro, Sofija, Torino; D.Pharm. Madrid, Strasbourg; D.Sc. Techn.TH Darmstadt, ETH Zurich.

He received the: Swiss Marcel Benoist Award (1923); Italian Cannizzaro Award (1935); Nobel Prize, joint with W. N. Haworth, (1937); Officier de La Legion d'Honneur (1954); Honorary Citizen of Teufenthal (his native village).

Karrer was appointed to Fellowships or Memberships to many academies, including: Chemical Society, London; Society of Chemical Industry, London; Royal Society of Edinburgh from the UK; Indian Chemical Society; Indian National Academy of Sciences; National Institute of Sciences of India; Indian Association for the Cultivation of Science from India; Swiss Academy of Medical Sciences, Swiss Society of Natural Sciences, Swiss Chemical Society, Swiss Societies of Chemical Industry, MicroPhysiology, Physiological Chemistry and Pharmacology from Switzerland. He was also appointed Honorary, Associate or Corresponding National Academies Membership of the following countries: United States of America; Belgium; Brazil; France; Italy; Romania; Spain; Sweden. Similarly of: Scientific Academies of: Bologna; Haarlem; Halle; Munchen; Paris; Roma;

Torino. Also of Chemical Societies of: Austria; Belgium; France; Germany; the Netherlands; Poland. The Pharmaceutical Societies of: Brazil, Italy, Spain.

He served as President of the Swiss Chemical Society (1924–26); the Society of Natural Sciences of Zurich (1928–30); the Swiss Foundation for Scholarin Chemistry (1944–67). He was a member of the Committee of Helvetica Chimica Acta (1924–71). He served as Rector of the University of Zurich (1950–52); President of the I.U.P.A.C. Congress in Zurich (1955); Chairman of International Symposia on Vitamin A (1960) and Vitamin E (1962).

Karrer's citation for election (date: 01/05/1947) as Foreign Member of the Royal Society reads [13]:

> **'distinguished for his outstanding contributions to the chemistry of natural products, with special reference to the chemistry of the carotenoids, carbohydrates and vitamins'.**

He was proposed by:

A. V. Hill; E. D. Adrian; E. J. Salisbury; M. L. Oliphant; W. T. Asbury; Thomas R. Merton; E. L. Hirst; M. H. A. Newman; S. Goldstein; R. E. Stradling; A. C. Chibnall; C. H. Andrewes; N. Hamilton Fairley; R. A. Fisher; H. H. Read; W. Brown; E. C. Bullard.

Bibliography

1. Reproduced with permission.
2. From Nobel Lectures, Chemistry 1922–1941, Elsevier Publishing Company, Amsterdam, 1966.
3. The Nobel Population 1901–1937, E. Crawford, J. L. Heilbron and R. Ullrich. Office of the History of Science and Technology, University of California, 1987, ISBN: 0-918102-15-4.
4. Biographical Memoirs of Fellows of the Royal Society, 1978, vol. 24, pp. 245–321.
5. Who was who: 1971–80.
6. Zur Kenntnis aromatischer Arsenverbindungen 1. Ueber p-Nitroso phenylarsinsaure, Ber. Dt. Chem. Ges., 1912, 45, 2065–2068.
7. A. Wettstein, *Helv. chim. Acta*, 1972, 55, 313–328.
8. C. H. Eugster, *Chemie in unserer Zeit*, 1972, 6, 146–153.
9. Carotenoids: Volume 3 (Carotenoids) edited by G. Britton, Hanspeter Pfander and Liaaen-Jensen Birkhauser, 1st ed. 1998, ISBN: 3764358297.
10. Carotenoids Handbook edited by George Britton, Synnove Liaaen-Jensen and Hans P. Pfander Birkhauser, 1st ed. 2004, ISBN: 3764361808.
11. Fundamentals of Medicinal Chemistry by Gareth Thomas, John Wiley & Sons, 2003, ISBN: 0470843071.
12. Flavins and Flavoproteins 1999, edited by G. Ghisla Rudolf, Weber, 1999 ISBN: 3000051287.
13. Biosynthesis: Polyketides and Vitamins (Springer Desktop Editions in Chemistry) edited by F. J. Leeper and J. C. Vederas Springer, 1st ed. 2000, ISBN: 3540669698.
14. Sackler database, Royal Society.

1938

Richard Kuhn

"for his work on carotenoids and vitamins" [2]

Professor Richard Kuhn was awarded the Nobel Prize in Chemistry for 1938, for his work on carotenoids and vitamins. Owing to political conditions at the time, Professor Kuhn was prevented from accepting the prize at the time. He later received the gold medal and the diploma (1949) [2]:

> *...In 1930 Karrer clarified the constitution of carotene. The elementary composition of carotene, $C_{40}H_{56}$, had previously been ascertained by Willstätter. In 1931, R. Kuhn (at that time already Professor at Heidelberg), Karrer in Zurich, and Rosenheim in London discovered simultaneously and independently of each other the fact that the carotene in carrots consists of two separate components: one of these, β-carotene, rotates the plane of polarized light to the right, while the other, α-carotene is optically inactive. In 1933 Kuhn discovered a third carotene isomer which was called γ-carotene.... Kuhn's second great field of activity concerns the clarification of the Vitamin B complex. Kuhn has the great merit, together with von Szent-Györgyi and Wagner-Jauregg, of having been the first to isolate the extraordinarily important substance Vitamin B_2 (Lactoflavin or Riboflavin). He has made very important contributions to the elucidation of the chemistry of this substance...*

> *...Kuhn also had an important share in establishing the composition of Rodoxanthin and Astaxanthin as well as in discovering the connection of this latter carotenoid with the chromoproteids of the Crustaceans...*

> *...Kuhn's second great field of activity concerns the clarification of the Vitamin B complex. Kuhn has the great merit, together with von Szent-Györgyi and Wagner-Jauregg, of having been the first to isolate the extraordinarily important substance Vitamin B_2 (Lactoflavin or Riboflavin). He has made very important contributions to the elucidation of the chemistry of this substance...*

Kuhn was awarded the reserve prize in 1939. He was also nominated for the years 1932, 1935, 1936 and 1937 [3].

Richard Kuhn was born in Vienna on 3rd December 1900. His father, Richard Clemens Kuhn, was an engineer and his mother, Angelika, née Rodler, was a teacher. Kuhn married Daisy Hartmann (1928) and had two sons and four daughters. He died on 1st August 1967 [2,4].

The Early Years

Kuhn was educated at the local grammar school. For a period of eight years (1910–1918), Kuhn was a schoolmate of Wolfgang Pauli who was awarded the Nobel Prize for Physics (1945). He studied chemistry at Vienna University and his Ph.D. under R. Willstätter in Munich. He graduated (1922) following his thesis, *Über Spezifität der Enzyme* (On the specificity of enzymes).

The Career

Year	*Post*	*Institute*
1925	Lecturer in Chemistry	Munich University
1926	Professor for General and Analytical Chemistry	Eidgenössische Technische Hochschule, Zurich
1929	Principal of the Institute for Chemistry	Kaiser-Wilhelm-Institut renamed the Max-Planck-Institut (in 1950)
1937	Head of the Institute.	Kaiser-Wilhelm-Institut
1950	Professor of Biochemistry – in addition to head of the Institute	University of Heidelberg

Kuhn spent much of his career in Germany working on compounds containing double bonds. He had many other research interests including:

- the stereochemistry of aliphatic and aromatic compounds
- the syntheses of polyenes
- the synthesis of cumulenes
- the constitution and colour
- the acidity of hydrocarbons
- an investigation on the carotenoids
- work on the flavins
- a study of the vitamins, particularly vitamin B_2 (Lactoflavin or Riboflavin) (Approximately 1g of pure lactoflavin was produced from 5,300 litres skimmed milk)
- Isolating vitamin B_6 and determining its chemical composition and structure [5,6]
- work on enzymes discovering a third carotene isomer, which was called g-carotene (1933)
- a contribution to the structure of Rodoxanthin
- a contribution to the structure of Astaxanthin

Kuhn also discovered and determined the structure of the following carotenoids:

- physalien, from berries of species of Physalis, Helenien, Flavoxanthin
- violaxanthin from Viola tricolor

- crocetin (which is unstable) from saffron
- taraxanthin and cryptoxanthin from Zea Mays Rubixanthin.

These studies, often working with very small quantities of materials, led Kuhn and his group to further develop chromatographic methods.

Kuhn was editor of *Liebigs Annalen der Chemie*, and was Board Member of the "Badische Anilin- und Soda-Fabrik".

Kuhn was prevented from attending the Nobel Prize award ceremony and so he did not write or present a lecture.

For further references concerning Kuhn's work see for example [7–9].

Honours and Awards

Kuhn was awarded honorary degrees by the University of Vienna (1960), Technische Hochschule in Munich (1960), and the University of St. Maria, Brazil (1961).

Kuhn received numerous awards and prizes from many countries including the Nobel Prize for Chemistry (1938).

Kuhn was a member of many scientific academies and societies. He was the President of the Society of German Chemists, Vice-President of the Max Planck Society.

He was Visiting Research Professor for Physiological Chemistry at the University of Pennsylvania, Philadelphia.

Bibliography

1. Reproduced with permission.
2. From Nobel Lectures, Chemistry 1922–1941, Elsevier Publishing Company, Amsterdam, 1966.
3. The Nobel Population 1901–1950: A Census of the Nominators and Nominees for the Prizes in Physics and Chemistry, Elisabeth Crawford, Universal Academy Press, Inc. (2002), ISBN: 4-946443-70-3.
4. Who was who: 1961–1970.
5. Chem. Ber. 1938, 71, 1534.
6. Chem. Ber. 1939, 72, 309.
7. Carotenoids: Volume 3 (Carotenoids) edited by G. Britton, Hanspeter Pfander and Liaaen-Jensen, Birkhauser, 1st ed. 1998, ISBN: 3764358297.
8. Carotenoids Handbook edited by George Britton, Synnove Liaaen-Jensen and Hans P. Pfander, Birkhauser, 1st ed. 2004, ISBN: 3764361808.
9. Advanced Dairy Chemistry, Volume 3: Lactose, Water, Salts and Vitamins by Patrick F. Fox Springer, 2nd ed. 1992, ISBN: 0412630206.

1939

© Nobel Foundation [1]

Adolf Friedrich Johann Butenandt (0.5)

"for his work on sex hormones" [2]

The political situation of the time prevented Butenandt from accepting the prize in person. He received the gold medal and the diploma in 1949. Part of an unattributed comment during the 1939 ceremony placed Butenandt's achievements into perspective [2]:

> *...In 1932, Butenandt was able, from observations made in spectral analysis, and especially on the basis of the then established correct formula of cholesterol to draw up the formulae of the chemical structure of oestrone and oestriol. But there remained the important task of proving the chemical structure of the ring system as assumed by Butenandt. By breaking down the oestriol molecule stage by stage Butenandt proved that both oestrogenic hormones contained a phenanthrene core. At the same time he was able to obtain the same dimethyl phenanthrene from etiobilianic acid, a transformation product of cholic acid. He had thus confirmed the close relationship existing between the follicle hormones on the one hand and the bile acids and sterols on the other...*

Butenandt was nominated for the 1939 Nobel Prize by P. Karrer. He was also nominated for the years 1934, 1935, 1936 and 1938 [3].

Adolf Frederick Johann Butenandt was born on 24th March 1903 in Lehe, near Bremerhaven-Wesermünde, the second child and only son of a businessman, Otto Butenandt of Hamburg. Butenandt's elder sister, Anne Mathilde, died from scarlet fever at the age of 16. Butenandt married Erika von Ziegner (1931) who was a laboratory assistant that he met during his time in Göttingen. They had seven children and lived at Munich-Obermenzing. Butenandt died in Munich after a short illness on 18th January 1995 [2, 4, 5].

The Early Years

Butenandt attended a school in Lehe where he studied for nine years, in English as his first language. Although he was interested in both biology and chemistry he eventually chose chemistry, helping his teacher, Dr. Sauer, with school experiments. Butenandt set up his own experiments in the kitchen at home, ordering his chemicals from the nearby pharmacy. One order, for 1g of KCN, caused a major issue in his village resulting in a visit from the police. He spent one year (1917–18), during the First World War, working in field hospitals. The end of his school education (1921) coincided with the German depression, which left no money to finance Butenandt's university education. He nevertheless decided to enrol at Marburg, where he took part in athletics and other physical activities whilst specializing in chemistry with physics as his secondary subject. His liking for fencing resulted in a facial scar, forcing him to undertake his fencing secretly, when it became banned by law.

He completed his studies in Marburg in 1921, with the qualification of 'Verbands-Examination'. Crucially, this qualification was equivalent to the present-day first degree and was recognized by German universities. This allowed Butenandt to enrol in Gottingen University where he started his doctoral research under Adolf Windaus in the field of natural products. He obtained his D. Phil degree (1927) for studies on '*Ober die chemische Konstitution des Rotenons, des physiologisch wirksamen Bestandteils der Derris elliptica*'. Windaus then appointed Butenandt as a postdoctoral associate.

The Career

Year	*Post*	*Institute*
1927	Scientific Assistant	Institute of Chemistry, Göttingen
1931	Acting Head of the laboratories for organic and inorganic chemistry	Department of Biological Chemistry, University of Göttingen
1933	Professor of Organic Chemistry and Director of the Institute for Organic Chemistry	Institute of Technology at Danzig
1936	Director of the Kaiser-Wilhelm Institute and Honorary Professor	Berlin-Dahlem. The Institute moved to Tv̈bingen (1943–44) shortly before the end of the war, later moving to Munich University of Berlin
1945–1956	Professor of Physiological Chemistry	University of Tvbingen (in addition to his other duties)
1956	Professor of Physiological Chemistry and Director of the Max Planck Institute	University of Munich
1960–72	President of the Max Planck Society	Remained Director of the Institute but took leave of absence from the Chair at Munich.

Butenandt's interests in chemistry and biology resulted in a career spanning the two areas concentrating initially on the sex hormones, later moving into insect biochemistry. He determined the correct chemical structure of the female sex hormone in 1932, when he was only 28 years old. He was 32 when he completed the structure determination of the androgens and progestogens, and was awarded the 1939 Nobel Prize in Chemistry (jointly) when only 36 years of age.

Steroid Sex Hormones

- Isolation of pure crystallized oestrone (1929) independent of the Doisy group
- Isolation of pure androsterone (1931)
- Isolation of testosterone (1939) independent of the Ruzicka group
- Isolated progesterone from the corpus luteum (1934)
- Confirmed the structure of progesterone (from stigmasterol)

Insect Biochemistry

- Postulated the one gene one enzyme hypothesis (following on from the work concerning genes for the formation of insect dye pigments)
- Isolation and structure determination of moulting hormones
- Isolation and structure determination of pheromones by studying the sex attraction of male and female silk moths

Work on insect biochemistry was challenging. Karlson, Butenandt's collaborator, biographer and lifelong friend commented (1995):

> *...I was a participant in the research (on the moulting hormone). The bioassay concerned was based on the work of G. Fraenkel and involved use of ligated larvae of the blowfly Calliphore. The starting material consisted of pupae of the silkworm, and in the summer of 1953 we bought up all the silk cocoons available. We used cocoons of ... males for the moulting hormone... From 107 pupae (500 kg), we ended up with 25mg of the ciystalline hormone which we called ecdysone. The stucture when elucidated, came as a surprise. Ecdysone was another steroid hormone...*

A full list of Butenandt's publications appears in his biography [6].

Butenandt did not give an acceptance speech as he was not allowed to attend in 1939. By not receiving his prize within one year of its award, Butenandt forfeited the right to the prize money – although he did receive the citation and medal.

References [7,8] detail the recent thinking and data concerning the sex hormones.

Honours and Awards

Butenandt was presented with fourteen honorary degrees from the following Universities/Institutes: Graz; Leeds; Madrid; Munich; and two from Tübingen.

He received the following awards and medals: Emil-Fischer Commemorative Medal of the Society of German Chemists (1935); Nobel Prize for Chemistry (1939); Cams Medal from the German Academy of Science Leopoldina zu Halle (1943); Paul Ehrlich Prize, Frankfurt (1935); Member of the Order Pour le Mérite for Sciences and Arts. (1962); Medal of the Rising Sun, Class H, Japan (1963); Austrian Badge of Honour for Science and Art (1964); Commander of the Legion of Honour, France

(1969); Adolf von Harnack Gold Medal of the Max Planck Society (1973); Medal of Merit of the German Academy for Science Leopoldina (1978); Bavarian Maximilian Medal for Science and Art (1981); Great Cross of the Order of Merit of the German Federal Republic (the highest distinction Germany can bestow. Butenandt was only the second scientist to have received this honour (1985) and Honorary Member of the Berlin-Brandenburg Academy of Sciences (1994).

He was a Corresponding Member of the Academy of Sciences at Göttingen, Honorary Life Member of the New York Academy of Sciences, and Honorary Member of the Japanese Biochemical Society, the Deutsche Akademie der Naturforscher Leopoldina, Halle, and the Austrian Academy of Sciences. He was elected Foreign Member of the Royal Society (1968).

Butenandt's citation for election (date: 25/04/1968) as Foreign Member of the Royal Society reads [9]:

> **'Adolf Butenandt of Munich distinguished for his outstanding work in the field of steroid sex hormones and on chemical aspects of insect physiology.'**

No proposers are listed in the Royal Society's database.

Bibliography

1. Reproduced with permission.
2. From Nobel Lectures, Chemistry 1922–1941, Elsevier Publishing Company, Amsterdam, 1966.
3. The Nobel Population 1901–1950: A Census of the Nominators and Nominees for the Prizes in Physics and Chemistry, Elisabeth Crawford, Universal Academy Press, Inc. (2002), ISBN: 4-946443-70-3.
4. Biographical Memoirs of Fellows of the Royal Society, 1998, vol. 44, pp. 77–92.
5. Who's who: 1995.
6. P. Karlson: Adolf Butenandt, 1990, Stuttgart, Wissenschaftliche Verlagsgesellschaft mbH.
7. Encyclopedia of Hormones, Three-Volume Set by Helen Henry and Anthony Norman, Academic Press, 2003, ISBN: 0123411033.
8. Peptide Hormone Secretion/Peptide Hormone Action: A Practical Approach 2 Volume Set (Practical Approach Series) by K. Siddle and J. C. Hutton, Oxford University Press, USA, 2005, ISBN: 0199630720.
9. Sackler database, Royal Society.

1939

© Nobel Foundation [1]

Leopold Stephen Ruzicka (0.5)

"for his work on polymethylenes and higher terpenes" [2]

Ruzicka was awarded half of the Nobel Prize in Chemistry for 1939, for his work on polymethylenes and higher terpenes. Owing to the war conditions Professor Ruzicka attended the ceremony later (1945), having been presented with the medal and citation on 16th January 1940 by the Swedish ambassador, Baron H. G. Beck-Friis. During the special ceremony, which took place in Munich, Beck-Friis remarked [2]:

...Professor Ruzicka. On behalf of the Nobel Foundation and the Royal Swedish Academy of Sciences I convey to you the most cordial congratulations - to which I may be permitted to add my own - and at the same time express my hope that you will be able to extend your research successfully to ever-increasing fields in the service of mankind. With this hope I have the honour to present you with the Certificate of the Nobel Prize for Chemistry for the year 1939 and with the Nobel Medal...

Commenting on Ruzicka's scientific achievements, Beck-Friis remarked:

...The higher terpenoids or polyterpenes, which occur in great variety in the vegetable kingdom, were originally the main subject of Ruzicka's investigations, and the extremely difficult determination of their structure has been made possible through his exceptional experimental skill. The pioneering work of Ruzicka and his co-workers has led to thorough investigations of a very large number of important polyterpenes...

Ruzicka was nominated for the chemistry Nobel Prize by M. Auméras, G. Charrier, F. Fichter, J. Fontes, G. Fromageot, P. Karrer, R. Locquin, L. Meunier, F. de Mira, A. Paris and P. Ruggli. He was also nominated for the years 1931, 1935, 1936, 1937 and 1938 [3].

Leopold Stephen Ruzicka was born on 13th September 1887 (some say 3rd September 1887), in Vukovar (a small Croatian town on the Danube), the son of Stjepan Ruzicka, a cooper, and Ljubica, née Sever. He first married Anna Hausmann, the daughter of a railway official (1912), and then Gertrud Acklin, the mother of a grown-up son (July1951). He had no children. Ruzicka died on 26th September 1976, in Mammern on Lake Constance, Switzerland [2,4].

The Early Years

Ruzicka moved to Osijek, on the river Drava, following the death of his father (1891), where he attended the primary school and the gymnasium. Ruzicka's interests were in physics and mathematics. Although there was no chemistry in the curriculum, Ruzicka decided to study this subject as he had an interest in natural products and their composition.

On completion of his schooling, Ruzicka enrolled in the Technische Hochschule at Karlsruhe (1906) to study chemistry, where his mentors were Christian Bunte, Carl Engler, Fritz Haber and Hermann Staudinger. In less than two years, Ruzicka had completed the practical course and begun work on his doctorate, working under Staudinger on ketenes. Two years of research led to the "Dipl. Ing" (for work on reaction kinetics of ketenes), and two weeks later "Dr. Ing" (for work on phenyl methyl ketene), which was completed in 1910. Unfortunately, Ruzicka's specialization in organic chemistry to the neglect of other areas, resulted in problems during his doctoral examination in physical chemistry. At Harber's insistence, Ruzicka failed to pass 'summa cum laude'.

Staudinger appointed Ruzicka as his Assistant (a post he held until 1916), and together they started to work on the active constituents of Dalmatian insect powder which they named the pyrethrins. Staudinger moved to the Swiss Federal Institute of Technology (E.T.H.) in 1912 and was followed by the newly married Ruzicka. Ruzicka published his first paper with Staudinger in 1911 [5].

The collaboration of Ruzicka with Staudinger covered many areas of research including:

- the active constituents of the insecticidal plant Pyrethrum cinerariifolium
- the isolation and structure determinaof the pyrethrins – published in 1924 (some of this work later proved to be incorrect – it was nevertheless pioneering for its day)
- some work on the terpenes which formed the basis of his career

The pyrethrins and their synthetic analogues are used as insecticides. Ruzicka used cockroaches (Blatta Germanica) for his testing – a ready supply of which were available from the laboratory building.

The Career

Year	*Post*	*Institute*
1918	Privatdozent	E. T. H. in Zurich
1920	Privatdozent	University of Zurich
1923	Titular (honorary) Professor	E. T. H. in Zurich
1926	Chemist at the industrial laboratories	M. Naef et Cie in Geneva
1927	Professor of Organic Chemistry	University of Utrecht
1929	Professor of General Chemistry	E. T. H. in Zurich
1957	retirement	

With the exception of a few years spent at the University of Utrecht, Ruzicka was to spend most of his career working in Switzerland. He acquired Swiss citizenship (1917). Amongst his collaborators for his second period at E. T. H. were Moses Wolf Goldberg (nominated for the Chemistry Nobel Prize for the years 1915, 1916*, 1921, 1922, 1924, 1927, 1928, 1929 *received after deadline), and Tadeus Reichstein (Nobel Laureate for Medicine and Physiology, 1950)

His main areas of work concerned:

- Diterpenes
- Macrocyclic compounds
- Monoterpenes
- Natural compounds from animals
- Natural compounds from plants
- Sesquiterpenes
- Steroids
- Triterpenes
- And some Miscellaneous investigations

Ruzicka formed associations with the Swiss chemical industry, most notably the perfume industry. In this capacity he worked on the perfume constituents of:

- Muscone
- Civetone
- Farnesol
- Nerolidol
- Jasmone

He also worked on:

The partial synthesis of androsterone (from 3α-cholestanol) in 1934
The partial synthesis of testosterone (1935)

Ruzicka was one of the editors of *Experientia* (launched in 1945 by A. Birkhäuser in Basle) until 1973. He was also on the editorial board of Helvetica Chimica Acta.

A full list of Ruzicka's papers can be found in the R.S. biography [4]. He published a total of 583 papers spanning 60 years, although the bulk (530) were published in the period 1930–50.

Ruzicka's Nobel lecture was titled 'Multimembered rings, higher terpene compounds and male sex hormones' and is dated 12th December 1945. His opening remarks were:

> *...This lecture should have been delivered in December 1939, but owing to a postponement of 6 years forced by the intervention of war, I have been able to use some results obtained in the meantime to round off the overall picture. Consequently I can now report on the results of 25 years of laboratory work in two fields of the chemistry of alicyclic compounds: multimembered rings and higher terpene compounds (polyterpenes), to which the third group named in the title, the male sex hormones, is closely related...*

References [6,7] detail the recent thinking and data concerning the sex hormones.

Honours and Awards

Ruzicka was conferred with eight honorary degrees (4 Science, 2 Medicine, 1 Natural Sciences, 1 Law) from the following Institutes: Harvard University (1936); Universities of Basel (1937); Zagreb (1940); Paris (1945); Bordeaux (1947); Praha (1948); Glasgow (1951); Genève (1958).

He was awarded the following awards and medals: Award Swiss Chemical Society (1918); Werner Award and Medal (1923); Swiss Chemical Society (1923); Le Blanc Médaille et Conférence, Société Chimique de France (1928); Pedler Lecture and Medal, Chemical Society, London (1931); Conference et Médaille de la Société Industrielle de Mulhouse (1935); Premio Cannizaro, Accademia dei Lincei (1936); Lavoisier Médaille et Conference,Société Chimique de France (1937); Marcel Benoist-Preis, Switzerland (1939); Nobel Prize in Chemistry (1939); Conference et Médaille de l'Université de Liege (1940); Conferenze e Premio Donegani, Academia dei Lincei (1948); Faraday Lecture and Medal, Chemical Society London (1958); Hanus Medal, Chemical Society Czechoslovakia (1966).

He held the following Fellowships/Memberships: Deutsche Akademie der Naturforscher, Leopoldina Halle/Saale (1932); Académie Royale des Sciences des Lettres et des Beaux-Arts Bruxelles (1936); American Academy of Arts and Sciences, Boston (1938); Yugoslav Academy of Sciences and Arts, Zagreb (1940); Royal Academy of Sciences, Amsterdam (1940); Royal Society London (1942); Pontificia Accademia Scientiarum, Roma (1942); Accademia delle Scienze Benedettina, Bologna (1943); National Academy of Sciences, Washington (1944); Royal Flemish Academy of Sciences, Literature, and Arts, Bruxelles (1945); Serbian Academy of Sciences, Beograd (1948); Accademia Nazionale dei Lincei, Roma (1948); Bayerische Akademie der Wissenschaften, München (1951); National Academy of Sciences of India, Allahabad (1957); Academy of Sciences of the USSR, Moskva (1958); Polish Academy of Sciences, Warszawa (1959); Oesterreichische Akademie der Wissenschaften, Wien (1967).

Ruzicka's citation for election (date: 18/06/1942) as Foreign Member of the Royal Society reads [8]:

> **'Distinguished for his work in Organic Chemistry, notably in each of the fields of chemistry, large ring carbon compounds and the chemistry of the steroids and sex hormones'.**

He was proposed by:

H. H. Dale; A. V. Hill; Thomas R. Merton; O. T. Jones; E. J. Russell; J. B. S. Haldane; W. H. Pearsall; H. Munro Fox; J. Herbert Parsons; C. M. Wenyon; C. R. Harington; S. P. Bedson; E. F. Armstrong; I. M. Heilbron; P. M. S. Blackett; A. M. Tyndall; J. E. Lennard-Jones; L. J. Mordell; H. T. Tizard; A. C. Egerton.

Bibliography

1. Reproduced with permission.
2. From Nobel Lectures, Chemistry 1922–1941, Elsevier Publishing Company, Amsterdam, 1966.
3. The Nobel Population 1901–1950: A Census of the Nominators and Nominees for the Prizes in Physics and Chemistry, Elisabeth Crawford, Universal Academy Press, Inc. (2002), ISBN: 4-946443-70-3.
4. Biographical Memoirs of Fellows of the Royal Society, 1980, vol. 26, pp. 411–501.
5. F. L. Ruzicka and H. Staudinger, Justus Leibigs Annalen Chem. 1911, 380, 278.

6. Encyclopedia of Hormones, Three-Volume Set by Helen Henry and Anthony Norman, Academic Press, 2003, ISBN: 0123411033.
7. Peptide Hormone Secretion/Peptide Hormone Action: A Practical Approach 2 Volume Set (Practical Approach Series) by K. Siddle and J. C. Hutton, Oxford University Press, USA, 2005, ISBN: 0199630720.
8. Sackler database, Royal Society.

1940

The prize money was with 1/3 allocated to the Main Fund and with 2/3 to the Special Fund of this prize section

1941

The prize money was with 1/3 allocated to the Main Fund and with 2/3 to the Special Fund of this prize section

1942

The prize money was with 1/3 allocated to the Main Fund and with 2/3 to the Special Fund of this prize section

1943

© Nobel Foundation [1]

George de Hevesy

"for his work on the use of isotopes as tracers in the study of chemical processes" [2]

In regard to de Hevesy's work, Professor A. Westgren, member of the Nobel Committee for Chemistry of the Royal Swedish Academy of Sciences, made the following statement, which was given on the radio on 10th December, 1944, with amendments and additions. The Nobel Prize in Chemistry 1943 was announced on November 9, 1944 [2]:

> *...He studied how physiological saline containing radioactive sodium which was injected into a human subject first spread into the blood and then slowly penetrated into the cells; he also studied the manner in which it is excreted. After 24 hours the blood corpuscles had lost approximately half their sodium content...Frogs and fishes swimming in water containing deuterium absorb it and, after about 4 hours, are in equilibrium with the medium as far as the deuterium is concerned. Heavy nitrogen and heavy oxygen have also been used in many investigations...*

De Hevesy was nominated for the 1943 Prize by L. Rolla. He was awarded the reserved prize in 1944. De Hevesy was nominated for the Physics Nobel Prize in 1942, and for the Chemistry Prize in the years 1924, 1927, 1929, 1933, 1934, 1935, 1936, 1939, 1940, 1941, 1942 and 1944 [3].

George de Hevesy was born in Budapest on 1st August 1885, the son of Louis de Hevesy, Court Counsellor, and Eugénie, née Baroness Schosberger. His father was the general manager of a mining company and also supervised the farms belonging to his mother. De Hevesy married Pia Riis (1924). They had one son and three daughters. De Hevesy died on 05th July 1966 [2,4,5].

The Early Years

De Hevesy matriculated from the school of the Holy Order of Joseph Calazanc, Budapest (Piarist) in 1903, where he concentrated his studies on mathematics and physics. Having spent a year at the

University of Budapest, he decided to become a chemical engineer. A few months after enrolling at the Technical High School in Berlin, he contracted pneumonia and was advised to spend some time studying in a warmer climate. De Hevesy chose the University of Freiburg, moving there in 1905 and graduating with his doctor's degree (Rer. Nat.), where the main subject was physics, with chemistry and mathematics as subsidiary subjects, in the autumn of 1908.

De Hevesy moved to the Technical High School in Zurich for two years to work under Richard Lorenz, who at the time, was an authority on the chemistry of molten salts. De Hevesy's work at Zurich concerned a study of the chemistry and electrochemistry of molten salts.

De Hevesy's time in Zurich was followed by a three month period working in Karlsruhe under Harber, on the emission of electrons accompanying the oxidation of liquid sodium-potassium alloys, which was met with little success. Shortly after arriving with Haber (3 months), de Hevesy moved to Manchester (1910) to study under Professor Ernest Rutherford. Three years later, de Hevesy moved to the Vienna Institute of Radium Research (1913) where he undertook the first radioactive-tracer experiment jointly with Frederic Paneth. He was drafted into the Austrian-Hungarian Army (1915). At the end of the war, he was taught in the University of Budapest (for 6 months) but left in the spring of 1919 to discuss the possible opportunities in Copenhagen following the decision to create the Niels Bohr Institute.

The Career

Year	*Post*	*Institute*
1918	Lecturer in Experimental Physics	University of Budapest
1919	Professor of Physical Chemistry	University of Budapest
1920		Neils Bohr Institute, Copenhagen
1926	Professor of Physical Chemistry	University of Freiburg
1930	Baker Lecturer	Cornell University, Ithaca, New York
1934		Niels Bohr Institute, Copenhagen
1943	Domiciled in Stockholm, Associate of the Institute for Research in Organic Chemistry	Continued to work part time at Niels Bohr Institute, Copenhagen and at the University of Stockholm
1949	Franqui Professor	University of Ghent
1952	Terminated links with the Bohr Institute	Continued association with the University of Stockholm in retirement

De Hevesy was appointed by the War Office in Vienna to be technical supervisor of an electrochemical works. This was followed by training at the arsenal in Vienna as a member of the Austrian-Hungarian Army. In the October 1918, de Hevesy was ordered to travel to Vienna with instructions to supervise and accelerate the coining of war medals. The war ended before he could undertake this task.

De Hevesy's scientific contributions include:

- a study of the chemical behaviour of molten salts
- some practical aspects of radiochemistry
- the use of radium and lead isotopes, an involvement in the first clinical use of isotopes
- artificially radioactive isotopes
- the use of fluorescent X-rays in analysis

- investigations into plant and animal physiology, using labelled atoms
- the effect of X-rays on the formation of nucleic acid in tumours
- discovery of Hafnium

Professor de Hevesy was the author of several important books on radiochemistry and many scientific papers. Several books have been written concerning his life, see for example [6,7].

De Hevesy's Nobel acceptance speech was titled 'Some applications of isotopic indicators' and is dated 12th December 1944. During this talk, he commented on a number of areas of his work including [2]:

- *Self-diffusion*
- *Arrhenius' theory of electrolytic dissociation*
- *Analytical chemistry*
- *Early biological application*
- *Heavy water*
- *Application of artificially radioactive isotopes*
- *Radio-phosphorus*
- *Permeability investigations*
- *Rate of formation (of labeled organic phosphorus compounds)*
- *Site of formation of phosphorus compounds in the organism – origin of yolk phosphatides*
- *Excretion studies*
- *Labellled red corpuscles.*

De Hevesy concluded:

> *...I have attempted to give a short review of the earliest applications of isotopic indicators and to discuss a few examples of their earlier and more recent employment. Their use may be much extended in the time to come...*

The use of isotopes to label selected atoms in molecules is now a common practice for certain applications, some of which are documented in the following text [8].

Honours and Awards

De Hevesy received honorary degrees from the following Institutes/Universities: Cape Town Sc.D.(1929); Uppsala, D.Phil. (1945); Freiburg, D.Phil. (1948); Ghent, D.Sc. (1950); Copenhagen, Phil.D. (1950); São Paulo, Med.D.(1954); Rio de Janeiro, Med.D. (1955); Turin, Med.D. (1957); Burlington, Jur.D. (1957); Freiburg, Med.D. (1959); Liege, D.Sc. (1959); London, D.Sc. (1960); Cambridge, D.Sc.(1964); Budapest, Dr. Ing. h.c. (1965).

He was awarded the following prizes/medals: Cannizzaro Prize, Rome (1929); Danish Cancer Research Prize (1942); Chemistry Nobel Prize (1943); Copley Medal, Royal Society (1949); Faraday Medal, Chemical Society (1950); Bruxelles Academy Medal (1950); Bailey Medal, College of Physicians, London (1951); Silvanus Thompson Medal, British Society of Radiology (1956); Medal of the City of Paris (1957); Atoms for Peace Award (1959); Rosenberger Medal of Chicago University (1961); the Niels Bohr Medal of the Danish Ing. Society (1961).

De Hevesy held Fellowships/Membership of: Danish Academy of Sciences (1925); Heidelberg Academy of Sciences (1926); Gothenburg Academy of Sciences (1928); Royal Society (1939); Swedish Academy of Sciences (1942); Hungarian Academy of Sciences (1945); Rome Academy of Sciences (Lincei) (1948); Genoa Academy of Sciences (1948); National Academy, Boston (1948); Bruxelles Academy of Sciences (1950); National Academy of India (1950); Hon. Member Halle Academy (1960); Pontificial Academy (1961); International Academy (1962) and Austrian Academy of Science (1965). He was also Honorary Fellow of: the Chemical Society, London (1939); Chemical Society, Helsingfors (1945); Royal Institution, London (1948); German Bunsen Society (1951); Swedish Medical Society (1952); Biochemical Society, Helsingfors (1955); Austrian Society of Physics and Chemistry (1955); Japan Chemical Society (1958); Society of Nuclear Medicine (1958); British Society of Radiology (1958); German Physiological Society (1960); Swedish Society for Radiobiology (1965).

De Hevesy's citation for election (date: 11/05/1939) as Foreign Member of the Royal Society reads [9]:

> **'George de Hevesy of Copenhagen distinguished for his work in experimental chemical physics, particularly the use for the solution of biological and chemical problems, of radioactive and isotopic substances as indicators, for his discovery of Hafnium, for his geochemical researches and for his work on isotopes and their separation'.**

He was proposed by:

W. H. Bragg; A. V. Hill; H. G. Lyons; Thomas H. Holland; A. D. Imms; D. M. S. Watson; W. W. C. Topley; M. Greenwood; F. G. Bartlett; J. Mellanby; F. E. Fritsch; C. K. Ingold; R. Whytlaw-Gray; H. O. Guy; G. B. Jeffrey; J. Proudman.

Bibliography

1. Reproduced with permission.
2. From Nobel Lectures, Chemistry 1942–1962, Elsevier Publishing Company, Amsterdam, 1964.
3. The Nobel Population 1901–1950: A Census of the Nominators and Nominees for the Prizes in Physics and Chemistry, Elisabeth Crawford, Universal Academy Press, Inc. (2002), ISBN: 4-946443-70-3.
4. Biographical Memoirs of Fellows of the Royal Society, 1967, vol. 13, pp. 125–166.
5. Who was who: 1961–70.
6. George de Hevesy 1885–1966, Festschrift, ISBN: 963 05 5230 2.
7. George de Hevesy: Life and Work by Hilde Levi Rhodos, Copenhagen, 1985, ISBN: 87 7245 054 1.
8. Isotopically Labelled Compounds – 1994 ed. J. Allen and R. Voges, J. Wiley and Sons, 1995, ISBN: 0-471-95143-9.
9. Sackler database, Royal Society.

1944

© Nobel Foundation [1]

Otto Hahn

"for his discovery of the fission of heavy nuclei" [2]

The Nobel Prize in Chemistry for 1944 was announced on 15th November 1945. At the 1945 ceremony, which took place on 10th December 1945, the then Chairman of the Nobel Committee for Chemistry of the Royal Swedish Academy of Sciences, A. Westgren observed [2]:

...The Academy of Sciences has decided to award to Professor Hahn the Nobel Prize in Chemistry for the year 1944 in reward for his discovery of the fission of heavy atomic nuclei. Professor Hahn has expressed his gratitude, but he has informed us that he is regrettably unable to attend this ceremony. Therefore the decision of the Academy cannot at present be implemented by the awarding of the prize...

Hahn was able to attend the 1946 ceremony, which took place on 13th December 1946. At that ceremony, A. Tiselius addressed the following words to Hahn [2]:

...Professor Hahn. At the Nobel ceremony held on 10th December, 1945, you were unfortunately unable to be here in person to receive your Nobel Prize. On that occasion, however, a detailed account of the results of your researches was given by the President of the Nobel Committee for Chemistry. I must therefore restrict myself today to the expression of our great pleasure at the fact that you are able to attend today in person to receive your prize and our congratulations. The discovery of the fission of heavy nuclei has led to consequences of such a nature that all of us, indeed the whole of humanity, look forward with great expectations, but at the same time with great dread, to further developments. I am convinced, Professor, that just as your great discovery has been a result of your far-reaching researches on atomic nuclei, irrespective of any eventual practical applications, the further ardent development of research in this field as a consequence of your work will be of particularly great pleasure to you. With regard to its practical application I am also sure that you share all our hopes that this application will serve in the end as a blessing to mankind...

Hahn was nominated for the 1944 Chemistry Prize by V. M. Goldschmidt and L. Mazza. He was awarded the reserved prize in 1945. Hahn was nominated for the Physics Nobel Prize for the years 1937, 1940, 1941, 1943, 1945, 1946 and 1947. He was also nominated for the chemistry prize for the years 1914, 1923, 1924, 1925, 1929, 1930, 1933, 1934, 1936, 1937, 1939, 1941, 1942, 1943, 1944, 1945 and 1946 [3].

Otto Hahn was born on 8th March 1879 in Bockgasse, Frankfurt-on-the-Main, Germany. His father, Heinrich Hahn, was a glazier, his mother Charlotte, née Stutzmann, was a young widow when they married. Charlotte's son Karl from her former marriage lived with them and their three sons Heinrich, Julius and Otto. Hahn married Edith, née Junghans (1913), and they had one son, Hanno (born in April 1922) who was killed in a car accident in August 1960 along with his wife. Edith never recovered from the shock, remaining an invalid for the rest of her life. Thereafter, Hahn cared for his grandson, Dieter, as well as his wife. Hahn fell as he was getting out of a car, fracturing one of the vertebrae in his neck. He died following this accident, in Gottingen on 28th July 1968. Edith died two weeks later [2,4,5].

The Early Years

Hahn's schooling started in the spring of 1885 when he entered the preparatory department of the Klinger Realschule. Hahn was naturally left handed but forced to write with his right hand by the school authorities. He was commended for his skill of writing with mirror script on transparent paper with his left hand whilst his right hand was injured. Hahn suffered several periods of ill health whilst young, including a period of diphtheria, severe inflammation of the lungs and chronic asthma, and often slept in a semi-upright position. He carried out chemistry experiments in his mother's wash-house, burning carbon in oxygen, making hydrogen and experimenting with sodium, yellow phosphorus and potassium perchlorate.

Matriculation from secondary high school with firsts in gymnastics, religious studies and singing (he was third out of ten for mathematics and chemistry) in 1897 was followed by chemistry studies at Marburg and Munich. He submitted his doctoral thesis to Professor Theodor Zincke (1901) at Marburg after which be obtained the post of Assistant in the Chemical Institute at Marburg, where he stayed for two years. This was followed by some work under Sir William Ramsay at University College London (autumn of 1904 to the summer of 1905). He discovered radiothorium at this time whilst working on the preparation of pure radium salts. Hahn then moved to the Physical Institute of McGill University, Montreal (autumn of 1905 to the summer of 1906) and from there to work under Rutherford. He discovered radioactinium whilst with Rutherford and investigated alpha-rays of radiothorium and radioactinium in conjunction with Rutherford.

The Career

Year	*Post*	*Institute*
1907	Lecturer, Chemical Institute	University of Berlin
1914–1918	War work	
1928	Director, Kaiser Wilhelm Institute	Berlin-Dahlem

Hahn was a reservist who was called up to an infantry regiment which moved up to the Western Front. He was given command of a machine gun section and was awarded the Iron Cross Class II for his part in the offensive. He took part in the fraternization between German and English troops on Christmas Eve and Christmas Day. Although ordered to the 126th Infantry Regiment as a Gas Pioneer (near Gheluvelt), plans were cancelled and gas was used in Ypres instead. He took part in the Western Front chemical warfare operations in the spring and summer of 1918.

Hahn collaborated extensively with Lise Meitner. Either independently or in collaboration with Meitner, Hahn worked on:

- the discovery of mesothorium
- applied radiochemistry
- investigations of radiocolloids
- the discovery of nuclear fission
- investigations on beta-rays, their absorbability, magnetic spectra, *etc.*
- the use of the radioactive recoil
- the discovery of protactinium
- the discovery of uranium Z
- the elaboration of the strontium method to determine the age of geological periods
- the processes of irradiating uranium and thorium with neutrons (Meitner and afterwards with Strassmann)
- the discovery of an artificially active uranium isotope (with Meitner)
- the fission of uranium and thorium (with Strassmann)
- investigations on the proof and separation of many elements

A full list of his papers appears in the Royal Society obituary [4]. Klemm and Strassmann wrote Hahn's biography in 1969 [6] and his autobiography was published in 1970 [7].

Hahn's Nobel acceptance speech was titled 'From the natural transmutations of uranium to its artificial fission' and is dated 13th December 1946. Hahn commented:

> *...Thus the behaviour of uranium on irradiation with neutrons of different velocities, both fast and slow, is very complicated. In addition to the natural splitting process, which continues during the irradiation at a speed independent of all the other reactions, the following occur:*
>
> - *Nuclear fission with formation of numerous artificial atoms of all elements between 30 and 64.*
> - *Emission of surplus neutrons during this fission process, making a chain reaction possible.*
> - *The resonance capture of a neutron with a definite energy by uranium 238, with formation of uranium 239, which in its turn is transformed into the elements neptunium and plutonium.*
> - *The giving up of a surplus neutron by the ^{238}U with formation of a ^{237}U, which also forms a neptunium isotope...*

For further information on radiochemistry, please see [8,9].

Honours and Awards

Hahn received a number of honorary degrees from various Institutes: Gottingen Dr. rer. (1949); Technischen Hochschule Dr. Ing. (1949); Frankfurt Dr. Phil (1949) and from Cambridge D.Sc. (1957).

He received many honours, awards and prizes including: the Emil Fischer Medal of the Society of German Chemists (1922); the Cannizzaro Prize of the Royal Academy of Science in Rome (1938); the

Copernicus Prize of the University of Konigsberg (1941); the Cothenius Medal of the Akademie der Naturforscher in Halle (1943); the Nobel Prize for Chemistry (1944); the Max Planck Medal of the German Physical Society, jointly with Lise Meitner (1949); the Goethe Medal of the city of Frankfurt-on-the-Main (1949); the Golden Paracelsus Medal of the Swiss Chemical Society (1953); the Grand Cross of the German Distinguished Services Order (1954); the Grotius Medal of the Hugo Grotius Foundation (1956); the Faraday Medal of the Chemical Society, London (1956); Pour le mérite, Friedensklasse (1957); the Grand Cross of the Federal Republic (1959); the Enrico Fermi Prize, jointly with Lise Meitner and Fritz Strassmann, (1966); the Harnack Medal in Bronze and Gold; the Wilhelm Exner Medal; the Ernst Reuter Medal; the Helmholtz Medal; the Senckenberg Silver Medal; the highest civic and state honours; the Freedom of the cities of Frankfurt, Gottingen and Berlin; an Officer of the French Legion of Honour.

Hahn became scientific member of the Kaiser Wilhelm Institute for Chemistry (1912) and Director from 1928. He was appointed Visiting Professor at Cornell University, Ithaca. He was President of the Kaiser Wilhelm Society (from 1st April 1946) and served as President of the Max Planck Society in Western Germany (from 28th February 1948), was appointed Honorary President of the same Society (May 1960). He held memberships of the Academies of: Allahabad; Berlin; Boston; Copenhagen; Göttingen; Halle; Helsinki; Madrid; Mainz; Munich; Lisbon; Rome (Vatican); Stockholm; Vienna. Hahn was elected to the Indian Academy of Sciences and as a Foreign Member of the Royal Society.

Additionally, he held memberships in the following societies: Consejo Superior de investigaciones Cientificas; Deutsche Bunsengesellschaft; Deutsche Physikalische Gesellschaft; Gauss-Gesellschaft; Vorstandsrates des Deutschen Museums; Gesellschaft Deutscher Chemiker; the Physikalischer Verein; the Physical Society; Real Sociedad Espanola de Fisica y Quimica; the Rumanian Physical Society; the Senckenburg Institute; Steuben-Schurz-Gesellschaft and Stifterverb; für die Deutsche Wissenschaft.

Hahn also became: a Hon. Fellow of University College London; Hon. President Studiengesellschaft zur Forderung der Kernenergieverwertung in Schiffbau und Schiffahrt; Vice-President Deutschen Atomkommission.

Hahn's citation for election (date: 09/05/1957) as Foreign Member of the Royal Society reads [8]:

> **'Professor Otto Hahn of Gottingen distinguished for his work on the chemistry of radioactive substances, for the discovery of several radioelements and especially for the discovery of the fission of uranium and thorium'.**

No proposers are listed in the Royal Society database.

Bibliography

1. Reproduced with permission.
2. From Nobel Lectures, Chemistry 1942–1962, Elsevier Publishing Company, Amsterdam, 1964.
3. The Nobel Population 1901–1950: A Census of the Nominators and Nominees for the Prizes in Physics and Chemistry, Elisabeth Crawford, Universal Academy Press, Inc. (2002), ISBN: 4-946443-70-3.
4. Biographical Memoirs of Fellows of the Royal Society, 1970, vol. 16, pp. 279–313.
5. Who was who: 1961–70.

6. A. Klemm and F. Strassmann, *Z. Natuvforsch.*, 1969, 24, 485.
7. Hahn and Otto, *My Life*, Macdonald & Co, London, 1970.
8. Radiochemistry and Nuclear Chemistry by Gregory Choppin, JAN RYDBERG and Jan-Olov Liljenzin Butterworth-Heinemann, 3rd ed. 2001, ISBN: 0750674636.
9. Nuclear and Radiochemistry: Fundamentals and Applications by Karl Heinrich Lieser, Wiley-VCH, 2nd ed. 2001, ISBN: 3527303170.
10. Sackler database, Royal Society.

1945

© Nobel Foundation [1]

Artturi Ilmari Virtanen

"for his research and inventions in agricultural and nutrition chemistry, especially for his fodder preservation method" [2]

Professor A. Westgren, then Chairman of the Nobel Committee for Chemistry of the Royal Swedish Academy of Sciences, said of Virtanen (on 10th December 1945) [2]:

> *...In his attempts to improve the availability of feeding stuffs in Finland, Virtanen also founded on purely theoretical data his method of preserving green fodder, a method which avoids protein loss and minimizes vitamin loss.*
>
> *It had long been known that the addition of organic or mineral acids to ensilage inhibited the breathing of the vegetable cells as well as all the fermentation processes. There have been frequent endeavours to turn this phenomenon to practical account. However, these attempts were carried out empirically and without serious study of the conditions for conserving the nutrient value of the fodder, or for using it as an animal feeding stuff. By his systematic, thorough studies Virtanen was the first to solve this problem.*
>
> *To preserve fodder Virtanen uses hydrochloric acid with the addition of sulphuric acid. After painstaking, protracted studies he determined the limiting degrees within which the acidity must be kept in order to achieve the required result. The following beneficial results are simultaneously obtained. The breathing of the plant cells, which consume readily soluble carbohydrates is reduced to a minimum, particularly if the fodder is well stacked. The lactic-acid fermentation ceases or remains at an insignificant level. The butyric-acid fermentation, which decreases the attractiveness of the fodder and lowers the quality of the milk, ceases. Protein decomposition, which is apt to cause heavy financial losses, ceases almost altogether. The content of vitamin A and carotin is maintained and vitamin B and vitamin C are also well preserved. Vitamin-rich milk and butter are very important for public health and AIV fodder provides an effective means of improving this supply of vitamins. Because of these fodders "summer milk can be produced the whole year round"...*

Virtanen was nominated for the 1945 Chemistry Prize by P. Eskola, H. von Euler-Chelpin, S. Kilpi, N. J. Toivonen, E. Tommila and P. Tuorila. He was also nominated for the years 1933, 1934, 1935, 1936, 1937, 1938, 1939, 1941, 1942, 1943 and 1944 [3].

Artturi Ilmari Virtanen was born in Helsinki, Finland on the 15th of January 1895, the son of Kaarlo Virtanen and Serafiina, née Isotalo. He married Lilja Moisio (1920). They had two sons, Kaarlo and Olavi. Virtanen died on 11th November 1973 [2,4,5].

The Early Years

Virtanen attended school at the Classical Lyceum at Viipuri, Finland. After matriculation (1913), he studied biology, chemistry and physics at the University of Helsinki. He graduated with his M.Sc. in 1916 and his D.Sc. in 1919. He worked as First Assistant at the Central Laboratory of Industries (1916–1917) and as a Chemical Assistant in the Government Butter and Cheese Control Station (1919) whilst undertaking his degrees. Virtanen went on to study physical chemistry under G. Wiegner in Zurich (1920), bacteriology under Chr. Barthel in Stockholm (1921) and enzymology under H. von Euler in Stockholm (1923–1924). Virtanen concentrated on aspects of biochemistry from 1923.

The Career

Virtanen's career spanned both theoretical investigations and the practical use of some of his findings.

Year	*Post*	*Institute*
1916–1917	First-assistant	Central Laboratory of Industries at Helsinki
1919–1920	Chemist	Valio, Finnish Cooperative Dairies' Association
1921	Director	Valio, Finnish Cooperative Dairies' Association
1931	Director and Professor of Biochemistry	Biochemical Research Institute at Helsinki and Finnish Institute of Technology at Helsinki
1939	Professor of Biochemistry	University of Helsinki

Virtanen's work included:

- establishing the link between cozymase in lactic and propionic acid fermentation
- the phosphorylation of sugar (1924)
- the decomposition of sugar as the first stage in different fermentation processes
- many fermentation experiments (with his collaborators)
- the mechanism of different bacterial fermentations
- the fermentation of dioxyaceton to glycerol and glyceric acid in the presence of phosphates by the effect of *Coli* bacteria (1929). This was the first sugar fermentation to be elucidated chemically

- the adaptive formation of enzymes
- the uptake of nutrients by cells,
- the relation between the protein content and enzymic activity of cells
- the biological nitrogen fixation which takes place in the root nodules of leguminous plants (1925)
- proof of the importance of active root nodules for the fixation of nitrogen – the involvement of the pigment leghaemoglobin
- the formation of vitamins in plants
- the ability of plants to utilize organic nitrogen compounds as their nitrogen source, the chemical composition of higher plants
- the isolation of a large number of new amino acids from different plants
- the isolation of numerous organic sulphur compounds were isolated from vegetables and fodder plants
- the creation of a theoretical basis for the preservation of fresh fodder
- the development of a practical method on this basis (the AIV method)

Virtanen's Nobel Prize acceptance speech is dated 12th December 1945 and titled 'The biological fixation of nitrogen and the preservation of fodder in agriculture, and their importance to human nutrition'. During his speech, Virtanen commented on three aspects of his work, namely:

- N-fixation in the root nodules of leguminous plants
- Preservation of protein-rich crops
- The influence of the AIV method on human and animal nutrition

In summary, Virtanen commented:

> *...In this way it was possible to adjust the butter acidity to pH 6–7, which guaranteed its storability. This simple method was of decisive importance in the storage of butter and the quality thereof. The discovery also made it possible to separate the bacteriological butter defects from the chemical and it has therefore resulted in ready controllability of butter quality.*
>
> *Ladies and gentlemen. I have now come to the end of my lecture. We have followed the fixation of atmospheric nitrogen and the formation of amino acids in the root nodules of legumes, the production of protein-rich legume crops, the preservation of them to form a high-grade animal feed, and the production of physiologically high-grade milk adapted to the improvement of human nutrition. These questions represent widely different problems within biochemical research but are bound up with one another organically and form a whole from the practical aspect...*

Modern methods of animal nutrition and feeding can be found in [6].

Honours and Awards

Virtanen was presented with honorary degrees from many Institutes including: The Finnish Institute of Technology; Tekniska Högskolan; the University of Paris; the University of Helsinki; the Justus-Leibig University.

He was awarded medals, prizes and other distinctions including: the Scheele Medal of the Swedish Chemical Society (1938); the Adelsköld Medal of the Swedish Academy of Science (1943); the Great Medal of the University of Ghent (1946); the Gadolin Medal of the Finnish Chemical Society (1949); the Kairamo Medal, Finland (1950); the Cajander Medal of the Finnish Society of Forestry (1955); the Emanuele Paterrio Medal of the Italian Chemical Society (1957); the Medal of the Agricultural Societies of Finland (1950); the Medal of Helsinki University (1960); the Medal of Faber Foundation, Vienna (1964); the Friesland Prize and Medal (1967); the First Atwater Memorial Lectureship and Prize, U.S.A. (1968); Siegfried Thannhauser Medal (1969); the First Cuenca Villoro Lectureship and Medal, Spain (1972); the Uovo d'Oro, Italy (1973).

Virtanen was a Fellow or Member of a number of Institutes/Societies including: the Accademia Pugliese delle Scienze, Bari; the Bavarian Academy of Sciences, Munich; the Finnish Academy of Science (Chairman 1941–45); the Engineering Academy of Denmark; the Leopoldina German Academy; the Norwegian Academy of Science; the Pontifical Academy of Science, Vatican City; the Royal Flemish Academy of Science; the Royal Society of Sciences of Uppsala; the Royal Swedish Academy of Agriculture; the Royal Swedish Academy of Engineering Sciences; the Royal Swedish Academy of Sciences of Stockholm; the U.S. National Academy of Sciences. He was also an honorary member of scientific societies throughout Europe and the United Slates including the Royal Society of Edinburgh, to which he was proposed by their Council as an Honorary Foreign Fellow (4th May 1959).

Virtanen also received civic honours including: being appointed a Commander of the Order of the White Rose of Finland (1943); granted the Cross of Liberty, 2nd Class (1945); appointed Commander of the Order of the Lion of Finland (1945); the Grand Cross in 1953. He was appointed a Commander of the Royal Order of Dannebrog by Denmark (1951) and he received the Grand Cross of the Order of Agricultural Merit of Spain (1951). He was also appointed a Foreign Member of the Order pour Ie Mérite für Wissenschaften mid Künste of Bonn, with the Golden Cross (1971).

Virtanen was a Member and President of the State Academy of Science and Arts in Finland (1948). He served on the State Nutrition Committee, the State Production Committee, the Committee of Chemical Industry and acted as Advisor to the Ministry of Supply. He was also Finland's representative on the League of Nations' Committee on Nutrition.

Bibliography

1. Reproduced with permission.
2. From Nobel Lectures, Chemistry 1942–1962, Elsevier Publishing Company, Amsterdam, 1964.
3. The Nobel Population 1901–1950: A Census of the Nominators and Nominees for the Prizes in Physics and Chemistry, Elisabeth Crawford, Universal Academy Press, Inc. (2002), ISBN: 4-946443-70-3.
4. Royal Society of Edinburgh year book, 1975, p. 55–57.
5. Who was who: 1971–80.
6. Basic Animal Nutrition and Feeding by Wilson G. Pond, David C. Church, Kevin R. Pond and Patricia A. Schoknecht, John Wiley & Sons, 5th ed. 2004 ISBN: 0471215392.

1946

© Nobel Foundation [1]

James Batcheller Sumner (0.5)

"for his discovery that enzymes can be crystallized" [2]

A. Tiselius, then a member of the Nobel Committee for Chemistry of the Royal Swedish Academy of Sciences, made the following observation of Sumner during the 1946 presentation ceremony [2];

...It was then natural to assume that these cell juices or cell extracts contained some substance which had the capacity of initiating and maintaining the reactions and guiding them into the paths they follow in the cell. These unknown active substances were called enzymes or ferments, and the investigation of their effects became one of the principal problems of chemistry during the first decades of this century, which for the rest it still is.

The important question of the nature of the enzymes remained unsolved, however, in spite of the energetic efforts of the research workers. It is manifestly a question of substances of complicated structures, which are present in such extremely small amounts that they, so to speak, slip through the fingers when one tries to grasp them. It is really remarkable to see how far it was possible to get in the study of the effects of the enzymes and the course of the enzymatic reactions, without knowing anything definite about the nature of these very active substances, nay, even without even being quite clear that they were substances which could be isolated in the pure form at all.

In 1926, however, in connection with his studies of a special enzyme "urease", James B. Sumner of Cornell University, Ithaca, U.S.A. succeeded in producing crystals which exhibited strikingly great activity. The basic material was the bean of a South American plant, Canavalia ensiformis, in America called the "jack bean", and the crystals had an activity that was about 700 times as great as that of bean flour. What was still more important was that it was possible to dissolve the substance and re-crystallize it several times without its activity being affected. The crystals proved to consist of a protein substance. Sumner expressed the opinion that in reality this protein substance was the pure enzyme...

Sumner was nominated for the 1946 Chemistry Prize by H. von Euler-Chelpin, T. Svedberg and A. Tiselius. He was also nominated for the years 1941, 1944 and1945 [3].

James Batcheller Sumner was born at Canton, Mass., on 19th November 1887, the son of Charles Sumner and Elizabeth Rand, née Kelly. His ancestors originated from Bicester, England and settled in Boston in 1636. Sumner was married three times. He married Bertha Louise Ricketts (1915), whom he later divorced, and had two sons and two daughters, one of whom died at an early age. He married Agnes Paulina Lundkvist (1931), and Mary Morrison Beyer (1943) by whom he had two sons, one of whom died also at an early age. Sumner died of cancer on 12th August 1955 [2,4].

The Early Years

Sumner attended Eliot Grammar School and then Roxbury Latin School. His interests were in chemistry and physics classes. Outside of the classroom his main interest was in fire-arms. Sumner was accidentally shot in the left arm whilst out grouse hunting, at the age of 17. As a consequence, he had to have his left arm amputated just below the elbow. Naturally left handed, he was forced to relearn many skills and forced himself to excel at many sports including billiards, clay-pigeon shooting, skating and skiing tennis.

He entered Harvard College (1906), where he specialized in chemistry and graduated in 1910. Sumner spent a short while working in his uncle's cotton knitting factory before accepting a teaching post at Mt. Allison College, Sackville, followed by an Assistantship in chemistry at Worcester Polytechnic Institute, Worcester (1911–1912) He left there in order to study biochemistry under Professor Otto Folin at Harvard Medical School. Sumner subsequently graduated with his Ph.D. (June 1914).

The Career

Year	*Post*	*Institute*
1914	Assistant Professor of Biochemistry	Cornell Medical School, Ithaca, NY, USA
1929	Professor of Biochemistry and Director of the Enzyme Research Laboratory	Cornell Medical School, Ithaca, NY, USA
1937	Guggenheim Fellowship	worked in the laboratory of Professor The Svedberg

Sumner's research work at Cornell concerned:

- analytical methods
- and the isolation an enzyme (urease) in its pure form (from the jack bean)-initially met with a mixed response

Sumner's Nobel Prize acceptance speech was dated 12th December 1946 and titled 'The chemical nature of enzymes'. Sumner commented:

...Why was it difficult to isolate an enzyme? Here I employ the word 'isolate' as meaning preparing in pure condition. It does not seem difficult to isolate and crystallize an enzyme now, but it was difficult 20 years ago. The reasons were several. One was the inertia of men's ideas. Another reason was the influence exerted by Wilistätter and his school, who held that enzymes were neither lipids, carbohydrates nor proteins and who believed enzymes to exist in excessively low concentrations in plants and animals. The misconception that colloidal substances had a chemistry different from that of crystalloids was another stumbling block. A yet more important reason why enzymes could not be isolated readily was because work with these substances requires a special technique. The enzyme is present in relatively low concentration in the raw material. It is often highly unstable. The raw material nearly always contains natural protectors which are left behind during the purification processes. In the absence of these natural protectors the enzyme may become destroyed by traces of heavy metals, by oxidation, by unfavorable pH, or even by autolysis...

...I wish to speak now about proofs of the identity of the crystals with urease. In cases of this sort one piece of experimental evidence is not sufficient to constitute a valid and satisfactory proof; one must submit many pieces of evidence. At this time I had no access to the ultra-centrifuge of Svedberg nor to the electrophoresis apparatus of Tiselius. However, I was able to offer evidence of another sort, as is shown below:

(i) *When the crystals separated there occurred a great increase in purity, namely from 700 to 14000-fold. Using other methods the increase in purity observed was very much less than this and at times there even occurred a decrease in purity.*
(ii) *When the distribution of urease was followed quantitatively it was found that of the urease passing into the filtrate as much as 40 per cent or more separated with the crystals.*
(iii) *Recrystallization increased the purity of the urease.*
(iv) *Adding traces of poisons, such as mercuric chloride or formaldehyde, to jack-bean meal inactivated the urease and, at the same time, prevented the appearance of the crystals.*

In obtaining the crystals I felt much the same as a person does who is trying vainly to place in position a piece of a machine. Suddenly the piece slides in as if covered with butter. One knows that it is now where it belongs.

The purification of proteins and enzymes is now common practice, see for example [5–7].

Honours and Awards

Sumner was awarded the Scheele Medal, Sweden (1937) and the Nobel Prize for Chemistry (jointly) in 1946.

He was elected to the USA National Academy of Sciences (1948).

Bibliography

1. Reproduced with permission.
2. From Nobel Lectures, Chemistry 1942–1962, Elsevier Publishing Company, Amsterdam, 1964.

3. The Nobel Population 1901–1950: A Census of the Nominators and Nominees for the Prizes in Physics and Chemistry, Elisabeth Crawford, Universal Academy Press, Inc. (2002), ISBN: 4-946443-70-3.
4. Who was who: 1951–1960.
5. Isolation and Purification of Proteins (Biotechnology and Bioprocessing Series) by Rajni Hatti-Kaul, CRC, 2003, ISBN: 0824707265.
6. Protein Purification Protocols (Methods in Molecular Biology) edited by Paul Cutler, Humana Press; 2nd Ed., 2004, ISBN: 1588290670.
7. Enzyme Assays: A Practical Approach (Practical Approach Series) by Robert Eisenthal and Michael J. Danson, Oxford University Press, USA, 2005, ISBN: 0199631433.

1946

© Nobel Foundation [1]

John Howard Northrop (0.25)

"for their preparation of enzymes and virus proteins in a pure form" [2]

During his presentation speech, Professor A. Tiselius, then a member of the Nobel Committee for Chemistry of the Royal Swedish Academy of Sciences, commented of Northrop [2]:

> *...Sumner's pioneer work was not immediately followed by similar work in other quarters, which might perhaps have been expected. About three years after Sumner's work had been published, however, Dr. John Northrop of the Rockefeller Institute at Princeton began to work on the purification of the protein-splitting enzymes met with in the digestive apparatus and gradually succeeded in obtaining a number of them in crystallized form, e.g. the pepsin met with in the gastric juice and the trypsin and chymotrypsin in the pancreas. Northrop and his collaborators, among whom should be mentioned in the first place Kunitz, also made extremely comprehensive studies of the homogeneity and purity of these purified enzymes, and in that connection gave further proof of their nature as protein substances. Exceedingly interesting results were attained also in the isolation of some protein substances which appeared to be the mother substances of these enzymes. On the whole Northrop used his purified material for detailed chemical studies to a greater extent than did Sumner, and his contributions in the matter of working out the most satisfactory conditions for the crystallization of enzymes have been of the greatest importance for subsequent research workers...*

Northrop was nominated for the 1946 Chemistry Prize by T. Svedberg and A. Tiselius. He was also nominated for the years 1940, 1941, 1942, 1943, 1944 and 1945 [3].

John Howard Northrop was born in Yonkers, New York, on 5th July 1891. His father, John I. Northrop, an Instructor at Columbia, was fatally injured in a laboratory accident shortly before his birth. His mother, Alice R. Northrop, taught botany at Hunter College, New York City.

Northrop married Louise Walker in 1917. Their only son, John, was an oceanographer and their only daughter, Alice, married Frederick C. Robbins, Nobel Laureate in Medicine, 1954. Northrop died on 27th May 1987 [2, 4].

The Early Years

Northrop entered Columbia University in 1908 to study zoology and chemistry, under T.H. Morgan and J.M. Nelson, following a public school education. He graduated with a Bachelor of Science degree (1912), a Master of Arts degree (1913) and received his Ph.D. in chemistry, in 1915. He was appointed W.B. Cutting Travelling Fellow, Columbia University, and spent the next year in Jacques Loeb's laboratory at the Rockefeller Institute.

The Career

Year	*Post*	*Institute*
1916	Staff member	Columbia University
1917	Associate	Columbia University
1917–1918	Captain	Chemical Warfare Service
1920	Associate Member	Columbia University
1924	Member	Columbia University
1949	Professor of Bacteriology	University of California
1958	Professor of Biophysics	University of California

Northrop's researches concerned:

- general studies of carbohydrates
- theories of the duration of life
- the discovery of a fermentation process for acetone
- a kinetic study of enzymes essential to life processes (with Loeb)
- the isolation of pure crystalline pepsin (1929). The techniques were later used by himself and other workers to crystallize carboxypeptidase, chymotrypsin, pepsinogen and trypsin
- studies of the proteins of viruses and antibodies
- isolation of a nucleoprotein which had bacteriophage activity
- the origin and relationship of viruses and the transforming principle

Northrop was a consultant to the National Defence Research Committee during World War II, when he studied the action of war gases and methods of detection. He developed an apparatus for their automatic detection and analysis.

Northrop was the author of *Crystalline Enzymes* (pub. 1939), 2nd Edition 1948, Columbia University Press. He edited the *Journal of General Physiology* (Rockefeller Institute) for some years and he wrote numerous papers on the agglutination of bacteria, the chemical nature of enzymes, kinetics of enzyme reaction and the physical chemistry of proteins.

Northrop's Nobel Prize acceptance speech is dated 12th December 1946 and titled 'The preparation of pure enzymes and virus proteins'. Northrop commented [2]:

> *...The relationship of the protein and the enzyme may be tested further by measuring the rate of diffusion of the preparation by means of protein determinations and also by enzyme activity*

measurements. The rate of diffusion was suggested by Arrhenius years ago as a means of determining the size of unknown substances. The method has been developed and used by von Euler.

Theoretically it is simple but experimentally it was difficult. These difficulties were overcome by Anson and myself, who developed a diffusion cell made of a porous disk across which diffusion takes place. This greatly accelerates and simplifies the experiment. The method is the only one, I believe, which permits the estimation of the molecular weight of unknown and impure substances.

The diffusion of pepsin, trypsin, chymotrypsin, chymotrypsinogen, and bacteriophage has been followed by this method. The results show that the protein and the active substance diffuse at exactly the same rate...

He summarized thus:

...The enzymes, pepsin, trypsin, chymotrypsin, carboxypeptidase, rihonuclease, and hexokinase have been isolated and crystallized. The precursors of pepsin, trypsin, and chyrnotrypsin have also been isolated and crystallized. A nucleoprotein which appears to be bacteriophage has been isolated but not crystallized.

The purity of the enzyme preparations, with the exception of carboxypeptidase, has been tested by means of solubility measurements, ultracentrifuge analysis (pepsin, ribonudease, hexokinase, and bacteriophage), and electrophoresis (pepsin, ribonuclease, hexokinase, chymotrypsinogen). They appear to be pure proteins by all these methods.

The relation of the enzymatic activity to the protein has been tested by diffusion measurements (pepsin, trypsin, chymotrypsm, chymotrypsinogen, bateriophage), denaturation of the protein (pepsin, trypsin, chymotrypsin, ribonuclease, hexokinase), hydrolysis of the protein (trypsin, ribonuclease, pepsin, bacteriophage), formation of the active enzyme from an inactive precursor (pepsin, trypsin, chymotrypsin), and by the formation and isolation of definite chemical derivatives of pepsin. These experiments confirm the conclusion that the enzymatic activity is a property of the protein molecule itself, and is not due to a non-protein impurity...

The purification of proteins and enzymes is now common practice, see for example [5–7].

Honours and Awards

Northrop received Honorary Doctor of Science degrees from the Universities of: Columbia (1937); Harvard (1936); Princeton (1940); Rutgers (1941); Yale. He also received an Honorary Doctor of Law degree from the University of California.

He was awarded: the Stevens Prize from Columbia (1931); the Chandler Medal (1936); the Daniel Giraud Elliot Medal (1939); the Nobel Prize for Chemistry, jointly (1946); the Certificate of Merit (U.S. Government) (1948); the Alex. Hamilton Medal (1961).

Northrop was an Honorary Fellow of the Chemical Society (London) and Member of the American Society of Biological Chemists; the National Academy of Sciences; the Halle Akademie der Naturforscher; the Société Pharmathique; the American Phylosophical Society and the Society of General Physiologists. He was also a Fellow of the World Academy and a Benjamin Franklin Fellow. He was a member of many other societies including the Sons of the American Revolution, the Delta Kappa Epsilon Fraternity and a member of Sigma Xi Phi Lambda Upsilon.

In addition, he held the following posts: DeLamar Lecturer at the School of Hygiene and Public Health, Johns Hopkins University (1937); Jesup Lecturer at Columbia University (1938); Hitchcock Lecturer, University of California (1939); Thayer Lecturer, Johns Hopkins University 1940).

Northrop was a consultant for OSRD (1941–45).

Bibliography

1. Reproduced with permission.
2. From Nobel Lectures, Chemistry 1942–1962, Elsevier Publishing Company, Amsterdam, 1964.
3. The Nobel Population 1901–1950: A Census of the Nominators and Nominees for the Prizes in Physics and Chemistry, Elisabeth Crawford, Universal Academy Press, Inc. (2002), ISBN: 4-946443-70-3.
4. Who was who: 1981–1990.
5. Isolation and Purification of Proteins (Biotechnology and Bioprocessing Series) by Rajni Hatti-Kaul, CRC, 2003, ISBN: 0824707265.
6. Protein Purification Protocols (Methods in Molecular Biology) edited by Paul Cutler, Humana Press; 2nd ed. 2004, ISBN: 1588290670.
7. Enzyme Assays: A Practical Approach (Practical Approach Series) by Robert Eisenthal and Michael J. Danson, Oxford University Press, USA, 2005, ISBN: 0199631433.

1946

© Nobel Foundation [1]

Wendell Meredith Stanley (0.25)

"for their preparation of enzymes and virus proteins in a pure form" [2]

A. Tiselius, then a member of the Nobel Committee for Chemistry of the Royal Swedish Academy of Sciences, observed of Stanley [2]:

> *...The problem which attracted his attention, namely the chemical nature of the viruses, was to a certain degree analogous to the problem of the enzyme just mentioned. As is well known, viruses are contagia which give rise to a large number of the best known illnesses in man, animals and plants, e.g. smallpox, infantile paralysis, influenza, foot-and-mouth disease, mosaic disease (on tobacco plants), etc. The virus particles are invisible in the microscope, and when Stanley began his work, they could only be identified by the symptoms of disease which they occasioned. Thus the problem was more difficult, inasmuch as the effect of the virus could not be as easily measured as that of an enzyme, where an exactly known chemical reaction can be employed. Stanley first tried to show the protein nature of viruses by studying how the virus of the tobacco mosaic disease was attacked by protein-splitting enzymes, but in 1934 he passed on to attempting to purify that virus by methods similar to those which Sumner and Northrop had employed so successfully for enzymes. In 1945, by using large quantities of infected tobacco leaves, he did succeed in producing small amounts of crystals which were extremely active, and which, after detailed investigation, proved to be the bearers of the virus's activity. Here, too, it was a matter of active protein substances. Subsequently it has been proved that nucleic acid also forms an important constituent of the latter...*

Stanley was nominated for the Prize by H. von Euler-Chelpin and H. Shapley. He was also nominated for the years 1938, 1939, 1941, 1942, 1943, 1944 and 1945 [3].

Wendell Meredith Stanley was born in Ridgeville, Indiana, USA on 16th August 1904, the son of James G. Stanley and Claire, née Plessionger. He married Marian Staples on 15th June 1929 and they had one son, Wendell M. Junior, and three daughters, Marjorie Jean, Dorothy Claire, and Janet Elizabeth. Stanley died in 1971 [2,4].

The Early Years

Stanley attended Earlham College prior to graduating with a Bachelor of Science degree (1926), Master of Science degree (1927) and a Ph.D. in chemistry (1929) from the University of Illinois. Stanley became a Research Associate and then an Instructor at the University of Illinois before leaving for Munich (latter half of 1930), as a National Research Council Fellow, where he worked under Wieland (until late 1931).

The Career

Year	*Post*	*Institute*
late 1931	Assistant	Rockefeller Institute
1937	Associate Member	Rockefeller Institute
1940	Member	Rockefeller Institute
1948	Professor of Biochemistry, Professor of Molecular Biology and Director of the Virus Laboratory (Chairman of the Department of Biochemistry from 1948–1953)	University of California
1958	Professor of Virology and Chairman of the Department	University of California

Stanley's work included:

- investigations on lepracidal compounds
- investigations on diphenyl stereochemistry
- the chemistry of the sterols
- the isolation of a nucleoprotein
- the preparation and investigation of influenza and similar viruses
- the development of the centrifuge-type influenza vaccine

Stanley was a recognized world authority on viruses and has written over 150 papers on the topic. He has also contributed chapters to several books.

Stanley's Nobel Prize acceptance speech was dated 12th December 1946 and titled 'The isolation and properties of crystalline tobacco mosaic virus'. Stanley commented [2]:

> *...The viruses have been separated as a special group of infectious, disease-producing agents by means of several general properties, no one of which is, however, exclusively characteristic of viruses. Nevertheless, no great amount of difficulty has been encountered in the segregation of the virus group. Viruses are characterized by their small size, by their ability to reproduce or multiply when within the living cells of a given host, by their ability of change or mutate during multiplication and by their inability to reproduce or grow on artificial media or in the absence of specific living cells. The sole means of recognizing the existence of a virus is provided by the multiplicating of the virus which is, of course, usually accompanied by manifestations of disease. Viruses spread from*

diseased to normal susceptible hosts by different methods. Some are transferred by direct contact, as when a diseased leaf is caused to rub against a healthy leaf by a gust of wind, or when a normal person or animal comes into direct contact with a diseased person or animal. Such viruses can usually be spread by indirect contact through the medium of non-specific animate or inanimate objects. Some viruses cannot be transferred by direct contact, but require an intermediate host such as a mosquito, louse or leaf-hopper. In some cases a highly specific intermediate host is necessary, and a more or less definite period of incubation within this host may be required before it can pass on the virus.

Reproduction, mutation and metabolic activity have long been regarded as unique and special properties of living organisms. When viruses were found to possess the ability to reproduce and to mutate, there was a definite tendency to regard them as very small living organisms, despite the fact that the question of metabolic activity remained unanswered. Because of their small size they could not be seen by means of the ordinary light microscope. Although this fact puzzled some investigators, it was pushed aside and for over thirty years interest in virus research was centered about the discovery of new viruses and on studies of the pathological manifestations of viruses...

He concluded his acceptance speech:

...The new field of virus research is really in its infancy and much remains to be accomplished. Certain basic and fundamental problems relating to the mode of virus reproduction and mutation have taken definite form. Solution of these problems should yield information of great value to biology, chemistry, genetics and medicine. This new field of research has been laid upon a very broad foundation, for the present status of virus research has been achieved by contributions from almost every branch of science. It is likely that a continuation and extension of the application of weapons from many fields of science will be necessary for the proper and full development of the new field of virus research...

The purification of proteins and enzymes is now common practice, see for example [5–7].

Honours and Awards

Stanley was presented with honorary D.Sc. degrees from many Universities, Colleges and Institutes including: Earlham (1938); Harvard (1938); Illinois (1959); Princeton (1947); Yale (1938). Honorary D.Sc. degrees from: Pittsburgh (1962); Gustavus Adolphus College (1963); Toledo University (1968); Butler (1968). He was also presented with honorary Doctor of Law degrees from: the Universities of California (1946) and Indiana (1951); the Jewish Theological Seminary of America (1953); Mills College (1960). Stanley had also been presented with Doctor, *honoris causa*, University of Paris (1947).

He received many awards and prizes including: the American Association for the Advancement of Science Prize (1937); the Rosenberger Medal (University of Chicago); the Alder Prize (Harvard); the Scott Award (City of Philadelphia, 1938); the Gold Medal of the American Institute of New York (1941); the Copernican Citation (1943); the Nichols Medal (American Chemical Society, 1946); the Gibbs Medal (American Chemical Society, 1947); the Franklin Medal and Presidential Certificate of Merit (1948); the Modern Medicine Award (1958); the American Cancer Society's Medal for Distinguished Service in Cancer Control (1963).

Stanley was a member of many scientific societies and served on the councils of many academic, medical and other learned societies including the Rockefeller Institute for Medical Research.

He was a consultant and advisor to the United States government and the World Health Organization. He was Director-at-large of the American Cancer Society and a Member of the Board of Scientific Counsellors of the National Cancer Institute.

Stanley was a Visiting Professor of Chemistry at Earlham College (1941).

He presented many named lectures including: Messenger at Cornell University (1942); Vanuxem, Princeton (1942); Bergen, Yale (1944); Edgar Fahs Smith, University of Pennsylvania (1947); Silliman, Yale (1947); Lower, Cleveland Academy of Medicine (1948); Biology Colloquium (1950); Hanau Loeb, St Louis University (1956); Walter R. Bloor, University of Rochester (1956); R.A.F. Penrose Jr. (1957); John Wesley Powell (1957); E.C. Franklin Memorial, University of Kansas (1958); Frank Billings (1958); Shannon (1959); Stanley P. Black (1960); Martin Memorial (1960); Annual Distinguished Lecturer of the American Society of Hematology (1962); Arthur Holly Compton (1963); Mike Hogg (1963); Reynolds (1963); Christian Herter (1964); James Ewing (1964); the Donald Johnson in cancer research at the University of Michigan (1964).

Bibliography

1. Reproduced with permission.
2. From Nobel Lectures, Chemistry 1942–1962, Elsevier Publishing Company, Amsterdam, 1964.
3. The Nobel Population 1901–1950: A Census of the Nominators and Nominees for the Prizes in Physics and Chemistry, Elisabeth Crawford, Universal Academy Press, Inc. (2002), ISBN: 4-946443-70-3.
4. Who was who: 1971–1980.
5. Isolation and Purification of Proteins (Biotechnology and Bioprocessing Series) by Rajni Hatti-Kaul, CRC, 2003, ISBN: 0824707265.
6. Protein Purification Protocols (Methods in Molecular Biology) edited by Paul Cutler, Humana Press; 2nd ed. 2004, ISBN: 1588290670.
7. Enzyme Assays: A Practical Approach (Practical Approach Series) by Robert Eisenthal and Michael J. Danson, Oxford University Press, USA, 2005, ISBN: 0199631433.

1947

© Nobel Foundation [1]

Sir Robert Robinson

"for his investigations on plant products of biological importance, especially the alkaloids" [2]

A. Fredga, then a member of the Nobel Committee for Chemistry of the Royal Swedish Academy of Sciences, observed of Robinson during his presentation speech [2]:

> *...By your very important and very numerous investigations, you have gradually changed our ideas on fundamental questions. As a student of molecular architecture you have, with eminent success, pursued the line of work emerging from Kekulé and Couper, and you have thrown light upon the formation of complicated structures within the living plant. Among organic chemists, you are today acknowledged as a leader and a teacher, second to none. In recognition of your services to Science, the Royal Academy has decided to bestow upon you the Nobel Prize for Chemistry for your investigations on plant products of biological importance and especially for your outstanding work on the structure and the biogenesis of complicated alkaloids...*

Robinson was nominated for the 1947 Chemistry Prize by H. H. Dale, H. Gault, W. N. Haworth, E. L. Hirst and L. Ruzicka. He was also nominted for the years 1928, 1933, 1935, 1936, 1938, 1939, 1940, 1941, 1942, 1943, 1944, 1945 and 1946 [3].

Robert Robinson was born at Rufford Farm, near Chesterfield, Derbyshire on 13th September 1886, the eldest child of William Bradbury Robinson, a surgical dressing manufacturer and inventor, and Jane, née Davenport (died 1950). William Robinson was first married to Elizabeth, née Lowe (died 1871) by whom there were eight surviving children.

Robinson then married Gertrude Maud Walsh (1912), a fellow student at Manchester University, by whom there was one son and one daughter. Gertrude died in 1954. Robinson, subsequently, married

Stearn Sylvia Hillstrom (née Hershey) in 1957. Robinson suffered from failing eyesight towards the end of his life and became virtually blind for the last few years of his life. He died on 8th February 1975 [2, 4, 5].

The Early Years

Robinson attended Mrs. Wilke's Kindergarten, the Chesterfield Grammar School and Fulneck School at Pudsey near Leeds (started at the age of 12), prior to entering Manchester University in 1902. Early records from the University describe Robinson as 'a good worker but messy'. During his undergraduate days his father built an extension to the family home at New Brampton. Robinson used this as his laboratory to synthesize terebic acid whilst still an undergraduate. Robinson graduated with a B.Sc. degree (1906) and entered Perkin's laboratory for his D.Sc., which he gained in 1910. Robinson's D.Sc. work concentrated on natural products starting with work on brazilin and the related haematoxylin. His first paper, published with Perkin (1905), concerned ethyl-piperonylacetate [6].

The Career

Year	*Post*	*Institute*
1912	First Professor of Pure and Applied Organic Chemistry	University of Sydney
1915	Heath Harrison Professor of Organic Chemistry	University of Liverpool
1920	Director of Research	British Dyestuffs Corporation
1921	Professor of Chemistry	St. Andrews
1922	Chair in Organic Chemistry	Manchester University
1928	Chair in Organic Chemistry	University of London
1930	Waynflete Professor of Chemistry	Oxford University
1955	Emeritus Professor and Honorary Fellow of Magdalen College (on his retirement)	Oxford University
1955	Director and Chemical Consultant	Shell Chemical Company

In the Royal Society Obituary, Todd and Cornforth have described Robinson's work in the following sequence [4]:

- Brazilin and haematoxylin
- Alkaloids
- Anthocyanidins and anthocyanins
- Anthoxanthins
- Theoretical chemistry
- Antimalarials
- Steroids and synthetic oestrogens
- Aliphatic acids
- Penicillin
- Speculations on biosynthesis
- Iridoids, origins of petroleum

Robinson completed the total synthesis of many natural products. A comprehensive account of this work appears in the above reference. Some of the more important syntheses were listed in Robinson's citation for election to Fellowship of the Royal Society – see later in honours and awards.

Robinson published 718 scientific papers, mainly in the Journal of the Chemical Society. He also published 32 patents. A comprehensive list of these publications can be found in [4]. He published the first volume of his autobiography in 1976 [7], the second volume was not finished. A biography of Robinson was published in 1990 [8].

Robinson's Nobel Prize acceptance speech, dated 12th December 1947 was titled 'Some polycyclic natural products'. His opening remarks [2]...

> *...In dealing with this subject I do not propose to attempt any full description of a part of my researches in the fields of the alkaloids, plant colouring matters, or on the methods of synthesis of steroids, but rather to indicate some connecting threads and perhaps, in some places, to go a little behind the scenes. In 1905, I was privileged to be given a place in the private laboratory of my revered teacher, Professor W. H. Perkin, Jr. at the University of Manchester. The synthesis of limonene had just been completed and sylvestrene was the next objective. Such work involved long and laborious preparations and would not have suited my case. It was indeed a fortunate chance that Perkin introduced me instead to the catechol derivatives which, together with their cousins of the meta and para series, have befriended so many organic chemists, myself among them, by reason of the numerous colour reactions that they exhibit and by the versatility of their transformations...*

Robinson went on to describe some work concerning:

- Brazilin
- Anthocyanins
- Alkaloid and
- Synthesis in the steroid groups

He concluded:

> *...I conclude with an expression of deepest gratitude and appreciation to all of my numerous co-workers; any success which I have had was due to their unsparing efforts. Though it might be invidious to mention individuals, yet I may be allowed to say how much I owe to the constant help of my wife, not quite my first, but much my most consistent collaborator, and over the longest period of years...*

His papers are lodged with the Royal Society. Such a profound volume of work has influenced generations of organic chemists. Perhaps his work in the alkaloids [9,10] may be referenced here as a token of his total work.

Honours and Awards

Robinson was presented with honorary degrees from the following Universities and Institutes: LL.D. Brussels; D.Pharm. Madrid and Paris; D.Sc. Belfast, Bristol, Cambridge, Dunelm, Liverpool, London, Nottingham, Oxford, Sheffield, Strathclyde, Sydney, Wales, Delhi, Hokkaido and Zagreb; LL.D. Birmingham, Edinburgh, Glasgow, Liverpool, Manchester and St. Andrews.

He received the following awards, medals or prizes: Priestley Medal (American Chemical Society); Longstaff, Faraday and Flintoff Medals (Chemical Society); Franklin Medal (Franklin Institute of Philadelphia); Hofmann Medal (German Chemical Society); Copley, Davy and Royal Medals (Royal Society); Albert Medal (Royal Society of Arts); Paracelsus Medal (Swiss Chemical Society); the Nobel Prize for Chemistry (1947).

He held Honorary Fellowship, Membership, Associate Membership or Corresponding Membership with many societies including: Australian Academy of Science; Biochemical Society; Society of Analytical Chemistry; Society of Endrocrinology; Indian Academy of Sciences; Indian National Academy of Science; Indian Chemical Society; Institution of Chemical Engineers; Parliamentary and Scientific Committee; Royal College of Obstetricians and Gynaecologists; Royal College of Physicians (Edinburgh); Royal College of Surgeons; Royal Society; Royal Society of Edinburgh; Royal Society of New South Wales; Royal Society of New Zealand. Robinson also held Foreign Fellowship/ Memberships with National Academies of: United States of America; Belgium; Denmark; France; Ireland; Japan; Norway; Rumania; Sweden; U.S.S.R.; New York Academy of Science, American Academy of Arts and Sciences, American Philosophical Society. Scientific Academies of Haarlem, Halle, Milan, Munich, Naples and Rome. He was honorary member of: the Chemical Societies of Belgium, France, Poland, Japan, Chemists' Club of New York and Honorary Fellow Weizmann Institute of Science, Rehovoth.

His public honours included: Knighthood (1939); Order of Merit (1949); Commandeur de la Legion d'Honneur (France); Order of the Rising Sun (Japan); Freeman of the Borough of Chesterfield (1947);

Robinson was President of the British Chess Federation (1950–1953).

He served as Vice President of the Royal Society (1944–1945); President of the Royal Society (1945–50); President of the Chemical Society (1939–41); President of the Society of Chemical Industry (1958–59); President of the British Association for the Advancement of Science (1955). He was an Honorary Fellow of Magdalen College, Oxford (from 1956), of which he had formerly been a Professorial Fellow. The Chemical Society honoured Robinson by establishing a Robert Robinson Lectureship (in 1962), to be delivered biennially in lieu of the usual Presidential Address.

Robinson's citation for election (date: 13/05/1920) as Fellow of the Royal Society reads [11]:

'Professor of Organic Chemistry. Distinguished for his researches in organic chemistry. Author of more than 50 papers in the Journal of the Chemical Society dealing mainly with the constitution of the natural colouring matters and of the alkaloids. The following may be specially mentioned:-'The Constitution of Brazilin and Haematoxylin' (1908); 'Brazilein, Haematein and their derivatives' (1908); 'The Constitution of Trimethylbrazilein a- and beta-Anhydrotrimethyl-brazilone and the Corresponding Haematoxylin Derivatives' (1909); 'Experiments on the Constitution of the Aloins' (1909); 'Synthesis of Cotarnic Acid' (1909); 'Strychnine and Berberine' (1910); 'Synthesis of Nitro-gnoscopine' (1910); 'Synthesis and Resolution of Gnoscopine' (1911); 'Iso-narcotine' (1912); 'Harmine and Harmaline' (1912–1913); 'The Mechanism of Benzoin Synthesis' (1913); 'Synthesis of beta-gnoscopine' (1914); 'Mechanism of the Alkylation of Ethyl Aceto-acetate' (1916); 'A Theory of the Mechanism of the Phyto-chemical Synthesis of Alkaloids' (1917); 'A Synthesis of Iso-brazilein (1918). Has been actively engaged during the war on investigations on the preparation of drugs, particularly of beta-eucaine hydrochloride, and on the problems submitted by the Chemical Warfare Committee'.

He was proposed by:

H. B. Dixon; W. H. Perkin; A. Lapworth; William A. Tilden; Edward Thorpe; P. F. Frankland; F. S. Kipping; James J. Dobbie; J. Thorpe; A. G. Perkin; A. G. Green; A. W. Crossley; E. C. C. Baly.

Bibliography

1. Reproduced with permission.
2. From Nobel Lectures, Chemistry 1942–1962, Elsevier Publishing Company, Amsterdam, 1964.
3. The Nobel Population 1901–1950: A Census of the Nominators and Nominees for the Prizes in Physics and Chemistry, Elisabeth Crawford, Universal Academy Press, Inc. (2002), ISBN: 4-946443-70-3.
4. Biographical Memoirs of Fellows of the Royal Society, 1976, vol. 22, pp. 415–527.
5. Who was who: 1971–1980.
6. R Robinson and W. H. Perkin Jr Ethyl-piperonylacetate. Proc. Chem. Soc. 1905, 21, 287.
7. Sir Robert Robinson: Memoirs of a minor prophet 70 years of organic chemistry autobiography, Elsevier, 1976, ISBN: 0-444-41459-2.
8. Robert Robinson: Chemist Extraordinary by Trevor I Williams, Clarendon Press, 1990, ISBN: 0-19-858180-7.
9. The Alkaloids, vol 1–31, Ed. Arnold Brossi, Adademic Press Inc. Volume 31 was published in 1987, ISBN: 0-12-469531-0.
10. Alkaloids – Nature's Curse or Blessing? by Manfred Hesse, Wiley-VCH, 2002, ISBN: 3-906390-24-1.
11. Sackler database, Royal Society.

1948

© Nobel Foundation [1]

Arne Wilhelm Kaurin Tiselius

"for his research on electrophoresis and adsorption analysis, especially for his discoveries concerning the complex nature of the serum proteins" [2]

In his presentation speech Professor A. Westgren, then Chairman of the Nobel Committee for Chemistry of the Royal Swedish Academy of Sciences, commented of Tiselius [2]:

> *...Tiselius has made many discoveries of far-reaching effect by applying his method of electrophoresis; that globulin, a protein of blood serum, was not an entirely homogeneous substance, had already been supposed: Tiselius succeeded in separating this seroglobulin into three distinct parts, each comprising slightly different groups of molecules. This finding is at the root of research work of the utmost importance for practical medicine, which was carried out in the United States during the last World War - research work aimed at dividing human blood plasma into fractional parts. If American scientists had not had at their disposal Tiselius' method as a control, they would probably have failed when they tried to solve that problem. In the course of their research work on electrophoresis, Tiselius and his collaborators also carried out experiments of great medical value on the antibodies of a protein nature which are formed in the blood during immunization...*

Tiselius was nominated for the 1948 Chemistry Prize by E. Havinga, B. Kamienski, L. C. King, B. Riegel and J. Roche [2]. He was also nominated for the year 1947 [3].

Arne Wilhelm Kaurin Tiselius was born 10th August 1902 in Stockholm, the son of Hans Abraham J. Tiseliusand, Rosa, née Kaurin. He had one sister. Tiselius married Ingrid Margareta (Greta) Dalén (1930). They had two children Eva (1932) and Per (1934). Tiselius died in Stockholm, Sweden on 29th October 1971 [2, 4, 5].

The Early Years

The family moved to Gothenburg following the premature death of Hans Tiselius (1906). Tiselius attended the local grammar school (Vasa Läroverk). Tiselius excelled at chemistry so much so that his chemistry and biology teacher, Dr. Ludwig Johansson, gave him a private key to the school laboratory so that Tiselius could carry on his private studies after nomal lessons. Tiselius attended the University of Uppsala in 1921 where he specialized in chemistry, however, his M.A. was in chemistry, physics and mathematics (1924). He was appointed Research Assistant in the Svedberg's laboratory (1925), and published a joint paper (Tiselius' first) in 1926 [6], concerning an electrophoresis method which allowed for the migration of a protein boundary to be followed by ultraviolet absorption. Tiselius' first experiments concerning the sedimentation of haemoglobin were suggested by Svedberg who gave Tiselius a free hand. Tiselius commented [4]:

> *...electrophoresis appeared so much simpler from a technical standpoint, suitable for a young man to play around with more or less on his own. Moreover, the excellent instrument workshop greatly facilitated the testing of new ideas...*

Tiselius obtained his doctor's degree on a thesis "The moving-boundary method of studying the electrophoresis of proteins" in 1930, and was appointed docent at the University of Uppsala followed by a docent fellowship two years later. Tiselius began some work on zeolites. He studied the diffusion of water vapour and other gases as well as kinetics of diffusion and adsorption into unhydrated zeolite crystals at Princeton University (under Hugh S. Taylor) during his tenure of a Rockerfeller Fellowship (1934–1935). Tiselius returned to Uppsala in 1935 and resumed his studies on electrophoresis.

The Career

Year	*Post*	*Institute*
1938	Research Professor of Biochemistry	University of Uppsala, Sweden
1968	Retirement	

Tiselius' return to the University of Uppsala prompted Svedberg to try and establish a post for Tiselius at the University. By the end of 1937 Svedberg had found a financial sponsor (Major Herbert Jacobsson and Mrs. Karin Jacobsson), so that a special Chair in Biochemistry was established for Tiselius. Royal Assent for the appointment was granted on 20th May 1938, and the inaugural lecture 'Biology and protein research' was delivered by Tiselius on 5th November 1938.

Tiselius made developments to a number of established methods during the course of his work including:

- electrophoresis
- chromatography
- phase partition
- gel filtration

He used these methods to investigate large molecular weight substances such as:

- proteins
- enzymes

- polysaccharides (dextran)
- nucleic acids

Tiselius instigated a cross-cultural Nobel Symposium, the first of which was held in 1968. The purpose of this was 'cultural, social and scientific aspects of the predicament of mankind'. In his opening address to the first Nobel Symposium, held outside of Stockholm (September 1969), Tiselius remarked that the title of the symposium was to be "The place of value in a world of facts". The issues under discussion were the use of science to further mankind.

Tiselius' Nobel Prize acceptance speech was dated 13th December 1948 and was titled 'Electrophoresis and adsorption analysis as aids in investigations of large molecular weight substances and their breakdown products'. He commented:

> *...For proteins and similar substances the immediate problem is to find suitable adsorbents. Above all, a relatively quick and reversible adsorption process is required for chromatographic analyses, and for this reason many adsorbents commonly used in biochemistry cannot be used for these substances. The advantages in the use of the chromatographic procedure - in the form just described or with some similar method - are however so great that it appears worth-while to search for such adsorbents. The advantage of using an adsorption column in comparison to the simple one-stage operation ('batch process') is almost comparable to the benefits accrued from the use of a distillation column instead of a simple evaporation procedure for the fractionation of fluid mixtures. Just recently we have studied an adsorption phenomenon previously hardly known, which we have called salting out adsorption, which possibly may fulfil the recently mentioned requirement for many proteins and similar substances. The adsorbent used in the column is a finely divided material which in water or weak salt solutions does not adsorb proteins or does so only very slightly...*

He concluded:

> *...It is perhaps too early yet to predict what practical value this method may attain - in conjunction with the recently described methods of observation. It is certain, however, that the need for specific and gentle methods of separation in the high molecular field is great, especially with respect to the enormous multitude of substances occurring in nature which are similar chemically but dissimilar with respect to function and activity...*

Electrophoresis is still a preferred method of separation for molecules of biological importance [7–9].

Honours and Awards

Tiselius was presented with honorary degrees from various institutes and universities including: Berkeley; Bologna (1955); California (1964); Cambridge (1949); Charles (1968); Glasgow (1956); Gustavus Adolphus College (1963); Lyons (1962); Madrid (1957); Michigan (1967); Oslo (1961); Oxford (1958); Paris (1948); Prague; Stockholm (1948); St. Peter.

He was presented with many awards and honours including: Nobel Prize for Chemistry (1948); Paterno (1954); Hoffmann (1955); Franklin (1956); Mookerjee (1959); Karrer (1961); Messel (1962). His civic honours included: Commander Stella Polaris 1st Class; Grand Officer Order of Merit Italian Republic; Commander Legion of Honour, France.

Tiselius was a member of the following Academies: Swedish Royal Academy of Sciences; Royal Society of Sciences of Uppsala; Royal Society of Gothenburg; Academy of Engineering Sciences, Stockholm; a member of the Danish Acad. Sciences; the Pontificia Scientiarum Academia, Vatican; the Polska Akademia Nauk, Warsaw; the World Acad. Art and Sciences. He was Honorary Fellow of: Royal Institute of Chemistry, London; Inst. Chem. of India; Indian Acad. Science, Bangalore. Tiselius held Honorary Membership of: the Finnish Acad. Sciences; the Harvey Soc.; the New York Acad. Sciences; the New York Acad. of Medicine; the Royal Inst. of Great Britain; the Chemical Soc., London; the Amer. Soc. of Europ. Chemists, New York; the Real Sociedad Espaflola de Fisica y Quimica; the Consejo Superior de Investigaciones Cientificas, Madrid; the Koninki. Nederlandse Chemische Verenigung, Haag; the Schweiz. Chem. Ges.; Hon. M. Nat. Acad. Science of India, Allahabad; the Franklin Inst., Philadelphia; the Soc. Chim. de France; the Sv. Lakaresallsk.; the Amer. Soc. Biol. Chemists; the Acad. Republ. Populare Romine, Bucharest. He was a Foreign Member of: the National Acad. of Sciences, Washington; the Acad.-Nazion. de Quaranta, Rome; the American Acad. of Arts and Sciences; the Royal Society, London; the Amer. Philosoph. Soc., Philadelphia; the Soc. of Chem. Ind., London. He was also a Corresponding Member of: the Soc. Philomat., Paris; the Academie des Ciencias de Lisboa; the Académie des Sciences, Paris.

Tiselius was: Chairman of the Swedish Natural Science Research Council (1946–50); a Member of the Nobel Committee for Chemistry (since 1946), Chairman (1965–71); Vice-president of the Nobel Foundation (1947–60), President (1960–64); Chairman of the Research Committee of the Swedish Cancer Society (1951–1955); President of the International Union of Pure and Applied Chemistry (1951–55); President of the Royal Swedish Academy of Sciences (1956–57); Head of the Nobel Institute, Royal Swedish Academy of Sciences (from 1968); President of the Royal Society of Sciences of Uppsala (1960–61); Member Science Advisory Council of the Swedish Government (1963–71) and a Member of the Wallenberg Foundation (1966–71).

Tiselius' citation for election (date: 09/05/1957) as Foreign Member of the Royal Society reads [10]:

> **'Professor Arne Tiselius of Uppsala distinguished for his outstanding contributions in the field of physiochemical analytical procedures and their application to labile molecules of high molecular weight and biological importance'**

No proposers appear in the Royal Society online database.

Bibliography

1. Reproduced with permission.
2. From Nobel Lectures, Chemistry 1942–1962, Elsevier Publishing Company, Amsterdam, 1964.
3. The Nobel Population 1901–1950: A Census of the Nominators and Nominees for the Prizes in Physics and Chemistry, Elisabeth Crawford, Universal Academy Press, Inc. (2002), ISBN: 4-946443-70-3.
4. Biographical Memoirs of Fellows of the Royal Society, 1974, vol. 20, pp. 401–428.
5. Who was who: 1971–1980.
6. A. Tiselius and T. Svedberg, J. Am. Chem. Soc, 1926, 48, 2272–2278.
7. Electrophoresis in Practice: A Guide to Methods and Applications of DNA and Protein Separations by Reiner Westermeier, John Wiley & Sons, 4th ed. 2005, ISBN: 3527311815.

8. Capillary Electrophoresis of Nucleic Acids, Vol. II: Practical Applications of Capillary Electrophoresis (Methods in Molecular Biology) edited by Keith R. Mitchelson and Jing Cheng, Humana Press, 1st ed. 2001, ISBN: 0896037657.
9. Adsorption: Theory, Modeling, and Analysis (Surfactant Science Series) by Jozsef Toth, CRC, 1st ed. 2002, ISBN: 0824707478.
10. Sackler database, Royal Society.

1949

© Nobel Foundation [1]

William Francis Giauque

"for his contributions in the field of chemical thermodynamics, particularly concerning the behaviour of substances at extremely low temperatures" [2]

A. Tiselius, then a member of the Nobel Committee for Chemistry of the Royal Swedish Academy of Sciences, observed of Giauque during his presentation speech [2]:

...There is a general law that all spontaneous chemical and physical processes are associated with an increase in entropy. We must conclude, therefore, that the world is becoming more and more disordered - a conclusion which is naturally only proved for molecular processes.

There is another very important law stating that the entropy is zero for a crystallized substance at a temperature equal to the absolute zero, $-273.16°$.

This law - the third law of thermodynamics - first formulated by the German Nobel Prize winner, Walther Nernst, has only now been fully proved as a result of Giauque's work. By means of this law we can now calculate not only differences in entropy but the entropy itself for a considerable number of elements and chemical compounds. Thus, if we want to calculate the free energy necessary for the formation of, for example, a certain organic substance made up of carbon, hydrogen, oxygen, and nitrogen, we look up the entropies of these elements in a table and determine in addition the heat of formation and the entropy of the compound. The problem of calculating the chemical affinity from the heat of formation or the heat of reaction is therefore solved in principle.

Behind the data used in such calculations and behind the final proof of the third law lie, however, extensive studies of the properties of substances at temperatures approaching absolute zero. It is here that this year's Nobel Prize winner has made his greatest contributions to chemistry. By supreme experimental skill he has overcome the many great difficulties which must obviously be inherent in investigations under such extreme conditions. He has also worked out new methods for experiment in this field, of which his magnetic cooling method deserves special mention. This method has made it possible to reach temperatures nearer to absolute zero than was possible by any earlier technique...

...William Francis Giauque, has added considerably to our knowledge of the entropy of chemical substances, particularly at very low temperatures...

Giauque was nominated for the 1949 Chemistry Nobel Prize by G. Hägg, G. A. Lander, T. Svedberg and A. Tiselius. He was nominated for the Physics Prize for 1940 and for the Chemistry Prize for the years 1936, 1937, 1942, 1947 and 1948 [3].

William Francis Giauque was the eldest of three children of William Tecumseh Sherman Giauque and Isabella Jane, née Duncan. Although his parents were U.S. citizens, Giauque was born in Niagara Falls, Canada (on 12th May 1895). Giauque married Muriel Frances Ashley (1932). They had two sons, William Francis Ashley Giauque and Robert David Ashley Giauque. Giauque died on 28th March 1982 [2,4].

The Early Years

Giauque's early education took place mainly in public grammar schools in Michigan. The family relocated back to Niagara Falls following the death of his father (1908). Giauque secondary school education was in the Niagara Falls Collegiate Institute. Giauque planned a career as an electrical engineer, and so sought jobs in the power plants at Niagara Falls. Additionally, his income provided much needed funds for the family finances. He successfully applied for a post in the laboratory of the Hooker Electrochemical Company in Niagara Falls. Captivated by the issues and problems that he experienced at the Hooker plant H, Giauque decided to become a chemical engineer. Two years later, Giauque left the Hooker plant and enrolled in the College of Chemistry at the University of California. The influence of Professor Gilbert N. Lewis and his co-workers further altered Giauque's thinking, coupled with his interest in and aptitude for fundamental research. Giauque graduated with the highest honours in a B.Sc. degree (1920). This was followed by a University Fellowship (1920–1921) and a James M. Goewey Fellowship (1921–1922). His postgraduate research, under Professor G.E. Gibson, was concerned with the relative entropies of glycerine crystals and glass. He received his Ph.D. in chemistry, with a minor in physics, in 1922.

The Career

Year	*Post*	*Institute*
1922	Instructor of Chemistry	University of California
1927	Assistant Professor	University of California
1930	Associate Professor	University of California
1934	Professor of Chemistry	University of California
1939–1945	War work	Government
1962	Emeritus	

Much of Giauque's early career involved investigations into the third law of thermodynamics, the principal objective for which was to determine that it is a basic natural law. His work included:

- accurate entropy determinations from low temperature measurements
- quantum statistics as a means of determining entropies and other thermodynamic properties

- determining molecular energy levels from band spectra (and other sources)
- investigations of the entropy of oxygen from its incompletely interpreted band spectrum
- entropy of oxygen from low-temperature heat capacity measurements (Dr. H.L. Johnston)
- the discovery of oxygen isotopes 17 and 18. This led to a re-evaluation of the atomic weight scale
- Investigation of the effect of magnetic fields on the entropies of paramagnetic substances
- the invention of the adiabatic demagnetization method of producing temperatures considerably below 1° absolute (in collaboration with Dr. D.P. MacDougall)

Giauque's scientific work has been described in some 75 papers. His Nobel prize acceptance speech is dated 12th December 1949 and is titled 'Some consequences of low temperature research in chemical thermodynamics'.

...The basic purpose which underlies most of the work in the Low Temperature Laboratory, of the University of California, is the study of entropy. It is easy for a chemist to write an equation for a desired reaction, but this does not mean that the reaction will actually take place. If one knows the heats of formation and entropies of the substances concerned in the equation, the so-called free energy or thermodynamic potential of the reaction may be calculaled. A knowledge of the free energy change permits chemists to determine all reactions which are thermodynamically possible and the extent to which they are possible. When it has been found that a certain reaction should go, but experiment fails, then a catalyst can be sought or the conditions can be altered so as to secure a practicable rate of reaction, If the free energy shows that the reaction is thermodynamically impossible, a search for a catalyst is futile...

...The third law of thermodynamics, which developed from the Nernst Heat Theorem, states that all perfect crystalline substances approach zero entropy as the absolute zero of temperature is approached. According to this statement a knowledge of the heat capacity, to sufficiently low temperatures, will permit the calculation of the absolute entropy of a substance, It was this possibility which interested me in low temperature research...

...The commonest question asked in the early days of this work was: "How do you know it gets cold?" This was a fair question. Obviously no one had ever made thermometers which were calibrated at temperatures that had never been produced. Since even helium gas has negligible pressure, at the low temperatures obtained, a gas thermometer is useless.

Temperature can only be measured by some property of a substance which varies with temperature. In this case the magnetic susceptibility increases as temperature decreases...

...If a substance becomes a permanent magnet 'at very low temperatures', it is easy to measure the strength of the magnet. This may be done by slowly warming the magnetic material. The apparatus shown then acts as an electric generator with no moving parts except the atoms. As the paramagnetic substance is warmed, any change in the magnetic moment will generate a potential in the coil, which can be quantitatively measured by a sensitive galvanometer and used to determine the total magnetic induction.

The resistance of the wire in measuring coils drops to such low values at the temperatures of liquid helium that very large numbers of turns can be used. The sensitivity which is obtained with coils at these low temperatures is such that it is necessary to compensate for the small fluctuations in the Earths magnetic field.

These are some examples of the type of things that are to be found by those who inquire into the subject of entropy. We consider it a rich field for further investigation...

Perhaps low temperature studies are now studied by as many physicists as chemists, which is reflected in the literature [5–7].

Honours and Awards

Giauque received an Honorary Sc.D. from Columbia University and an Honorary LL.D. from the University of California.

He was awarded the: Chandler Medal; Elliott Cresson Medal from the Franklin Institute; Nobel Prize for chemistry (1949); Willard Gibbs Medal (1951); Gilbert Newton Lewis Medal (1956).

He was a Member of the: American Chemical Society; American Philosophical Society; Institut International du Froid; National Academy of Sciences. He held Fellowships with the American Academy of Arts and Sciences and the American Physical Society.

He was Faculty Research Lecturer of the University of California in 1948.

Bibliography

1. Reproduced with permission.
2. From Nobel Lectures, Chemistry 1942–1962, Elsevier Publishing Company, Amsterdam, 1964.
3. The Nobel Population 1901–1950: A Census of the Nominators and Nominees for the Prizes in Physics and Chemistry, Elisabeth Crawford, Universal Academy Press, Inc. (2002), ISBN: 4-946443-70-3.
4. Who was who: 1981–1990.
5. Experimental Techniques in Low-Temperature Physics (Monographs on the Physics and Chemistry of Materials, 59) by Guy K. White and Philip Meeson, Oxford University Press, USA, 4th Ed., 2002, ISBN: 019851428X.
6. Low-Temperature Physics by Christian Enss and Siegfried Hunklinger, Springer, 1st Ed., 2005, ISBN: 3540231641.
7. Matter and Methods at Low Temperatures by Frank Pobell, Springer, 2nd Ed., 2002, ISBN: 3540585729.

1950

Otto Paul Hermann Diels (0.5)

"for their discovery and development of the diene synthesis" [2]

During his presentation speech, Professor A. Fredga, then a member of the Nobel Committee for Chemistry of the Royal Swedish Academy of Sciences, observed of Diels [2]

...Diene synthesis occurs between two different molecules of more or less complex structure. One of them, the diene, contains a chain of four carbon atoms, linked by two double bonds and - in the centre of the chain - a single bond. The first syllable of diene is the Greek word for "two", and refers to the two double bonds. The other molecule must contain two carbon atoms linked by a double bond, and must also satisfy certain other structural requirements. This component is usually called the dienophil or philodiene, i.e. the diene-loving, partner. In diene synthesis the two chains attach themselves to four or two carbon atoms to form a ring, which thus contains six atoms; this number is particularly favourable to cyclization. At the same time the double bonds are broken, so that only one remains in the new system.

It has long been known that molecules with double bonds readily become attached to one another; today we also know that this process usually results in very long atom chains, known as giant molecules. Diene synthesis is a special case, where, owing to certain geometrical and arithmetical relations, growth of the atom chain by attachment of new molecules ceases. Instead, the chain catches its own tail, so to speak, and forms itself into a ring which can grow no further...

...Professor Diels, Professor Alder. More than two decades have passed since the day when you sent in the first paper on "Syntheses in the Hydroaromatic Series" to the Editor of the Annals. Much has happened since then-in Science, as well as in the world.

The reciprocal addition of the unsaturated compounds was at that time still somewhat mysterious. The chemistry of the high polymer compounds belonged for the most part to the future. You found it necessary to stress that the new substances which you obtained so easily were not molecular compounds but were really new molecules, stable and firmly bonded. New simple bonds had come into being of the old, well-known kind. This identification has paved the way to a right understanding of the polymerization phenomena of the unsaturated compounds, and through it you have very greatly advanced the development of high polymer chemistry and technics...We have been told that you, Professor Diels, were in the past years unable to do any experimental work. And yet your thoughts played keenly upon scientific problems and we hope that you will soon have the opportunity of approaching the problems experimentally as well...

Diels was nominated for the 1950 Chemistry Prize by C. Dufraisse and H. Wieland. He was also nominated for the years 1932, 1933, 1934, 1935, 1936, 1937, 1942, 1943, 1944 and 1949 [3].

Otto Paul Hermann Diels was born in Hamburg, Germany, on 23rd January 1876. He married Paula Geyer (1909) and had three sons and two daughters. Two of their sons were killed in action during World War II. He died on 7th March, 1954 [2, 4].

The Early Years

Diels attended school at the Joachimsthalsches Gymnasium (1882 to 1895) in Berlin prior to enrolling at Berlin University in 1895. His family moved to Berlin when Diels was two years old, upon the appointment of his father to a professorship. Diels studied chemistry under Emil Fischer and graduated in 1899. He was appointed Assistant at the Institute of Chemistry at Berlin University upon graduation.

The Career

Year	*Post*	*Institute*
1904	Lecturer	Institute of Chemistry at Berlin University
1906	Professor	Institute of Chemistry at Berlin University
1913	Head of Department	Institute of Chemistry at Berlin University
1914	Professor and Director of the Institute of Chemistry	University of Kiel
1945	Retirement	

Diels started his career in inorganic chemistry. He discovered carbon suboxide and spent sometime looking at its properties. He is best known, however, for his work in organic chemistry where he worked on:

- the use of selenium as a specific reagent for the dehydrogenation of hydroaromatic compounds (1927)
- elucidating structures of some of the steroids

- 3′-methyl-1,2-cyclopentenophenanthrene which he obtained dehydrogenating cholesterol with selenium
- the Diels-Alder reaction (1928) - also known as the diene synthesis.

The Diels-Alder reaction is a conjugate addition reaction of a conjugated diene to a alkene. For example the reaction of 1,3-butadiene with ethane to form cyclohexene as in the example below.

It is a generic reaction in that many dienes, and compounds containing double bonds can be used.

Diels was the author of the *Einführung in die organische Chemie* (1907). He had many papers published, mostly in German scientific periodicals such as *Justus Liebigs Annalen der Chemie*.

Diels was unable to present his Nobel Prize acceptance speech due to illness. He wrote a manuscript titled 'Description and importance of the aromatic basic skeleton of the steroids'. He commented on the use of selenium in the structure elucidation of cholesterol:

> *...Then, fortunately, I hit upon the happy idea of replacing sulphur by selenium, because the action of this element is much more gentle than that of sulphur, and because it does not have the tendency to enter organic molecules.*
>
> *This new dehydrating agent was an unexpected success.*
>
> *By heating melted cholesteryl chloride or cholesterol with selenium powder until evolution of hydrogen selenide had ceased, and by fractionating the reaction product I obtained, as main product, a large yield of a beautifully crystallized aromatic hydrocarbon with the composition $C_{18}H_{16}$ and melting point 124–125°C. This hydrocarbon has loomed very large in the literature on organic chemistry during the past two or three decades.*
>
> *Since this wholly unexpected result, obtained by a method - 'selenium dehydration" - which I had recently discovered, aroused much interest everywhere, it was only natural that, like a great many chemists throughout the world, I should have attempted to explain the structure of this hydrocarbon $C_{18}H_{16}$, which was obviously the aromatic basic skeleton of cholesterol. Strangely enough, however, the task was by no means easy, despite the fact that this was an aromatic system.*
>
> *And so it was that just before my own investigation had been brought to a successful and decisive conclusion, Harper, Kon, and Ruzicka forestalled me by establishing the identity of the γ-methylcyclopentenophenanthrene, which they had produced by synthesis, with the hydrocarbon $C_{18}H_{16}$ which I had obtained by dehydrating cholesterol with selenium.*
>
> *Shortly after this I myself again proved the identity of the two products by converting them into the same tribromide...*

It is somewhat unusual for an acceptance speech to detail work that was not the subject of the award. In this case, Diels was unable to attend the Nobel Prize ceremony at which Alder gave a comprehensive treatise of the Diels-Alder reaction.

Further references to the use of the Diels-Alder reaction can be found in [5–7].

Honours and Awards

Diels was presented with the honorary degree of Doctor of Medicine from the University of Kiel.

He was awarded: Gold Medal at the St. Louis (USA) International Exhibition (1904); the Adolf v. Baeyer Medallion (1930); the Nobel Prize for Chemistry, jointly with K. Alder (1950); the Grosskreuz des Verdienstordens der Bundesrepublik (1952);

He was a Member of the Academies of Halle, Munich, and Göttingen.

Bibliography

1. Reproduced with permission.
2. From Nobel Lectures, Chemistry 1942–1962, Elsevier Publishing Company, Amsterdam, 1964.
3. The Nobel Population 1901–1950: A Census of the Nominators and Nominees for the Prizes in Physics and Chemistry, Elisabeth Crawford, Universal Academy Press, Inc. (2002), ISBN: 4-946443-70-3.
4. Who was who: 1951–1960.
5. Classics in Total Synthesis II: More Targets, Strategies, Methods by K. C. Nicolaou and S. A. Snyder, John Wiley & Sons, 2003, ISBN: 3527306846.
6. Strategic Applications of Named Reactions in Organic Synthesis by Laszlo Kurti and Barbara Czako, Academic Press, 2005, ISBN: 0124297854.
7. Organic Synthesis: Concepts and Methods by Jürgen-Hinrich Fuhrhop and Guangtao Li, Wiley-VCH, 3rd ed. 2003, ISBN: 3527302735.

1950

© Nobel Foundation [1]

Kurt Alder (0.5)

"for their discovery and development of the diene synthesis" [2]

During his presentation speech Professor A. Fredga, then a member of the Nobel Committee for Chemistry of the Royal Swedish Academy of Sciences, said of Alder [2]:

...Diene synthesis occurs between two different molecules of more or less complex structure. One of them, the diene, contains a chain of four carbon atoms, linked by two double bonds and-in the centre of the chain-a single bond. The first syllable of diene is the Greek word for "two", and refers to the two double bonds. The other molecule must contain two carbon atoms linked by a double bond, and must also satisfy certain other structural requirements. This component is usually called the dienophil or philodiene, i.e. the diene-loving, partner. In diene synthesis the two chains attach themselves to four or two carbon atoms to form a ring, which thus contains six atoms; this number is particularly favourable to cyclization. At the same time the double bonds are broken, so that only one remains in the new system.

It has long been known that molecules with double bonds readily become attached to one another; today we also know that this process usually results in very long atom chains, known as giant molecules. Diene synthesis is a special case, where, owing to certain geometrical and arithmetical relations, growth of the atom chain by attachment of new molecules ceases. Instead, the chain catches its own tail, so to speak, and forms itself into a ring which can grow no further... Professor Diels, Professor Alder. More than two decades have passed since the day when you sent in the first paper on "Syntheses in the Hydroaromatic Series" to the Editor of the Annals. Much has happened since then - in Science, as well as in the world...

...The reciprocal addition of the unsaturated compounds was at that time still somewhat mysterious. The chemistry of the high polymer compounds belonged for the most part to the future. You found it necessary to stress that the new substances which you obtained so easily were not molecular

compounds but were really new molecules, stable and firmly bonded. New simple bonds had come into being of the old, well-known kind. This identification has paved the way to a right understanding of the polymerization phenomena of the unsaturated compounds, and through it you have very greatly advanced the development of high polymer chemistry and technics...

Alder was nominated for the 1950 Chemistry Prize by C. Dufraisse and H. Wieland. He was also nominated for the years 1936, 1942, 1943, 1944 and 1949 [3].

Kurt Alder was born in Königshütte, Upper Silesia, on the 10th of July 1902, the son of Joseph and Maria Alder. Alder died on 20th June 1958 [2,4].

The Early Years

Alder's childhood and school years were spent in Königshütte until after the First World War. Thereafter, the family left the region because of the political climate.

Alder enrolled at Berlin University (1922) and continued his studies at the University of Kiel, graduating with a Ph.D. (1926) working under Diels, the thesis for which was titled *Über die Ursachen der Azoester-reaktion* (On the causes of the azoester reaction).

The Career

Year	*Post*	*Institute*
1930	Reader in Chemistry	Kiel University
1936	Head of Department in the science laboratories	I. G. Farbenindustrie
1940	Professor of Experimental Chemistry and Chemical Technology and Principal of the Institute of Chemistry	Cologne University

Alder declined the invitation to return to Berlin University (1944) and also declined an invitation from the University of Marburg (1950), preferring to remain in Cologne.

Whilst part of Alder's time was devoted to exploring the extent of the Diels-Alder reaction, he worked with other colleagues on other areas of research. These included:

- the behaviour of double bonds in stressed carbon rings
- intermolecular rearrangements
- selection phenomena particularly in unsaturated systems during chemical reactions
- Investigations into various aspects of stereochemistry

Alder published about 150 papers, mainly in *Angewandte Chemie, Chemische Berichte der Deutsches Gesellschaft and Justus Liebig's Annalen der Chemie*,

Alder's Nobel Prize acceptance speech, dated 12th December 1950 is titled 'Diene synthesis and related reaction types'. During the course of this speech, Alder comprehensively covered many aspects of the Diels-Alder reaction during his presentation where he grouped the various topics in the following way:

- *The precursors*
- *The general formula*
- *The diene component*
- *The philodiene partner*
- *The role of maleic and fumaric acid*
- *Substituting addition*
    - *to the double bond*
    - *in the allyl position*
- *Combination of diene synthesis and substituting addition*
    - *the divinylmethane type*
    - *the styrene type*
- *Diene synthesis as a stereochemical phenomenon*
- *The 'cis' principle*
- *Retention of the diene configuration*
- *Diene synthesis and free rotation*
- *The universal orientation formula*
- *Diene synthesis with asymmetrical addenda*
- *Diene synthesis and substituting addition as a special case among general reaction types*

Alder concluded:

> *...It is not surprising that processes of this type should not be confined to pure carbon systems but that other arrangements of atoms with suitable electron configurations show a tendency towards similar conversions. Cases of this type have become known in increasing numbers, so far primarily with the elements oxygen and nitrogen. Thus, instead of the arrangement C=C, –C=C, the groups C=C–C=O and N=C–CN can also act as dienes, whereas in addition to the classical philodienes, >C=C< and –C=C–, combinations such as >C=O, –N=N–, –CN, and –NO are capable of fundamentally the same effects. This second group is rounded off by molecular oxygen. Developments during the past few years have shown ever more clearly the similarity of this substance to maleic anhydride in its reactions with dienes and olefines. The merging of diene synthesis and of substituting addition, on the one hand, with the apparently completely heterogeneous phenomena of autoxidation, on the other, into a homogeneous group of processes will presumably be only one stage in this development. It is very probable that in the majority of cases the additions of diazomethane, of diazoacetic acid, of organic azides, etc., to unsaturated systems, are triggered off and are determined, as far as their course is concerned, by resonance phenomena between the π-electron systems of the two partners. The task now is to investigate such effects of unsaturated substances upon one another from aspects which have become evident for the first time during the study of diene synthesis as the ideal type of this reaction...*

A simple schematic of the Diels-Alder reaction can be found in the entry for Otto Diels (1950), further references to the use of the Diels-Alder reaction can be found in [5–7].

Honours and Awards

Alder received an Honorary M.D. from the Medical Faculty of the University of Cologne and an Honorary Doctorate from the University of Salamanca.

He was awarded the Emil Fischer Memorial Medal from the Association of German Chemists (1938) and the Nobel Prize for Chemistry, jointly with Otto Diels (1950).

He was: Member of the Kaiserlich Leopold.-Karol.-Deutsche Akademie der Naturforscher (Imperial Leopold.-Karol.-German Academy of Natural Philosophers) in Halle (1938); a Honorary Member of the Real Sociedad Española de Física y Química (1953); Corresponding Member of the Mathematisch-naturwissenschaftliche Klasse (1955). Alder was also an Honorary Adviser to the Consejo Superior de Investigaciones Cientificas (1953).

Bibliography

1. Reproduced with permission.
2. From Nobel Lectures, Chemistry 1942–1962, Elsevier Publishing Company, Amsterdam, 1964.
3. The Nobel Population 1901–1950: A Census of the Nominators and Nominees for the Prizes in Physics and Chemistry, Elisabeth Crawford, Universal Academy Press, Inc. (2002), ISBN: 4-946443-70-3.
4. Who was who: 1951–1960.
5. Classics in Total Synthesis II: More Targets, Strategies, Methods by K. C. Nicolaou and S. A. Snyder, John Wiley & Sons, 2003, ISBN: 3527306846.
6. Strategic Applications of Named Reactions in Organic Synthesis by Laszlo Kurti and Barbara Czako, Academic Press, 2005, ISBN: 0124297854.
7. Organic Synthesis: Concepts and Methods by Jürgen-Hinrich Fuhrhop and Guangtao Li Wiley-VCH, 3rd ed. 2003, ISBN: 3527302735.

1951

Edwin Mattison McMillan (0.5)

"for their discoveries in the chemistry of the transuranium elements" [2]

During his presentation speech at the 1951 Nobel Prize awards ceremony, Professor A. Westgren, then Chairman of the Nobel Committee for Chemistry of the Royal Swedish Acadeny of Sciences, made the following comments of McMillan [2]:

...The first transuranium element of which there was definite proof was produced by McMillan and Abelson in May 1940 at the University of California, by irradiating uranium with neutrons with the aid of the cyclotron built by Lawrence. It was obtained as a disintegration product of a beta-radiating uranium isotope, which has a half-life of 23 minutes. Hahn and Meitner had also discovered this body, but their preparation was too weak for its daughter-product to be demonstrated. The Americans were able to investigate this thoroughly, and showed that it forms an isotope of element 93, that is to say, a transuranium element. They called it neptunium after the planet Neptune, whose orbit lies next outwards after Uranus in the solar system. By irradiating uranium with rapid neutrons or with heavy-hydrogen nuclei, deuterons, other neptunium isotopes were soon produced in Berkeley...

...Dr. McMillan. In 1934 Fermi showed that nuclear transmutations could be brought about by irradiating the heaviest elements with neutrons. Research into the reactions thus produced has, however, met with certain difficulties, and it took longer than was expected for the existence of the transuranium elements to be proved. You were the first to succeed in this enterprise. By your discoveries you have opened a field of research in which vast and fundamentally important scientific and technical gains have been made. Later, by your work on the accelerator problem you have also actively furthered the progress in this domain of chemistry...

Edwin Mattison McMillan was born on 18th September 1907, at Redondo Beach, California. He was the son of Edwin Harbaugh McMillan (a physician) and his wife, Anne Marie McMillan, née Mattison.

McMillan married Elsie Walford Blumer (1941). They had three children, Ann Bradford (1943), David Mattison (1945) and Stephen Walker (1949). McMillan died on 7th September 1991 [2,3].

The Early Years

McMillan grew up in Pasadena, California and attended local schools there. He entered the California Institute of Technology graduating with a B.Sc. degree (1928) and M.Sc. degree (1929). He then moved to Princeton University where his took a Ph.D. (1932), the subject for which was in the field of molecular beams. After graduation, he returned to California, and joined the University of California as a National Research Fellow, where he studied the measurement of the magnetic moment of the proton by a molecular beam method.

The Career

Year	*Post*	*Institute*
1934	Research Associate in the Radiation Laboratory	University of California
1935	Instructor, Department of Physics	University of California
1936	Assistant Professor, Department of Physics	University of California
1941	Associate Professor, Department of Physics	University of California
Nov 1940–Sept 1945	War work	
1946	Professor, Department of Physics	University of California
1954	Associate Director, Radiation Laboratory	University of California
1958	Deputy Director, then Director	University of California
1971	Director	Lawrence Berkeley Laboratory
1973	Retired	

As a staff member at the University of California, McMillan studied nuclear reactions and their products whilst also helping in the design of equipment, particularly cyclotrons. This area of research continued until the outbreak of the war during which he was seconded from the University. McMillan's war work started in November 1940 and ended in September 1945. He worked in three Institutes during this period:

- Radiation Laboratory, Massachusetts Institute of Technology (1940–1941)
- U. S. Navy Radio and Sound Laboratory, San Diego (1941–1942)
- Manhattan District, Los Alamos(1942–1945)

McMillan's most notable achievements were:

- the discovery of neptunium
- the partial identification of plutonium
- the development of the synchroton
- the development of the synchro-cyclotron

McMillan's Nobel Prize acceptance speech is dated 12th December 1951 and is titled 'The transuranium elements: early history'. He recounted his part of the story which took place prior to his secondment for the war effort. The rest of the story can be found in Seaborg's acceptance speech, with whom McMillan shared the 1951 Nobel Prize for Chemistry.

> *...In this talk I shall tell of the circumstances that led to the discovery of neptunium, the first element beyond uranium, and the partial identification of plutonium, the next one beyond that. The part of the story that lies before 1939 has already been recounted here in the Nobel Lectures of Fermi and Hahn; I played no part in that and shall not repeat it now. Rather I shall start with the discovery of fission by Hahn and Strassmann. News of this momentous discovery reached Berkeley early in 1939...*
>
> *...I decided to do an experiment of a very simple kind. When a nucleus of uranium absorbs a neutron and fission takes place, the two resulting fragments fly apart with great violence, sufficient to propel them through air, or other matter, for some distance. This distance, called the "range", is a quantity of some interest, and I undertook to measure it by observing the depth of penetration of the fission fragments in a stack of thin aluminium foils. The fission fragments came from a thin layer of uranium oxide spread on a sheet of paper, and exposed to neutrons from a beryllium target bombarded by 8 MeV deuterons in the 37-inch cyclotron. The aluminum foils, each with a thickness of about half a milligram per square centimeter, were stacked like the pages of a book in immediate contact with the layer of uranium oxide. After exposure to the neutrons, the sheets of aluminum were separated and examined for radioactivity by means of an ionization chamber. The fission fragments of course are radioactive atoms, and their activity is found where they stop...*
>
> *...The natural supposition was that plutonium followed chemically the properties of uranium and neptunium, with the lower state still more stable than it is in neptunium. However I did not participate in the final proof of this, since I left Berkeley in November 1940, to take part in the development of radar for national defense. The rest of the story belongs to Seaborg, who took up the work after my departure, and I shall let him tell it in his own way...*

Further texts concerning the actinides/transuranium elements can be found in [4, 5, 6].

Honours and Awards

McMillan received a Honorary Doctorate in Science by the Rensselaer Polytechnic Institute (1961) and by the Gustavus Adolphus College (1963).

He was awarded: the 1950 Research Corporation Scientific Award (1951); the Nobel Prize for Chemistry, jointly with Seaborg (1951); the Atoms for Peace Award jointly with Professor V. I. Veksler (1963); the Alumni Distinguished Service Award (1966); the Centennial Citation (1968).

Professor McMillan was a Fellow of the American Academy of Arts and Sciences and of the American Physical Society. He was a Member of the American Philosophical Society and of the National Academy of Sciences.

He served on the General Advisory Committee to the Atomic Energy Commission (1954–1958); he was appointed to the Commission on High Energy Physics of the International Union of Pure and Applied Physics (1960); he was a member of the scientific committee of Stanford Linear Accelerator Centre (1962–66) and trustee of the University Research Association (1969–84).

Bibliography

1. Reproduced with permission.
2. From Nobel Lectures, Chemistry 1942–1962, Elsevier Publishing Company, Amsterdam, 1964.
3. Who was who: 1991–95.
4. Americium and Curium Chemistry and Technology ed Norman M. Edelstein, James D. Navratil and Wallace W. Schultz, D Reidel Publishing Co, 1985, ISBN: 90-277-2097-5.
5. Transplutonium Elements – Production and Recovery by James D. Navratil and Wallace W. Schultz, Americal Chemical Society, 1981, ISBN: 0-8412-0638-4.
6. Lanthanides and Actinides by Simon Cotton, McMillan Education, 1991, ISBN: 0-333-53603-7.

1951

Glenn Theodore Seaborg (0.5)

"for their discoveries in the chemistry of the transuranium elements" [2]

A. Westgren, then Chairman of the Nobel Committee for Chemistry of the Royal Swedish Acadeny of Sciences, said of Seaborg during his presentation speech [2]:

> *...In 1940 McMillan and Seaborg and their fellow-workers had already reported that when neptunium disintegrates it gives rise to an element 94. By analogy with the way in which names had been found for neptunium and uranium, this second transuranium element was called plutonium, after the planet Pluto, which has its orbit next outside that of Neptune. The first isotope of this element, which has a half-period of 24,000 years and thus is relatively stable, is what is called an atomic fuel. This plutonium isotope reacts with slow neutrons in the same way as the uranium isotope* ^{235}U*, that is to say, when it is split it develops great energy and gives off neutrons. In this way it came to play an important part in the atomic-bomb project during the war, and methods were developed for its production on a large scale...*

> *...After these problems, conditioned by the war, had been solved, Seaborg, as leader of a comprehensive group of able colleagues, completed the studies of the transuranium elements. In doing this, he has written one of the most brilliant pages in the history of the discovery of chemical elements...*

> *...Dr. Seaborg. At a time when the possibilities of finding new elements appeared to be exhausted, you have produced a whole row of them and thus extended the Periodic System beyond the limits which, one might say, Nature seemed to have established. With great skill you have studied the chemical characteristics of the new-found elements, and so made clear their atomic structure...*

Glenn Theodore Seaborg was born in Ishpeming, Michigan, on 19th April 1912, the son of Herman Theodore and Selma Erickson Seaborg. He married Helen Lucille Griggs (1942), who was at that time, secretary to Ernest O. Lawrence (Nobel Laureate for Physics 1939). They had six children: Peter

(1946), Lynne (1947), David (1949), Stephen (1951), John Eric (1954), and Dianne (1959). Seaborg died on the 25th February 1999 [2–4].

The Early Years

Seaborg graduated from the David Starr Jordan High School in Los Angeles (1929) as valedictorian of his class. The family moved to California when Seaborg was 10 years of age. He entered the University of California (1929) and studied chemistry, graduating with a Ph.D. in 1937. After graduation he was appointed the personal laboratory assistant of the late G. N. Lewis (1937 to 1939), with whom he published a number of scientific papers.

The Career

Year	*Post*	*Institute*
1939	Instructor in Chemistry	University of California
1941	Assistant Professor	University of California
1942–1946	War work – leave of absence	
1945	Professor of Chemistry	University of California
1946	Additional responsibility for direction of nuclear chemical research at the Lawrence Radiation Laboratory	Atomic Energy Commission by the University of California
1954–1961	Associate Director	LRL
1958	Chancellor	University of California
1961–1968	Chairman of the Atomic Energy Commission	US Government
1971	University Professor of Chemistry	University of California

Seaborg was based at the Metallurgical Laboratory of the University of Chicago for his war work (1942–1946) where he headed the plutonium work of the Manhattan Project.

His main career scientific contributions included:

- co-discovery of plutonium
- co-discovery of all further transuranium elements through to element 102
- the identification of more than 100 isotopes of various elements
- author of the actinide concept of heavy element electronic structure
- demonstrated that the heavy elements form a "transition" series of actinide elements analogous to the rare-earth series of lanthanide elements
- developed instrumentation
- developed methodology for the study of elements and isotopes

Seaborg published approximately 200 scientific papers, including a number of comprehensive reviews and compilations in scientific publications. He also authored/co-authored several books on chemistry and the elements.

Seaborg's Nobel Prize acceptance speech, dated 12th December 1951 is titled 'The transuranium elements: present status'. Seaborg commented:

> *...Beginning at essentially the point where Professor McMillan left off in his lecture, I shall first describe our work on plutonium. The beginning of this can be considered to have taken place in the fall of 1940 when I asked a graduate student, Arthur C. Wahl, to consider the possibility of studying the tracer chemical properties of element 93 as a thesis problem, a suggestion which he was happy to accept. This, and related work on element 94, was carried on in collaboration with Dr. Joseph W. Kennedy who, like myself, was at that time an instructor in the Department of Chemistry at the University of California. After Dr. McMillan's departure from Berkeley in November, 1940, and his gracious assent to our continuation of the search for, and possible identification of element 94, our group turned its major efforts to this problem.*
>
> *A bombardment of uranium oxide with the 16-MeV deuterons from the 60-inch cyclotron was performed in December, 1940. Alpha radioactivity was found to grow into the chemically separated element 93 fraction, and this alpha-activity was chemically separated from the neighboring elements, especially elements 90 to 93 inclusive, in experiments performed during the following months. These experiments, which constituted the positive identification of element 94, showed that this element has at least two oxidation states, distinguishable by their precipitation chemistry, and that it requires stronger oxidizing agents to oxidize element 94 to the upper state than is the case for element 93. The particular isotope identified has been shown to be of mass number 238...*
>
> *...The chemical properties of elements 93 and 94 were studied by the tracer method at the University of California for the next year and a half. These first two transuranium elements were referred to by our group simply as "element 93" and "element 94" until the spring of 1942, at which time the first detailed reports concerning these elements were written. In order to write the report on the chemical properties, it became necessary to have chemical symbols for the two elements. It was remembered that McMillan had suggested the name "neptunium" (symbol Np) for element 93 after Neptune, the planet immediately beyond Uranus, which gives its name to uranium, and, therefore, it was thought proper that element 94 should assume the name 'plutonium" (symbol Pu) after the next planet Pluto. These names and symbols have been adopted officially by the International Union of Chemistry and will be used here...*
>
> *...The name "berkelium" (symbol Bk) was suggested for element 97 after the city where the work was done, in analogy with the naming of its chemical homologue terbium, which was named after the village of Ytterby here in Sweden where the rare earth minerals were first found. The name "californium" (symbol Cf) was suggested for element 98 in honor of the university and state where the work was done; thus the name does not reflect its strong homology to dysprosium. These names are used in the present talk including the various slides...*

Further texts concerning the actinides/transuranium elements can be found in [5,6,7].

Honours and Awards

Seaborg was presented with honorary degrees from the following Universities and Institutes:

Doctor of Humane Letters from Northern Michigan College (1962); Doctor of Public Administration from the University of Puget Sound (1963); Doctor of Public Service from George Washington

University (1962); Doctor of Law from the University of Michigan (1958) and University of Massachusetts (1963). Doctor of Science the University of Denver (1951); Gustavus Adolphus College (1954); the Northwestern University (1954); the University of Notre Dame (1961); Ohio State University (1961); Florida State University (1961); University of Maryland (1961); the Drexel Institute of Technology (1962); Georgetown University (1962); University of the State of New York (1962); Temple University (1962); Tulane University (1962); Mundelein College (1963); Trinity College (1963).

He was awarded the following medals and prizes: the American Chemical Society's Award in Pure Chemistry (1947); the John Ericsson Gold Medal by the American Society of Swedish Engineers (1948); the Nichols Medal of the New York Section of the American Chemical Society (1948); the Nobel Prize for Chemistry, jointly with McMillan (1951); the John Scott Award and Medal of the City of Philadelphia (1953); the Perkin Medal of the American Section of the Society of Chemical Industry (1957); the Atomic Energy Commission's Enrico Fermi Award (1959); the Priestly Memorial Award (1960); the Franklin Medal of the Franklin Institute, Philadelphia (1963); the Charles Lathrop Parsons Award (1964); Chemistry Pioneer Award (1968); the Gold Medal Award (1973); the Arches of Science Award (1968); the John R. Juebler Award (1978); the Priestley Medal (1979); the Henry DeWolf Smyth Award (1982); the Actinide Award (1984) and the Great Swedish Heritage Award (1984).

Seaborg was an Honorary Fellow of the Chemical Society of London and of the Royal Society of Edinburgh (elected 6th July 1959) and Foreign Member of the Royal Society (1985). He is a Fellow of the American Association for the Advancement of Science (President 1972, Chairman 1973); the American Institute of Chemists; the American Physical Society; the California Academy of Sciences and the New York Academy of Sciences. He was a Member of: the American Academy of Arts and Sciences; the National Academy of Sciences; the Royal Society of Arts of England; the Royal Swedish Academy of Engineering Sciences.

He was named as one of America's 10 outstanding young men by the U.S. Junior Chamber of Commerce (1947), and named Swedish American of the Year by Vasa Order of America, Stockholm (1962).

Seaborg was a member of: the AEC's first General Advisory Committee (1946–1950); the scientific advisory board of the Robert A. Welch Foundation (1957); the board of directors of the national Educational television and radio center (1958–1964, 1967–70); the President's Science Advisory Committee (1959 to 1961); the National Committee for chemical education material study (1959–74); the Federal council for science and technology (1961–71); the Federal radiation council (1961–69); the National Aeronautics and space council (1961–71); the board of trustees of the Pacific Science Centre Foundation (1962 onwards); the National Committee on America's Goals and Resources, National Planning Association (1962–64); the Presidents Committee on Manpower (1964–69); the Council on Foreign Relations (1965 onwards); the National Council on Marine Resources and Engineering Development (1966–71); the board of governors of the American-Swedish Historical Foundation (1972).

Part of Seaborg's citation for election (date: 27/06/1985) as Foreign Member of the Royal Society [8] is reproduced below:

> **'In the summer of 1940, E.M. McMillan and P.H. Abelson, using the 60 inch Berkeley cyclotron, discovered that the neutron bombardment of uranium yielded a new element which proved to be element 93. It was named neptunium. In December 1940, Glenn T. Seaborg and his associates took the next step and bombarded uranium (as the oxide) with 16-Mev deuterons and succeeded in**

isolating the unstable element-93 fraction of the resulting products, which decayed according to expectation, by emitting a beta particle to yield element 94, with a mass number of 238. This element was given the name plutonium. Further research with plutonium led to the isolation of isotope Pu-239 which was found to be a potential source of nuclear energy. This resulted in Seaborg's translation to the Manhatten project at the Metallurgical Laboratory of the University of Chicago as head of a group responsible for the isolation of plutonium on an industrial scale. By 1944, Seaborg recognized that the fourteen elements beyond [sic] actinium probably constituted an "ectinide series" analogous to the rare-earth series of lanthanide elements. The "actinide concept" proved to be crucial in the experimental work on the actinides beyond plutonium, and enabled Seaborg and his collaborators in Berkeley after the war to anticipate the chemistry of the heavier elements synthesized during the next decade: americium (95), curium (96), berkelium (97), californium (98), einsteinium (99), fermium (100), mendelevium (101), and nobelium (102).'

Bibliography

1. Reproduced with permission.
2. From Nobel Lectures, Chemistry 1942–1962, Elsevier Publishing Company, Amsterdam, 1964.
3. Royal Society of Edinburgh database.
4. Who was who: 1996–2000.
5. Americium and Curium Chemistry and Technology ed Norman M. Edelstein, James D. Navratil and Wallace W. Schultz, D Reidel Publishing Co, 1985, ISBN: 90-277-2097-5.
6. Transplutonium Elements – Production and Recovery by James D. Navratil and Wallace W. Schultz, Americal Chemical Society, 1981, ISBN: 0-8412-0638-4.
7. Lanthanides and Actinides by Simon Cotton, McMillan Education, 1991, ISBN: 0-333-53603-7.
8. Sackler database, Royal Society.

1952

© Nobel Foundation [1]

Archer John Porter Martin (0.5)

"for their invention of partition chromatography" [2]

During his presentation speech, A. Tiselius, then a member of the Nobel Committee for Chemistry of the Royal Swedish Academy of Sciences, commented of chromatography in general and of Martin in particular [2]:

> *...The method of Martin and Synge, in different forms, has already found extensive application in all branches of chemistry and important discoveries have been made with it. New and interesting substances have been traced and isolated with its help. Metabolic pathways in the organism can be studied and formerly unknown intermediary products identified. This has been done, for example, with studies of the way in which the green leaves of plants build up starch out of the carbon dioxide of the air. With Martin and Synge's method, Calvin and his co-workers in Berkeley have been able to identify some of the most important links in this process, perhaps the most important chemical reaction on our planet...*

> *...For their part, Martin and his co-workers have recently expanded the method's field of usefulness in other directions, of which experiments with gases and vapours on the one hand and experiments concerning the separation of mixtures of very large molecules on the other are attracting special interest...*

Archer John Porter Martin was born in London on 1st March 1910, the son of Dr W.A.P. Martin and Mrs L.K. Martin. Martin's father was a general medical practitioner. Martin married Judith Bagenal (1943) and they had one son and three daughters. Archer J.P. Martin died in Herefordshire on 28th July 2002 [2,3,4].

The Early Years

Martin's childhood was spent in North London. He did not read until he was eight years old. He went to Oaklands School in Crouch End for a year from the age of nine, after which the family moved to

Bedford. He attended Bedford School (1921 to 1929) prior to entering Peterhouse College, Cambridge University with the intention of becoming a chemical engineer. The influence of Professor J.B.S. Haldane, then Reader of Biochemistry at Cambridge, was sufficient for Martin to specialize in biochemistry. He graduated in 1932 and spent a further year in the Physical Chemistry Laboratory. He then transferred to the Dunn Nutritional Laboratory, Cambridge (1933–1938), completing his Ph.D, the subject of which was antipelagra factor in pigs, under Sir Charles Martin (1936).

The Career

Year	*Post*	*Institute*
1932	Chemist	Dunn Nutritional Laboratory, Cambridge
1938	Chemist	Wool Industries Research Association, Leeds
1946	Head of the biochemistry division	Boots Pure Drug Co., Nottingham
1948	Staff	Lister Institute for Preventative Medicine of the Medical Research Council
1950	From 1952 Head of Physical Chemistry Division	National Institute of Medical Research at Mill Hill
1956	Chemical consultant	
1959	Director	Abbotsbury Laboratories Ltd
1965–1973	Extrordinary Professor	Technological University of Eindhoven
1970	Consultant	Wellcome Research Laboratory
1973–74	Professorial Fellow	Sussex University
1974–1979	Robert A Welch Professor of Chemistry	University of Houston Texas
1979–80	Consultant	École Polytechnique, Lausanne
1980–1984	Invited Professor	École Polytechnique, Lausanne

Martin's academic work involved:

- a method of detecting pyro-electricity by observing the attraction of a metal plate for crystals that had been immersed in liquid air
- ultraviolet adsorption spectra
- work on liquid-liquid partition with R.L.M. Synge (1941)
- the development of paper chromatography (1944)
- the attempted isolation of Vitamin E
- the pathological effects of prolonged Vitamin E deficiencies
- chromatography
- the B2 group of vitamin deficiencies in pigs (with others)

The Wool Industries Research Association

- the felting of wool (initially with R.L.M. Synge, then Consden and Gordon)
- on amino-acid analysis

- the development method of partition chromatography
- the development method of gas-liquid chromatography (with A.T. James)

Martin's Nobel Prize acceptance speech is dated 12th December 1952 with the title 'The development of partition chromatography'. During his speech, Martin commented [2]:

> *...The subject of partition chromatography has now an extensive literature and I feel on an occasion such as this it is better to describe the development of the method rather than to summarize the present position or to attempt to put forward any of the more recent advances, which have been described elsewhere. Perhaps indeed, if enough histories, written while the ideas are still fresh in the minds of the people concerned, are available for a variety of discoveries or inventions, it may eventually be possible to lay down some of the principles required to facilitate the obtaining of fruitful results in scientific research in general. Clearly also the background of knowledge at the time the advance was made will be best understood if the history is as recent as possible.*
>
> *Partition chromatography resulted from the marrying of two techniques, that of chromatography and that of countercurrent solvent extraction. All of the ideas are simple and had peoples' minds been directed that way the method would have flourished perhaps a century earlier. In fact the minds of laboratory workers seem to have been closed to countercurrent procedures, which were adopted in industry, e.g. in distillation and lixiviation, long before they were generally used in the laboratory. In industry the use of the countercurrent principle led to great economies in heat and solvents, and its value was obvious: in the laboratory such economies are of comparatively little importance and the more subtle point that by its aid a vast improvement in certain difficult separations could he effected was not appreciated. Indeed it is only within the last twenty years that high efficiency distillation apparatus has been available in the laboratory and twenty-five years elapsed after Tsvett's classical work on chromatography before the latter method was put to general use. Synge and I were fortunate to work at a time when the need for new methods was apparent to many people, so that our methods have gained general acceptance in ten years...*

It is difficult to imagine separation techniques without the various aspects of chromatographic separation. The references [5–7] attempt to provide examples of current techniques.

Honours and Awards

Martin was presented with an Honorary LL.D. from Glasgow University (1973) and an Honorary D.Sc. from Leeds University (1968).

Martin received the following: Tswett Medal; the Kolthoff Medal; the Berzelius Medal of the Swedish Medical Society (1951); the Nobel Prize for Chemistry, jointly with Synge (1952); the John Scott Award (1958); the John Price Wetherill Medal (1959); the Franklin Institute Medal (1959); the Leverhulme Medal (1963); Koltoff Medal (1969); the Rising Sun Medal (1972); the Fritz-Pregl Medal (1985).

He was a Fellow of the Royal Society (1950), a Fellow of the Royal Society of Chemistry and a Millennium Fellow of the Royal Society of Chemistry (2000).

He was made a Companion of the British Empire in 1960.

He was appointed to deliver special lectures (as "buitengewoon hoogleraar") at the Technological University of Eindhoven, the Netherlands (1963).

Martin's citation for election (date: 16/03/1950) as Fellow of the Royal Society reads [8]:

> **'Distinguished as a biochemist. Was responsible with R. L. M. Synge, for the development of partition chromatography in the general field of amino acid analysis of proteins. His introduction, with Consden and Gordon, of filter paper (strip) chromatography, was an outstanding advance which has been extended in recent years to nearly all families of low-molecular organic substances. Martin has evolved several other ingenious methods for the separation of amino acids and peptides, which he has used with success in investigations in the structure of wool keratin; he has also made fundamental contributions to the study of felting in wool.'**

He was proposed by:

From personal knowledge. A. C. Chibnall; C. R. Harington; F. G. Young; Robert Hill; Malcolm Dixon; Charles J. Martin; Joseph Needham; A. R. Todd; Charles S. Hanes.

Bibliography

1. Reproduced with permission.
2. From Nobel Lectures, Chemistry 1942–1962, Elsevier Publishing Company, Amsterdam, 1964.
3. Biographical Memoirs of Fellows of the Royal Society, 2004, vol. 50, pp. 157–170.
4. Who's who: 2002.
5. Chromatography: Concepts and Contrasts by James M. Miller, Wiley-Interscience; 2nd ed. 2004, ISBN: 0471472077.
6. Modern Practice of Gas Chromatography edited by Robert L. Grob and Eugene F. Barry, Wiley-Interscience, 4th ed. 2004, ISBN: 0471229830.
7. Drug Monitoring and Clinical Chemistry, Volume 5 (Handbook of Analytical Separations) by Georg Hempel, Elsevier Science, 2004, ISBN: 0444509720.
8. Sackler database, Royal Society.

1952

Richard Laurence Millington Synge (0.5)

"for their invention of partition chromatography" [2]

During his presentation speech, A. Tiselius, then a member of the Nobel Committee for Chemistry of the Royal Swedish Academy of Sciences, commented of chromatography in general and of Synge in particular [2]:

...Partition chromatography has had other extremely important applications when it has been used as a means of studying the structure of giant molecules. It has been possible in this manner to attack problems of the structures of proteins and carbohydrates with considerably greater prospects for success. The intact molecule is too complicated for us to be able to penetrate its structure by chemical methods. If one breaks down these large molecules, a mixture of molecular fragments of different sizes and different chemical natures results. A separation of such extremely complicated mixtures of rather similar substances and an identification of every fragment is a way which allows us to draw conclusions about the structure of the original molecule in the same manner that an archeologist among the ruins of an ancient temple can find archeological details which make it possible for him to some extent to reconstruct the building in its original condition. As a guide in such work, the method of Martin and Synge is of the greatest value. Synge has shown this in some very beautiful investigations on the structure of gramicidin, an antibiotic active against certain bacteria. In work such as this, it is of special importance that the method is suitable for the isolation not only of the smallest building blocks (the amino acids) but also of larger fragments (peptides)...

...It is as if one should have a puzzle to put together: if by chance several pieces happen to hang together, the problem immediately becomes more simple. The young English chemist Sanger has recently succeeded in putting together an unusually difficult puzzle of this sort; from the mixtures which were separated by Martin and Synge's method, among others, he has been able to get an almost complete picture of the structure of the insulin molecule-a result which perhaps more than any other shows the method's great scope and principal significance...

Richard Laurence Millington Synge was born in Great Barrow, Chester on 28th October 1914, the son of Laurence Millington Synge who worked on the Liverpool Stock Exchange, and Katharine Charlotte, née Swan. He married Ann Stephen (1943), daughter of the late Adrian and Karin Stephen. They had seven children Mary, Charlotte, Alexander Millington, Patrick Millington, Matthew Millington, Elizabeth and Jane. Synge died on 18th August 1994 [2, 3, 4].

The Early Years

Synge was educated by his mother until he was nine years of age, at which point he was sent to a boarding school to prepare for Winchester College (1928–1931). Although Synge studied mainly classics at Winchester, his interest in biochemistry – a new subject for that time – was aroused by one of the Winchester teachers, the Rev. S. A. McDowall. Synge won an exhibition scholarship to read classics at Trinity College, University of Cambridge, but switched to science, winning the science prize in 1933. He completed Part I of the Natural Sciences Tripos (1935) studying chemistry, physics and physiology. He studied biochemistry for Part II (1936) during which he chose to work on proteins. He wrote an unpublished history of the countercurrent principles used in distillation titled '*Fractional distillation and the countercurrent principle*' during his Part II. Synge was awarded a first class honours degree in Natural Sciences. He studied as a research student under Mr. N.W. Pirie in the University Biochemical Laboratory (1936–1939). Wanting to increase his knowledge of carbohydrates, Synge spent some time under D. J. Bell, the results of which were three joint papers [5,6,7]. He spent two years (during 1939–1941) at the Wool Industries Research Association in Leeds. He graduated from Cambridge University with a Ph.D. degree in 1941.

The Career

Year	*Post*	*Institute*
1943	Research Assistant, Biochemistry Department	Lister Institute, London
1946–1947	Visiting Scientist	with Professor Tiselius at Uppsala
1948–1967	Head of the Department of Protein Chemistry and for the last year Deputy Director	Rowett Institute, Aberdeen
1958–59	On assignment	Ruakura Animal Research Station, Hamilton, New Zealand
1967	A.R.C. Food Research Institute	Norwich
1968–1984	Honorary Professor of Biology	University of East Anglia

General Academic work:

- The separation of amino acids by means of differential solubilities of their acetyl derivatives between cholorform and water
- The study of partition coefficients of acetyl amino acids
- Separations by liquid-liquid partition

- He built the first continuous flow steady state countercurrent apparatus (with A.J.P. Martin)
- He demonstrated the first working partition chromatography system (May 1941)
- The use of partition chromatography in the sequencing of the residues in the pentapeptide Gramicidin S
- A demonstration of paper chromatography
- Various analytical problems concerning the larger peptide molecules
- Work concerning antibiotic peptides
- Work on electrokinetic ultrafiltration of various polysaccharides (with Tiselius)

At the Rowett Institute

- The digestion of proteins by the ruminant animal and its associated micro-organisms
- Peptides, proteins and other components of plant material
- Physico-chemical methods for the purification of intermediates in the metabolism of proteins

At the Ruakura Animal Research Station, New Zealand

The isolation of the toxic fungal component sporidesmin (with E.P. White).

Synge's papers are deposited in Trinity College, Cambridge

He was editor of the Geological Chemistry section of the *American Chemical Abstracts* from Dec 1963, and a member of the editorial board of the *Biochemical Journal* (1949–55). Synge published 133 papers/review articles.

Synge's Nobel Prize acceptance speech was dated 12th December 1952 and titled 'Applications of partition chromatography'. Synge commented [2]:

> *...While still at school I heard tell of the science of biochemistry, which should draw together two aspects of nature into a common study. I had long been fascinated by living things (particularly the wild plants), and more recently by chemistry. This last was probably more for the appeal to the senses of the chemist's activities than for the knowledge of the nature of things which chemistry imparts. It was particularly reading in the newspaper Sir Frederick Hopkins' presidential address to the British Association for the Advancement of Science at Leicester in 1933 that impressed on me the idea that living things must have wonderfully precise and complicated working parts on the molecular scale and that the biochemist had the best chance of finding out how these are put together and do their work...*

> *...I will discuss in most detail the application of partition chromatography to the study of proteins. Partly because of intense interests of biochemists in proteins and partly because the earliest work with the method was in the field, the results obtained with partition chromatography have been especially striking here, although the method has similar capabilities in other branches of biochemistry...*

> *...When the effects with small molecules are better understood we shall be much better able to understand from the structure of the larger molecules what their function in terms of such molecular forces. The chemistry really will begin to merge with biology as closely as it has already merged*

with physics. Partition chromatography may well be remembered for this contribution long after it is quite obsolete as an analytical method...

It is difficult to imagine separation techniques without the various aspects of chromatographic separation. The references [8–10] attempt to provide examples of current techniques.

Honours and Awards

Synge was presented with Honorary D.Sc. degrees from the University of East Anglia (1977) and the University of Aberdeen (1987). He was also presented with an Honorary Ph.D. from the University of Uppsala.

Synge received the Nobel Prize for Chemistry, jointly with Martin (1952)

He was elected a Fellow of the Royal Society (1950) and the Royal Institute of Chemistry (1952). Synge was an Honorary Member of the American Society of Biological Chemists and of the Royal Society of New Zealand. He was appointed an Honorary Fellow, Trinity College in 1972.

Synge was treasurer of the Norwich Peace Council.

Synge's citation for election (date: 16/03/1950) as Fellow of the Royal Society reads [8]:

'Distinguished as a biochemist. Was the first to show the possibility of using counter-current liquid-liquid extraction in the separation of N-acetylamino acids. In collaboration with A.J.P. Martin this led to the development of partition chromatography, which they have applied with conspicuous success in problems related to the composition and structure of proteins, particularly wool keratin. Synge's recent work on the composition and structure of gramicidins is outstanding and illustrates vividly the great advances in technique for which he and Martin are responsible.'

He was proposed by:

From Personal Knowledge. A. C. Chibnall; N. W. Pirie; C. R. Harington; F. G. Young; Robert Hill; Malcolm Dixon; Charles J. Martin; Joseph Needham; A. R. Todd; Charles S. Hanes; F. C. Bawden.

Bibliography

1. Reproduced with permission.
2. From Nobel Lectures, Chemistry 1942–1962, Elsevier Publishing Company, Amsterdam, 1964.
3. Biographical Memoirs of Fellows of the Royal Society, 1996, vol. 42, pp. 453–479.
4. Who was who: 1991–1995.
5. D. J. Bell and R. L. M. Synge, J. Chem. Soc., 1937, 1711–1718.
6. D. J. Bell and R. L. M. Synge, J. Chem. Soc., 1937, 833–836.
7. D. J. Bell and R. L. M. Synge, J. Chem. Soc., 1937, 836–838.
8. Chromatography: Concepts and Contrasts by James M. Miller, Wiley-Interscience; 2nd ed. 2004, ISBN: 0471472077.
9. Modern Practice of Gas Chromatography edited by Robert L. Grob and Eugene F. Barry, Wiley-Interscience, 4th ed. 2004, ISBN: 0471229830.
10. Drug Monitoring and Clinical Chemistry, Volume 5 (Handbook of Analytical Separations) by Georg Hempel, Elsevier Science, 2004, ISBN: 0444509720.
11. Sackler database, Royal Society.

1953

© Nobel Foundation [1]

Hermann Staudinger

"for his discoveries in the field of macromolecular chemistry" [2]

During his presentation speech, A. Fredga, then a member of the Nobel Committee for Chemistry of the Royal Swedish Academy of Sciences, commented of Staudinger [2]:

> *...In the early 1920's Professor Staudinger expressed the view that a molecule could be very large, almost arbitrarily large in fact, that such macromolecules could very easily, sometimes with apparent spontaneity, be formed of 10,000 or 100,000 atoms and that the particles of colloidal solutions were in many cases actual molecules of this type.*
>
> *Professor Staudinger. More than thirty years ago you expressed the view that a chemical molecule can attain an almost arbitrary size and that such macromolecules are of great importance in our world. Your view was based on logical reasoning. You drew attention to the fact that what are termed high polymers are formed when for some reason or another an anticipated ring closure fails to occur. You thus submitted an argument which an organic chemist cannot ignore. Moreover, in extensive and painstaking series of studies you have provided experimental proof.*
>
> *It is no secret that for a long time many colleagues rejected your views which some of them even regarded as abderitic. This was understandable perhaps. In the world of high polymers almost everything was new and untried. Longstanding, established concepts had to be revised or new ones created. The development of the macromolecular science does not present a picture of a peaceful idyll...*

Hermann Staudinger was born in Worms on the 23rd of March 1881, the son of Dr. Franz Staudinger. He married Magda Woit (1928), who was his collaborator and his co-worker for many years, jointly writing numerous publications. Staudinger died in 1965 [2,3].

The Early Years

Following matriculation from school (1899), Staudinger studied at the University of Halle and then at Darmstadt and Munich. He graduated from Halle in 1903 and qualified for inauguration as an academic lecturer under Professor Thiele at Strasbourg University in spring 1907.

The Career

Year	*Post*	*Institute*
1907	Lecturer	Strassburg University
1908	Professor of Organic Chemistry	Institute of Chemistry of the Technische Hochschule in Karlsruhe
1912	Ordinary Professor	Eidgenössische Technische Hochschule in Zurich
1926	Professor of Chemistry	University of Freiburg i. Br.
1940	Additional post of Principal	Research Institute for Macromolecular Chemistry
1951	Resigned from his post as Principal	
1951	Honorary appointment as Head	State Research Institute for Macromolecular Chemistry
1956	Retirement	

Staudinger's research interests involved work on:

- aliphatic diazo-compounds
- autoxidation
- cellulose
- coffee aroma
- explosions
- insecticides
- ketenes
- macromolecular compounds
- oxalyl chloride
- rubber and isoprene
- synthetic pepper.

His main research area was macromolecules, the study of which was published in approximately 500 papers. A further 120 papers were published on cellulose and 50 on rubber and isoprene.

In addition to his papers, Staudinger published many books including:

Anleitung zur organischen qualitativen Analyse (Introduction to organic qualitative analysis), published by Springer, Berlin, 1st Edition 1923, 5th Edition 1948, 6th Edition 1955; *Die hochmolekularen organischen Verbindungen, Kautschak und Cellulose* (The high-molecular organic compounds, rubber and cellulose), published by Springer, Berlin, 1932; *Die Ketene* (Ketenes), published by Enke, Stuttgart, 1912; *Fortschritte der Chemie, Physik und Technik der Makromolekularen Stoffe* (Progress of the chemistry, physics and technique of the macromolecular substances), jointly with Professor Vieweg and Professor Röhrs, Volume 1, 1939, Volume II, 1942, publisher Lehmann, Munich; *Makromolekulare Chemie und Biologie* (Macromolecular chemistry and biology), publisher Wepf & Co., Basle, 1947; *Organische Kolloidchemie* (Organic colloid chemistry), published by Vieweg, Braunschweig, 1st Edition 1940, 3rd Edition 1950; *Tabellen zu den Vorlesungen über allgemeine und anorganische Chemie* (Tables for the lectures on general and inorganic chemistry), published by Braun, Karlsruhe, 1st Edition 1927, 5th Edition 1947; *Vom Aufstand der technischen Sklaven* (The uprising of the technical slave), published by Chamier, Essen-Freiburg, 1947. He wrote *Arbeitserinnerungen* (Working memoirs) in 1961 published by Dr. A. Hüthig, Heidelberg.

He edited the periodical *Die Makromolekulare Chemie* (Macromolecular chemistry), published by Dr. A. Hüthig, Heidelberg and Wepf & Co., Basle (from September 1947).

Staudinger's Nobel Prize acceptance speech is dated 11th December 1953 and is titled 'Macromolecular Chemistry'. During the course of the speech, Staudinger set the scene for polymer research in general, and his own work in particular [2]:

> *...In common with all organic compounds, the structure of the organic macromolecular compounds (inorganic macromolecular compounds are not discussed in the following) involves in addition to carbon atoms, chiefly hydrogen, oxygen, and nitrogen atoms which in accordance with the laws of Kekulé's structural theory are bound by chief valences. The only difference between macromolecules and the small molecules of low molecular substances is one of structural size. If it is desired to lay down a boundary between macromolecular and low molecular compounds - there are of course transitions linking the two groups - the substances with a molecular weight greater than ten thousand, i.e. the molecules of which consist of one thousand and more atoms, may he classified as macromolecular. Beyond roughly this size, characteristic macromolecular properties occur. So far no upper limits can he given for the size of the macromolecules. Macromolecular compounds with a molecular weight of several millions are known...*

He went on to classify various polymer types:

Classification of macromolecular substances

- *Substances occurring in nature*
    - *Hydrocarbons - rubber, guttapercha, balala*
    - *Polysaccharides - celluloses, starches, glycogens, mannans, pectins, polyuronic acids, chitines*
    - *Polynucleotides (nucleic acids)*
    - *Proteins and enzymes*
    - *Lignins and tans (transition from low- to macromolecular substances)*

- *Conversion products of natural substances*
    - *Vulcanized rubber, rayon, cellophane, cellulose nitrate, leather, lanital, galalith, etc.*

- *Synthetic materials*
  - *Plastics (polyplastics) formed by*
    - *polymerization - buna, polystyrene, poiymethacrylic ester*
    - *polycondensation - bakelite, nylon, Perlon, Terylene*
    - *polyaddition - polyurethane*

He concluded:

> *...In this way macromolecular chemistry appears today to fit between low molecular organic chemistry and cytology. It is the connecting link between them, growing systematically out of low molecular chemistry but, with the incomparably larger wealth of its chemical scope, forming living matter. In addition, over and above the quantitative laws of pure chemistry, macro-molecular chemistry makes use of a number of qualitative correlations:*
>
> *those of shape and of the associated configurational scope, up to the level of the "atomos" of living substance, on which the game of Life ensues.*
>
> *In the light of this new knowledge of macromolecular chemistry, the wonder of Life in its chemical aspect is revealed in the astounding abundance and masterly macromolecular architecture of living matter...*

Polymer chemistry is at the heart of modern plastics. By way of an illustration of the types of materials being prepared to suit current needs, the texts indicated will provide the interested reader with an entry into the subject [4,5,6].

Honours and Awards

Staudinger received various honorary degrees from European Institutes and Universities including: Dr. Ing. of the Technische Hochschule Karlsruhe; Dr. rer. nat. of the University of Mainz; Dr. (C) of the University of Salamanca; Dr. of the University of Strasbourg; Dr. Chem. of the University of Torino; Dr. sc. techn. of the Eidgenössische Technische Hochschule, Zurich;

Staudinger received many awards and prizes including the Cannizzaro Prize of the Reale Accademia Nazionale dei Lincei in Roma (1933) and the Nobel Prize for Chemistry (1953).

He was a Member of the Institut de France and the Society of Macromolecular Chemistry in Tokyo. He was also Member or Honorary Member of many Chemical Societies including the Science Academies of Göttingen, Heidelberg, Halle and Munich. He was also a Honorary Member of the Union of Finnish Chemists.

Bibliography

1. Reproduced with permission.
2. From Nobel Lectures, Chemistry 1942–1962, Elsevier Publishing Company, Amsterdam, 1964.

3. Who was who: 1961–1970.
4. Synthetic Rubbers – Their chemistry and their technology by D. C. Blackley, Applied Science Publishers, 1983, ISBN: 0-85334-152-4.
5. Handbook of Polyethylene – Structures, Properties and Applications by Andrew J. Peacock, Marcel Dekker Inc, 2000, ISBN: 0-8247-9546-6.
6. Modern Ployesters ed John Scheirs and Timothy E. Long, John Wiley and Sons Inc, 2003, ISBN: 0-471-49856-4.

1954

© Nobel Foundation [1]

Linus Carl Pauling

"for his research into the nature of the chemical bond and its application to the elucidation of the structure of complex substances" (and Peace prize in 1962) [2]

During his presentation speech, G. Hägg, then a member of the Nobel Committee for Chemistry of the Royal Swedish Academy of Sciences, commented of Pauling [2]:

...In the late nineteen thirties Pauling and his colleagues had already begun to determine with X-rays the structure of amino acids and dipeptides, that is to say, compounds of relatively simple structure containing what may be called fragments of proteins. From this was obtained valuable information - about atomic distances and bond directions. These values were supplemented by the determination of the probable limits of variation for distances and directions.

On this basis Pauling deduced some possible structures of the fundamental units in proteins, and the problem was then to examine whether these could explain the X-ray data obtained. It has thus become apparent that one of these structures, the so-called alpha-helix, probably exists in several proteins.

How far Pauling is right in detail still remains to be proved, but he has surely found an important principle in the structure of proteins. His method is sure to prove most productive in continued studies.

Professor Pauling. Since you began your scientific career more than thirty years ago you have covered a diversity of subjects ranging over wide fields of chemistry, physics, and even medicine. It has been said of you that you have chosen to live "on the frontiers of science" and we chemists are keenly aware of the influence and the stimulative effect of your pioneer work.

Wide though your field of activity may be, you have devoted the greater part of your energy to the study of the nature of the chemical bond and the determination of the structure of molecules and crystals...

Linus Carl Pauling was born in Portland, Oregon, on 28th February, 1901, the first child of a druggist, Herman Henry William Pauling and his wife, Lucy Isabelle, née Darling. Pauling had two younger sisters Pauline Darling (1902) and Lucile (1904). Pauling married Ava Helen Miller of Beaver Creek, Oregon (1923, d 1981). They had four children, Linus (Carl) Jr. (1925), Peter Jeffress (1931), Linda Helen (Kamb) (1932) and Edward Crellin (1937). Pauling died on his ranch at Big Sur, California on 19th August 1994 [2–4].

The Early Years

Pauling attended the public elementary and high schools in the town of Condon until the family moved to Portland in 1910. Pauling learnt to read books at an early age. He read the Bible and Darwin's theory of evolution before his ninth birthday. Pauling later commented that one of his favourite books was the Encyclopaedia Brittanica. Pauling entered the Oregon State College in 1917. His mother wanted him to leave college to earn money for the support of the family as money was in short supply following the death of his father. Fortunately, he was offered a full-time post as Instructor in Qualitative Analysis in the Chemistry Department (1919). The following year Pauling resumed his studies graduating with a B.Sc. in chemical engineering. He started his doctoral work (1922) under Professor Roscoe G. Dickinson and Richard C. Tolman. He published his first paper in 1923 on the crystal structure of magnesium stannide [5]. In his final year, he was asked to teach introductory chemistry for the young ladies on the home economics course. Ava Helen Miller was one of the students. They subsequently married and remained so for almost sixty years. Pauling was awarded the Ph.D. (summa cum laude) in chemistry, with minors in physics and mathematics in 1925.

The Career

Year	*Post*	*Institute*
1925	Research Associate	California Institute of Technology
1926	Fellow of the John Simon Guggenheim Memorial Foundation	European Universities
1927	Assistant Professor of Chemistry	California Institute of Technology
1929	Associate Professor	California Institute of Technology
1931	Professor and Chairman of the Division of Chemistry and Chemical Engineering	California Institute of Technology
1936	Director	Gates and Crellin laboratories of Chemistry
1958	Retirement	
1969–74	Professor of Chemistry	Stanford University

Pauling worked with Sommerfeld, Schrödinger, and Bohr during his Guggenheim Fellowship (1926–27). He was a National Research Fellow in Chemistry (1925–1926).

Pauling's many research papers (over 350) reported work in the following areas:

- molecular structure
- the nature of the chemical bond
- the experimental determination of the structures of some crystals by x-ray diffraction (1922)
- theoretical work on the nature of the chemical bond
- the interpretation of crystal structures in terms of the radii and other properties of atoms
- the properties of atoms
- the application of quantum mechanics to physical and chemical problems
- the rotational motion of molecules in crystals
- the structure of metals and intermetallic compounds
- the theory of ferromagnetism
- the structure of proteins
- the structure of antibodies and the nature of serological reactions
- the structure and properties of hemoglobin and related substances
- the molecular theory of general anesthesia

Pauling wrote several books, including:

- *General Chemistry* (1947, 1953) - which was translated into nine languages
- *No More War!* (1958, 1959, 1962)
- *The Nature of the Chemical Bond*
- *The Structure of Molecules and Crystals (1939, 1949, 1960).*

Pauling's Nobel Prize acceptance speech is dated 11th December 1954 and titled 'Modern structural chemistry'. Pauling chose to review theories concerning the nature of the chemical bond for his presentation. He observed [2]:

> *...After the discovery of quantum mechanics in 1925 it became evident that the quantum mechanical equations constitute a reliable basis for the theory of molecular structure. It also soon became evident that these equations, such as the Schrödinger wave equation, cannot be solved rigorously for any but the simplest molecules. The development of the theory of molecular structure and the nature of the chemical bond during the past twenty-five years has been in considerable part empirical-based upon the facts of chemistry - but with the interpretation of these facts greatly influenced by quantum mechanical principles and concepts...*
>
> *...These early applications of quantum mechanics to the problem of the nature of the chemical bond made it evident that in general a covalent bond, involving the sharing of a pair of electrons between two atoms, can be formed if two electrons are available (their spins must be opposed, in order that the bond be formed), and if each atom has available a stable electronic orbital for occupancy by the electrons.*
>
> *The equivalence of the four bonds formed by a carbon atom, which had become a part of chemical theory, was not at first easily reconciled with the quantum mechanical description of the carbon atom as having one 2s orbital and three 2p orbitals in its outer shell. The solution to this difficulty was obtained when it was recognized that as a result of the resonance phenomenon of quantum mechanics a tetrahedral arrangement of the four bonds of the carbon atom is achieved. The carbon atom can be described as having four equivalent tetrahedral bond orbitals, which are hybrids of the s*

and p orbitals. Further study of this problem led to the discovery of many sets of hybrid bond orbitals, which could be correlated with bond angles, magnetic moments, and other molecular properties. In particular it was found that sp^3, dsp^2, and d^2sp^3 hybrid orbitals correspond respectively to the tetrahedral, square planar, and octahedral configurations of inorganic complexes that had been discovered by Werner. Conclusions as to the utilization of atomic orbitals in bond formation can be drawn from experimental values of magnetic moments. For example, the theory of the dsp^2 square complexes of bipositive nickel, palladium, and platinum requires that these substances be diamagnetic. The square complexes of bipositive palladium and platinum had been recognized by Werner and their structure verified by Dickinson (J. Am. Chem. Soc., 44 (1922) 2404); but the assignment of the square configuration to the complexes of nickel which are diamagnetic had not been made until the development of the new theory...

Many books have been written about Pauling, references to some of which are included in the bibliography [6–9]. There are also many books that have been written concerning the nature of the chemical bond. These texts cover a wide age range from [10] which is for the younger reader to classic texts [11–13].

Honours and Awards

Pauling received Honorary Doctorates from: Oregon State College (1941); Brooklyn Polytechnic Institute (1955); Reed College (1959); and the Universities of Cambridge (1947), Chicago, Humboldt (Berlin) (1959), Liège (1955), London (1947), Lyon (1970), Melbourne (1964), Montpellier (1958), New Brunswick, Oxford (1948), Paris (1948), Princeton (1946), Tampa (1949), Toulouse (1949), Yale (1947), York (Toronto) (1966) and Warsaw (1969).

He received many awards and prizes, perhaps the most notable of which are: William H. Nichols Medal (1941); J. Willard Gibbs Medal (1946); Davy Medal (1947); Theodore William Richards Medal (1947); Presidential Medal for Merit (1948); Gilbert Newton Lewis Medal (1951); Nobel Prize for Chemistry (1954); Thomas Addis Medal (1955); Amedeo Avogadro Medal (1956); Pierre Fermat Medal (1957); Paul Sabatier Medal (1957); International Grotius Medal (1957); Nobel Prize for Peace (1962); Linus Pauling Medal (1966); International Lenin Peace Prize (1971); Martin Luther King Jr Medical Award (1972); National Medal of Science (1975); Lomonosov Gold Medal (1978); Vannevar Bush Medal (1989).

Pauling was a member of numerous professional societies in the U.S.A., many European countries, Chile, India and Japan. He was also elected to the U.S. National Academy of Sciences, the Royal Society of London, the Academic Française des Sciences and the Akademie Nauk S.S.R.

He was elected Rationalist of the Year (1960) and Humanist of the Year (1961).

Pauling delivered many named lectures starting with the George Fisher Baker lectures at Cornell University.

Pauling's citation for election (date: 27/05/1948) as Foreign Member of the Royal Society reads [14]:

'distinguished for his researches on valency and on structural organic and inorganic chemistry'.

He was proposed by:

R. Robinson; E. J. Salisbury; E. D. Adrian; S. Chapman; M. L. Oliphant; W. N. Haworth; W. E. Garner; J. H. C. Whitehead; J. D. Bernal; B. N. Wallis; H. D. Kay; C. Lovatt Evans; A. C. Chibnall; Charles H. Kellaway; A. C. Hardy; C. F. A. Pantin; H. H. Read; A. E. Trueman; W. Brown; J. R. Merton; A. C. Egerton.

Bibliography

1. Reproduced with permission.
2. From Nobel Lectures, Chemistry 1942–1962, Elsevier Publishing Company, Amsterdam, 1964.
3. Biographical Memoirs of Fellows of the Royal Society, 1996, vol. 42, pp. 315–338.
4. Who was who: 1991–1995.
5. L. Pauling, *J. Am. chem. Soc.*, 1923, 45, 2777–2780.
6. Linus Pauling A Life in Science and Politics by Ted Goertzel and Ben Goertzel, BasicBooks a division of HarperCollins, 1995.
7. Linus Pauling: Scientist and Peacemaker ed Clifford Mead and Thomas Hager, Oregon State University Press, 2001, ISBN: 0-87071-489-9.
8. Linus Pauling: A man and his science by Anthony Serafini, 1989, ISBN: 0-671-65311-3.
9. Linus Pauling Scientist and Advocate by David Newton, Facts on File Inc., 1994.
10. Linus Pauling and the Chemical Bond (Unlocking the Secrets of Science) (Library Binding) by Susan Zannos, Mitchell Lane Publishers, 2003, ISBN: 1584151234.
11. The Chemical Bond in Inorganic Chemistry: The Bond Valence Model (International Union of Crystallography Monographs on Crystallography) by I. David Brown, Oxford University Press, 2002, ISBN: 0198508700.
12. The Chemical Bond: A Fundamental Quantum-Mechanical Picture (Springer Series in Chemical Physics) by Tadamasa Shida, Springer, 1st ed. 2004, ISBN: 3540206388.
13. How Chemical Bonds Form and Chemical Reactions Proceed by Victor Gankin and Yuriy Gankin, Inst of Theoretical Chemistry Inc, 1998, ISBN: 0966414306.
14. Sackler database, Royal Society.

1955

© Nobel Foundation [1]

Vincent du Vigneaud

"for his work on biochemically important sulphur compounds, especially for the first synthesis of a polypeptide hormone" [2]

A. Fredga, then a member of the Nobel Committee for Chemistry of the Royal Swedish Academy of Sciences, commented of du Vigneaud [2]:

Underneath the brain, there is a small, well-protected gland, the pituitary gland. In man it is about as big as a bean. There are secreted several hormones, that is, substances which regulate important physiological functions. In spite of its small size, the pituitary gland is made up of several distinct parts with different functions. We are interested here in the posterior lobe, which contains two substances called oxytocin and vasopressin. The former stimulates the contractions of the uterus and also the lactation, the latter raises the blood pressure and regulates the function of the kidneys. As early as in 1933, when rather impure preparations from the posterior lobe were used in experiments, du Vigneaud found a high percentage of sulphur, which seemed to be correlated to the physiological activity.

Using the experimental methods, which the development of science has put at his disposal and making the best of his own intimate knowledge of the organic chemistry of sulphur, du Vigneaud has step by step forced his way. Both hormones were isolated in a state of purity, and it was found that they are built up from amino acids in the same way as proteins, but with a far lower molecular weight. Such compounds are, as distinguished from real proteins, called polypeptides. The nature of the amino acids and their positions in the molecule could be determined. The sulphur is present in cystine. The two hormones have a very similar structure; both contain eight amino acids, connected to a chain, which at one point is closed to a ring. The molecule has some resemblance to a figure six or nine, where the loop contains five amino acids and the "tail" three. Two sulphur atoms, linked to each other, form a part of the ring.

The design of the molecule was thus known. It remained to build it up by synthesis and check the correctness of the design. That was perhaps the most difficult part of the work. The interest was first concentrated on the synthesis of oxytocin...

Professor du Vigneaud. Some years ago you gave a survey of your most important works and you called it: "A trail of research in sulphur chemistry and metabolism". At an early stage you found that the element sulphur has a special fascination, and you started out to follow its winding trail into the unknown. It has led you to many important discoveries and recently to the first synthesis of a polypeptide hormone. We are sure it will lead you further still.

In the structure of these hormones there is nothing mysterious, nothing thrilling. They are built up like proteins, and the building-stones, the eight amino acids, are those found in common proteins. The conspicuous properties must be due to the sequence, the arrangement. This is a very important step on the way, leading to the understanding of the mode of action of such hormones...

Vincent du Vigneaud was born in Chicago, Illinois, on 18th May 1901, the son of Alfred J. du Vigneaud, an inventor and machine designer, and his wife Mary Theresa. Du Vigneaud married Zella Zon Ford (1924). They had a son, Vincent, Jr. (1933) and a daughter, Marilyn Renée (1935). Du Vigneaud died on 11th December 1978 [2].

The Early Years

Du Vigneaud studied at the University of Illinois under Professor C.S. Marvel, taking his B.Sc. degree in 1923 and M.Sc. in 1924. He was Assistant Biochemist to Dr. W.G. Farr at the Philadelphia General Hospital (1924–1925) and also worked on the Staff of the Graduate School of Medicine of Pennsylvania University. His Ph.D. thesis, under Professor J.R. Murlin at the School of Medicine of Rochester University, was submitted in 1927. This was followed by a period as a Fellow of the National Research Council. In this capacity, he worked with J.J. Abel at Johns Hopkins University Medical School, George Barger at Edinburgh University Medical School, and Charles R. Harington at London University College Hospital.

The Career

Year	*Post*	*Institute*
Early 1930s	Physiological Chemical Staff member	The University of Illinois
1932	Head of the Biochemistry Department	George Washington University School of Medicine
1938	Professor and Head of the Biochemistry Department	Cornell University Medical College

Du Vigneaud investigated many biologically important sulphur containing compounds, including:

- amino acids
- biotin (also known as vitamin H)

- coenzyme R
- numerous works on cystine
- the sulphur of insulin
- methionine
- first synthesis of a polypeptide hormone, oxytocin
- penicillin
- peptides
- vasopressin

He also undertook more general work, including:

- intermediary metabolism
- metabolism of one-carbon compounds
- the biological importance of the methyl group and of its migrations in the organism
- transmethylation
- transsulphuration

A Trail of Research in Sulphur Chemistry and Metabolism and Related Fields was the subject of the Messenger Lectures that he delivered in 1950 at Cornell University, and published in 1952.

Du Vigneaud's Nobel Prize acceptance speech is dated 12th December 1955 and titled 'A trail of sulfa research: from insulin to oxytocin'. He commented [2]:

> *...As already mentioned, we felt that synthetic proof of the structure we had arrived at had to be adduced; that is, of course, if the synthesis were within the range of possibility. I might point out that the synthesis of this structure meant the synthesis of an octapeptide of eight different amino acids, one of them being cystine. In addition, this desired octapeptide contained three amide groups and was a cyclic polypeptide for which no known synthetic methods were available - a cyclic pentapeptide with a tripeptide side chain. To add to the challenge was the fact that we knew the final product was relatively unstable and easily inactivated. Gentle means would therefore have to be employed.*
>
> *The clue to the synthetic approach to this compound rested on our work with Sealock, carried out in the 1930s, on the reduction of oxytocin and the oxidation of the reduced oxytocin without appreciable inactivation at either step, which has already been discussed. On the basis of our postulated structure for oxytocin, the reduction of oxytocin and subsequent oxidation could be interpreted as involving the opening and closing of the 20-membered ring. The reduced oxytocin would then have a linear structure containing two sulfhydryl groupings in place of the disulfide linkage in oxytocin. Furthermore, if the proposed structure for oxytocin were correct, reduction of oxytocin with sodium in liquid ammonia followed by addition of benzyl chloride should give rise to the S,S′-dibenzyl derivative of reduced oxytocin.*
>
> *Since benzylation of reduced oxytocin had led to inactivation in the earlier study, the expectation was that this benzylated derivative of reduced oxytocin would be biologically inactive. With what we now knew of the structures involved, we could see no reason why treatment of the S,S′-dibenzyl derivative of reduced oxytocin with sodium in liquid ammonia should not lead to the biologically active, reduced oxytocin; oxidation of the sulfhydryl form by aeration should then lead to the regeneration of oxytocin itself, if our concepts and line of reasoning were valid...*

The structure of oxytocin is:

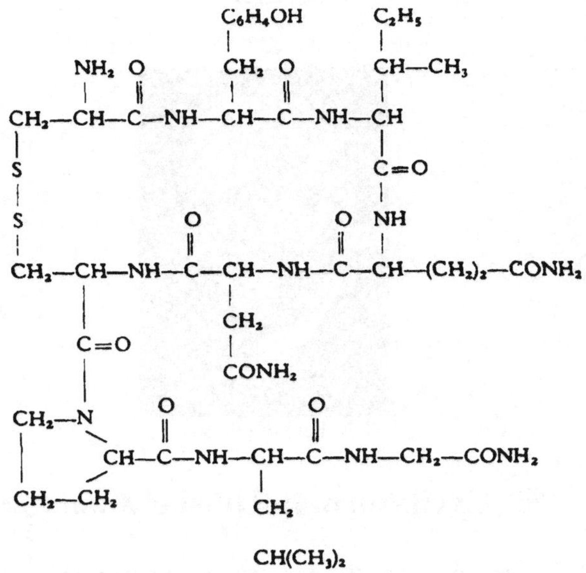

Modern methods of peptide synthesis can be found in [3–5].

Honours and Awards

Du Vigneaud was presented with honorary doctorates from the Universities of New York (1955), Yale (1955) and Illinois at Urbana-Champaign (1960).

He was presented with the: Nobel Prize for Chemistry (1955); Chandler Medal of Columbia University (1955); Willard Gibbs Medal of the American Chemical Society (1956).

He held many Honorary Fellowships including those of the: Royal Society of Edinburgh (elected 2nd July 1951); Chemical Society and Royal Institute of Chemistry, London. He was elected to membership of many scientific academies and institutes. He was also on the Board of Trustees of the National Institute of Arthritis and Metabolic Diseases, and the Rockefeller Institute.

Du Vigneaud held many lectureships in universities in the United States and England, including the Liversidge Lectureship at Cambridge, and was a Visiting Lecturer of the American Swiss Foundation for Scientific Exchange in Switzerland (1947).

Bibliography

1. Reproduced with permission.
2. From Nobel Lectures, Chemistry 1942–1962, Elsevier Publishing Company, Amsterdam, 1964.
3. Amino Acid and Peptide Synthesis (Oxford Chemistry Primers, 7) by John Jones, Oxford University Press, USA, 2nd Ed 2002, ISBN: 0199257388.
4. Chemistry of Peptide Synthesis by N. Leo Benoiton, CRC, 2005, ISBN: 1574444549.
5. Peptide Synthesis And Applications (Methods in Molecular Biology) edited by John Howl, Humana Press, 2005, ISBN: 1588293173.

1956

Sir Cyril Norman Hinshelwood (0.5)

"for their researches into the mechanism of chemical reactions" [2]

During his presentation speech, A. Ölander, member of the Nobel Committee for Chemistry of the Royal Swedish Academy of Sciences, commented of Hinshelwood [2]:

> *...Thirty years ago Hinshelwood studied a number of chemical reactions which allowed him to draw important conclusions concerning the collisions between molecules, which set them in such vibration that they became unstable...*
>
> *...I will first mention the combustion of hydrogen with oxygen. This important reaction was studied both by Hinshelwood and his team in Oxford and by Semenov and his team in Leningrad. Of course also many other scientists have contributed to the final elucidation, but the present prize-winners have indicated the principles guiding the work. Another technically important chain reaction is the combustion of carbon monoxide, not to mention the combustion of hydrocarbons...*
>
> *...Sir Cyril Norman Hinshelwood, Academician Nikolai Nikolaevic Semenov. More than half a century has elapsed since the first Nobel Prize in Chemistry was given to van't Hoff for his discovery of the laws of chemical dynamics. Some of the greatest advances in chemical kinetics since that time have emerged from your researches and they have inspired a great number of scientists to continued fruitful studies. Your results are of equally great importance to technology and to the more theoretical aspects of chemistry...*

Cyril Norman Hinshelwood was born in London on 19th June 1897, the only child of Norman Macmillan Hinshelwood, a chartered accountant, and Ethel, née Smith. Hinshelwood never married, and died in London on 9th October 1967 [2–4].

The Early Years

Hinshelwood's parents moved to Canada when he was a small child. Whilst in Canada, Hinshelwood attended a Montreal kindergarten and, for a short time, a national school in Toronto. He suffered from ill health as a child, and at his father's suggestion, he was brought back to England by his mother. His father died shortly thereafter (1904). His English education started in ca 1904 when he attended Westminster City School. Whilst at Westminster, he won a Brackenbury Scholarship to Balliol College, but the war prevented him from taking it up immediately. During the latter part of the First World War, Hinshelwood worked in the Department of Explosives, Queensferry Royal Ordinance Factory (1916–18), leaving as Deputy Chief Chemist in the main laboratory. He went up to Balliol College, University of Oxford (1919) to do the shortened post-war honours course in chemistry. He gained his Master of Arts degree (1920) with Distinction and was made a Fellow of Balliol College. His Ph.D. degree followed shortly thereafter.

The Career

Year	*Post*	*Institute*
1920	Fellow of Balliol College	University of Oxford
1921	Tutor of Trinity College	University of Oxford
1921 (until 1964)	Fellow of Trinity College	University of Oxford
1937	Dr. Lee's Professor of Chemistry	University of Oxford
1964	Retirement	
1964	Senior Research Fellow	Imperial College of Science, Technology and Medicine

Hinshelwood studied aspects of:

- antibiotics
- the effects of biological responses to environmental changes to bacteria
- chemical changes in the bacterial cell
- heterogeneous gas reactions
- homogeneous gas reactions
- the mechanism of chemical change
- molecular kinetics
- therapeutic agents

He published several books including: *Thermodynamics for Students of Chemistry* (1926); *The Kinetics of Chemical Change* (1st ed 1926, 4th ed 1940); *The Chemical Kinetics of the Bacterial Cell* (1946); *The Structure of Physical Chemistry* (1951).

Hinshelwood's Nobel Prize acceptance speech was dated 11th December 1956, the title of which was 'Chemical kinetics in the past few decades'. Of his early interest in chemical kinetics, Hinshelwood commented [2]:

> *...There are many ways in which the mysteries of chemical change might begin to impress themselves on a young chemist. I will mention one: it belongs to practical life and is a real example.*

Explosives have to be tested for stability and this is done by measuring their decomposition at a suitably high temperature. Phenomena of great interest appear, revealing a remarkably large influence of temperature, a great acceleration of the rate if the solid liquefies, and various striking catalytic effects. Many major problems present themselves, which can be roughly summed up as those involving the dependence of chemical change upon the energy and the environment of the molecules. It was obvious that these problems offered many mysterious and fascinating aspects. In fact they prove to contain more than enough for a lifetime's study...

In summing up the subject, Hinshelwood observed:

...There are many other problems which can be envisaged in terms of the reaction pattern of the cell. Cells contain numerous specific enzymes. The way in which these are built up as the cells multiply has some curious aspects, of which the explanation is certainly linked with the mode of coordination of the reaction pattern. When cells are provided with nutrient they grow and divide. The division seems to be determined by the moment at which a more or less critical amount of deoxyribose nucleic acid has been built up. Here is a fact relating kinetic phenomena closely to structural considerations.

I cannot give you a full and coherent account of these things, but I mention them to show that there are plenty of mysteries left to explore, and also to suggest that kinetics can play a part in explaining some phenomena of that fascinating entity, the living cell. Even if I am wrong about this, it will have been worth the trouble to prove it.

One of my colleagues, Professor G. Temple, has drawn an interesting distinction between the "classical" and the "romantic" in scientific research.

The former is exemplified by the demonstration of clear-cut and unquestionable mathematical propositions, or the devising and application of crucial logical tests of alternatives, the latter by the exploration of the unknown and the formulation of provisional hypotheses from incomplete data which need continual revision as the search proceeds. There is little doubt to which class the study of chemical kinetics belongs...

For further information concerning chemical kinetics, please see [5–7].

Honours and Awards

He received the degree of D.Sc./Sc.D., honoris causa from the Universities of Bristol, Cambridge, City University, Dublin, Hull, Leeds, London, Ottawa, Sheffield, Southampton and Wales; Doctor of Civil Law, honoris causa from Oxford.

Hinshelwood was awarded many Medals and prizes including: Avogadro Medal, Accademia dei Lincei, Rome (1956); Copley medal (1962); Davy Medal (1942); Faraday Medal (1953); Guldberg Medal, Oslo University (1952); Lavoisier medal of the French Chemical Society (1935); Leverhulme Medal, Royal Society (1960); Longstaff Medal, Chemical Society (1948); Nobel Prize for Chemistry jointly with N N Semenov (1956); Royal Medal (1947); Society of Chemical Industry Medal (1961). Also US Medal for Freedom (1947).

Honorary Foreign Fellow or Member of: Academy of Science, USA; Academy of Sciences of the U.S.S.R; Accademia del Lincei; Accademia dei XL; American Academy of Arts and Sciences;

American Philosophical Society; Bologna Academy; Ligurian Academy; Manchester Literary and Philosophical Society; National Academy of Sciences, U.S.A.; New York Academy; Pontifical Academy in Rome; Real Academia de Ciencias, Spain; Royal German Academy; Royal Society of Edinburgh; Spanish Academy of Science. He was also a Member of the Chemical Societies of: Belgium; French; Italy; London and Switzerland; Spanish, Physics and Chemistry Society; Royal Scientific Society, Uppsala. He was a Fellow of the Royal Society.

He was appointed: a Knight Bachelor (1948); a Grand Officer of the Order of Merit, Italy (1956); Order of Merit (1960). He was admitted to freedom and livery of Goldsmiths Company without fine or fees (30th April 1952).

Hinshelwood was: President of the Chemical Society (1946–48), a Vice-President (1940–43, 1948–64), Chairman of the Publications Committee 1944–53); President of the Faraday Society (1961–63); President of Science Masters Association; President of the Royal Society (1955–60), Foreign Secretary (1950–55), Council (1933–34, 1949–50); President of Oxford Branch of Modern Languages Association (1939); Member of Gas Council research committee (1950); President of Classical Association (1959); on Council of Assistants of Goldsmiths Company (1960); Chairman of education committee of Goldsmiths Company (1964); Chairman of the Council of Queen Elizabeth College, London (1964–1967); President of British Association for the Advancement of Science (1964–1965); served on Courtauld Foundation grants committee; Leverhulme Trusts grants committee; served on Standing Committee for Museums and Galleries; served on Advisory Committee of the National Gallery; Trustee of the British Museum (1963). He was an honorary Counsellor of the Higher Council of Scientific Research, Madrid, and a Delegate of Oxford University Press (1934–1967).

His war work resulted in memberships of various boards including: Chemical Defence Board of the Ministry of Supply (1940–1945) - Respirator Sub Committee; Chemical Advisory Board (1946–1948, 1949–1952); Chairman of Defence Equipment Committee (1948–1951); Chairman of Committee on Pyrotechnics, Initiators, Fuses and Incendiaries; Committee of Propellants and Internal Ballistics; Liquid Fuels Committee of the Munitions Chemistry Advisory Board; Fuel Research Council of Department of Scientific and Industrial Research and Coal Research Board; Physical and Inorganic Chemistry Committee of Ministry of Supply.

Hinshelwood presented many Lectures including: Bakerian (1946); Boyle Lecture (1949); Faraday (1952); the Fourth Coal Science Lecture on 'The Carbon of Coal' (1955); the Eddington Memorial Lecture (1961).

Posthoumously he held an exhibition of his oil paintings at Goldsmiths Hall (20th March 1968).

As detailed above, Hinshelwood received a vast array of honours: civil, scientific and academic. He valued them greatly, but tried to adopt a stoical attitude when a thief stole many of his most important medals from his London flat, several years before he died [3].

Hinshelwood's citation for election (date: 2/05/1929) as Fellow of the Royal Society reads [8]:

'Distinguished for his work in Physical Chemistry. By his discovery of a number of homogeneous gas reactions, and by his interpretation of his experimental results he has made important additions to our knowledge of the mechanism of chemical change. He has also worked on surface catalysis, and on the comparison of heterogeneous and homogeneous gas reactions. He has published over 30 papers on these and kindred subjects, and 'The Kinetics of Chemical Change in Gaseous Systems' (Oxford, Clarendon Press)'

He was proposed by:
N. V. Sidgwick; H. Hartley; F. A. Lindemann; T. R. Merton; D. L. Chapman; F. G. Donnan.

Bibliography

1. Reproduced with permission.
2. From Nobel Lectures, Chemistry 1942–1962, Elsevier Publishing Company, Amsterdam, 1964.
3. Biographical Memoirs of Fellows of the Royal Society, 1973, vol. 19, pp. 375–431.
4. Who was who: 1961–1970.
5. Introduction to Chemical Kinetics by Margaret Robson Wright, John Wiley & Sons, 2004, ISBN: 0470090596.
6. Chemical Kinetics and Reaction Dynamics by Paul L. Houston, McGraw-Hill, Science/Engineering/Math, 1st ed. 2001, ISBN: 0072435372.
7. Chemical Kinetics & Catalysis by Richard I. Masel, Wiley-Interscience, 2001, ISBN: 0471241970.
8. Sackler database, Royal Society.

1956

© Nobel Foundation [1]

Nikolay Nikolaevich Semenov (0.5)

"for their researches into the mechanism of chemical reactions" [2]

During his presentation speech, A. Ölander, then a member of the Nobel Committee for Chemistry of the Royal Swedish Academy of Sciences, commented of Semenov [2]:

> *...There are many reactions which do not start at once when the substances have been mixed, but after a while. A number of explosions behave that way and are therefore extremely unpleasant. Semenov showed that the concept of chain reaction could also explain this behaviour.*
>
> *I shall conclude with an example showing how these researches have a direct bearing upon our daily life. I have just mentioned that the combustion of hydrocarbons is a chain reaction. When internal combustion engines were built with higher and higher compression in order to get more energy out of them, the result was that the air-gasoline mixture burned much more rapidly than was appropriate, it exploded. But the rapid combustion could be regulated by using gasoline containing suitable hydrocarbons or with suitable additions. Such gasoline is said to have a high octane number. The vast amount of work carried out in recent years to augment the effect of internal combustion engines in this way is founded, as far as its chemical aspects are concerned, on the fundaments laid down by these prize-winners...*
>
> *...Nikolaevic Semenov. More than half a century has elapsed since the first Nobel Prize in Chemistry was given to van't Hoff for his discovery of the laws of chemical dynamics. Some of the greatest advances in chemical kinetics since that time have emerged from your researches and they have inspired a great number of scientists to continued fruitful studies. Your results are of equally great importance to technology and to the more theoretical aspects of chemistry...*

Nikolai Nikolaevic Semenov was born in Saratov on 3rd April 1896 (some say 16th April 1896), son of Nikolai Aleksandrovitch Semenov and Helena Aleksandrovna, née Dmitrieva. Semenov had a younger sister, Ksenia. He married Natalya Nikolaevna Semenova and they had one son and one daughter. He died on 25th September 1986 [2–4].

The Early Years

A year after Semenov's birth, the family moved to the small town of Voisk, some 60 miles northeast of Saratov, where they spent the winter months and Shiroky Buyerak, where they spent the summer months. The lack of a secondary school in Voisk prompted the family move to Kuybyshev (formally Samara) sometime in the period 1909–1912. Semenov attended the Real school, studying chemistry and physics in particular, as he had developed a friendship with Vladimir Ivanovich Karmilov, his physics teacher. Semenov had a home laboratory at this time and accompanied by his school friends, performed experiments, some of which involved explosives. Semenov enrolled in Petersburg University at the age of 17, graduating with a high honours physics degree in 1917. He moved to the University and Technological Institute at Tomsk shortly after graduation, returning to Petrograd in 1920, where he worked on aspects of electron impact and electrical discharges, whilst taking charge of the electron phenomena laboratory.

The Career

Year	*Post*	*Institute*
1920	Lecturer in the electron phenomena laboratory	Leningrad Physico-Technical Institute
1928	Professor	Leningrad Physico-Technical Institute
1931	Director of the Institute of Chemical Physics	U.S.S.R. Academy of Sciences
1944	Professor	Moscow State University

Semenov was appointed an Academician in 1932.

Semenov's work included:

- the application of the chain theory to various reactions
- a study of combustion processes
- he proposed a theory of degenerate branching
- an analysis of dielectric breakdown
- investigations on electron phenomena
- the propagation of explosive waves
- phenomena associated with the induction periods of oxidation processes
- the mechanism of chemical transformation
- aspects of molecular physics

Semenov wrote two books which were considered seminal works at the time. They were:

Chemical Kinetics and Chain Reactions (published in 1934, English ed 1935), *Some Problems of Chemical Kinetics and Reactivity* (pub 1954, revised in 1958, translated into Chinese, English and German editions).

Semenov's Nobel Prize acceptance speech is dated 11th December 1956, the title of which was 'Some problems relating to chain reactions and to the theory of combustion', during which he commented [2]:

...Our investigations can be divided into two different groups, which are however closely linked with each other:

(1) the application of chemical kinetics to problems relating to the theory of combustion and explosion processes;
(2) the mechanism of chemical reactions, particularly chain reactions

Every combustion or explosion process is above all a chemical reaction between the components of the combustible mixture, accompanied by the liberation of heat and various kinds of motion in gases. For this reason, all phenomena which occur in the combustion process are closely linked with the ideas and laws of chemical kinetics, and above all with the velocity of chemical reaction under non-isothermal conditions.

Chain propagation (a reaction between molecules and radicals) usually occur at a fairly high velocity, at least where the temperature is relatively high. The slowest stage is the initiation of a chain which is often combined with increased activation energy. For this reason, the reaction rate is in fact higher with artificial chain initiation (e.g. in the case of photochemical processes or by the action of additives) than for any molecular reaction. If no such artificial chain initiation occurs, then considerable significance must be attributed to the formation of free radicals by molecules participating in the reaction. The easiest way to form free radicals lies in the decomposition of the weakest bond of the molecule. The extensive investigations into elementary decomposition reaction carried out recently by M. Szwarc and other authors (to a certain extent in our institute also) showed that, in some cases, the decomposition of molecules occurs more readily in radicals than in saturated fresh molecules. It follows from this that thermal decomposition of such substances proceeds by a radical-chain mechanism.

For chain processes, the reaction of free radicals with the molecules is characteristic (the reaction of chain propagation). Such reactions belong to the simplest because here only one bond is disrupted and one bond is generated afresh. Therefore, the investigation of these simple reactions offers the best opportunity for breaking fresh ground in the development of the theory of reactivity and it can be applied subsequently to other types of reaction also.

It is one of the main tasks of chemistry to create a possibility of rationally regulating the rate and direction of a chemical change. Unfortunately, chemistry and the chemicals industry are in this respect lagging behind radio technology, electronics, atomic engineering, etc., where technology is based far more on theoretical anticipation and calculation. The chain reaction theory shows the first possibilities in this respect also. That the chain reaction includes the stages of initiation of the chain and its termination, which are greatly dependent upon the average condition and upon external influences, opens up the possibility of regulating the rate and sometimes even the yield of the process. The knowledge of the reaction of chain propagation often makes it possible to control the process in the desired direction.

Accordingly, the chain theory provides fresh prospects of perfecting established industrial processes and others which are new and in the development stage, and is already to a certain extent being applied; this is particularly true with polymerization processes...

For further information concerning chemical kinetics, please see [5–7].

Honours and Awards

Semenov received Honorary Doctorates (all in science) from the following Universities/Institutes: Oxford (1960); Brussels (1962); Milan (1964); Budapest (1965); London (1965); Prague (1965); Humboldt Universitat, Berlin (1973); Wroclaw (1975).

He was presented with many awards and prizes including: the State Prize (1941 and 1949); seven times various ranks of the Lenin Order culminating with that of Hero (over the period 1945 and 1981); the Nobel Prize for Chemistry, joint (1956); the Order of Labour, Red Banner (1956); the Hero of Socialist Labour (accompanied by the Golden Star, Hammer and Sickle, and Order of Lenin) (1966 and 1976); the Lenin Prize (1976); the Order of the October Revolution (1986).

He became a Corresponding Member of the U.S.S.R. Academy of Sciences (1929) and held many Honorary Fellowships/Memberships of Academies and Learned Societies including: the Chemical Society (1949); the Royal Society (1958); the Indian Academy of Science (1959); the Hungarian Academy of Science (1961); the New York Academy of Sciences (1962); the U.S.

National Academy of Science (1963); the Rumanian Academy of Science (1965); the Czechoslovakian Academy of Science (1965); the Royal Society of Edinburgh (1966); the Akademie der Wissenschaften, Berlin (G.D.R.) (1966); the Bulgarian Academy of Sciences (1969).

Semenov was Chairman of the All-Union Society for Propagation of Political and Scientific Knowledge.

Semenov's citation for election (date: 24/04/1958) as Foreign Member of the Royal Society reads [8]:

> **'Nikolai Semenov of Moscow distinguished for his work on chemical kinetics, especially for his pioneer work on chain reactions'.**

No proposers are listed in the Royal Society database.

Bibliography

1. Reproduced with permission.
2. From Nobel Lectures, Chemistry 1942–1962, Elsevier Publishing Company, Amsterdam, 1964.
3. Biographical Memoirs of Fellows of the Royal Society 1990, vol. 36, pp. 525–546.
4. Who was who: 1981–1990.
5. Introduction to Chemical Kinetics by Margaret Robson Wright, John Wiley & Sons, 2004, ISBN: 0470090596.
6. Chemical Kinetics and Reaction Dynamics by Paul L. Houston, McGraw-Hill, Science/Engineering/Math, 1st ed. 2001, ISBN: 0072435372.
7. Chemical Kinetics & Catalysis by Richard I. Masel, Wiley-Interscience, 2001, ISBN: 0471241970.
8. Sackler database, Royal Society.

1957

© Nobel Foundation [1]

Lord (Alexander R.) Todd

"for his work on nucleotides and nucleotide co-enzymes" [2]

During his presentation speech, Professor A. Fredga, then member of the Nobel Committee for Chemistry of the Royal Swedish Academy of Sciences, commented of Todd [2]:

> *...Through Sir Alexander's work a solid foundation is laid for the future development in this field. Starting from this work, other scientists have advanced very fascinating theories as to the arrangement of the chains; it seems that they may be coiled up as a helix with the bases inside. This model can perhaps explain how a nucleic acid chain can bring about the formation of another similar chain or even of a protein. We are here approaching very fundamental biological questions.*
>
> *The synthetic methods have also been successfully applied to the preparation of low-molecular nucleotide coenzymes, for instance the cozymase, which plays a part in alcoholic fermentation and other biochemical processes. The ways are now open for synthetic preparation of the different types occurring in nature. It is also possible to synthesize coenzymes with slightly modified structure and study the effect of these modifications on the activity, and hence gain better insight into the mode of action of the enzymes.*
>
> *Sir Alexander Todd. Some fifteen years ago you started your work in nucleotide chemistry. You saw the great importance of this topic and you did not underrate the difficulties. Today the chemical structure of these compounds is established - in any case in its outlines - and a solid foundation is laid for future work by biochemists and biologists. Results of utmost interest have already been reported and others will follow...*

Alexander Robertus Todd was born in Cathcart to the south of Glasgow on 2nd October 1907, the elder son of Alexander Todd, a businessman and his wife Jean Ramsay, née Lowrie. Todd married Alison Sarah (1937), daughter of Nobel Prize winner Sir Henry Dale (PRS 1940–1945), and they had a son, Alexander Henry, and two daughters, Helen Jean and Hilary Alison. Alison died in 1987, Todd died on 10th January 1997 [2–4].

The Early Years

He was educated at Allan Glen's School from the age of eleven, and Glasgow University (1924). He won the James Black Medal and the Roger Muirhead Prize during his first year at university, which gave him a scholarship for the remainder of his time as an undergraduate. He graduated with a B.Sc. degree (first class honours) in 1928 and, for a short time, undertook some research under T.S. Patterson with a Carnegie research scholarship which paid £100 per year. He published two papers with Patterson [5–6]. From there he moved to the University of Frankfurt-on-Maine under W. Borsche, obtaining a Ph.D. (Dr.Phil.nat.) in 1931 for a thesis on the chemistry of the bile acids. Todd then enrolled at Oriel College, Oxford University (1931–1934), funded by an 1851 Exhibition Senior Studentship, to work on anthocyanins and other dyes under Sir Robert Robinson (1947 Nobel Chemistry Laureate) graduating with a Ph.D. in 1933. He was subsequently awarded a D.Sc. degree from Glasgow University.

The Career

Year	*Post*	*Institute*
1934	Assistant in Medical Chemistry	Edinburgh University
1935	Beit Memorial Research Fellow	Edinburgh University
1936	Staff member	Lister Institute of Preventive Medicine, Chelsea
1937	Reader in Biochemistry	University of London
1938	Sir Samuel Hall Professor of Chemistry and Director of the Chemical Laboratories	University of Manchester
1944–71	Professor of Organic Chemistry and Fellow of Christ's College	Cambridge University

Todd was elected the 32nd Master of Christ's College, Cambridge University in 1963. He held the post until he was 70 (1978).

Todds research interests included:

- the constituents of cannabis and related plants
- the chemistry of natural products of biological importance
- studies of nucleotide and nucleotide coenzymes
- insect colouring compounds
- factors influencing obligate parasitism
- mould products
- the chemistry of vitamins B_1, E and B_{12},
- nucleoside synthesis
- phosphorylation and nucleotide synthesis

- nucleic acids
- aphid pigments

Todd published his autobiography in 1983 [7].

Todd's Nobel Prize acceptance speech is dated 11th December 1957, the title for which is 'Synthesis in the study of nucleotides'. During the course of this presentation, Todd remarked [2]:

> *...In retrospect, the path which led me to the nucleotides and to the methods of study which I employ is relatively clear. After my undergraduate work in Glasgow, I had my first introduction to natural product chemistry in Frankfurt where I worked for two years with the late Professor Borsche on bile acids. Thereafter at Oxford I had the privilege of working on the anthocyanins and other natural colouring matters with Robert Robinson. It was from him that I learnt that synthesis is not merely a means of confirming a structure arrived at by degradative studies. Synthesis and degradation are complementary tools in structural work, and, properly used, synthesis can serve not only in the elucidation of structure, but can in many instances open up wider vistas regarding the significance and function of biologically important compounds. This view of synthesis in organic chemistry I have held ever since and it has played a major role in determining the pattern of all my subsequent researches...*

> *...I decided that we should seek to clarify the nucleotide field beginning with the simplest units - the nucleosides. To do so we applied primarily the method of synthesis since the amount of preliminary information available from earlier work had at least given sufficient indication of the nature of the nucleosides to make such an attack appropriate. This phase of our work, although providing, I believe, an interesting example of the power of synthetic methods in structural work, would take an entire lecture to describe in itself and I shall not, therefore, discuss it here...*

> *...Today the nucleotides occupy a prominent place in chemical, biochemcial and biological research and new vistas are opening before us which may in a relatively short time lead to a far deeper understanding of the mechanisms of the living cell than seemed possible only a few years ago. And this is surely a matter of profound importance to humanity in its ceaseless struggle against disease. We still have far to go, but if, in a small way, our chemical researches contribute to such an understanding, then my colleagues and I are more than satisfied. I specially include my colleagues in this statement, for the work I have attempted to review would not have been possible without the untiring efforts of a large number of brilliant and enthusiastic students and co-workers, among whom I would mention particularly Lythgoe, Baddiley, Atherton, Kenner, Michelson, Brown, Clark, and Webb. It has been a privilege to be associated with them, and to all of them I owe a deep debt of gratitude...*

The synthesis of nucleotides is now established as a field of biochemistry in its own right. Lord Todd was a pioneer in this endeavour, indeed there are many literature examples of modern biochemists following his example, see for example [8–10].

Honours and Awards

Todd was presented with over 30 honorary doctorates from many Universities including: Hon.D.Sc. London (1958), Madrid (1959), Aligarh (1960), Exeter (1960), Leicester (1960), Wales (1961), Sheffield (1961) and Yale (1961); Dr.rer.nat. Kiel and Hon.LL.D. Glasgow and Melbourne in (1960).

He was awarded many medals and prizes including: the Cannizzaro Medal of the Italian Chemical Society; the Copley Medal (1970); the Davy Medal (1949); the Lavoisier Medal of the French Chemical Society; the Meldola Medal of the Royal Institute of Chemistry and the Society of Maccabeans; the Royal Medal of the Royal Society (1955); the Nobel Prize for chemistry (1957); the Paul Karrer Medal (1962); the Stas Medal (1962); the Longstaff Medal (1963); the Copernicus Medal (1979); the Lomonosov Medal (1979); the Hanbury Medal (1986).

Todd was elected Hon. Member, New York Academy of Sciences (1959) and Hon. Fellow of the Royal Australian Chemical Institute (1960). He was also a Honorary Member of the French, German and Spanish chemical societies. He was a Fellow of the Royal Society (President 1975–1980) and a member of the Deutsche Akad. Naturforscher Leopoldina, Halle, (1959). He was a Foreign Member of: the American Academy of Arts and Sciences; the Austrian Academy of Sciences; the National Academy of Sciences; the Spanish Council of Scientific Investigation.

Todd was Bakerian Lecturer of the Society of Chemical Industry (1954), Pedler and Tilden Lecturer Chemical Society and Tercentenary Lecturer of the Royal Society (1960).

He was Visiting Professor at: the California Institute of Technology (1938); the University of Chicago (1948); Sidney University (1950); Massachusetts Institute of Technology (1954); the University of California (1957).

Todd was: Chairman of the British Government's Advisory Council on Scientific Policy (1952); President of the Chemical Society, London (1960–1962); Master of the Worshipful Company of Salters (1961–1962); President of the International Union of Pure and Applied Chemistry; Chairman of the British National Committee for Chemistry (1963–65); a Director of Fisons Ltd (1963–78) and contributed to the Trend Report as a member of the Committee of Enquiry into the Organisation of Civil Science (1961); Chair of the Royal Commission on Medical Education (1965–1968); President of the Society of Chemical Industry (1981–82); President of the British Association for the Advancement of Science (1970); President of the Royal Society (1975–1980); one of the ten Founding Trustees (and later Honorary Fellow) of Churchill College; Chancellor of Strathclyde University (1965–91); Managing Trustee of the Nuffield Foundation (in 1950) – Deputy Chairman and Chairman until 1979 and then Chairman of the Ordinary Trustees (until his death); Todd held a great many evinent positions including: Chairman of the Croucher Foundation (based in Hong Kong).

He was knighted in 1954 (some say 1952), raised to the peerage (March, 1962) being created Baron Todd of Trumpington and was awarded the Order of Merit (1977). He was also presented with the Order of the Rising Sun (1978).

Todd's citation for election (date: 19/03/1942) as Fellow of the Royal Society reads [11]:

> **'Professor of Organic Chemistry. Distinguished as an organic chemist and biochemist. Early work in the cholic acid series was followed by the synthesis (with Robertson) of the chief diglucosidic anthocyanins and later (with Taistrick and Robinson) the constitutions of some metabolic products of species of Helminthosporium were determined, partly by synthesis. He has developed remarkable new syntheses of vitamin B1 and established the structure of thiochrome. Equally significant has been his contribution to the chemistry of vitamin E and the vitamin B1complex especially the liver filtrate factor. He has established the constitution of constituents of hashish and has synthesized reference compounds.'**

He was proposed by:

Robert Robinson; J. C. G. Ledingham; I. M. Heilbron; S. Sugden; Warrington Yorke; Harold King; N. V. Sidgwick; C. N. Hinshelwood; E. J. Bowen; Joseph Barcroft; Harold Hartley; J. Kenner; J. L. Simonsen; F. L. Pyman.

Bibliography

1. Reproduced with permission.
2. From Nobel Lectures, Chemistry 1942–1962, Elsevier Publishing Company, Amsterdam, 1964.
3. Biographical Memoirs of Fellows of the Royal Society, 2000, vol. 46, pp. 515–532.
4. Who was who: (1996–2000).
5. The action of phosphorus pentachloride on ethyl tartarate Barton and Patterson, J. Chem. Soc. 1929, 1768.
6. The influence of solvents and of other factors on the rotation of optically active compounds part 28. The rotation dispersion of mannitol and some if its derivatives. Barton and Patterson, J. Chem. Soc, 1929, 2876.
7. A time to remember: the autobiography of a chemist, Cambridge University Press, 1983, ISBN: 0 521 25593 7.
8. Organic Chemistry of Enzyme-Catalyzed Reactions by Richard B. Silverman, Academic Press, 2nd ed. 2002, ISBN: 0126437319.
9. Evaluation of Enzyme Inhibitors in Drug Discovery: A Guide for Medicinal Chemists and Pharmacologists (Methods of Biochemical Analysis) by Robert A. Copeland, Wiley-Interscience, 2005, ISBN: 0471686964.
10. Reagents for Glycoside, Nucleotide, and Peptide Synthesis (Hdbk of Reagents for Organic Synthesis) edited by David Crich, John Wiley & Sons, 2005, ISBN: 047002304X.
11. Sackler database, Royal Society.

1958

© Nobel Foundation [1]

Frederick Sanger

"for his work on the structure of proteins, especially that of insulin" [2]

Sanger is the only chemist to be awarded the Nobel Prize for chemistry twice. For that reason the entry here deals with his career up to the award in 1958. The entry for his second award, shared with Paul Berg (0.5) and Walter Gilbert (0.25) in 1980, contains further information.

During his presentation speech, Professor A. Tiselius, then a member of the Nobel Committee for Chemistry of the Royal Swedish Academy of Sciences, commented of Sanger [2]:

> *...Thus, Sanger could give the exact sequence of the 31 amino acids in one chain and the 20 in the other. Already earlier he had shown that the two chains are held together to form an insulin molecule with the aid of two bridges of sulphur atoms. The exact positions of these bridges were determined in a way similar to that used for determination of the structure of the chains.*
>
> *The structure of insulin had thus been established - a remarkable achievement, indeed. Insulin is a protein, and this was the first time one had succeeded in determining the structure of a substance belonging to this very important group.*
>
> *The significance of Sanger's work reaches, however, much further. The procedure he has used so successfully can be applied to proteins in general, in attempts to determine their structure. Already now, many research workers are engaged in such investigations, and important results appear to be on the way in the exploration of how proteins play their role as key substances in the chemical processes of Life.*
>
> *Doctor Frederick Sanger. It sometimes happens that an important scientific discovery is made so to say "overnight" - if the time is ripe and the necessary background is there. Yours is not of that kind. The first successful determination of the structure of a protein is the result of many years of persistent and zealous work, in which the final solution of the problem has been approached step by*

step. You knew when you began to look into the structure of the insulin molecule 15 years ago that the problem was a formidable one. So did the whole scientific world. Those who knew you, were confident, however, that you would ultimately succeed...

Frederick Sanger was born on 13th August 1918 in Rendcombe, Gloucestershire, the second son of Frederick Sanger M.D., and his wife Cicely. Sanger married Margaret Joan Howe (1940). They have two sons and one daughter, Robin (1943), Peter (1946) and Sally Joan (1960). He is retired and lives in Cambridgeshire, England [2,3].

The Early Years

Sanger was educated at Bryanston School and at St. John's College, Cambridge, graduating with a B.A. degree in natural sciences (1939). He worked under Dr. A. Neuberger on the metabolism of the amino acid lysine (1940–1943) graduating with a Ph.D. degree in 1943. Sanger has carried out research in the Department of Biochemistry at Cambridge University since 1940.

The Career

Year	*Post*	*Institute*
1944	Beit Memorial Fellowship for Medical Research	Department of Biochemistry at Cambridge
1951	Member of the External Staff	Medical Research Council
1961–1958	Head of the Division of Protein Chemistry	M.R.C. Laboratory for Molecular Biology at Cambridge

Since the completion of his Ph.D. in 1943, Sanger has concentrated his efforts on the structure of proteins leading to the determination of the structure of insulin.

Sanger's 1958 Nobel Prize acceptance speech is dated 11th December 1958 with the title 'The chemistry of Insulin'. During his speech, Sanger commented [2]:

...Since the work on insulin has extended over about 12 years it will be necessary to give a somewhat simplified account and to omit most of the work that did not contribute directly to the main problem, the determination of the structure of a protein.

In 1943 the basic principles of protein chemistry were firmly established. It was known that all proteins were built up from amino acid residues bound together by peptide bonds to form long polypeptide chains. Twenty different amino acids are found in most mammalian proteins and by analytical procedures it was possible to say with reasonable accuracy how many residues of each one was present in a given protein. Practically nothing, however, was known about the relative order in which these residues were arranged in the molecules. This order seemed to be of particular importance, since although all proteins contained approximately the same amino acids they differed markedly in both physical and biological properties. It was thus concluded that these differences were dependent on the different arrangement of the amino-acid residues in the molecules. Although very little was known about amino acid sequence, there was much speculation in this field. The most widely discussed theory was that of Bergmann and Niemann who suggested that the amino acids were arranged in a periodic fashion, the residues of one type of amino acid occurring at regular intervals along the chain. On the other extreme there were those who suggested that a pure protein was not a chemical individual in the classical sense but consisted of a random mixture of similar individuals...

...The determination of the structure of insulin clearly opens up the way to similar studies on other proteins and already such studies are going on in a number of laboratories. These studies are aimed at determining the exact chemical structure of the many proteins that go to make up living matter and hence at understanding how these proteins perform their specific functions on which the processes of Life depend. One may also hope that studies of proteins may reveal changes that take place in disease and that our efforts may be of more practical use to humanity...

The structure of insulin is:

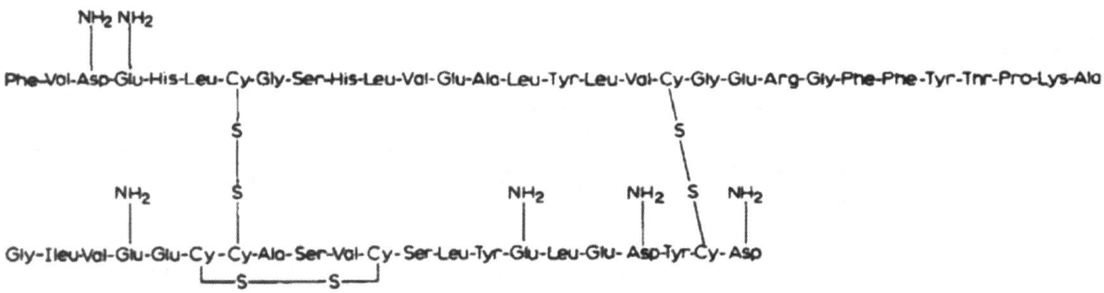

(taken from the 1958 acceptance speech)

Pioneering work of this kind always stands the test of time. Protein structure determinations are now a matter of routine, whereas in the 1950s, that was far from the case. The selected references [4–6] are provided as examples of the depth now associated with protein structure determination.

Honours and Awards

Sanger was awarded: the Corday-Morgan Medal and Prize of the Chemical Society (1951); the Nobel Prize for Chemistry (1958); the Nobel Prize for Chemistry, jointly (1980).

He is: an Honorary Foreign Member of the American Academy of Arts and Sciences; an Honorary Member of the American Society of Biological Chemists; a Member of the Academies of Science of Argentina and Brazil; an Honorary Member of the Japanese Biochemical Society; a Corresponding Member of the Association Qulmica Argentina. He was elected a Fellow of the Royal Society (1954) and a Fellow of King's College, Cambridge (1954).

Sanger's citation and proposers for election to the Royal Society do not appear on their Sackler database.

Bibliography

1. Reproduced with permission.
2. From Nobel Lectures, Chemistry 1942–1962, Elsevier Publishing Company, Amsterdam, 1964.
3. Royal Society Yearbook 2005.
4. Insulin and Related Proteins – Structure to Function and Pharmacology edited by Matthias Federwisch, Markus Leyck Dieken and Pierre De Meyts, Springer, 1st ed. 2002, ISBN: 1402006551.
5. Cytochrome P450: Structure, Mechanism, and Biochemistry by Paul R. Ortiz de Montellano, Springer, 3rd ed. 2004, ISBN: 0306483246.
6. Introduction to Protein Structure by C. Branden, Garland, 2nd ed. 1999, ISBN: 0815323050.

1959

© Nobel Foundation [1]

Jaroslav Heyrovsky

"for his discovery and development of the polarographic methods of analysis" [2]

During his presentation speech A. Ölander, then a member of the Nobel Committee for Chemistry of the Royal Swedish Academy of Sciences, commented of Heyrovski [2]:

...The striving of analysts is not only to develop methods yielding accurate results, but even more important for practical work is that the analysis can be carried out rapidly, that it can be done using as small samples as possible, and that very small percentages of various substances can be detected and ascertained.

Polarography is one of these micro-methods which are available to the modern analyst...

...Professor Heyrovsky. You are the originator of one of the most important methods of contemporary chemical analysis. Your instrument is extremely simple, only falling droplets of mercury, but you and your collaborators have shown that it can be used for the most diverse purposes.

Several years elapsed before the polarographic method was noticed outside your own country. But after that, its significance has ever increased, not suddenly, attracting the attention of outsiders, but steadily your method has won the confidence of analytical chemists...

Jaroslav Heyrovsky was born in Prague, Czech Republic on 20th December 1890, the fifth child of Professor Leopold Heyrovsky and his wife Clara, née Hanlová. He married Marie Koranová (1926) and they had a daughter Jitka (Czech for Judith), and a son, Michael. Heyrovsky died in Prague on 27th March 1967 [2–4].

The Early Years

Heyrovsky obtained his early education at the Akademicke Gymnasium (1901–1909), where his favourite subjects were mathematics and physics. He then enrolled in the Phylosophical Faculty of the Czech University, Prague where he read chemistry, physics and mathematics. His strict father gave permission for the young Heyrovsky to study in London, thus Heyrovsky attended the lectures of Sir William Ramsay and others at University College, London (1910 onwards). He graduated with a B.Sc. degree from the University of London (1913) and shortly thereafter was appointed Demonstrator for the year 1913–1914. War broke out during a visit to his parents, preventing Heyrovsky from returning to England. J. S. Stérba-Bohm made facilities available at the University of Prague, allowing Heyrovsky to work on his doctorate at the Chemical Institute.

He was called up for Military Service (1915) in the Austro-Hungarian Army but found to have a weak physical condition, accordingly, he was posted to a military hospital, where he became a dispensing chemist and radiologist for the remainder of the war. There was sufficient time to continue his research and perform his military duties as he found the time to submit a thesis with the title *'The electro-affinity of aluminium'* to the Philosophical Faculty of the Czech University in Prague. He was awarded the Ph.D. on 26th September 1918. The University of London awarded him a D.Sc. degree in 1921.

The Career

Year	*Post*	*Institute*
1920	Assistant to Professor B. Brauner	Institute of Analytical Chemistry of the Charles University, Prague
1922	Lecturer	Institute of Analytical Chemistry of the Charles University, Prague
1924	Assistant Professor	Institute of Analytical Chemistry of the Charles University, Prague
1926	the first Professor of Physical Chemistry	Institute of Analytical Chemistry of the Charles University, Prague
1950	Director	Polarographic Institute[a]

[a] subsequently incorporated into the Czechoslovak Academy of Sciences (1952).

Heyrovsky was given a Rockerfeller Fellowship in 1926 allowing Heyrovsky to travel to the Sorbonne in Paris and work with G. Urbain.

Heyrovsky's work included:

- the determination of the electrode potential of aluminium
- studies of electrochemistry
- the invention of polarography
- the study and interpretation of electrocapillary curves

Heyrovsky published 193 papers, a comprehensive list for which appears in one of the references [3].

Heyrovsky's Nobel Prize acceptance speech is dated 11th December 1959 with the title 'The trends in polarography', during which he commented [2]:

> *...The reason why I keep some 38 years to the electrochemical researches with the dropping mercury electrode is its exquisite property as electrode material. Its physical conditions of dropping*

as well as the chemical changes during the passage of the electric current are well defined, and the phenomena displayed at the dropping mercury electrode proceed with strict reproducibility. Owing to the latter property the processes at the electrode can be exactly expressed mathematically. According to the registering apparatus called "polarograph", which draws automatically curves characteristic of the electrode processes, the electrochemical studies with the dropping mercury electrode and the analytical methods developed on these investigations have called "polarography"...

...Contrary to the kinetic currents very much higher are the polarographic currents due to catalytic processes in the evolution of hydrogen, especially those due to SH groups. For example, cystine and proteins in NH_3 buffers containing cobalt ions give a 500 to 700 times higher "catalytic" wave as cystine without the catalyst. Of these reactions the most noteworthy is the serological reaction of Brdička, to which many publications relate. This is characterized by the "protein double wave" which has been found to depend on the pathological state of the patient. All inflammations cause an increase of the protein wave. In non-feverish state it is cancer which is invariably manifested by an elevated protein wave. After the removal of the cancerous centre, the reaction becomes normal; metastases elevate it. It has been ascertained that the proteins responsible for the cancer reaction are mucoproteins, belonging to the γ-2-globulins. Electrophoresis shows that the mucoproteins are akin to the serum component MP-2. On the other hand, liver diseases lower the protein wave, so that in liver troubles the polarographic test is specific...

...All the above-mentioned polarographic methods are used for analytical determinations, qualitative and quantitative, in micro-doses as well as macro-doses of inorganic or organic depolarizers. If the substance to be determined is not a depolarizer, it may be chemically converted into a depolarizer and so determined. For example benzene is no depolarizer, its traces in air are nox and its microdetermination wanted. The smallest traces of benzene are easily converted by nitration into nitrobenzene and after alkalization exactly determined in expired air, in blood or urine...

Further data concerning polorography and its applications may be obtained from [5,6,7].

Honours and Awards

Heyrovsky received honorary doctorates from the following Universities: Aix-Marseille (1959); Technical University, Dresden (1955); Frankfurt am Main; Paris (1960); Prague (1966); Vienna and Warsaw (1956). Ill health prevented Heyrovsky from attending the ceremony in Frankfurt so the University sent a delegation to Prague where the ceremony took place in 1966.

He was presented with many awards and prizes including: the State Prize of Czechoslovakia, First Grade (1951); the Order of the Czechoslovak Republic (1955); the Nobel Prize for Chemistry (1959); the Golden Plaquet (1962).

Heyrovsky was an Honorary Fellow or Member of: the Societies for Analytical Chemistry India and London; the American Academy of Arts and Sciences, Boston, Mass (1933); the Chemical Societies of Austria, Czechoslovakia, England, India and Poland; the Hungarian Academy of Sciences (1955); the Indian Academy of Sciences, Bangalore (1955); the International Society for Research on Nutrition and Vital Substances; the Polarographic Society of Australia, Japan and London; the Polish Academy of Sciences, Warsaw (1962); the Verein Osterreichischer Chemiker. He was elected a Foreign

Member of the Royal Danish Academy of Sciences (1962) and the Royal Society (1965). He was also a Corresponding Member of the German Academy of Sciences (1955).

He was a Member of the Czechoslovak Academy of Sciences; the Danish Academy of Sciences; the Deutsche Akademie der Wissenschaften (Berlin); the Leopoldina (Halle-Saale) (1956); the German Academy of Natural Scientists; the Polish Academy of Sciences; the Academy of Sciences of the USSR.

Heyrovsky was also: Vice-President of the International Union of Physics (1951 to 1957); President and first Honorary Member of the Polarographic Society, London; a Fellow of the University College London (1927); a Founding Member of the University of Peace, Huy (Belgium).

He was Carnegie Visiting Professor (1933) and lectured in: the People's Republic of China (1958); England (1946); Sweden (1947); in U.A.R. (Egypt) (1960 and 1961); the United States of America (1933); the USSR (1934).

Heyrovsky's citation for election (date: 08/04/1965) as Foreign Member of the Royal Society reads [8].

> **'Professor Jaroslav Heyrovsky of Prague distinguished as the founder of the science of polarography and for his many contributions to electrochemistry.'**

No proposers are listed in the Sackler database of the Royal Society.

Bibliography

1. Reproduced with permission.
2. From Nobel Lectures, Chemistry 1942–1962, Elsevier Publishing Company, Amsterdam, 1964.
3. Biographical Memoirs of Fellows of the Royal Society, 1967, vol. 13, pp. 167–191.
4. Who was who: 1961–1970.
5. Polarography and Other Voltammetric Methods (Analytical Chemistry By Open Learning) by Tom Riley, Arthur Watson, Arthur M. James, John Wiley & Sons, 1987, ISBN: 0-471-91395-2.
6. Principles of Polarography by Ramesh Chandra Kapoor, B. S. Aggarwal, John Wiley & Sons, 1991, ISBN: 0470217324.
7. Polarography and Allied Techniques by V.S. Rao, Sangam Books Ltd., 2002, ISBN: 8173714029.
8. Sackler database, Royal Society.

1960

© Nobel Foundation [1]

Willard Frank Libby

"for his method to use carbon-14 for age determination in archaeology, geology" [2]

During his presentation speech A. Westgren, then Chairman of the Nobel Committee for Chemistry of the Royal Swedish Academy of Sciences, commented thus on the issue of carbon dating [2]:

> *...The Royal Swedish Academy of Sciences this year has decided to award with the Nobel Prize for Chemistry a scientific feat which is less a direct improvement of our material living standards, but more a means of widening and deepening our knowledge in different scientific fields. Professor Willard Libby has been selected to be the prize winner for his method of age determination of materials of biological origin by use of carbon-14 as a measurer of time. His method has obtained widespread use and has become indispensable in archaeology, geology, geophysics and other sciences. Fortunately, it is so simple - which is probably not always the case with chemical research distinguished with the Nobel Prize - that everyone should be able to understand the conditions and principles for its execution.*
>
> *Carbon-14 is a kind of carbon, an isotope of carbon with an atomic weight of 14, which is found in the carbon dioxide of the air. It is formed high up in the atmosphere by cosmic radiation coming from outer space. How that occurs, we can ignore. The newly formed carbon-14 has high energy at the moment of its formation, so that it rapidly oxidizes to carbon dioxide, which spreads out and distributes itself evenly in the atmosphere.*
>
> *The ratio of carbon-14 in the carbon dioxide of the atmosphere is very low. In about one million million of these carbon atoms, there is only one which has an atomic weight of 14. But nevertheless, this ratio can be determined, for carbon-14 is a radioactive isotope and manifests itself by its radiation. It is converted into nitrogen by the emission of an electron which can be detected by a sensitive apparatus. The disintegration is such a slow procedure, however, that 5,600 years are required to convert half of these atoms into nitrogen. After another 5,600 years, there is still one*

quarter left, and after an equally long period of time one eighth, etc. Carbon-14 is thus said to have a half - life of 5,600 years...

Willard Frank Libby was born in Grand Valley, Colorado on 17th December 1908, the son of Ora Edward Libby and his wife Eva May, née Rivers. He married Leonor, née Hickey of King City, California (1940, marriage dissolved 1966). They had twin daughters Janet and Susan (1945). Libby married Leona Marshall (1966). He died on 8th September 1980 [2,3].

The Early Years

Libby attended grammar and high schools near Sebastopol, California (1913 to 1926) prior to enrolling at the University of California in 1927. He graduated with a B.Sc. in 1931 and a Ph.D. degree in 1933.

The Career

Although Libby was awarded a Guggenheim Memorial Foundation Fellowship (1941) and made arrangements to work at Princeton University but his plans were interrupted by the outbreak of the Second World War. Libby joined the Manhattan District Project at Columbia University taking leave of absence from the Department of Chemistry, California University, till 1945.

Year	*Post*	*Institute*
1933	Instructor in the Department of Chemistry	California University (Berkeley)
1938	Assistant Professor of Chemistry	California University (Berkeley)
1945	Associate Professor of Chemistry	California University (Berkeley)
1945	Professor of Chemistry in the Department of Chemistry and Institute for Nuclear Studies (now the Enrico Fermi Institute for Nuclear Studies)	Chicago University
1954	Member	US Atomic Energy Commission
1959	Professor of Chemistry	University of California at Los Angeles
1962	Director of the Institute of Geophysics and Planetary Physics	University of California at Los Angeles

Libby's carrer has involved studies of:

- carbon-14 (radiocarbon) and its use in dating archaeological artifacts
- hot atom chemistry
- isotope tracer studies
- radiochemistry
- tracer techniques
- natural tritium
- the use of tritium in geophysics and hydrology

Libby published a book concerning carbon dating called *Radiocarbon Dating* which was published by the University of Chicago Press (1952). A 2nd edition appeared in 1955.

He served as a member of the Editorial Board of the *Proceedings of the National Academy of Sciences* (from 1960) and a member of the Editorial Board of *Science* (from 1962) and of *Space Life Science* (from 1970).

Libby's Nobel Prize acceptance speech is dated 12th December 1960 and titled 'Radiocarbon dating'. In his opening remarks, Libby commented [2]:

> *...Radiocarbon dating had its origin in a study of the possible effects that cosmic rays might have on the earth and on the earth's atmosphere. We were interested in testing whether any of the various effects which might be predicted could actually he found and used. Initially the problem seemed rather difficult, for ignorance of billion-electron-volt nuclear physics (cosmic ray energies are in this range) was so abysmal at the time and incidentally fourteen years later still is so abysmal, that it is nearly impossible to predict with any certainty the effects of the collisions of the multibillion-volt primary cosmic radiation with air.*
>
> *However, in 1939, just before the war, Professor Serge Korff of New York University and others discovered that the cosmic rays produce secondary neutrons in their initial collisions with the top of the atmosphere. The neutrons were found by sending counters, designed to be sensitive to neutrons, up to high altitudes and they were found to have an intensity which corresponded to about two neutrons being generated for each square centimeter of the earth's surface per second. Whereas it was extremely difficult to predict the types of nuclei that might be produced by the billion-volt primary cosmic rays, the neutrons being secondaries were in the million-volt energy range and therefore subject to laboratory tests. So at this point the question was "What will million-electron-volt neutrons do if liberated in the air?" The answer to this question was already available - in fact, Professor Korff noted in one of the papers announcing the discovery of the neutrons that the principal way in which the neutrons would disappear would be to form radiocarbon...*

Libby went on to explain how this phenomenon can be used in radiocarbon dating:

> *...We are in the radiocarbon-dating business as soon as this has been said, for it is clear from the set of assumptions that have been given that organic matter, while it is alive, is in equilibrium with the cosmic radiation; that is, all the radiocarbon atoms which disintegrate in our bodies are replaced by the carbon-14 contained in the food we eat, so that while we are alive we are part of a great pool which contains the cosmic-ray produced radiocarbon. The specific activity is maintained at the level of about 14 disintegrations per minute per gram by the mixing action of the biosphere and hydrosphere. We assimilate cosmic-ray produced carbon-14 atoms at just the rate that the carbon-14 atoms in our bodies disappear to form nitrogen-14. At the time of death, however, the assimilation process stops abruptly. There is no longer any process by which the carbon-14 from the atmosphere can enter our bodies. Therefore, at the time of death the radioactive disintegration process takes over in an uncompensated manner and, according to the law of radioactive decay, after 5,600 years the carbon that is in our bodies while we are alive will show half the specific carbon-14 radioactivity that it shows now. Since we have evidence that this has been true for tens of thousands of years, we should expect to find that a body 5,600 years old would be one-half as radioactive as a present-day living organism. This appears to be true. Measurements of old artifacts of historically known age have shown this to be so within the experimental errors of measurement...*

Carbon dating is an analytic technique that continues to be used today for certain applications, perhaps the most well known in recent times is that of dating the Turin Shroud. Further texts concerning this technique and this application can be found in the references [4,5].

Honours and Awards

Libby was presented with many awards and prizes including: the Research Corporation Award (1951); the Chandler Medal of Columbia University (1954); the American Chemical Society Award for Nuclear Applications in Chemistry (1956); the Elliott Cresson Medal of the Franklin Institute (1957); the Willard Gibbs Medal Award (1958); the Albert Einstein Medal Award (1959); the Nobel Prize in Chemistry (1960); the Day Medal of the Geological Society of America (1961).

Libby was elected to Fellowships and Memberships of numerous learned societies in the United States including the American Academy of Arts and Sciences, and was also Member of the Bolivian Society of Anthropology and the Heidelberg Academy of Sciences. He was a Foreign Member of the Royal Swedish Academy of Sciences (1960).

Libby served on many US agencies/committees including: Member of the Committee of Senior Reviewers of the Atomic Energy Commission (1945 to 1952); Member of the Commission's General Advisory Committee (1950 to 1954 reappointed June 1960); Member of the Advisory Board of the Guggenheim Memorial Foundation (re-elected in May, 1960, for a further term of four years; Member of the Plowshare Advisory Committee from 1959; Member of the National Science Foundation's General Commission on Science, the Federal Government and the Academic Institution; Member of the Advisory Council to the Coordinator of Atomic Activities of the State of California (from 1959); Member of the Air Resources Board, California (1967–72); Member of the President's Task Force on Air Pollution (1969–70).

Libby was Visiting Professor at the University of Colorado (1967–70) and of the University of S. Florida (1972).

He was: a Guggenheim Memorial Foundation Fellow (1941, 1951 and 1959–1962); a Research Associate of the Carnegie Institute of Washington Geophysical Laboratory (1954 to 1959); Consultant to the Office of Civil and Defence Mobilization (1959–1961); a Director of the Douglas Aircraft Company (1963).

Bibliography

1. Reproduced with permission.
2. From Nobel Lectures, Chemistry 1942–1962, Elsevier Publishing Company, Amsterdam, 1964.
3. Who was who 1971–1980.
4. Nuclear and Chemical Dating Techniques – Interpreting the Environmental Record by Lloyd A Currie, American Chemical Society, 1982, ISBN: 0-8412-0669-4.
5. Shroud, Carbon Dating and Calculus of Probabilities by Bruno Barberis, Pietro Savarino, St Pauls Publishing, 1998, ISBN: 0854395377.

1961

© Nobel Foundation [1]

Melvin Calvin

"for his research on the carbon dioxide assimilation in plants" [2]

During his presentation speech K. Myrbäck, then a member of the Nobel Committee for Chemistry of the Royal Swedish Academy of Sciences, commented of Calvin [2]:

> *...A question that had occupied scientists for more than a century, was: "What is the primary product of the assimilation; what first happens to the carbon dioxide taken up by the plant?" Calvin demonstrated that the primary reaction is not, as had been assumed previously, a reduction of carbon dioxide as such, but a fixation of carbon dioxide to a substance, the carbon dioxide acceptor, occurring in the plant. Calvin was able to show that the product formed in this fixation reaction is a organic compound known as phosphoglyceric acid.*
>
> *This discovery was of fundamental importance for the development that followed. The primary product of assimilation was recognized as being a compound, well-known from earlier work as an intermediary product of the biological degradation of carbohydrates, and not some previously unknown compound; phosphoglyceric acid had been identified as a breakdown product of sugar as early as 1929 by Ragnar Nilsson here in Stockholm. Calvin's identification of the primary assimilation product with phosphoglyceric acid led to the very important conclusion that there is an intimate connection between photosynthesis and carbohydrate metabolism as a whole.*
>
> *Calvin's subsequent investigations mapped out the path between the primary product and the end products of assimilation, the various carbohydrates. What had formerly been assumed to be a reduction of carbon dioxide was shown to be a reduction of phosphoglyceric acid.*
>
> *Professor Melvin Calvin. Your investigations on plant photosynthesis have shed light on a field of biochemistry which was, until recently, veiled in obscurity. You have tracked the various steps of the path of carbon in photosynthesis and created a clear picture of this complicated sequence of reactions, reactions of immense importance for life on our planet...*

Melvin Calvin was born in St. Paul, Minnesota on the 8th April 1911 the son of Russian emigrant parents Rose and Elias Calvin. He married Genevieve, née Jemtegaard, (1942) and they had two daughters, Elin and Karole, and one son, Noel. Calvin died on 8th January 1997 (his wife died in 1987) [2–4].

The Early Years

Calvin graduated with a B.Sc. chemistry degree from the Michigan College of Mining and Technology (now Michigan Technological University) in 1931. His doctoral studies were under Professor George A. Glocker at the University of Minnesota, where he studied the electron affinity of halogens and graduated with a Ph.D. in 1935.

Calvin then moved to the University of Manchester in England to work under Professor Michael Polanyi (1935–1937), where he worked on coordination catalysis, particularly metalloporphyrins. His academic career started on his return to the United States (in 1937).

The Career

Year	*Post*	*Institute*
1937	Instructor	University of California at Berkeley
1941	Assistant Professor	University of California at Berkeley
1945–47	Associate Professor	University of California at Berkeley
1946	Director of the big-organic chemistry group whilst retaining his posts within the University of California	Lawrence Radiation Laboratory renamed Laboratory of Chemical Biodynamics (in 1960)
1947	Professor	University of California at Berkeley
1963–80	Professor of Molecular Biology	University of California at Berkeley
1967–80	Associate Director	Lawrence Radiation Laboratory
1971	University Professor of Chemistry	University of California at Berkeley

Calvin's research interests were broad. He worked on many issues including work concerning:

- cancer
- development of techniques for the use of carbon-14
- catalytic behaviour of coordination compounds
- chemical methods for extracting oxygen from air
- colour
- electron affinity of halogen atoms
- ligand complexing of metal ions
- oil from plants
- organic compounds in meteorites and moon rocks
- organic evolution
- origin of life
- paramagnetism of the triplet state

- photosynthesis
- photosynthetic carbon dioxide reduction
- the Rh negative factor

His work for NASA included:

- plans to protect the moon against biological contamination from the earth in the first lunar (Apollo) landings
- procedures to protect the earth from possible lunar pathogens on and in the returning Apollo spacecraft
- strategies for the search for organic and biological compounds in returned lunar samples
- plans for the search for biological compounds for life on other planets

His publications totalled nearly six hundred papers and seven books, one of which was '*Chemical evolution; molecular evolution towards the origin of living systems on the earth and elsewhere*'. (Hardcover Publisher: Oxford University Press, New York, Pub. (1969)). He published his autobiography in 1998 [5].

Calvin had a laboratory built for him at the Berkeley campus, now called the Calvin Laboratory (1964).

He was US Chairman of the Joint US - USSR Editorial Board for the four-volume series entitled '*Foundations of Space Biology and Medicine*'.

His Nobel Prize acceptance speech was dated 11th December 1961 and titled 'The path of carbon in photosynthesis'. In setting the scene Calvin commented [2]:

> *...The speculations on the way in which carbohydrate was built from carbon dioxide began not long after the recognition of the basic reaction and were carried forward first by Justus von Liebig and then by Adolf von Baeyer and, finally, by Richard Willstätter and Arthur Stall into this century. Actually, the route by which animal organisms performed the reverse reaction, that is, the combustion of carbohydrate to carbon dioxide and water with the utilization of the energy resulting from this combination, turned out to be the first one to be successfully mapped, primarily by Otto Meyerhof and Hans Krebs...*

He went on to cover various aspects of the elucidation of photosynthesis and included topics such as:

- The design of the experiment
- Plant material
- Apparatus
- Early analytical methods
- Paper chromatographic methods
- The development of the carbon reduction cycle
- Labeling of hexose
- Origin of phosphoglyceric aced
- Radioactivity determination of pentose and heptose
- Identification of the two-carbon acceptor
- Carbon-14 saturation experiments

- The photosynthetic carbon cycle
- Quantum conversion in photosynthesis

Many workers over the years have determined various aspects of the photosynthetic pathway. Some of their findings are contained in [6–8].

Honours and Awards

Calvin received Honorary D.Sc. degrees from: Brooklyn Polytechnic Institute (1969); Columbia University (1979); Northwestern University (1961); the University of Notre Dame (1965); University of Nottingham (1959); Michigan College of Mining and Technology (1955); Oxford University (1959); Rijksuniversiteit-Gent (1970)

He was presented with many awards and prizes including: the Nobel Prize for Chemistry (1961); the Davy Medal (1964); the Virtanen Medal (1978); the Gibbs Medal (1977); the Priestley Medal (1978); the American Institute of Chemists Gold Medal (1979); the National Medal of Science (1989); the John Ericsson Award in renewable energy (1991).

Calvin held Fellowships, Honorary Fellowships and/or Membership in numerous Institutes and learned societies including: the American Academy of Arts and Sciences; the American Philosophical Society; the German Academy of Scientists, Leopoldina; the National Academy of Sciences; the Royal Netherlands Academy of Sciences and Letters; the Royal Society.

He was a President of American Chemical Society and the American Society of Plant Physiology.

Calvin served on many government bodies including: the President's Science Advisory Committee (1963–1966); the President's Advisory Group on Major Advances in Science and Technology (1975); the Energy Research Advisory Board of the Department of Energy (1981 to 1985); president of the Pacific division of the AAAS and other committees of the National Academy of Sciences (including the Committee on Science and Public Policy); the Joint Commission on Applied Radioactivity of the International Union of Pure and Applied Chemisa member of the US Commission of the International Union of Biochemistry; the Commission on Molecular Biophysics of the International Organization for Pure and Applied Biophysics.

He was Bakerian lecturer of the Royal Society (1965).

Calvin's citation for election (date: 23/04/1959) as Foreign Member of the Royal Society reads [9]:

> **'Professor Melvin Calvin of Pasadena distinguished for his researches into the mechanism of photosynthesis.'**

No proposers are listed in the Sackler database.

Bibliography

1. Reproduced with permission.
2. From Nobel Lectures, Chemistry 1942–1962, Elsevier Publishing Company, Amsterdam, 1964.
3. Biographical memoirs, Proceedings of the American Philosophical Society, vol. 144, no. 4, (December 2000), 454–457.
4. Who was who: 1996–2000.
5. Calvin, Melvin, Following the Trail of Light - A Scientific Odyssey. Oxford University Press, Oxford, 1998.

6. Molecular Mechanisms of Photosynthesis by Robert E. Blankenship, Blackwell Science, 1st ed. 2002, ISBN: 0632043210.
7. Photosynthesis by D.W. Lawlor BIOS Scientific Publ, 3rd ed. 2000, ISBN: 1859961576.
8. Photosynthesis: Regulation Under Varying Light Regimes by V. S. Rama Das, Science Publishers Inc, 2004, ISBN: 1578083435.
9. Sackler database, Royal Society.

1962

© Nobel Foundation [1]

Max Ferdinand Perutz (0.5)

"for their studies of the structures of globular proteins" [2]

During his presentation speech G. Hägg, then a member of the Nobel Committee for Chemistry of the Royal Swedish Academy of Sciences, commented of Perutz [2]:

...In 1937 Max Perutz performed some experiments in Cambridge to find out whether it might be possible to determine the structure of haemoglobin by X-ray diffraction, since no other method could be imagined for this purpose. Sir Lawrence Bragg, who tirelessly continued the work begun jointly with his father, in 1938 became the head of the Cavendish Laboratory in Cambridge. When he saw the results obtained by Perutz, he encouraged him to continue and has ever since lent a very efficient support. Haemoglobin belongs to the proteins which play such an enormous part in life processes, and which are a basic material in living organisms. Haemoglobin is a component of the red blood corpuscles. It contains iron which can take up oxygen in the lungs and later give it up to the body's other tissues. Haemoglobin is counted among the globular proteins, whose molecules are nearly spherical. It was chosen for the initial attempt, partly because it could develop good crystals, and partly because the haemoglobin molecule is quite small for a protein molecule...

...But even if the path was now open for a direct structure determination of haemoglobin and myoglobin, there was still an enormous amount of data to be processed. Myoglobin, the smaller of the two molecules, contains about 2,600 atoms, and the positions of most of these are now known. But for this purpose, Kendrew had to examine 110 crystals and measure the intensities of about 250,000 X-ray reflections. The calculations would not have been practicable if he had not had access to a very large electronic computer. The haemoglobin molecule is four times as large, and its structure is known less thoroughly. In both cases, however, Kendrew and Perutz are currently collecting and processing an even greater number of reflections in order to obtain a more detailed picture...

Max Ferdinand Perutz was born in Vienna on 19th May 1914, the youngest child of Hugo Perutz and Adèle 'Dely', née Goldschmidt. Until the outbreak of the Second World War the Perutz family were textile manufacturers who introduced mechanical spinning and weaving into the Austrian monarchy. Whilst this made their fortune in the 19th century the family was dispossessed of the business during the war. Perutz married Gisela Peiser (24th March 1942) and they had two children Vivien (1944) and Robin (1949). Perutz died in Cambridge, England, on 6th February 2002 [2–4].

The Early Years

Perutz attended the Theresianum, a grammar school which was formally an officers academy (during the days of Maria Theresia). The parental career choice was for Perutz to study law and then enter the family business. However, a school teacher's enthusiasm for chemistry caught the imagination of the young Perutz and he decided to concentrate on chemistry thereafter.

He entered Vienna University (1932) where he developed an interest in organic biochemistry. He decided to study under J.D. Bernal at the Cavendish Laboratory of Cambridge University for his doctorate, enrolling in 1936. He has remained at Cambridge thereafter.

The Career

Year	*Post*	*Institute*
1937	Assistant to Sir Lawrence Bragg	Cavendish Laboratory, Cambridge University
War years	Interned as an "enemy alien"	Released for war work
1945	ICI Fellow	Cavendish Laboratory, Cambridge University
1947–1957	Head	MRC Unit for the Study of Molecular Structure of Biological Systems, Cambridge
1954–1968	Reader in Chemistry – additional post to the Head of MRC unit	Davy Faraday Research Laboratory, Royal Institution, London
1957–1979	Director Chairman (1967)	MRC Laboratory of Molecular Biology, Cambridge
1963–69	Chairman	European Molecular Biology Organisation
1973–1979	Fullerian Professor of Physiology – additional post to MRC work	Davy Faraday Laboratory, Royal Institution

Amongst Perutz's many research interests were:

- molecular biology
- structure of haemoglobin
- chemistry of proteins in general

He had one interest outside of molecular biology, that of glaciers. He studied their crystal texture and mechanism of flow. He was a keen climber which may have provided the stimulation.

The facilities at Cambridge in the 1940's were such that the X-ray work was carried out at the Cavendish Laboratories and the protein chemistry was done at Keilin's Molteno Institute. Perutz moved between the two Institutes on a bicycle.

Perutz worked with Kendrew (from 1946), Crick (in 1948), Watson (Visitor in 1951) and Sanger. All of whom went on to be Nobel Laureates.

Perutz's Nobel acceptance speech is dated 11th December 1962 with the title 'X-ray analysis of haemoglobin' During the course of the speech Perutz commented [2]:

> *...I started my X-ray work on crystalline proteins in the Cavendish Laboratory at Cambridge under J. D. Bernal in 1937, just after he and Dorothy Hodgkin had demonstrated that protein crystals can be made to yield sharp X-ray diffraction patterns which extend to spacings of the order of interatomic distancesa. Soon afterwards Sir Lawrence Bragg was appointed Cayendish Professor. My X-ray diffraction pictures of haemoglohin at once fired his imagination and enthusiasm. He was fascinated by the idea that the powers of X-ray analysis might be extended to the giant molecules which form the catalysts of living cells. Bragg's determined support made it possible for me, and later for both Kendrew and myself, to carry on with our studies over the many years that were needed to develop X-ray methods suitable for such highly complex structures...*

He continued:

> *...Verterbrate haemoglobin is a protein of molecular weight 64,500. Four of its 10,000 atoms are iron atoms which are combined with protoporphyrin to form four haem groups. The remaining atoms are in four separate polypeptide chains, each containing just over 140 amino acid residues. In human haemoglobin their sequences are known...*

He went on to discuss the structure of haemoglohin in detail:

> *...The haemoglohin molecule is assembled by first matching each chain with its symmetrically related partner, then inverting the pair of white chains and placing it over the top of the pair of black ones. The resulting arrangement is tetrahedral, the four subunits forming a compact spheroidal molecule with the haem groups arranged in separate pockets on the surface of the molecule. The wide distances separating the haem groups was perhaps the greatest surprise the structure presented to us, for one would have expected the chemical interaction between them to be due to their close proximity. As it stands, the structure of oxyhaemoglobin leaves its physiological properties unexplained...*

Haemoglobin has the following structure:

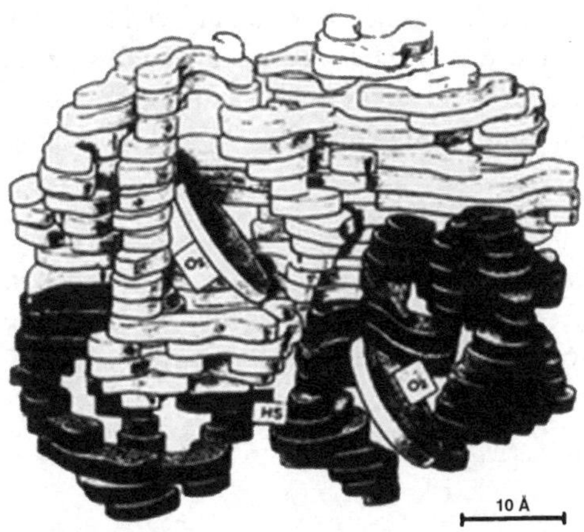

Complete haemoglobin molecule. The haem groups are indicated by gray dots.

Reproduced from Perutz's Nobel Prize acceptance speech.

Modern protein sequence analysis uses a variety of techniques including X-ray analysis further details for which may be obtained, for example, in the references [5,6,7].

Honours and Awards

Perutz was presented with many Honorary degrees including: in medicine from the University of Rome (1988); in philosophy from Universities in Vienna (1965), Saltsburg (1972) and Wales (1995); in sciences from the universities of Edinburgh (1965), East Anglia (1967), Cambridge (1981), York (1990), Oxford (1993), and Paris (1993).

He received many awards and prizes including: the Ehrenzeichen für Wissenschafte und Kunst (1966); the Nobel Prize for Chemistry (1962); the Royal Medal (1971); the Copley Medal (1979); the Actonian Prize (1984); the Lewis Thomas Prize (1997); Pour le Mérite from the FRG (1988).

Perutz was elected a Fellow of the Royal Society and a Honorary Fellow of the Royal Society of Edinburgh (1976). He was also: a Honorary Member of the American Academy of Arts and Sciences (1976); Corresponding Member of the Austrian Academy of Sciences (1963); a Member of the Akademie Leopldina, Halle (1964); a Foreign Member of the American Philosophical Society (1968), the Royal Netherlands Academy (1972), the French Academy of Sciences (1976), the Bavarian Academy of Sciences (1983), the National Academy of Sciences of Rome (1983), the Accademia dei Lincei of Rome (1984) and the Academy of Science of the DDR (1985). Perutz was also elected a Foreign Associate of the National Academy of Sciences of the USA (1970) and a Member of the Pontifical Academy of Sciences of Rome (1981). He was created a Millenium Fellow by the Royal Society of Chemistry (in 2000) and was an Honorary Fellow, Peterhouse, Cambridge (1962).

He held various 'named lectureships' including: Lectures: Croonian (1968); Humphry Davy (1988); Medawar (1992).

Perutz was made Companion of the British Empire (1963), Companion of Honour (1975) and Order of Merit (1988).

Perutz's citation for election (date: 18/03/1954) as Fellow of the Royal Society reads [8]:

> **Distinguished for his researches into the structure of protein molecules. In particular haemoglobin. His expert skill in manipulating protein crystals and his exhaustive studies of X-ray diffraction by haemoglobin in its various shrinkage states, and of its optical and chemical properties, form a firm basis for the analysis of its structure which has already yielded important and unique results, and holds out much promise for the future. Has also done much to put knowledge of glacier flow on a firmer basis by sponsoring theoretical studies of ice movement under gravity, and measurements of the plastic properties of ice, and by mapping, the flow at all depths of an actual glacier. List of publications attached.**

He was proposed by:

From personal knowledge. J. D. Bernal; W. L. Bragg; Kathleen Lonsdale; Dorothy Hodgkin; J. M. Robertson; W. T. Astbury; E. C. Bullard; C. R. Harington; L. R. Wager; D. Keilin; A. C. Chibnall; A. V. Hill; F. J. W. Roughton; F. P. Bowden; A. R. Todd; A. J. P. Martin.

Bibliography

1. Reproduced with permission.
2. From Nobel Lectures, Chemistry 1942–1962, Elsevier Publishing Company, Amsterdam, 1964.

3. Biographical Memoirs of Fellows of the Royal Society, 2004, vol. 50, pp. 227–256.
4. Who's who: 2002.
5. Methods in Protein Sequence Analysis-1986 ed Kenneth A. Walsh, Humana Press, 1987, ISBN: 0-89603-118-7.
6. Modern Protein Chemistry ed Gary C. Howard and William E. Brown, CRC Press, 2002, ISBN: 0-8493-9453-8.
7. Techniques in Protein Chemistry (multi-volume set published under the auspices of the Protein Society) ed Daniel R. Marshak, Academic Press, 1996, ISBN: 0-12-473556-8.
8. Sackler database, Royal Society.

1962

John Cowdery Kendrew (0.5)

"for their studies of the structures of globular proteins" [2]

G. Hägg, then a member of the Nobel Committee for Chemistry of the Royal Swedish Academy of Sciences, commented of Kendrew during his presentation speech [2]:

...About ten years later, John Kendrew joined Perutz' research group, and the task allotted to him was to try to determine the structure of myoglobin. Myoglobin is another globular protein, closely related to haemoglobin, but with a molecule only a quarter as large. It is found in the muscles, and enables oxygen to be stored there. Particularly large amounts of myoglobin are found in the muscular tissues of whales and seals, which need to be able to store large quantities of oxygen when diving...

...But even if the path was now open (after developing techniques of heavy atom incorporation) for a direct structure determination of haemoglobin and myoglobin, there was still an enormous amount of data to be processed. Myoglobin, the smaller of the two molecules, contains about 2,600 atoms, and the positions of most of these are now known. But for this purpose, Kendrew had to examine 110 crystals and measure the intensities of about 250,000 X-ray reflections. The calculations would not have been practicable if he had not had access to a very large electronic computer...

...Doctor Kendrew and Doctor Perutz. One of you recently said that today's students of the living organism do indeed stand on the threshold of a new world. You have both contributed very efficiently to the opening of the door to this new world and you have been among the first to obtain a glimpse of it. Through your combined efforts there is now in view, as it has been stated by yourself, a firm basis for an understanding of the enormous complexities of structure, of biogenesis and of the functions of living organisms both in health and disease...

John Cowdery Kendrew was born on 24th March 1917, in Oxford. His father Wilfrid George Kendrew, was Reader at the University of Oxford and his mother, Evelyn May Graham, née Sandberg) was an art historian. Kendrew married Elizabeth Jarvie, (née, Gorvin), the widow of a close friend, in St. Mary's the Less in 1948. They divorced in 1956. Kendrew died on 23rd August 1997 [2–4].

The Early Years

Kendrew attended school at the Dragon School, Oxford (1923–1930) and Clifton College, Bristol (1930–1936). He went up to Trinity College, Cambridge in 1936 as a Major Scholar in chemistry and attended lectures by Lord Rutherford, Sir Joseph Thomson and Sir Gowland Hopkins. He graduated with a B.A. in 1939. His was involved in various projects during the war including (from 1940–45):

- the Department of Physical Chemistry at Cambridge under the supervision of Dr. E.A. Moelwyn-Hughes working on reaction kinetics
- the Air Ministry Research Establishment (later Telecommunication Research Establishment) working on radar applications as a junior research officer
- the staff of Sir Robert Watson-Watt, Scientific Adviser to the Air Ministry in 1940 and for the rest of the war was engaged in operational research at the Royal Air Force headquarters, successively in Coastal Command, Middle East, and South East Asia (where he was Scientific Adviser to the Allied Air Commander-in-Chief)

By the end of the war, Kendrew held the honorary rank of Wing Commander R.A.F (1944).

Returning to the Cavendish Laboratory, Cambridge University (1946), Kendrew started to collaborate with Perutz under Sir Lawrence Bragg, and graduated with his Ph.D. degree in 1949 and his Sc.D. in I962.

The Career

Year	*Post*	*Institute*
1946–1962	Founder member, rising to Deputy Director	Medical Research Council Unit at the Cavendish Laboratory
1954–68	Additional post of Visiting Reader	Davy-Faraday Laboratory of the Royal Institution in London
1962	Deputy Chairman, and Director of its Division of Structural Studies	Medical Research Council Laboratory of Molecular Biology[a]
1975–82	Director General	European Molecular Biology Laboratory

[a] The Medical Research Council Laboratory of Molecular Biology achieved independent status from the Medical Research Council Unit at the Cavendish Laboratory

Kenrew's career interests were:

- protein structure
- X-ray analysis of myoglobin – partial structure (1957), almost complete (1960)
- haemoglobin

His papers are deposited in the Bodleian Library, Oxford.

Kenrew was a founder and Editor-in-Chief of the *Journal of Molecular Biology* from 1959–1987. He appeared with Perutz in 'Eye on Research', a BBC television science series. Kendrew also wrote a book called *'The Thread of Life'* (1966) London:Bell. He was a Director of Academic Press (1970–88).

Kendrew's Nobel Prize acceptance speech is dated 11th December 1962 and titled 'Myoglobin and the structure of proteins' [2].

> *...The whole diffraction pattern of a myoglohin crystal consists of at least 25,000 reflexions. In 1955, when the three-dimensional work began, no computers existed which were fast enough to calculate Fourier syntheses containing so many terms; besides, the method was unproved and it seemed advisable to test it first on a smaller sample of data...*

> *...We actually decided to undertake the solution of the structure in three stages; the first, completed in 1957, involved 400 reflexions and gave a resolution of 6 Å; the second (1959) included nearly 10,000 reflexions and gave a resolution of 2 Å; the third (not yet complete) includes all the observable reflexions - about 25,000 and gives a resolution of 1.4 Å. It may be recalled that polypeptide chains pack together at centre-to-centre distances of 5 to 10 Å; that atoms (other than hydrogen) of neighbouring groups in Van der Waals contact, or brought together by hydrogen bonds or charge interactions, lie 2.8 to 4 Å apart; and that the separation between covalently bonded atoms is 1 to 1.5...*

> *...What is the nature of the molecule which has emerged with progressively increasing clarity from successive Fourier syntheses? Some 118 out of the total of 151 amino acid residues make up 8 segments of right-handed α-helix, of lengths ranging from 7 to 24 residues. These segments are joined by 2 sharp corners (containing no non-helical residues) and 5 non-helical segments (of I to 8 residues); there is also a non-helical tail of 5 residues at the carboxyl end of the chain. The whole is folded in a complex and unsymmetrical manner to form a flattened, roughly triangular prism with dimensions about 45 × 35 × 25 Å. The whole structure is extremely compact; there is no water inside the molecule, with the probable exception of a very small number (less than 5) of single water molecules presumably trapped at the time the molecule was folded up; there are no channels through it, and the volume of internal empty space is small...*

Modern protein sequence analysis uses a variety of techniques including X-ray analysis, further details for which may be obtained, for example, in the references [5,6,7].

Honours and Awards

Kendrew received honorary degrees from a number of universities including: Honorary D.Sc. from the Universities of Keele (1968), Exeter (1982) and Buckingham (1983); Honorary Doctorate from the Universities of Pécs (1975), Madrid (1987), Siena (1991) and Santiago (1991).

He was presented with many awards and prizes including: the Nobel Prize for Chemistry (1962); the Royal Medal (1965); the William Procter Prize (1988); Member of the Order of the Madara Horseman (1980).

Kendrew was elected a Fellow of the Royal Society (1960) and a Honorary Member of the American Society of Biological Chemists (1962). He was a Member of the National Academy of Sciences; an Honorary Fellow of the Weizmann Institute (1969); a Member of the Biophysical Society council (1962); a Fellow of the Institute of Biology (1963); a Foreign Honorary Member of the American Academy of Arts and Sciences (1964); a Fellow of the American Association for the Advancement of Science (1966, Life Member 1968); a Foreign Member Max-Planck-Gesellschaft

(1969); a Foreign Associate of the US Academy of Sciences (1972); a Member of the Academic Advisory Council University of Buckingham (1973); a Foreign Member Bulgarian of the Academy of Sciences (1979); a Foreign Fellow of the Indian National Science Academy (1988).

He became: a Fellow of Peterhouse in 1947 (also Director of studies; Librarian; Praelector; Steward; wine steward and custodian of the paintings and portraits); Deputy to the Chief Scientific Adviser, Ministry of Defence, part-time (from 1960); Honorary Secretary of the British Biophysical Society; President of the International Union for Pure and Applied Biophysics (1969–72); Member of the Confederation of Science and Technology Organisations for Development (1981–85); Secretary General of the International Council of Scientific Unions (1974–80, first President 1983–88, Past President 1988–90); Member of the BBC Scientific Advisory Group; Trustee of the British Museum (1974–79); Founder of the European Molecular Biology Laboratory; President of St. John's College Oxford (post held for 7 years from 1987); Trustee of the International Foundation for Science (1975–78); Member of the Weizmann Institute's Board of Governors (1964 until his death); Chairman of the Weizmann Institute's Scientific and Academic Advisory Committee; Member of the Executive of the Weizmann Institute Foundation (from 1966); Chairman of the Defence Scientific Advisory Council (1971–74); President of the British Association for the Advancement of Science (1973–74); Member of Council of UN University (1980–86, Chairman 1983–85); Life Member of the University of Reading Society (1986); Co-opted member of the Governing Body of Queen Elizabeth House (Dec 1982–July 1987).

He was made Companion of the British Empire (1963) and knighted (KBE) in 1974.

Kendrew's citation for election (date: 24/03/1960) as Fellow of the Royal Society reads [8]:

> **'Kendrew's field of research has been the investigation of the structure of protein molecules by X-ray analysis, and he has played a large part in the development of the Medical Research Council's Unit at Cambridge which has added so much to our knowledge of biological molecules. He has recently achieved the three-dimensional analysis of myoglobin, the first protein structure to be determined by direct X-ray methods and far more complex than any molecular hitherto analysed. Great difficulties were encountered at each stage, in purifying and crystallising the protein, finding a favourable crystalline form, and preparing a series of isomorphous crystalline heavy atom derivatives. He has also done pioneer work in using electronic computing machines for crystallographic computations. During the War he was Officer-in-Charge, Operational Research Section, and Scientific Adviser to the Allied Air Commander-in Chief, South East Asia (Honorary Wing Commander).'**

No proposers listed.

Bibliography

1. Reproduced with permission.
2. From Nobel Lectures, Chemistry 1942–1962, Elsevier Publishing Company, Amsterdam, 1964.
3. Biographical Memoirs of Fellows of the Royal Society, 2001, vol. 47, pp. 311–332.
4. Who was who: 1996–2000.
5. Methods in Protein Sequence Analysis - 1986 ed. Kenneth A. Walsh, Humana Press, 1987, ISBN: 0-89603-118-7.
6. Modern Protein Chemistry ed. Gary C. Howard and William E. Brown, CRC Press, 2002, ISBN: 0-8493-9453-8.
7. Techniques in Protein Chemistry (multi-volume set published under the auspices of the Protein Society) ed. Daniel R. Marshak, Academic Press, 1996, ISBN: 0-12-473556-8.
8. Sackler database, Royal Society.

1963

Karl Ziegler (0.5)

"for their discoveries in the field of the chemistry and technology of high polymers" [2]

During his presentation speech A. Fredga, then a member of the Nobel Committee for Chemistry of the Royal Academy of Sciences, commented of free radical polymerization in general and Ziegler's work in particular [2]:

...Plastics consist of very large molecules or macromolecules often forming long chains of thousands of atoms. They are made by joining together normal size molecules constituting the basic units. These molecules must be reactive, but some outside help is also necessary to make them combine. This outside assistance often used to be supplied by free radicals, added to trigger off the reaction of polymerization. The term "free radical" may conjure up political connotations, and indeed free radicals have much in common with revolutionaries: they are full of energy, difficult to control, and have an unpredictable outcome. Thus, free - radical reactions give polymer chains with branches and other anomalies.

However, Professor Ziegler has found entirely new methods of polymerization. Studying organometallic compounds, he discovered that organoaluminium compounds, which are easy to prepare, are particularly suitable for work on the industrial scale. Peculiar electrical forces operate around an aluminium - carbon bond in a hydrocarbon chain: reactive molecules are drawn in and sandwiched between the carbon atom and the aluminium atom, thus increasing the length of the chain. All this happens much more quietly than in free - radical reactions. When the chain is long enough, we detach the aluminium and thus stop the further growth of the molecule. The combination of aluminium compounds with other metallic compounds gives Ziegler catalysts. These can be used to control polymerizations and to obtain molecular chains of the required length. However, many systematic experiments - and indeed some accidental findings - were necessary to reach this stage. Ziegler catalysts, now widely used, have simplified and rationalized polymerization processes, and have given us new and better synthetic materials...

Karl Ziegler was born in Helsa near Kassel in Germany, on 26th November 1898, the son of Karl (a Lutherian Minister) and Luise Ziegler. He married Maria Kurtz (1922) and they had a daughter Marianne Witte and a son, Erhart. Ziegler died in Mulheim, Germany on 12th August 1973 [2–4].

The Early Years

Ziegler graduated in 1920 under Prof. von Auwers at the University of Marburg/Lahn, qualifying as a lecturer in 1923.

The Career

Year	*Post*	*Institute*
1923	Lecturer	University of Marburg
1925	Lecturer	University of Frankfurt/Main
1926	Lecturer	University of Heidelberg
1927	Professor of Chemistry	University of Heidelberg
1936	Professor of Chemistry and Director of the Chemical Institute	University of Halle/Saale
1943–69	Director	Max-Planck Institute for coal research
1949	Honorary Professor (whilst remaining Director)	Technical High School in Aachen
1963	Retirement – but continued to work at the Institute as an Honorary Fellow	

Ziegler's work covered many fields of research, including:

- alkali organic compounds
- trivalent carbon radicals
- the syntheses of multi - membered ring systems
- the syntheses and reactions in the chemistry of organic aluminium compounds
- the application of the techniques of physical chemistry to the study of the mechanism of organic reactions
- the discovery of organometallic mixed catalysts for the polymerization of olefins (*e.g.* the synthesis of high - density polyethylene)

Ziegler was instrumental in the foundation of the Gesellschaft Deutscher Chemiker after the war. He was President for five years and also President of the Deutsche Gesellschaft für Mineralölwissenschaft und Kohlechemie (1954 until 1957).

His Nobel Prize acceptance speech, dated 12th December 1963 with the title 'Consequences and development of an invention', detailed his discovery of the catalysts that bear his name and their use in the manufacture of polyethylene [2]. He commented:

> *...The new development had its inception near the end of 1953, when I, together with Holzkamp, Breil and Martina, observed - during only a few days of an almost dramatic course of events that*

ethylene gas will polymerize very rapidly with certain catalysts that are extremely easy to prepare, at 100, 20 and 5 atmospheres and, finally, even at normal pressure, to a high molecular weight plastic...

...The catalyst is prepared simply by simultaneously pouring, with exclusion of air, two liquid materials into about two litres of a gasoline - like hydrocarbon, after which ethylene is introduced, while stirring. The gas is absorbed quickly; within an hour one can easily introduce 300–400 litres of ethylene into the two litres of liquid. At the same time, a solid substance precipitates, in such a way that after approximately one hour the material becomes doughy and can scarcely be stirred any more. If the brown catalyst is then destroyed, by the addition of some alcohol and by the introduction of air, the precipitate becomes snow - white and can he filtered off. In its final state it will accumulate, in amounts of 300–500g, as a dry, white powder...

...The results of this experiment greatly surprised us, and, later on, many others, since up to that time ethylene had been considered extremely difficult to polymerize. The polythene of the Imperial Chemical Industries, a product which had been known for some seventeen years, was being prepared under pressures of 1000–2000 atmospheres, and at a temperature of 200°C. Our experiment thus destroyed a dogma. It led, in addition, to a polyethylene which differed quite markedly from the high - pressure product. Low - pressure polyethylene not only has a better resistance to elevated temperatures and a higher density, but is also more rigid. This is easily demonstrated by holding in one hand two similar objects made of the two materials, and press - ing them together. Low - pressure polyethylene can be drawn without difficulty to form fibres or ribbons of high tensile strength. This cannot be done at all with high - pressure polyethylene, or at best only an indication of drawing is obtained. We established these facts immediately after our discovery, with test specimens which were still quite primitive.

The differences can be attributed to the fact that in our process molecules of ethylene are joined together linearly, without interruption, whereas in the high - pressure process chain growth is disturbed, so that a strongly branched molecule results. The low - pressure process found immediate acceptance in industry. By 1955, 200 metric tons of this new type of plastic had been produced; in 1958 it was 17,000 tons, and in 1962 some 120,000 tons. The much higher figures occasionally cited for this and other plastics have resulted from the confusion of available, but unused, capacities with actual production. The increase in dimensions can be indicated by comparison of our first test specimens, prepared ten years ago with rather primitive means, with containers that are twenty cubic metres in capacity, the largest now being made from polyethylene. A subsequent figure shows the lightness of the material, since a very large container can easily be carried by only a few men.

The catalyst employed in the experiment described was prepared by mixing aluminium triethyl, or diethyl aluminium chloride, with titanium tetrachloride. However, this is only one example, taken from the countless series of "organometallic mixed catalysts". Most generally they will form, as we found, whenever standard organometallic compounds, preferably those of aluminium, but also many of other metals, are brought into contact with compounds of certain heavy metals. Those of titanium, zirconium, vanadium, chromium, molybdenum, cobalt and the like are especially effective...

Homogeneous catalysts are used extensively for the modern production of some polymers, see for example [5,6,7].

Honours and Awards

Ziegler received Honorary Doctorates from: The Technische Hochschule (1951); the University of Giessen (1958); the University of Heidelberg (1958); the Technische Hochschule, Darmstadt (1968).

He was presented with many awards and prizes including: the ComMedal of the Society of German Chemists (1935); the Carl Duisberg Award of the German Chemical Society (1953); the Carl Engler Medal of the German Society of Mineralogy and Coal Chemistry (1958); the Lavoisier Medal of the French Chemical Society (1955); the Nobel Prize for Chemistry, joint (1963); the Swinburne Medal of the Plastics Institute, London (1964); the International Synthetic Rubber Medal of 'Rubber and Plastics Age' (1967); the Order of Merit for Science and Technology (1969); the Carl Dietrich - Harries Award of the German Rubber Society (1971); the Wilhelm Exner Medal of the Austrian Industrial Association (1971); Liebig Medal from the Verein Deutscher Chemiker; the Siemens Ring from the Werner von Siemens Foundation.

Ziegler was a Member, Honorary Member, or Foreign Member of various German and foreign scientific societies and academiesas well as: Honorary Member of the Chemical Society of Japan (1959); the Society of Chemical Industry (1959), New York; the Academy of Sciences (1966); the Société de Chimie Industrielle (1966); Gesellschaft Deutscher Chemie (1968); Foreign Member of the Royal Society (1971). He was also Honorary Fellow of the Royal Society of Edinburgh (1972).

He was Visiting Professor at Chicago University (1936), a Honorary Senator of the Max-Planck Gesellschaft (1969) and founder President (1970–1972) of the Rheinisch-Westfalische Akademie der Wissenschaften.

Ziegler's citation for election (date: 22/04/1971) as Foreign Member of the Royal Society reads [8]:

> **'Professor Kurt Ziegler of Mulheim - Ruhr distinguished for his discovery of organometallic catalysts for the low pressure polymerization of olefins, and for his contributions in organic chemistry.'**

No proposers are listed in the Sackler database.

Bibliography

1. Reproduced with permission.
2. From Nobel Lectures, Chemistry 1963–1970, Elsevier Publishing Company, Amsterdam, 1972.
3. Biographical Memoirs of Fellows of the Royal Society, 1975, vol. 21, pp. 569–584.
4. Who was who: 1971–1980.
5. Catalytic Chemistry by Bruce C. Gates, John Wiley and Sons Inc. 1992, ISBN: 0-471-51761-5.
6. Ziegler Catalysts by G. Fink, R. Mülhaupt and H. H. Brintzinger, Springer - Verlag, 1995, ISBN: 3-540-58225-8.
7. Homogeneous Catalysis by Piet W.N.M. van Leeuwen, Kluwer Academic Publishers, 2004, ISBN: 1-4020-1999-8.
8. Sackler database, Royal Society.

1963

© Nobel Foundation [1]

Giulio Natta (0.5)

"for their discoveries in the field of the chemistry and technology of high polymers" [2]

During his presentation speech A. Fredga, then a member of the Nobel Committee for Chemistry of the Royal Academy of Sciences, commented of Natta [2]:

...Isotactic polymers show very interesting characteristics. Thus, while ordinary hydrocarbon chains are zigzag-shaped, isotactic chains form helices with the side groups pointing outwards. Such polymers give rise to novel synthetic products such as fabrics which are light and strong at the same time, and ropes which float on the water, to mention only two examples...

...Nature synthesizes many stereoregular polymers, for example cellulose and rubber. This ability has so far been thought to be a monopoly of Nature operating with biocatalysts known as enzymes. But now Professor Natta has broken this monopoly...

...Professor Natta. You have succeeded in preparing by a new method macro molecules having a spatially regular structure. The scientific and technical consequences of your discovery are immense and cannot even now be fully estimated. The Swedish Royal Academy of Sciences wishes to express its appreciation by awarding you the Nobel Prize. Please accept the best wishes of the Academy. I would also like to express the admiration of the Academy for the intensity with which you are continuing your work in the face of difficulties...

Giulio Natta was born at Imperia on 26th February 1903. He married Rosita, née Beati, they had one son and one daughter. Natta died on 2nd May 1979.

The Early Years

Natta graduated with a degree in Chemical Engineering from the Polytechnic of Milan (1924) and passed the relevant teaching examinations there in 1927.

The Career

Year	*Post*	*Institute*
1925	Assistant Professor in Analytical Chemistry	Polytechnic of Milan
1933	Professor and Director of the Institute of General Chemistry	Pavia University
1935	Professor in physical chemistry	University of Rome
1936	Professor and Director of the Institute of Industrial Chemistry	Polytechnic of Turin
1938	Professor and Director of the Department of Industrial Chemistry	Polytechnic of Milan

Natta's extensive work on synthetic polymers included:

- a study for the production of synthetic rubber in Italy
- a study of solids by means of X-rays and electron diffraction
- a study of catalysts and the structure of some high organic polymers (the latter from 1934)
- a mechanistic study of oxosynthesis
- an investigate into the polymerisation of olefins (1938)
- a study of the selective hydrogenation of unsaturated organic compounds

Throughout his work, Natta used analysis techniques that allowed for work concerning:

- the kinetics of ploymerisation reactions (1938)
- kinetic research on methanol synthesis
- the use of X-ray techniques to investigate the polymers that he produced

Having studied butadiene in its various forms, he was the first to accomplish a physical separation of butadiene from 1-butadiene (using extractive distillation). Using the results from these and other experiments, Natta produced a range of new types of polymers including:

- stereospecific polymerization, by extending Ziegler's work on organometallic catalysts (1953)
- discovering new classes of polymers with a sterically ordered structure, for example
    - isotactic, di-isotactic and syndiotactic polymers
    - linear non branched olefinic polymers
    - copolymers with an atactic (or sterically nonordered) structure
- thermoplastic material ie isotactic polypropylene which was produced commercially
    - as a synthetic fibre
    - as a monofilament
    - as packing film
- new elastomers
    - by polymerization of butadiene into *cis*-1,4 polymers with a very high degree of steric purity
    - by copolymerization of ethylene with other α-olefins (propylene)
- crystalline alternating copolymers of different couples of monomers
- various sterically ordered polymers of non-hydrocarbon monomers.

Natta developed techniques that allowed for the synthesis of asymmetric polymers, i.e. optically active macromolecules, which he produced from optically inactive monomers. These are of great interest as the resultant polymers are similar to natural biological processes.

Natta published over 700 papers, of which about 500 concern stereoregular polymers. He also wrote a large number of patents that were filed in many different countries.

Natta's Nobel Prize acceptance speech is dated 12th December 1963 with the title 'From the stereospecific polymerization to the asymmetric autocatalytic synthesis of macromolecules'. Natta started his speech by placing his work in the context of previous work concerning polymers and then outlines his three-fold plan of research, namely [2]:

..Our efforts were then directed to three main fields of research:

(i) *To investigate the structures of the new polymers in order to establish the relationships existing between chemical structure, configuration, and conformation of the macromolecules in the crystalline state*
(ii) *To find the conditions that allowed the synthesis of olefinic polyhydrocarbons having a determined type of steric structure, with high yields and high degree of steric regularity, as well as to study the reaction mechanism, and regulation of the molecular weight.*
(iii) *To attempt the synthesis, possibly in the presence of nonorganometallic catalysts, of stereoregular polymers corresponding to other classes of monomers having a chemical nature different from that of α-olefins...*

One of the many issues Natta went on to discuss was that of stereospecific polymerization produced from each class of monomers. He went on to comment on:

- *Crystalline structure of high stereoregular polymers, covering*
    - *α-olefins*
    - *ditactic polymers*
    - *stereoregular polymers of conjugated diolefins*
- *Stereospecificities in polymerization processes of hydrocarbon monomers*
- *Stereospecificity in polymerization of nonhydrocarbon monomers*
- *Asymmetric synthesis of high polymers and interpretation of stereospecific polymerization*

Homogeneous catalysts are used extensively for the modern production of some polymers, see for example [3,4,5].

Honours and Awards

Natta was presented with honorary degrees from various Universities and Institutes including: the Catholic University of Louvain, Belgium (1965); the University of Genoa (1964); Mainz University (1963); the Polytechnic Institute of Brooklyn, New York (1964); ESPI, University of Paris (1971); Turin University (1962).

He received many awards and prizes including: the first International Gold Medal of the Synthetic Rubber Industry (1961); the STAS Medal (1962); the Gold Medal from the Society of Plastic Engineers (New York, 1963); the Perrin Medal from the French Chemical Physical Society (1963); the Lavoisier Medal from the Chemical Society of France (1963); the Perkin Gold Medal of the English Society of Dyers and Colourists (1963); the Gold Medal of the Union of Italian Chemists (1964); the Gold Medal "Lomonosov" of the Moscow Academy of Sciences (1969); the Medal "Leonardus Vincius Florentinus Doctor Ingenieurs" of FIDIIS, Paris (1971); the "Carl-Dietrich-Harries-Plakette, of the Deutsche Kautschuk Gesellschaft, Frankfurt/Main (1971).

Natta also received: a gold medal from the town of Milan (1960); a gold medal from the President of the Italian Republic (1961); a gold medal from the Milan district (1962) and the John Scott award from the Board of Directors of the City Trust of Philadelphia.

Amongst his many Fellowships and Memberships of Institutes and learned societies, Natta was a "National Member" of the Academia dei Lincei (1955); a Honorary Life Member of the New York Academy of Sciences (1961) of which he had been a Fellow since 1958; a Honorary Member of the Austrian (1960), Belgian and Swiss (1963) Chemical Societies, the Industrial Chemical Society of Paris (1966) and of the Chemical Society of London (1970). He was also a Member of: the Istituto Lombardo di Scienze e Lettere; the Accademia delle Scienze of Turin; the National Academy of XL, Rome (1964); the International Academy of Astronautics, Paris (1965). He was also a Foreign Member of the Academy of Sciences of Moscow, USSR (1966) and an associated Foreign Member of the Académie des Sciences de l'Institut de France (1964).

Natta was Honorary President of the Italian Section of the Society of Plastics Engineers (SPE) and an Honorary Member of the Rotary Club.

Bibliography

1. Reproduced with permission.
2. From Nobel Lectures, Chemistry 1963–1970, Elsevier Publishing Company, Amsterdam, 1972.
3. Catalytic Chemistry by Bruce C. Gates, John Wiley and Sons Inc, 1992, ISBN: 0-471-51761-5.
4. Ziegler Catalysts by G. Fink, R. Mülhaupt and H. H. Brintzinger, Springer-Verlag, 1995, ISBN: 3-540-58225-8.
5. Homogeneous Catalysis by Piet W.N.M. van Leeuwen, Kluwer Academic Publishers, 2004, ISBN: 1-4020-1999-8.

1964

© Nobel Foundation [1]

Dorothy Crowfoot Hodgkin

"for her determinations by X-ray techniques of the structures of important biochemical substances" [2]

During his presentation speech G. Hägg, then a member of the Royal Academy of Sciences, observed of Hodgkin [2]:

...However, even today structure determination by X-ray methods does not yield a direct route from the experimental data to the structure. In complicated cases the scientist only obtains a result after considerable mental effort, in which chemical knowledge, imagination and intuition play a significant part. In addition, the experimental data often have to be processed using different mathematical treatments, which must be varied according to the circumstances. Add to this the fact that the more complicated the structure, the greater becomes the volume of experimental data which must be amassed and processed. For relatively simply built compounds it was possible to carry out the calculations with pencil and paper. Nowadays it is nearly always necessary to use electronic computers, and their arrival has made an enormous difference to the possibility of carrying out structure determinations. However, it is not usually possible to just feed in the experimental data, and get out the figures which give the final structure; the scientist's ability to handle the data is still of vital importance. It is in this respect that Mrs. Hodgkin has shown such exceptional skill.

Mrs. Hodgkin has carried out a large number of structure determinations, primarily of substances which are of importance biochemically and medically, but two of these substances deserve especial mention. These are penicillin and vitamin B12, whose structures have become completely and definitely known through her efforts...

...Professor Hodgkin. You have for many years directed your efforts towards the determination of crystal structures by means of X-ray diffraction techniques. You have solved a large number of structural problems, the majority of great importance in biochemistry and medicine, but there are

two landmarks which stand out. The first is the determination of the structure of penicillin, which has been described as a magnificent start to a new era of crystallography. The second, the determination of the structure of vitamin B12, has been considered the crowning triumph of X-ray crystallographic analysis, both in respect of the chemical and biological importance of the results and the vast complexity of the structure...

Hodgkin was born in Cairo on 12th May 1910, the daughter of John and Grace Mary, née Hood, known as Molly Crowfoot. Her father worked in the Egyptian Education Service. She had two sisters Joan (1912) and Elisabeth Grace (Betty) (1912). Hodgkin married Thomas Lionel Hodgkin (1937), who died in 1982. They had three children, Luke Howard (1938), Prudence Elizabeth (1943), John Robin Tobias (1946). Hodgkin died on 29th July 1994 at her home, Crab Mill, and is buried in the local church cemetery in Ilmington [2, 3, 4].

The Early Years

Hodgkin spent very few childhood years with her parents as John Crowfoot, her father, was Director of Education in the Sudan (from 1916). Hodgkin and her sisters lived with a to nanny, Katie Stevens, on the South Downs, thereafter moved to Beccles in Suffolk (1920) and, shortly to the Old House in Geldeston, a Norfolk village. She became interested in chemistry and in crystals at about the age of 10. A family friend, Dr. A. F. Joseph, helped her to analyse ilmenite whilst on a trip to visit her parents. She was given the published version of *On the Nature of Things* and *Old Trades and New Knowledge* two lectures given by Sir William Bragg (1923 and 1925) which fired her interest in chemistry. However, in the 1925 school report from the Sir John Leman School, Beccles, Criss Deeley (one of Hodgkin's tutors) found Hodgkin to be 'good on the whole' in chemistry, but placed Hodgkin ninth in the class of fourteen! (p. 31 from [5]).

Hodgkin was accepted to read chemistry at Somerville College, Oxford University, from October 1928 attending lectures by Sir Robert Robinson and Sir Cyril Hinschelwood, graduating with a chemistry degree in 1932. She worked on thallium dialkyl halides for part II of her degree (under H. M. Powell) and also spent a summer visiting Victor Goldschmidt's laboratory in Heidelberg. Her postgraduate research was under J. D. Bernal at Cambridge University (Newnham College).

Following her postgraduate work, Hodgkin returned to Oxford University and continued her work on X-ray analysis.

The Career

Year	*Post*	*Institute*
1946	University Lecturer and Demonstrator	Oxford University
1956	University Reader in X-ray crystallography	Oxford University
1960	Wolfson Research Professor of the Royal Society	Oxford University
1977	Retirement	

For most of her working life, Hodgkin was Official Fellow and Tutor in Natural Science at Somerville College (University of Oxford), responsible mainly for teaching chemistry for the women's colleges. Margaret Thatcher was one of her students at Somerville College. Hodgkin was later awarded

Honorary Fellow of Somerville College Oxford, Linacre College Oxford, Girton College Cambridge, Newnham College Cambridge. She was also a Fellow of Wolfson College Oxford (1977–82).

Her work on X-ray analysis covered many substances, most of which were of biological significance and include:

- cobyric acid
- insulin
- penicillin (from 1942) and other antibiotics
- sterols
- vitamin B_{12} (from 1948)
- proteins

Her Nobel Prize acceptance speech was dated 11th December 1964 and had the title 'The x-ray analysis of complicated molecules'. During her speech she commented [2]:

...There have been some practical results following from our knowledge of the structure of the molecules so far described. Penicillin V has been synthesised by Sheehan and Henery-Logan. The coenzyme, a few years ago obtainable with difficulty in milligram quantities, can now be prepared very easily from cyanocobalamin - Dr. Lester Smith has made crystals 0.5 cm across. Soon, and no doubt very beautifully, cyanocobalamin itself will be synthesised by R.B.Woodward. But microorganisms are even more efficient than chemists at synthesis of molecules of this magnitude and will most likely continue to provide the main supplies of these compounds to be used in medicine. What we most hope to gain from knowledge of the structure and synthesis of these molecules is a complete understanding of their biogenesis and the part they play in metabolism. This should enable natural processes to be controlled when they go astray.

I should not like to leave an impression that all structural problems can be settled by X-ray analysis or that all crystal structures are easy to solve. I seem to have spent much more of my life not solving structures than solving them. I will illustrate some of the difficulties to be overcome by considering our efforts to achieve the X-ray analysis of insulin.

Insulin is a molecule of weight about 6000, larger than any so far described, though small if considered as a protein and compared with myoglobin and haemoglobin. Although its complete chemical structure is now known from Sanger's researches, it is quite unclear what exactly the molecule does that makes it so necessary to life. Our hope, following the kind of reasoning outlined above, is that a complete knowledge of the molecular geometry, how the peptide chains fit together within the molecule and the molecules within crystals, may make it possible for us to understand and control its behaviour. It crystallises in a number of different modifications and their very different degrees of complexity seem significant. In acid insulin salts the molecule appears to be dimeric, the two insulin molecules in one dimer being related to one another by two-fold axes of symmetry. Crystallographic two-fold axes also relate insulin molecules in a cubic metal-free form of insulin first observed by Abel in 1927. These crystals have so far only been obtained as extremely small rhombic dodecahedra; their habit and symmetry suggest that here six insulin dimers may be grouped in a larger aggregate, with symmetry 23, reminiscent of structures proposed for the smaller viruses. But it is difficult to get sufficient X-ray diffraction effects from these crystals to check the hypothesis. In all insulin crystals and in solutions which contain adequate zinc, or some other similar bivalent metal ions, a definite aggregate of six insulin molecules appears; having, indeed, the molecular weight 36000 first recorded by Svedberg in 1935. This insulin hexamer corresponds with the unit cell in the rhombohedral crystals first investigated and is the asymmetric unit in the monoclinic form which appears in the presence of phenol...

Honours and Awards

Hodgkin was presented with many Honorary Degrees including: Honorary DL. from Dalhousie; Honorary D.Univ from York and Zagreb; D.Sc./Sc.D. from the Universities of Cambridge, Leeds, Manchester, Oxford and others; Honorary LL.D from Bristol University.

Hodgkin was presented with many medals and prizes, including: the Royal Medal (1956); the Nobel Prize for Chemistry (1964); the Baly Medal (1971); the Copley Medal (1976); the Gold Medal of the Royal Society of Medicine (1977); the Longstaff Medal (1978); the Lomonosov Gold Medal of the USSR Academy of Sciences (1983); the Dimitrov Peace Prize, Bulgaria (1984); the International Lenin Peace Prize (1987).

She gave the Tercentenary (1960) and Bakerian (1972) Lectures of the Royal Society.

Hodgkin was appointed to Fellowships/Memberships in many Institutes and Centres of Learning including: Fellow of the Royal Society (1947); the Australian Academy of Science (1968); the Akademy Leopoldina (1968). She was Honorary Fellow of: the Royal Australian Chemical Institute; the Bangladesh Physical Society; the Ghana Academy of Sciences; the Indian Academy of Sciences; the Netherlands Academy of Science; of the Royal College of Physicians (1977); of the Royal Society of Chemistry (1980); of the Royal Society of Edinburgh (1970). She was a Member of the Norwegian Academy of Sciences and Letters and of the Yugoslav Academy of Sciences and Arts; Honorary Member of the Irish Academy and of the Royal Institution (1988); a Foreign Honorary Member of the American Academy of Arts and Sciences; the Akademi Leopoldina. Hodgkin was also Foreign Member of the Royal Netherlands Academy of Sciences (1956); the American Academy of Arts and Sciences, Boston (1958); the USSR Academy of Sciences. She was Corresponding Member of the Puerto Rico Academy of Atrs and Sciences and of the Bavarian Academy of Sciences as well as Foreign Associate of the National Academy of Sciences, Washington.

Hodgkin was awarded the Order of Merit in 1965 and was presented with the Freedom of Beccles (1965).

She was Vice-President of the Medical Aid Committee for Vietnam (1965), President (1971); Chancellor, Bristol University (1970–1988); President, International Union of Crystallography (1972–1975); President, Pugwash conferences on Science and World Affairs (1976–1988); President of the British Association for the Advancement of Science (1977–1978).

Hodgkin's citation for election (date: 20/03/1947) as Fellow of the Royal Society reads [9]:

'Dr. Hodgkin has been engaged for thirteen years on researches in crystal chemistry, chiefly in the field of the sterols and proteins. Her studies have thrown much light on the stereo-chemical structure of the sterol skeleton and, in one case, cholesteryl-iodide, a complete structure has been worked out. In proteins, she has established the general form and the leading inter-atomic distances in crystalline insulin and she has also carried out work on virus structures. During the war she has shown that it is possible to derive the complete structure of organic substances of unknown chemical composition by X-ray methods alone. Her contributions to this field has [sic] placed her in the forefront of X-ray crystal analysts'.

She was proposed by:

From Personal Knowledge. J. D. Bernal; I. M. Heilbron; W. T. Astbury; J. Monteath Robertson; H. W. Florey; F. Simon; W. L. Bragg; W. Hume-Rothery; Kathleen Lonsdale.

Bibliography

1. Reproduced with permission.
2. From Nobel Lectures, Chemistry 1963–1970, Elsevier Publishing Company, Amsterdam, 1972.
3. Biographical Memoirs of Fellows of the Royal Society, 2002, vol. 48, pp. 179–219.
4. Who was who: 1991–1995.
5. Dorothy Hodgkin: A Life by Georgina Ferry, Granta Books London, 1998, ISBN: 1-86207-167-5.
6. Structure determination by X-ray crystallography by M.F.C. Ladd and R.A. Palmer, 3rd ed. Plenum Press, ISBN: 0-306-44290-6.
7. Time-resolved Diffraction ed J.R.Helliwell and P.M. Rentzepis, Oxford Science Publications, 1997, ISBN: 0-19-850032-7.
8. A Practical Guide to X-Ray Crystallography of Biomacromolecules by Albrecht Messerschmidt, Wiley-VCH for publication January 14 2007, ISBN: 3527313966.
9. Sackler database, Royal Society.

1965

© Nobel Foundation [1]

Robert Burns Woodward

"for his outstanding achievements in the art of organic synthesis" [2]

During his presentation speech A. Fredga, then a member of the Nobel Committee for Chemistry of the Royal Swedish Academy of Sciences, commented of Woodward [2]:

> *...Professor Woodward's research work covers vast and various fields in Organic Chemistry. A leading feature is that the problems have been extremely difficult and that they have been solved with brilliant mastery. He has attacked them with a maximum of theoretical knowledge, a never-failing practical judgement and, not least, a genial intuition. He has, in a conspicuous way, widened the limits for what is practically possible. As a stimulating example he has exerted a profound influence on the organic chemistry of today...*

Robert Burns Woodward was born in Boston on 10th April 1917, the only child of Arthur Chester and Margaret Burns Woodward. He married Irja Pullman in 1938, and had two daughters, Siiri Anne (1939) and Jean Kirsten (1944). He married Eudoxia Muller in 1946 and also had two children, Crystal Elisabeth (1947), and a son, Eric Richard Arthur (1953). Both marriages ended in divorce. Woodward died of a heart attack on 8th July 1979 at his home on Memorial Drive, Cambridge [2,3,4].

The Early Years

Woodward attended local public schools in Quincy, a suburb of Boston, for both his primary and secondary education. His first interests were in mathematics and literature of all types. His father died in 1933. As a schoolboy, Woodward used to devise syntheses for a variety of natural products. He liked to say that his synthesis of quinine was planned when he was 12 years old.

In the autumn of 1933, at the age of 16, Woodward enrolled at MIT to study chemistry. He quickly made it clear that he wanted to spend all of the time as an undergraduate in the library and in the laboratory, and to take his final examinations without attending the set courses, and to forget about compulsory courses in physical education. Woodward was expelled from MIT for failing to attend the physical education course at the end of the autumn term of 1934. He nevertheless published his first paper that year, with W. J. Hall [5]. He was allowed to re-enrol in the fall term of 1935 and took his BS degree in 1936 - but only after a brief but highly intensive effort in the gymnasium in order to fulfill one of the requirements for graduation.

He went on to receive his doctorate degree also aged 20. Woodward was a Instructor during the summer of 1937 at the University of Illinois and returned to MIT, as a Research Assistant to E. P. Kohler, just before the latter's death.

The Career

Year	*Post*	*Institute*
1938	Junior Fellow	Harvard
1941	Instructor	Harvard Chemistry Department
1944	Assistant Professor	Harvard
1946	Associate Professor	Harvard
1950	Full Professor	Harvard
1953	Morris Loeb Professor	Harvard
1960	Donnor Professor of Science	Harvard
1963	Director	Woodward Research Institute, Basel
1973–74	Todd Professor and Fellow Christ's College	Cambridge, England

Woodward elucidated the structures of many natural products and developed the synthesis of a number of others including:

Structure Elucidations:

Penicillin (1945); Strychnine (1948); Patulin (1949); Terramycin (1952); Aureomycin (1952); Cevine (1954); Magnamycin (1956); Gliotoxin (1958); Oleandomycin (1960); Streptonigrin (1963) and Tetrodotoxin (1964).

Syntheses:

Quinine (1944); Patulin (1950); Cholesterol (1951); Cortisone (1951); Lanosterol (1954); Lysergic acid (1954); Strychnine (1954); Reserpine (1956); Ellipticine (1959); Chlorophyll (1960); Mycarose and Cladinose (1962); Tetracycline (1962); Colchicine (1963); Cephalosporin C (1965); Vitamin B_{12} (with A. E. Eschenmoser); Prostaglandin F_{2a} (1973) and Marasmic acid (1976). Progress towards the synthesis of erythromycin was the text of one of his lectures. The synthesis was completed after his death and written up by 49 authors in three papers [6,7,8].

General:

Structure and ultraviolet spectra of ketones (1941), Structure and aromaticity of ferrocene and other metallocenes (1952), Structure and rotary dispersion of ketones (the octant rule ((1961), conservation of orbital symmetry (1965).

Biogenetic Theory:

Indole alkaloids (1948), Steroids (1953), Macrolides (1956)

Synthetic Methods:

Polypeptides (1947), peptides (1961)

The structure below is of vitamin B_{12}

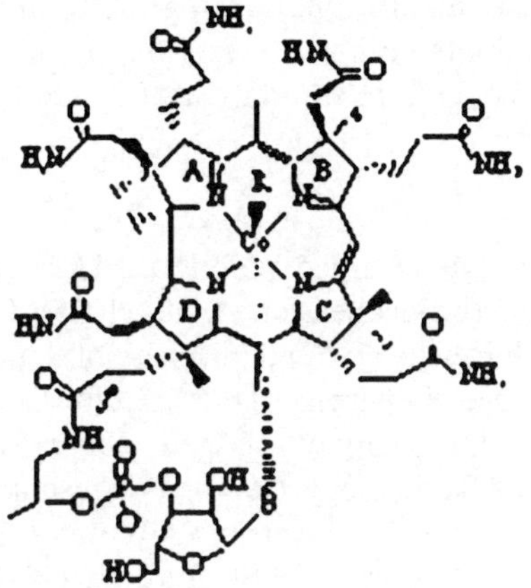

During the course of his career, Woodward published a total of 196 publications, of which 85 or so are full papers, the remainder being preliminary communications, lecture texts and reviews. Much of his achievements are documented in published lectures, short communications and through the theses and laboratory notebooks of his Harvard students and collaborators.

He co-founded *Tetrahedron* and *Tetrahedron Letters* (with Sir Robert Robinson) in 1958, and served as co-chairman until his death.

Woodward's Nobel Prize acceptance speech is dated 11th December 1965 with the title 'Recent advances in the chemistry of natural products' During his speech he commented [2]:

> *...I have chosen to present an account of an entirely new and hitherto unreported investigation which, I hope, will illuminate many facets of the spirit of contemporary work in chemical synthesis.*
>
> *Cephalosporin C, a product of the metabolism of Cephalosporium acremonium, was isolated in 1955 by Newton and Abraham in an investigation notable for its perspicacity as well as its painstaking attention to detail. The investigation of the structure of the metabolite was successfully concluded in 1961 through studies in which both chemical and X-ray crystallographic techniques were employed. The molecular array*

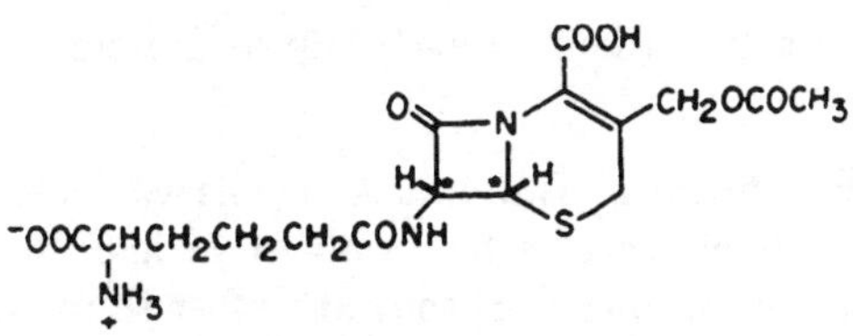

The structure of Cephalosporin C

thus laid bare strikes one at once as having affinities with a hitherto well-known class of substances which has constituted one of the most challenging and recalcitrant synthetic objectives of our generation. I refer of course to the penicillins, of which penicillin G (II) - one of the earliest known and one which has been widely used in medicine-may serve as an example...

Woodward was an extremely prolific synthetic organic chemist and thus made contributions to all aspects of organic synthesis. Some of the syntheses mentioned above have been reproduced in the collections of classic syntheses [9]. The Woodward-Hoffmann rules, which were developed with R. Hoffmann (Chemistry Nobel Laureate 1982) during the total synthesis of vitamin B_{12}, now form part of most standard undergraduate courses – see [10]. Woodward's use of functional protection during some of his work is also covered in standard texts – see for example [11].

Honours and Awards

The following Universities and Colleges awarded Woodward Honorary Doctorates: Wesleyan University (1945); Manchester University (1954); Bucknell University (1955); University of New Brunswick (1956); Yale University(1956); Harvard University (1957); University of Southern California (1959); University of Chicago (1961); New England College of Pharmacy (1961); Colby College (1963); University of Cambridge (1964); Brandeis University (1965); University of Glasgow (1966); Stonehill College (1966); University of Sheffield (1966); Israel Institute of Technology (1966); Polytechnic Institute of Brooklyn (1967); Eidgenössische Technische Hochschule (1967); University of Western Ontario (1968); Columbia University (1969); University of Louvain (1970); University of St. Andrews (1976); The University of London (1976).

Woodward was presented with the following Awards: the John Stott Medal (1945); the Baekeland Medal (1955); the Ledie (1955); the Research Corporation Award (1955); the Nichols Medal (1956); the Synthetic Organic Chemistry Award (1957); the Richards Medal (1958); the Davy Medal (1959); the Roger Adams Medal (1961); the Pius XI Gold Medal (1961); the Acientific Achievement Medal (1961); the Priestley Medallion (1962); the Stas Medal (1962); the Gold Medal for Creative Research in Synthetic Organic Chemistry (1962); the National Medal of Science (1964); the Kirkwood Medal (1965); the Nobel Prize in Chemistry (1965); the Willard Gibbs Medal (1967); the Lavoisier Medal (1968); The Order of the Rising Sun second class (1970); the Hanbury Memorial Medal (1970); the Pierre Bruylants Medal (1970); the Scientific Achievement Award (1971); the Dr. B. C. Law Gold Medal (1972); the Arthur C. Cope Award joint with R. Hoffmann (1973).

He held Honorary Fellowships/Memberships of many Academies and learned societies including:

Member of the National Academy of Sciences; Fellow of the American Academy of Arts and Sciences; Honorary Member of the Royal Institution of Great Britain; Foreign Member of the Royal Society (1956).

He was a member of: the Committee on Medical Research, and of the Office of Scientific Research and Development (1944–45); on the War Production Board (1944–45); a Director, Pfizer and Co Inc (1951–70); a Member of the Corporation, Massachusetts Institute of Technology (1966–71); a Director of Ciba-Geigy Ltd, Basel (1970–1979); the Board of Governors Weizmann Institute of Science (1968–1979).

Woodward's's citation for election (date: 26/04/1956) as Foreign Member of the Royal Society reads [12]:

'Professor Robert Burns Woodward of Harvard distinguished for outstanding investigations in organic chemistry with special reference to the synthesis of important natural products such as quinine, various alkaloids and antibiotics, cortisone and other steroids.'

No proposers are listed in the Sackler database.

Bibliography

1. Reproduced with permission.
2. From Nobel Lectures, Chemistry 1963–1970, Elsevier Publishing Company, Amsterdam, 1972.
3. D Barton and H Wasserman, Tetrahedron Letters, 1979, 38, i–x.
4. Biographical Memoirs of Fellows of the Royal Society 1981, 27, 628–695.
5. W. J. Hall and R. B. Woodward, Ind. Eng. Chem. Anal. Ed., 1934, 6, 478.
6. J. Amer. Chem. Soc. 1981, 3210.
7. J. Amer. Chem. Soc. 1981, 3213.
8. J. Amer. Chem. Soc. 1981, 3215.
9. Classics in Total Synthesis: Targets, Strategies, Methods by K. C. Nicolaou and E. J. Sorensen, Wiley-VCH 1996, ISBN: 3527292314.
10. Elements of Molecular Symmetry by Yngve Öhrn, Wiley-Interscience, 2000, ISBN: 0471363235.
11. Protective Groups in Organic Synthesis by Theodora W. Greene and Peter G. M. Wuts, Wiley-Interscience, 3rd ed. 1999, ISBN: 0471160199.
12. Sackler database, Royal Society.

1966

© Nobel Foundation [1]

Robert S. Mulliken

"for his fundamental work concerning chemical bonds and the electronic structure of molecules by the molecular orbital method" [2]

During his presentation speech Professor Inga Fischer-Hjalmars, from the University of Stockholm, commented of Mulliken and his work [2]:

...but Mulliken has developed the general scheme of an elegant method to combine the theoretical computations with experimental information from small molecules. This kind of calculation is exceedingly illuminating in several connections, for instance for the interpretation of measurements. Just as for small molecules the method has also been used to gain information about molecules, inaccessible to experiments, such as compounds of importance for life processes. In such cases the theoretical results cannot be directly compared with measurements, but they can suggest new kinds of experiments...

...Professor Mulliken, it is now more than forty years since you started to investigate what the electron is doing in the molecule. Your deep understanding of the physical laws, governing the behaviour of the electron, combined with your intimate knowledge of the problems of Chemistry have greatly advanced our understanding of the properties of molecules. Especially, by the development of the molecular-orbital method you have given us a powerful key to the mechanism of the chemical bond. Your thorough penetration of the inherent possibilities of the method and your unfailing enthusiasm in guiding the development have very lately pushed the progress to the point of accurate, quantitative applications to chemical problems. By these achievements you have opened a road of exciting possibilities for future, theoretical investigations of molecular properties...

Robert Sanderson Mulliken was born in Newburyport, Massachusetts, on 7th June 1896, the son of Samuel Parsons Mulliken, Professor of Organic Chemistry at MIT, and Katherine W. Mulliken. He married Mary Helen von Noè on 24th December 1929. They had two daughters Lucia Maria (Mrs. John P. Heard) and Valerie Noè. Mulliken died on 31st October 1986 [2–4].

The Early Years

Mulliken's interest in chemistry, particularly theoretical chemistry, developed at an early age. When 17 years of age (1913), Mulliken read Bohr's quantum theory of the atom. He chose to present an essay entitled *'Electrons-what they are and what they do'* at his high school graduation ceremony [5]. He took his B.Sc. degree at the Massachusetts Institute of Technology (1917) and his Ph.D. at the University of Chicago (1921). His studies were interrupted during the War years when he became a Private, First Class, in the Chemical Warfare Service, US Army (in 1918). His first paper on *'Relation between Intensity of Tyndall Beam and Size of Particles'* was published during his doctorate years [6].

The Career

Year	*Post*	*Institute*
1917–18	Junior Chemical Engineer	Bureau of Mines, US Department of Interior, Washington DC
1919	Assistant in Rubber Research	New Jersey Zinc Company, Pennsylvania
1926	Assistant Professor of Physics	Washington Square College, New York University
1928	Associate Professor of Physics	University of Chicago
1931–61	Professor of Physics (and Chemistry, 1961)	University of Chicago
1942–1945	Director	Editorial Work and Information, Plutonium Project, University of Chicago
1956–61	Ernest de Witt Burton Distinguished Service Professor	University of Chicago
1961–83	Distinguished Service Professor of Physics and Chemistry	University of Chicago
1964–72	Distinguished Research Professor of Chemical Physics	Florida State University (January to March)
1973	Emeritus Professor	

Mulliken spent periods of time visiting universities through scholarships and fellowships including: National Research Council Fellow, University of Chicago and Harvard University (1921–25); Guggenheim Fellow, Germany and Europe (1930 and 1932–33); Fulbright Scholar, Oxford University (1952–54); Visiting Fellow, St. John's College, Oxford (1952–53).

Mulliken undertook work on a variety of subjects including:

- computations on conjugated organic molecules
- work on dipole moments, electronegativities, and valence-state energies
- a study of isotopes and on diatomic band spectra
- the use of interatomic overlap integrals as measures of bond energies
- the theoretical interpretation of absolute intensities of electronic spectra
- aspects of molecular-orbital theory

- the quantum-mechanical statement of the concept of hyperconjugation
- the structure and spectra of molecular complexes
- a theoretical systematization of the electronic states of molecules
- work on valence theory and molecular structure

Mulliken published several books including his autobiography [7] and: *Diatomic molecules: Results of ab initio calculations;* William Draper Harkins: 28th December 1873-7th March 1951 and *Spectroscopy, molecular orbitals, and chemical bonding*: Nobel lecture.

His Nobel Prize acceptance speech is dated 12th December 1966 with the title 'Spectroscopy, molecular orbitals, and chemical bonding'. During the course of his presentation, and to set the scene, Mulliken posed the question [2]:

> *...Let me now ask, what is a molecular orbital? A really adequate answer is unavoidably technical. However, in an effort to make matters as clear as possible shall begin this lecture by reviewing a number of things which may be regarded as uninteresting old history, or else as boringly well known, at least by physical scientists. For this approach I beg your indulgence and ask your forgiveness...*

Later on in his presentation he went on to answer his previously posed question,

> *...Now to attempt an answer to the question posed earlier, an* ***orbital*** *means, roughly,* ***something like an orbit****; or, more precisely,* ***something as much like an orbit as is possible in quantum mechanics****. Still more precisely, the term "orbital" is simply an abbreviation for* ***one-electron orbital wavefunction*** *or, preferably,* ***for one-electron orbital eigen-function****. The last-mentioned expression refers to any one of the so-called characteristic solutions or eigen-functions of Schrödinger's quantum-mechanical wave equation* ***for a single electron*** *in an atom or molecule.*
>
> *According to a picturesque expression once used by Van Vleck, a set of orbitals represents a housing arrangement for electrons. A very strict rule (Pauli's exclusion principle) applies to every orbital, whether atomic or molecular, namely that not more than two electrons can occupy it. In other words, it can be empty, or it can hold one electron, or it can hold two electrons. Every electron has a spin, like the earth on its axis; if there are two electrons in an orbital, their spins are oppositely directed.*
>
> *An* ***atomic orbital*** *(abbreviated AO) is best taken to be an eigen-function of a one-electron Schrödinger equation which is based on the attraction of the nucleus for the electron we are considering plus the average repulsion of all the other electrons. Following Hartree, this average field is called a* ***self-consistent field*** *because of the way the calculations are made; namely, the orbital for each electron in turn is calculated assuming all the other electrons to be occupying appropriate orbitals.*
>
> *A* ***molecular orbital*** *(abbreviated MO) is defined in exactly the same way, except that its one-electron Schrödinger equation is based on the attractions of two or more nuclei plus the averaged repulsions of the other electrons.*
>
> *An orbital (either atomic or molecular) is, strictly speaking, just a mathematical function in ordinary 3-dimensional space. When an electron is occupying an orbital, the form of the orbital tells us, among other things, what fraction of time the electron in it can spend in different regions of space around the nucleus, or nuclei. Each orbital* ***favours*** *some particular regions of space and disfavours others, yet all the orbitals in a given atom or molecule extend at least to some small*

extent throughout all regions of the atom or molecule. Orbitals differ most strikingly from the orbits of Bohr theory in the fact that they do not specify detailed paths for electrons. They do, however, give some information about average speeds as well as positions of electrons.

I have always felt that a ***true*** *AO or MO (sometimes I have called it a best AO or MO) for an electron is one which corresponds to a self-consistent field as above described. Commonly, however, the word orbital is used to refer to mathematical functions which are only* ***approximations*** *to these true AO's or MO's. The main reason for this fact is that until recently we have had in most cases, especially for MO's, only a rather roughly approximate knowledge of the true forms...*

Quantum mechanics is as much a subject of physics as it is of chemistry, many texts have been written on the subject from both perspectives, see for example [8–10].

Honours and Awards

Mulliken received Honorary Sc.D. degrees from: Columbia University (1939); Marquette University (1967); Cambridge University (1967). He was also presented with an Honorary Ph.D. from University of Stockholm.

He was presented with many Awards and Prizes including: the Bronze Medal Award, University of Liege (1948); the Theodore William Richards Gold Medal, Northeastern Section of the American Chemical Society (1960); the Gilbert Newton Lewis Gold Medal, California Section of the American Chemical Society (1960); the Peter Debye Award, California Section of the American Chemical Society (1963); the John Gamble Kirkwood Award, New Haven Section of the American Chemical Society (1964); the Willard Gibbs Medal Award, Chicago Section of the American Chemical Society (1965); Gold Medal Award for Scientific Achievement, City College Chemistry Alumni Association, and 15th Bicentennial Lecturer, City College of New York (1965); the Nobel Prize in Chemistry, Stockholm, Sweden (1966); the Chicagoan of the Year Award, Chicago Press Club (1966).

Mulliken held Fellowships/Memberships with many Institutes and Academies including: Fellowship in the American Physical Society, the American Academy for the Advancement of Science and the International Academy of Quantum Molecular Science; Membership of the American Academy of Arts and Sciences, the American Chemical Society, the American Philosophical Society, the National Academy of Sciences, the Cosmos Club (Washington, D.C.) and the Quadrangle Club (Chicago, Illinois). He was: Honorary Fellow of the Chemical Society of Great Britain and the Indian National Academy of Science; Foreign Member of the Royal Society of London; Honorable Member of the Société de Chimie Physique, the Royal Irish Academy and the Chemical Society of Japan; Corresponding Member of the Société Royale des Science de Liege.

Mulliken was a Scientific Attaché, U.S. Embassy in London (1955).

He also visited many Institutes and was a 'named' Lecturer for many Societies including: Fisher Baker Lecturer, Cornell University (1960); Indian Atomic Energy Establishment, Bombay (1962); Indian Institute of Technology, Kanpur (1962) and Silliman Lecturer, Yale University (1965). He was Visiting Professor in Bombay (1962) and Kanpur (1962); Jan van Geuns Visiting Professor, University of Amsterdam (1965).

Mulliken's citation for election (date: 20/04/1967) as Foreign Member of the Royal Society reads [11]:

'Professor Robert Sanderson Mulliken of Chicago distinguished for his pioneering studies in the application of quantum mechanics to molecular structure.'

No proposers are listed in the Sackler database.

Bibliography

1. Reproduced with permission.
2. From Nobel Lectures, Chemistry 1963–1970, Elsevier Publishing Company, Amsterdam, 1972.
3. Biographical Memoirs of Fellows of the Royal Society, 1990, vol. 35, pp. 327–354.
4. Who was who: 1981–1990.
5. Chemistry 1967, 40, 13–15.
6. R. C. Tolman, R. H. Gerke, A. P. Brooks, A. G. Herman, R. S. Mulliken and D. DeW. Smyth, J. Amer. Chem. Soc 1919, 41, 575–87.
7. Robert S. Mulliken. Life of a scientist autobiography (ed B. J. Ransil), New York:Springer-Verlag ISBN: 1-0-3875037-57.
8. Valency and Bonding: A Natural Bond Orbital Donor-Acceptor Perspective by Frank Weinhold and Clark R. Landis, Cambridge University Press, 2005, ISBN: 0521831288.
9. Introduction to Quantum Mechanics (2nd Edition) by David J. Griffiths, Prentice Hall, 2nd ed. 2004, ISBN: 0131118927.
10. Mathematical Foundations of Quantum Mechanics (Dover Books on Mathematics) by George W. Mackey, Dover Publications, 2004, ISBN: 0486435172.
11. Sackler database, Royal Society.

1967

Manfred Eigen (0.5)

"for their studies of extremely fast chemical reactions, effected by disturbing the equlibrium by means of very short pulses of energy" [2]

During his presentation speech H.A. Ölander, then a member of the Nobel Committee for Chemistry of the Royal Swedish Academy of Sciences, commented of Eigen [2]:

> *...Eigen has invented several methods, however. If, say, a solution of acetic acid is subjected to a high-tension electric pulse, more molecules of this substance are dissociated than else would be the case in an aqueous solution. That takes a certain length of time. When the electric pulse is turned off, the solution goes back to its former equilibrium; this also takes some time, and that relaxation can be recorded.*
>
> *The shock current caused by the application of the high-tension pulse will heat the solution a few degrees. Most chemical equilibria are slightly displaced when the temperature is changed, and the rapid establishment of the new equilibrium after heating can be recorded...*
>
> *Although Eigen starts his reactions in another way than that employed by Norrish and Porter, the instruments that record the fast reactions are largely identical for both research groups.*
>
> *The chief importance to chemists of the methods worked out by Eigen, Norrish and Porter is their usefulness for the most widely diverse problems. A great many laboratories round the world are now obtaining hitherto undreamt-of results with these methods, which thereby fill what used to be a severely-felt gap in the means of advance available to Chemistry.*
>
> *Professor Dr. Manfred Eigen. Although chemists had long been talking of instantaneous reactions, they had no way of determining the actual reaction rates. There were many very important reactions of this type, such as the neutralization of acids with alkalis. Thanks to you, chemists now have a*

whole range of methods that can be used to follow these rapid processes, so that this large gap in our chemical knowledge has now been filled...

Manfred Eigen was born in Bochum, Germany on 9th May 1927, the son of the chamber musician Ernst Eigen and his wife Hedwig, née Feld. Eigen married Elfriede, née Müller (1952). They have two children, Gerald (1952) and Angela (1960). Eigen can be contacted through the Max-Planck-Institut für biophysikalische Chemie, Göttingen [2,3].

The Early Years

Following his school years at the Bochum humanistic Gymnasium, Eigen enrolled at the University in Göttingen, where he read physics and chemistry (1945–50). His dissertation was under Arnold Eucken, on the specific heat of heavy water and aqueous electrolyte solutions. He obtained his doctorate (Dr. rer. Nat) in natural sciences in 1951. He spent a further two years at the university as a Assistant Lecturer in the Physical Chemistry Department under Ewald Wicke, after which he transferred to the Max-Planck-Institut für physikalische Chemie, which at that time had moved to Göttingen. His mentor during this time was Karl Friedrich Bonhoeffer, who later influence Eigen to work in biophysical chemistry.

The career

Year	*Post*	*Institute*
1953	Assistant	Max-Planck-Gesellschaft
1958	Research Fellow	Max-Planck-Gesellschaft
1962	Director of the independent Kinetics Department	Max-Planck-Institute for Physical Chemistry
1964	Director	Max Planck Institute for Physical Chemistry
1967	Managing Director of the Institute for a three year term	Max Planck Institute for Physical Chemistry
1971	Honorary Professor	Department of Medicine at the University of Göttingen
1987	Honorary Senator	Georg-August-Universität, Göttingen

In addition to studying fast ionic reactions, Eigen's work also covered:

- developing a series of measuring techniques involving times down to the order of a nanosecond (with Leo de Maeyer from 1954)
- thermodynamic properties of aqueous solutions and water
- the theory of electrolytes
- thermal conductivity
- sound absorption
- the neutralization rate
- a study of the anomalous conduction characteristics of protons in ice crystals
- the development of the theory of relaxation of multi stage processes

- studies on metal complex reactions
- a general theory of acid-base catalysis (which was verified experimentally)

In the latter phase of his career he concentrated on some biochemical issues such as:

- biological control and regulation processes
- biopolymerisation
- code transfer dynamics
- enzymes and lipid membranes
- hydrogen bridges of nucleic acids
- information storage in the central nervous system
- origin and evolution of life (theory and experiments)
- sequential kinetics

Eigen's work has been published in close to 100 papers. He has also written several books including: *The Hypercycle - A Principle of Natural Self-Organization* (with Peter Schuster), Springer-Verlag, Berlin, 1979; *Perspektiven der Wissenschaft,* Deutsche Verlagsanstalt, Stuttgart, 1988; *Steps Towards Life: A Perspective on Evolution*, Oxford University Press, Oxford, 1992 and *Laws of the Game: How the Principles of Nature Govern Chance* (with Ruthild Winkler), Princeton University Press, Princeton, 1993.

Eigen's Nobel Prize acceptance speech is dated 11th December 1967 with the title 'Immeasurably fast reactions'. During the course of his presentation, Eigen posed the question [2]:

> *...We generally describe as "fast" anything that takes place quickly compared to the rate of resolution of our sense perceptions. However, since our perceptions are in turn based on chemical processes - including the charge neutralization reactions that have just been mentioned - these processes must necessarily be "fast" - indeed "extremely fast".*

But what does "immeasurable" mean? If we wish to measure the rate of a chemical reaction, we must do two things: (1) induce the reaction; (2) follow the course of the reaction in time.
Chemists generally induce a reaction by mixing the reagents with each other. This must be done very quickly, and here the first difficulty appeared:...
He went on to comment on some of his work which he presented under the following headings:

- *A trip to the seaside*
- *Back to physics*
- *Historical review*
- *New paths*
- *Things get even simpler*
- *On the way to technical perfection*
- *Applications*
    - *For the inorganic chemist: aperiodic system of reaction rates*
    - *For the organic chemist: unused possibilities*
    - *For the biochemist: intelligent molecules*
- *Where now*
- *Epilogue*

The 'where now' section is reproduced below in its entirety as it illustrates Eigen's thinking at the time of his award:

...We are just beginning to understand how molecular reaction systems have found a way to "organize themselves". We know that processes of this nature ultimately lead to the life cycle, and that (for the time being?) Man with his central nervous system, i.e. his memory, his mind, and his soul, stands at the end of this development and feels compelled to understand this development. For this purpose he must penetrate into the smallest units of time and space, which also requires new ideas to make these familiar concepts from physics of service in understanding what has, right into our century, appeared to be beyond the confines of space and time...

Further information on general aspects of fast chemical reactions and the equipment needed for them see references [4–6].

Honours and Awards

Eigen holds many Honorary Degrees including: D.Sc. h.c Washington University, St. Louis, USA (1966); Harvard University, USA (1966); University of Chicago, USA (1966); University of Nottingham, UK (1968); University of Hull, England (1976); University of Cambridge, UK (1982); Utah State University, Logan, USA (1990). He received the degree of: D.Ph. h.c. from the Hebrew University, Jerusalem, Israel (1973). He was also Honorary Doctor of: the University of Bristol (1978); University of Debrecen, Hungary (1982); the University of Bielefeld (1985). He was also presented with Dr. rer. nat. h.c. of the Technische Universität Munich (1983).

Eigen has been presented with many awards and prizes including: the Bodenstein Prize of the Deutsche Bunsengesellschaft (1956); the Otto Hahn Prize for Chemistry and Physics (1962); the Kirkwood Medal of the American Chemical Society (1965); the Harrison Howe Award of the American Chemical Society(1966); the Nobel Prize in Chemistry, jointly (1967); the Linus Pauling Medal of the American Chemical Society (1967); the Carus Medal of the Deutsche Akademie der Naturforscher Leopoldina, Halle (1968); the Carus Prize of the Town of Schweinfurt (1968); the Paracelsus Medal of the Schweizerische Chemische Gesellschaft (1968); the Keilin Medal of the Biochemical Society, UK (1969); the Intra-Science Award of the Intra-Science Research Foundation, Santa Monica, USA (1971); the Austrian Decoration for Arts and Sciences (1976); the Faraday Medal of the Chemical Society, London (1977); the Prize of the Country of Niedersachsen (1980); the Medal of the Government of Niedersachsen in Hanover (1982); the Paul Ehrlich and Ludwig Darmstädter Prize (1992); the Hermann von Helmholtz-Medal of the Berlin-Brandenburgische Akademie der Wissenschaften, Berlin (1994); the Literature Prize of the Fonds der Chemischen Industrie, Frankfurt/Main (1994); the Karl Küpfmüller-Ring of the Technische Hochschule Darmstadt (1994); the Merck-Prize, Darmstadt (1996); the Diesel Medal (1996).

He is a Member of the following Institutes/Academies: Honorary Fellow of the Indian Chemical Society (1974); the Royal Society of Chemistry, London (1981) and the Royal Society of Edinburgh, Scotland (1983). He is a Foreign Honorary Member of the American Academy of Arts and Sciences, Boston, USA (1964) and the Weizmann Institute of Science, Israel (1969). He is a Foreign Member: of the Royal Danish Academy of Sciences (1971); the Royal Society, London (1973); the USSR Academy of Sciences now the Russian Academy of Sciences (1976); the Science Academy in the Institut de France, Paris (1978). He is Honorary Member of: the American Association of Biological Chemists (1966); the Österreichische Akademie der Wissenschaften (1970); the Deutsche Bunsengesellschaft (1979); the Hungarian Academy of Sciences (1991). He is a Member of: the Deutsche Akademie der Naturforscher Leopoldina, Halle (1964); Member of the Akademie der Wissenschaften, Göttingen (1965); the Popal Academy of Sciences, Rome (1981); the Salk Institute's Society of Honorary Fellows

(1985); the Berlin-Brandenburgische Akademie der Wissenschaften (1990); the Croatian Academy of Sciences and Arts (1992); the Society for the Development in the Field of Medicine (1992). He is: a Corresponding Member of the mathematisch-naturwissenschaftliche Klasse der Bayerischen Akademie der Wissenschaften, Munich (1972) and of the Senckenbergische Naturforschende Gesellschaft, Frankfurt (1977); a Foreign Associate of the National Academy of Sciences, Washington, USA (1966); a Fellow of the Royal Society of Arts, London (1972); a Associate of the Académie Royale des Sciences, des Lettres et des Beaux Arts, Belgium (1977) and of the Berlin-Brandenburgische Akademie der Wissenschaften, Berlin (1993).

He is also: Andrew D. White Professor at Large, Cornell University, Ithaca, USA (1965); Titular Professor at the Technische Hochschule Braunschweig (1965); appointed to the Scientific Council of the German Federal Republic (1967); Palmes Académiques (1972); a Member of the Orden pour le mérite (1973); Fogarty Lecture in Bethesda, USA (1995); Walker-Ames Professor at the University of Washington, Seattle, USA (1996).

There is no entry in the Sackler database of Eigen's citation concerning his election as Foreign Member of the Royal Society [7].

Bibliography

1. Reproduced with permission.
2. From Nobel Lectures, Chemistry 1963–1970, Elsevier Publishing Company, Amsterdam, 1972.
3. Who's who: 2005.
4. Chemical Reaction Engineering, 3rd Edition by Octave Levenspiel, Wiley, 3rd ed. 1998, ISBN: 047125424X.
5. The Mechanisms of Fast Reactions in Solution by E. F. Caldin, Ios. Pr. Inc., 2001, ISBN: 1586031031.
6. Fundamentals of Chemical Reaction Engineering by Mark E. E. Davis and Robert J. J. Davis, McGraw-Hill Science/Engineering/Math, 1st ed. 2002, ISBN: 007245007X.
7. Sackler database, Royal Society.

1967

© Nobel Foundation [1]

Ronald George Wreyford Norrish (0.25)

"for their studies of extremely fast chemical reactions, effected by disturbing the equlibrium by means of very short pulses of energy" [2]

During his presentation speech, H.A. Ölander, then a member of the Nobel Committee for Chemistry of the Royal Swedish Academy of Sciences, commented of Norrish [2]:

...Ever since the 1920's, Professor Norrish has been studying reaction kinetics and he was one of the leading scientists in this field. A younger associate joined him in the late 1940's in the person of George Porter. They decided to make use of a flash lamp, the kind you have seen photographers use. The only difference was that they made their lamp thousands of times more powerful. Indeed, subsequent refinements have led to the construction of such lamps with an effect greater than the total effect which the whole city of Stockholm consumes on a winter afternoon with the lights turned on and the factories still humming before closing time - and that is 600000 kilowatts. There is just one catch, however; this enormous effect in the lamp lasts no more than one millionth of a second or so. Still, in this way much if not most of a substance in a tube next to the flash lamp can be converted into an activated form, or the molecules broken up so as to yield atom groups with a high reactivity. It then becomes possible to study these newly formed molecules spectroscopically, but since they react so readily, this must be made extremely fast. Thanks to modern electronic equipment, however, these rapid processes can be recorded.

The new method developed by Norrish and Porter enabled them to study at first hand many fast reactions which one had previously only guessed that they took place. I cannot begin to enumerate even a sample of the reactions which Norrish and Porter, not to mention a great many other scientists, have investigated with this method. Suffice it to say that, in an earlier day, the study of these short-lived high-energy molecules and their chemical characteristics could hardly even have been contemplated as a wild dream...

...Professor Ronald George Wreyford Norrish, Professor George Porter. Photo-reactions have been studied by chemists for more than two hundred years, but the detailed knowledge of the

behaviour of the activated molecules was meagre and most unsatisfactory. By your flash photolysis method you have provided us with a powerful tool for the study of the various states of molecules and the transfer of energy between them...

Ronald George Wreyford Norrish was born in Cambridge, England on the 9th November 1897, the son of Herbert and Amy Norrish. Amy died in 1905, his father re-married some three years later. Norrish married Annie Smith (1926). They had twin daughters. Norrish died in Addenbrooke's hospital on 7th June 1978 [2–4].

The Early Years

Norrish won a scholarship to the Perse Grammar School, a local boarding school in Cambridge. He passed his entrance scholarship to Emmanuel College, Cambridge to read natural sciences in 1915, but left to take up a commission in the Royal Field Artillery, for service in France (1916).

He was captured by the Germans in March 1918 and spent the rest of the war in Germany as a prisoner of war, first at Rastatt and later at Graudenz in Poland. He was repatriated in 1919 and returned to Emmanuel College to complete his studies. Norrish graduated with a 1st class honours degree in Natural Sciences (tripos I in 1920, and tripos II in 1921).

The Career

Year	*Post*	*Institute*
1925	Demonstrator and Research Fellow of Emmanuel College	Cambridge University
1930	Humphrey Owen Jones Lecturer in Physical Chemistry in the Department of Chemistry	Cambridge University
1937	Director of Department and Professor	Cambridge University
1965	Retirement	

Norrish was a Fellow of Emmanuel College, Cambridge.

During the Second World War, Norrish was involved with a variety of research work connected with various ministries whilst still directing the Department of Physical Chemistry, and also teaching at the university. He was also involved in some of the government committees of the time.

Whilst Norrish is best known for his work on photochemistry he also studied:

- combustion
- co-polymerization
- the development of flash photolysis and kinetic spectroscopy
- general issues of photochemistry
- the importance of photochemistry and spectroscopy to chemical kinetics
- reaction kinetics short lived transients in chemical reactions

Norrish published some 246 papers in the scientific literature. His personal papers are deposited in Cambridge University Library.

Norrish's Nobel acceptance speech is dated 11th December 1967 with the title 'Some fast reactions in gases studied by flash photolysis and kinetic spectroscopy'. He started his presentation thus [2]:

> *...Realisation that free radicals and atoms take part in chemical reactions has focused attention on the processes of photo-chemistry which are not only paramount in the geochemistry of the upper atmosphere but are also basic to many reactions of organic chemistry involving free radicals and the triplet state; this realization also has led to the development of gas lasers, and to the exploration in detail of the intimate anatomy of reactions of pyrolysis, combustion and explosion.*
>
> *Classical photochemistry emerged in 1908 with the understanding by Stark of the distinction between primary and secondary photochemical processes, of which the former is the immediate result of the absorption of a light quantum by a molecule or atom and the latter the subsequent 'dark' reactions initiated by the products of the former. Into this simple pattern it has been possible to fit the whole gamut of photochemical phenomena – fluorescence, phosphorescence, photolytic and photosynthetic processes, photocatalytic and photosensitized reactions...*

He went on to describe some of his own work under the headings:

- Vibrational excitation by primary reaction
    - The flash photolysis of nitrosyl halides – vibrational relaxation
- Vibrational excitation by secondary reaction
    - The reactions of oxygen atoms
- Application of the adiabatic method
    - The study of explosive processes exemplified by the oxidation of hydrides
- A general mechanism for the combustion of hydrides

Norrish concluded:

> *...The opportunity for the application of the methods of flash photolysis to chemical kinetics, not only in the gas phase, but also to the study of photo-chemical reactions in solution is very great, and is increasing steadily as improvement in technique gives greater time resolution, and ever increasing accessibility to reactions of the* **'vacuum** *untraviolet'...*

Photochemistry over the last 30 years or so has continued to be developed and is now applied to many reaction types, see for example [5–8].

Honours and Awards

Norrish was presented with Honorary Doctorates from various Universities including: the Sorbonne (1958); Leeds (1965); Sheffield (1965); Lancaster (1968); Liverpool (1968); British Columbia (1969).

He received many awards and prizes including: the Meldola Medal of the Royal Institute of Chemistry (1926); the Davy Medal of the Royal Society (1958); the Lewis Gold Medal of the Combustion Institute (1964); the Liversidge Medal of the Chemical Society; the Faraday Medal of the Chemical Society (1965); Nobel Prize for Chemistry joint (1967); the Longstaff Medal (1969).

Norrish was: a Fellow of the Royal Society; a Fellow of the Royal Society of Arts; an Hon. Fellow of the Royal Society of Edinburgh; held honorary membership of the Belgian Academy of Sciences, the Bulgarian Academy of Sciences, the Göttingen Academy of Science, the New York Academy of Science, the Polish Academy of Science, Polish Chemical Society, the Royal Society of Science of Liege, the Royal Society of Sciences in Uppsala, the Societe de Chimie Physique; was a Foreign Member of the Polish, the Bulgarian Academy of Sciences; he was a Corresponding Member of the Academy of Sciences in Göttingen and of the Royal Society of Sciences in Liege and Membre d'honneur of the Société de Chimie Physique in Paris.

He was awarded the Knights Cross of the Order of Poland Restituta (1974) and the Order of Cyril and Methodius of Bulgaria (1974).

Norrish served on the councils of: the Chemical Society; the Faraday Society (President in 1953–55); the Royal Institute of Chemistry (Vice President from 1957 to 1959); of the Senate of the University of Cambridge (1938–41). He served on the scientific advisory council of the Ministry of Supply (1942–45). He was President of the British Association Section B (Chemistry) in 1961. He was made Liveryman of the Worshipful Company of Gunmakers in 1961.

Norrish delivered the Liversidge Lecture to the Chemical Society in 1958, the Faraday Memorial Lecture to the Chemical Society in 1965, and the Bakerian Lecture to the Royal Society in 1966.

Norrish's citation for election (date: 7/05/1936) as Fellow of the Royal Society reads [9]:

'Meldola Medallist of the Institute of Chemistry. Distinguished for researches in physical chemistry with special reference to the mechanism of chemical change. He has made important advances in photo-chemistry, and in particular has investigated photokinetics, photosensitised explosions, and the relationship between molecular spectra and photochemical change.'

He was proposed by:

W. J. Pope; Rutherford; T. M. Lowry; W. H. Mills; Eric K. Rideal; Arthur Lapworth; Robert Robertson; R. H. Fowler; F. G. Donnan; W. B. Hardy.

Bibliography

1. Reproduced with permission.
2. From Nobel Lectures, Chemistry 1963–1970, Elsevier Publishing Company, Amsterdam, 1972.
3. Biographical Memoirs of Fellows of the Royal Society, 1981, vol. 27, pp. 379–424.
4. Who was who: 1971–1980.
5. Photochemistry and Reaction Kinetics ed. P. G. Ashmore, F. S. Dainton and T. M. Sugden, Cambridge University Press, 1967, Library of Congress Card Number 67-10157.
6. Organic Photochemistry 10 volume set ed. A. Padwa, Marcel Dekker Inc, 1989, ISBN: 0-8247-7920-7.
7. Organic and Inorganic Photochemistry ed. V. Ramamurthy and Kirk S. Schanze, Marcel Dekker Inc, 1998, ISBN: 0-8247-0174-7.
8. Chiral Photochemistry ed. Yoshihisa Inove and V. Ramamurthy, Marcel Dekker Inc, 2004, ISBN: 0-8247-5710-6.
9. Sackler database, Royal Society.

1967

© Nobel Foundation [1]

George Porter (0.25)

"for their studies of extremely fast chemical reactions, effected by disturbing the equlibrium by means of very short pulses of energy" [2]

During his presentation speech, H.A. Ölander then a member of the Nobel Committee for Chemistry of the Royal Swedish Academy of Sciences, commented thus of Porter [2]:

...The 1967 Nobel Laureates in Chemistry have opened up the whole of this vast field of reaction kinetics for research. They did so by applying the principle I have just mentioned: to start from a system in equilibrium and to disturb this equilibrium suddenly by one means or another...

...The new method developed by Norrish and Porter enabled them to study at first hand many fast reactions which one had previously only guessed that they took place. I cannot begin to enumerate even a sample of the reactions which Norrish and Porter, not to mention a great many other scientists, have investigated with this method. Suffice it to say that, in an earlier day, the study of these short-lived high-energy molecules and their chemical characteristics could hardly even have been contemplated as a wild dream...

George Porter was born near Doncaster in the West Riding of Yorkshire on the 6th December 1920, the son of John Smith Porter and Alice Ann Porter. He married Stella Jean Brooke on 25th August 1949 and they had two sons, John and Andrew. Porter died on 31st August 2002 in Canterbury [2–4].

The Early Years

Porter attended Thorne Grammar School (1931–1938) and Leeds University (1938–1941) as an Ackroyd Scholar. Whilst at Leeds one of his final year lecturers, M.G. Evans, inspired Porter in physical chemistry, in particular chemical kinetics. Indeed during his final year he took a special course in radio physics. Towards the latter part of his last undergraduate year, Porter became an

Officer in the Royal Naval Volunteer Reserve Special Branch, concerned with radar, serving in the Western approaches and in the Mediterranean Sea. He graduated with a B.Sc. later in that year.

His postgraduate research was under R.G.W. Norrish at Emmanuel College, Cambridge (1945), working on flow techniques of free radicals produced in gaseous photochemical reactions. One year into his doctorate work, Porter decided to initiate the free radical reactions using short bursts of high intensity light. The apparatus was constructed in early 1947, and together with Norrish, they applied this technique to the study of gaseous free radicals and to combustion. He graduated with an M.A., a Ph.D. and a Sc.D. from Cambridge.

The Career

Year	*Post*	*Institute*
1949	Demonstrator in Physical Chemistry	University of Cambridge
1952	Assistant Director of Research in Physical Chemistry	University of Cambridge
1954	Assistant Director	British Rayon Research Association, Manchester
1955	Professor of Physical Chemistry	University of Sheffield
1963	Firth Professor of Chemistry and Head of Chemistry Department	University of Sheffield
1963	Professor of Chemistry	Royal Institution
1966–1985	Director and Fullerian Professor of Chemistry	Royal Institution, London
1987–2002	Professor	Imperial College
1988–2002	Emeritus Professor	Royal Institution
1990–2002	Chairman of the Centre for Photomolecular Sciences	Imperial College, London
1990–1994	Gresham Professor of Astronomy	Gresham College

Porter's research interests concentrated on the development of flash-photolysis, and how the technique could be applied to solve or investigate other problems in the disciplines of biology, chemistry and physics. He also made significant input into free radical trapping and matrix stabilization.

Porter also found the time to use television and radio for science education. He wrote, presented or appeared in the following programmes:

- Eye on Research. '*Quick as a Flash*'. BBC Television (1960)
- '*A Candle to Nature*'. Pilot for 'Horizon' series. BBC Television. (1973)
- *Laws of Disorder*'. 10 programmes. BBC Television (repeated four times). (1965–66)
- '*Time Machines*', six Christmas Lectures from the Royal Institution. BBC (repeated twice). (1969–70)
- '*Controversy*', series of debates, chairman. BBC. (1971–75)
- '*The Natural History of a Sunbeam*' six Christmas Lectures from the Royal Institution. BBC Television.(1976–77)

- Judge. Young Scientists of the Year. BBC. (1966–81)
- Contributor, Open University course on '*Photochemistry: Light, Chemical Change and Life*', BBC Television.(1982)
- '*Man of Action*,' Radio. BBC Radio 4. (1984)
- '*Conversation Piece*,' Radio. BBC Radio 4. (1987)
- Dimbleby Lecture. BBC Television. (1988)
- '*Desert island Discs*' (with Sue Lawley). BBC Radio 4. (1990)

Porter's Nobel Prize acceptance speech is dated 11th December 1967 with the title 'Flash photolysis and some of its applications'. During his presentation, Porter commented [2]:

> *...My original conception of the flash photolysis technique was as follows: the transient intermediates, which were, in the first place, to be gaseous free radicals, would be produced by a flash of visible and ultraviolet light resulting from the discharge of a large condenser bank through an inert gas. The flash would be of energy sufficient to produce measurable overall change and of short duration compared with the lifetime of the intermediates. Calculation showed that an energy of 10 000 J dissipated in a millisecond or less, in lamps of the type which were being developed commercially at that time, would be adequate for most systems. The bank of condensers was given by my friends in the Navy, and, although I am grateful to them for saving us much expense, it consisted of a motley collection of capacitors which, owing to their high inductance, gave a flash of rather longer duration than was desirable. The detection system was to consist of a rapid-scanning spectrometer and much time was wasted in the development of this before I realised that to demand high spectral resolution, time resolution and sensitivity in a period of a few milliseconds was to disregard the principles of information theory...*

Photochemistry over the last 30 years or so has continued to be developed and is now applied to many reaction types, see for example [5–8].

Honours and Awards

Porter was awarded an Honorary D.Sc. degree from the Universities of: Sheffield (1968); Utah, Salt Lake City (1968); East Anglia (1970); Durham (1970); Leicester (1971); Leeds (1971); Heriot-Watt (1971); City (1971); Manchester (1972); St. Andrews (1972); London (1972); Kent (1973); Oxford (1974); Hull (1980); Instituto Quimico de Sarria, Barcelona (1981); Pennsylvania (1983); Coimbra, Portugal (1983); Lille (1984); Phillipines (1985); Notre Dame (1986); Bristol (1986); Reading (1986); Loughborough (1987); Brunel (1988); Cordoba, Argentina (1990); Liverpool (1990); Mangalore (1995); Buckingham (1995); Bath (1995). He was awarded: the Honorary D.Univ from the University of Surrey (1970) and the Open University (1983); the Laurea in Chemica h.c. University of Bologna (1988); the Honorary LL.D. Cantab. (1993).

Porter was presented with many medals and prizes including: the Corday Morgan Medal (1955); the Nobel Prize for Chemistry, joint (1967); the Silvanus Thompson Medal (1968); the Davy Medal of The Royal Society (1971); the Kalinga Prize (UNESCO) (1977); the Bose Medal (1977); the Robertson Memorial Lecture and Prize (1978); the Communications Award of the European Physical Society (1978); the Rumford Medal (1978); the Faraday Medal (1980); the Longstaff Medal (1981); the Melchett Medal (1987); the Porter Medal for Photochemistry 1988; the Society Medal of the Society of Chemical Industry (1991); the Michael Faraday Medal (1991); the Ellison-Cliffe Medal (1991) and the Copley Medal (1992).

Porter held Fellowships/Memberships in a number of Academies and Institutes including: Honorary Fellowship of the Institute of Patentees and Inventors (1970); the Royal Scottish Society of Arts (1975); Royal Society of Edinburgh (1983); The Royal Medical Society, Edinburgh (1987); Imperial College, London (1987); the Royal Society of Chemistry, London (1991); Science Museum of London (1991); Honorary Fellow, Society of Radiation Physics (1991); University of North Lancashire (1992) and of the Tata Institute, Bombay (1995). He was a Foreign Fellow of the Indian National Academy of Sciences (1986) and a Fellow of the Royal Society (1960 Council: 1966–67, 1985–90; PRS 1985–90). Porter held Honorary Membership of: the New York Academy of Sciences (1968); the Leopoldina Academy (1970); the Chemical Society, Spain (1978); the Société de Chimie Physique (1979); and he was the Honorary Member of: the Chemical Society of Japan (1982); the Association for Science Education (1985); the American Philosophical Society (1986); the Society for Free Radical Research (1986); the Foundation de la Maison de la Chimie, Paris (1987); the Indian Academy (1987); the Royal Institution (1988); the Japan Academy (1988); the Indian Science Academy, Bangalore (1989); the Royal Society of Medicine, Edinburgh (1989); the Soviet Academy of Sciences (1989); the Indian Academy of Sciences (1990); the Scientific Society of Argentina (1990); of KAST (Korean Academy of Science and Technology) (1995) and KIAS (Korean Institute for Advanced Studies) (1996). He was a Foreign Honorary Member of the American Academy of Arts and Sciences (1979) and an Honorary Foreign Member, Hungarian Academy of Sciences (1988). Porter was Foreign Member of the following institutions: the Institute of Fundamental Studies, Sri Lanka (1985); the Academy of Sciences, Lisbon (1983); the European Academy of Arts, Sciences and Humanities (1987); the Foreign Member of the Academia Nazionale Dei Lincei, Italy (1987). He was a Member of: the Pontifical Academy of Sciences (1974); the Academic International de Lutece (1979); the European Academy of Arts, Sciences and Humanities (1980); the Academia Europaea (1989). Porter was Foreign Associate of the National Academy of Sciences (1974); Corresponding Member of the Göttingen Academy of Sciences (1974); Foreign Corresponding Member of La Real Academia de Ciencias, Madrid (1978); Frew Fellow, Australian Academy of Science (1976).

He was awarded the following civic honours: Knight Bachelor (1972); Order of Merit (1989); Created Baron Porter of Luddenham (1990); Freeman and Liveryman, honoris causa, of the Salter's Company (1981); Master, Salters' Company 1993–94.

Porter held honorary professorships at: the University of Kent (1966); the Dalian Institute of Chemical Physics of the Academy of Sciences of China (1980); the Royal Institution (1988); Institute of Technology, Beijing, China (1986); Distinguished Visiting Professor University of Exeter (1991). He was an Honorary Fellow of: Emmanuel College, Cambridge (1967); of Queen Mary College, London (1986); King's College, London (1995).

He worked on many public bodies including: President, Comité International de Photobiologie (1968–72); President, the Chemical Society (1970–72); President, Faraday Division of the Chemical Society (1973–74); Trustee, British Museum (1972–74); President, The National Association for Gifted Children (1975–80); President, The Research and Development Society (1977–82); President, Association for Science Education (1985); President, British Association for the Advancement of Science (1985–86); President of the Royal Society of London (1985–90); Trustee, the Exploratory, Bristol (1986–); Trustee, Glynn Research Institute (1986); Chancellor, University of Leicester (1986–95); President, London International Youth Science Fortnight (1987–89); President, Scitech (1987–89); Vice President of the British Video History Trust (1987–); President, British Mathematics Olympiad (1991–); Member of the Cabinet office Advisory Council on Science and Technology

(1987–91); Trustee, National Energy Foundation (1989–91); President, National Energy Foundation (1989–91); Member of the House of Lords Select Committee on Science and Technology (1990–94); Second Warden, Upper Warden (1992/93) and Master (1993/94) Salters' Company; Almoner of Christ's Hospital (1994–).

Porter's citation for election (date: 24/03/1960) as Fellow of the Royal Society reads [9]:

'Distinguished for his investigation of unstable chemical intermediates and for kinetic studies of the very fast reactions in which they take part. The method of direct observation by flash photolysis which he devised and developed in collaboration at Cambridge constitutes a very important post-war development in physiochemical technique. The application of this and other direct methods by Porter and others has yielded some highly significant contributions to the fields of free radical spectroscopy, the kinetics of atomic recombination and radical reactions in gases, the occurrence and kinetic behaviour of the triplet state, the mechanisms of combustion and carbon formation and the detailed reactions of certain organic radicals in solution. A list of fifty-four papers and articles is appended.'

No proposers listed in the Sackler database.

Bibliography

1. Reproduced with permission.
2. From Nobel Lectures, Chemistry 1963–1970, Elsevier Publishing Company, Amsterdam, 1972.
3. Biographical Memoirs of Fellows of the Royal Society, 2004, vol. 50, pp. 257–284.
4. Who's who: 2002.
5. Photochemistry and Reaction Kinetics ed. P.G. Ashmore, F.S. Dainton and T.M. Sugden, Cambridge University Press, 1967, Library of Congress Card Number 67-10157.
6. Organic Photochemistry 10 volume set ed. A. Padwa, Marcel Dekker Inc., 1989, ISBN: 0-8247-7920-7.
7. Organic and Inorganic Photochemistry ed. V. Ramamurthy and Kirk S. Schanze, Marcel Dekker Inc., 1998, ISBN: 0-8247-0174-7.
8. Chiral Photochemistry ed. Yoshihisa Inove and V. Ramamurthy, Marcel Dekker Inc., 2004, ISBN: 0-8247-5710-6.
9. Sackler database, Royal Society.

1968

© Nobel Foundation [1]

Lars Onsager

"for the discovery of the reciprocal relations bearing his name, which are fundamental for the thermodynamics of irreversible processes" [2]

During his presentation speech, S. Claesson of the Royal Swedish Academy of Sciences, commented of Onsager [2]:

...Professor Lars Onsager has been awarded this year's Nobel Prize for Chemistry for the discovery of the reciprocal relations, named after him, and basic to irreversible thermodynamics. On hearing this motivation for the award one immediately gets a strong impression that Onsager's contribution concerns a difficult theoretical field. A closer study shows this indeed to be the case. Onsager's reciprocal relations can be described as a universal natural law, the scope and importance of which becomes clear only after being put in proper relation to complicated questions in border areas between physics and chemistry...

...Onsager's great contribution was that he could prove that if the equations governing the flows are written in an appropriate form, then there exist certain simple connections between the coefficients in these equations. These connections - the reciprocal relations - make possible a complete theoretical description of irreversible processes...

...The proof of the reciprocal relations was brilliant. Onsager started from a statistical mechanical calculation of the fluctuations in a system, which could be directly based on the simple laws of motion which are symmetrical with regard to time. Furthermore he made the independent assumption that the return of a fluctuation to equilibrium in the mean occurs according to the transport equations mentioned earlier. By means of this combination of macroscopic and microscopic concepts in conjunction with an extremely skilful mathematical analysis he obtained those relationships which are now called Onsager's Reciprocal Relations...

Lars Onsager was born in Oslo, Norway on 27th November 1903, the son of Erling Onsager (a barrister of the Supreme Court of Norway) and Ingrid, née Kirkeby. He married Margarethe Arledter in 1933. They had three sons Christian Carl, Erling Frederick and Hans Tanberg and a daughter Inger Marie. Onsager became a naturalized US citizen (1945). He died in Coral Gables, Florida on the 5th October 1976 [2–4].

The Early Years

Onsager's early education was with tutors Inga and Anna Platou in Oslo, followed by a year in a private school, and a few months of tutoring by his mother. Onsager then entered Frogner School, as the family had by then moved to Oslo. His early education was such that he moved up to the next academic year, completing his schooling in 1920. He enrolled in the Norges tekniske høgskole (autumn 1920) to study chemical engineering. He graduated with the degree of Ch.E. in 1925.

The Royal Society biography of Onsager, written by H. Christopher Longuet-Higgins and Michael E. Fisher, describe an incident that took place in 1925 thus [2]:

> ..."*One day in 1925 Pieter Debye was sitting in his office at the E.T.H. in Zurich when a visitor from Norway was announced. In came a tall young man, who walked silently across the room, bent over the desk and said solemnly: 'Professor Debye, your theory of electrolytes is incorrect.' Whereupon Debye, after begging the stranger to sit down, and inviting him to discuss his objections, offered him an Assistantship for the following year. The young man's name was Lars Onsager (Murphy & Cohen 1968)*"...

Onsager returned to Zurich the following year as a Research Assistant and stayed with Debye for two years. He later graduated from Yale University with a Ph.D.

The Career

Year	*Post*	*Institute*
1928	Associate in Chemistry	John Hopkins University
1928	Instructor in Chemistry	Brown University
1933	Sterling & Gibbs Fellow	Yale University
1934	Assistant Professor of Chemistry	Yale University
1940	Associate Professor of Chemistry	Yale University
1945	J. Willard Gibbs Professor of Theoretical Chemistry	Yale University
1972	Distinguished University Professor	University of Miami

Onsager obtained his Ph.D. whilst working at Yale University (1935), the subject of which was the mathematical background for his interpretation of deviations from Ohm's law in weak electrolytes. He was a Fulbright Scholar for the period 1951–1952, which he spent with Schoenberg at the Mond Laboratory in Cambridge, England. At the time this laboratory was a centre of excellence for studies of low temperature physics.

Onsager had many research interests, including:

- colloids
- dielectrics
- hydrodynamics and fractionation theory
- metals and superfluids
- order-disorder transitions

Onsager's Nobel Prize acceptance speech was dated 11th December 1968 with the title 'The motion of ions: principles and concepts'. Onsager commented [2]:

> *...Fortunately, my own efforts in the summer of 1923 had produced a modest but firm result. The relaxation effect ought to reduce the mobilities of anion and cation in equal proportion. Much to my surprise, the results of Debye and Huckel did not satisfy that relation, nor the requirement that whenever an ion of type A is 10 Å West of a B, there is a B 10 Å East of that A. Clearly, something essential had been left out in the derivation of such unsymmetrical results. The model used was this: one particular ion is constrained to move at a constant speed along a straight line in the solution; neighbouring ions respond to the fields in the distorted screening cloud, and in addition they mill around in random fashion according to the laws of Brownian motion. Recipe: Remove restraints on the central ion but retain an external force on it, let it execute its own thermal motion and respond to the fields of its neighbours, and recognize whatever external forces act on them. That done, the result for binary electrolytes became very simple: the relaxation effect reduces the migration velocity of every ion by a fraction which depends neither on its own mobility nor on that of the partner species (of opposite, numerically equal charge). Otherwise the effect is charge dependent and proportional to the square root of the one concentration just like the corrections to the equilibrium properties, but with a different coefficient. As to the electrophoretic effect, it was easy to show that plausible variations of the hydrodynamics near the center of the countercurrent system driven by a widely distributed force could not matter enough to affect the limiting law; Debye and Huckel had unjustly impugned their own result...*

He went on to comment:

> *...Going back to the time when I revised the theory of Debye and Hückel, the task was by no means easy. The key was a principle of superposition applied to the ion cloud around a pair. To begin with, it was a bit confusing that the force exerted by the external field on an ion as well as the interaction between the ions were proportional to the charge. In order to gain perspective I decided to ignore the relation between the charge and the driving force, and took a look at a more general problem. One constant field of force is acting on each kind of ion; what are the effects of the Coulomb interaction? The problem is in fact equivalent to that which arises in the most general case of diffusion and electrical conduction combined; the gradients of chemical potentials are equivalents of forces...*

Towards the end of his presentation, and by way of summarizing a point, Onsager observed:

> *...Estimates of the ionic mobilities vary over a considerable range; but in any event the positive ionic defect is much more mobile in the solid than in the liquid, and its mobility varies very little with the temperature. The mobilities of the bonding defects are more like those of ordinary ions in the liquid, and the temperature coefficients are similar. Nevertheless, the bonding defects determine the direction of polarization in pure ice, because they are much more numerous, possibly several pairs for each million molecules. As to the task of transporting a direct current, that is about equally*

distributed between bonding and ionic defects; each type carries about half an elementary charge. One kind of current may get out of step with the other for a short time; but this produces a polarization which equalizes the number currents...

A text concerning irreversible thermodynamics was published some 20 years after Onsager's Nobel award which summarized the development of the field [5].

Honours and Awards

Onsager held Honorary D.Sc. degrees from: Harvard University (1954); Brown University (1962); Rensselaer Polytechnic Institute (1962); Rheinisch-Westfahlische Technische Hochschule (1962); the University of Chicago (1968); Ohio State University (1969), Cambridge University (1970); Oxford University (1971). He was also presented with Doctor Technicae from Norges tekniske høgskole (1960).

He was presented with many awards and prizes including: the Rumford Medal (1953); the Lorentz Medal (1958); the Lewis Medal (1962); the Kirkwood Medal (1962); the Gibbs Medal (1962); the Richards Medal (1964); the Peter Debye Award in Physical Chemistry (1965); the Belfer Award in Science (1966); the Nobel Prize for Chemistry (1968); the National Science Medal (1969).

Onsager was elected Fellow or Member of a number of learned societies and Institutes including Fellow of: the American Physical Society (1933); the New York Academy of Sciences (1942); National Academy of Sciences (1947); the American Academy of Arts and Sciences (1953). He was a Member of: the American Physical Society (1928); the Sigma Xi Fraternity (Brown University Chapter in 1929); the Connecticut Academy of Arts and Sciences (1940); the Royal Norwegian Academy of Science (May 1953); the Royal Swedish Academy of Science (April 1957); Norwegian Academy of Technical Science (February 1958); the Royal Netherlands Academy of Science (1958); the American Philosophical Society (April 1959); the American Chemical Society (1962); the Deutsche Bunsen Gesellschaft (May 1969). He was a Foreign Member of: the Norwegian Academy of Sciences and Letters, Oslo (1938); the Royal Society of Sciences of Uppsala (1963); the Royal Society (1975). He was an Honorary Member of the Institute of Physics (1974).

He was Visiting Professor to a number of Universities including: California, San Diego (1961); Rockefeller (1967–68); Göttingen (1968); Lorentz Professor in Leiden (1970); Distinguished University Professor, Miami, Coral Gables (1972–76).

Onsager's citation for election (date: 24/04/1975) as Foreign Member of the Royal Society reads [6]:

'Professor Lars Onsager of Yale, U.S.A. distinguished for his fundamental contributions, through the application of powerful mathematical techniques, to statistical physics and chemistry.'

No proposers are listed in the Sackler database.

Bibliography

1. Reproduced with permission.
2. From Nobel Lectures, Chemistry 1963–1970, Elsevier Publishing Company, Amsterdam, 1972.
3. Biographical Memoirs of Fellows of the Royal Society, 1978, vol. 24, pp. 443–471.
4. Who was who: 1971–1980.
5. Kinetic Theory and Irreversible Thermodynamics by Byung Chan Eu, John Wiley and Sons Inc., 1992, ISBN: 0-471-61524-2.
6. Sackler database, Royal Society.

1969

© Nobel Foundation [1]

Derek H.R. Barton (0.5)

"for their contributions to the development of the concept of conformation and its application in chemistry" [2]

During his presentation speech Arne Fredga, then a member of the Nobel Committee for Chemistry of the Royal Swedish Academy of Sciences, observed of Barton and his work [2]:

...Metaphorically one could say that the molecule tries to arrange itself in the most comfortable way. It will avoid crowding and strain and must consider that certain groups may attract or repel each other. Often a great number of conformations are possible, but some are more stable than others. These are statistically favoured. A ring of six carbon atoms can have two conformations, known as the chair and boat forms, which easily interchange. At room temperature, a molecule changes its conformation about a million times in a second. One of the conformations is, however, strongly predominant (about 99%)...

...Professor Barton. In the classical work "The Conformation of the Steroid Nucleus" you have advanced the leading principles of conformational analysis. In this paper you have also drawn attention to the notable researches of Hassel, which have thrown considerable light on these more subtle aspects of stereochemistry. Your ideas were soon accepted and they play a fundamental role in organic chemistry of today. According to a prominent fellow scientist, your paper represents the first real advance in stereochemistry since the theory of Van't Hoff and Le Bel, i.e. since 1874. I have no objections...

Derek Harold Richard Barton was born in Gravesend, Kent on the 8th September 1918, the son of William Thomas and Maude Henrietta Barton, née Lukes. He married Jeanne Kate Wilkins in Harrow (1944) but this marriage was later dissolved. There was one son from this marriage, W.G.L. Barton. Barton married Christiane Cognet, a Professor of the Lycée Français de Londres (1969). Christiane died in 1992. Barton went on to marry Judith Von-Leuenberger Cobb (1993). He died of heart failure in College Station, Texas on 16th March 1998 [2–4].

The Early Years

Barton was educated at Gravesend County School for Boys (1926–29), King's School, Rochester (1929–32), Tonbridge School (1932–35) and Medway Technical College in Gillingham (1937–38) prior to enrolling at Imperial College, University of London (1938). As an undergraduate, he was awarded the Hofmann Prize prior to graduating with a first class honours degree in chemistry in 1940. This was followed by his Ph.D. in organic chemistry under Heilbron (1942) and a D.Sc. from the University of London in 1949.

The Career

Year	*Post*	*Institute*
1942	Research chemist	Government projects
1944	Research chemist	Albright and Wilson
1945	Assistant lecturer	Imperial College
1946	ICI research fellow	Imperial College
1949	Visiting Lecturer	Harvard University
1950	Reader	Birkbeck College
1953	Professor	Birkbeck College
1955	Regius Professor	University of Glasgow
1957	Professor	Imperial College
1970–78	Hofmann Professor of Organic Chemistry	Imperial College
1978	Emeritus Professor	Imperial College
1976	Director	Natural Products Institute (C.N.R.S.) at Gif-sur-Yvette, France
1986	Distinguished Professor of Chemistry	Texas A&M University
1995	Dow Distinguished Professor of Chemical Invention	Texas A&M University

During World War II (specifically 1942–44), Barton worked for Military Intellegence at Baker Street in London, where he is understood to have worked on developing invisible inks that could be used on human skin.

Barton published his observations concerning preferred conformations of molecules in a paper in *Experientia* (1950). It was to be the basis of his subsequent work on conformational analysis [5]. Barton also involved in many aspects of organic synthesis including:

- deoxydation of secondary alcohols
- phenol oxidation – particularly the biological importance
- rules for ortho-para-phenolic oxidative coupling
- the use a photochemical reaction in a synthesis of aldosterone – now known as the Barton reaction [6].

Barton published 1041 papers in the chemical literature. A full listing was provided by Scott [4]. In addition to the papers, Barton published many successful patents. He also published his biography in the series *Profiles, Pathways and Dreams* [7]. He was the Chairman of the executive board of editors for *Tetrahedron Publications* (from 1979 until his death).

Barton's Nobel Prize acceptance speech is dated 11th December 1969 with the title 'The principles of conformational analysis'. During his introduction, Barton commented [2]:

> *...The importance of conformational analysis in Chemistry became manifest during the decade immediately after the last World War. This lecture is, therefore, more of an account of chemical history than of recent advances,...*
>
> *...A constitutional formula has no stereochemistry. The necessity to consider the stereochemistry of molecules became self-evident when two or more distinctly different substances were found to have the same constitution. It was Le Bel and, especially, van't Hoff who, also in the nineteenth century, introduced the idea of configuration. One can consider that the configuration of a molecule of a given constitution represents the specification of the order of the bonds in space about the atoms, or groups in the molecule which give rise to stereoisomers. For the great majority of substances this means the specification of the order of the bonds about an asymmetric 'double bond" or about a "centre of asymmetry", normally a carbon atom substituted by four different groups. The first type of stereoisomerism is called "geometrical isomerism", the latter "optical isomerism"...*

During his conclusion, Barton suggested:

> *...We are now in a position with most substrates to specify the preferred conformation involved in this reaction (an enzymatic reaction). In the case of steroidal substrates the conformation can be described in a detail which will soon be quite exact. This knowledge must soon have important consequences in biology...*

A detailed knowledge of the stereochemistry of all organic chemical, expecially those of biological importance, is mandatory to their synthesis. Many texts have been written to help the chemist appreciate the importance of conformational analysis, see for example [8–10].

Honours and Awards

Barton was presented with Honorary Degrees from many Universities including: Montpellier (1962); Dublin (1963); St. Andrews (1970); Columbia University (1970); Coimbra (1971); Oxford (1972); Manchester (1972); S. Africa (1973); La Laguna-Tenerife (1974); Western Virginia (1974); Sydney (1976); Valencia (1979); Sheffield (1979); Western Ontario (1979); the Weizmann Institute of Science (1979); Frankfurt (1980); Metz (1981); Université de Lyon (1981); University of Uruguay (1983); Stevens Institute of Technology (1984); London (1984); Debrechen (1986); Université Paris-Sud (1988); Shanghai Institute of Organic Chemistry (1991); Universite D'Aix-Marseilie-flI (1993); Universida de Nova de Lisboa (1993); Universidad Nacional de Cordoba (1993); Universidad Nacional del Sur, Bahía Blanca (1993); Hong Kong (1994); Autonoma de Puebla (1995); de Salamanca, Salamanca (1995); de Concepcion (1995); Munich (1995); Belgrade (1997).

He received many awards and medals including the: Corday-Morgan Medal (1951); Fritzsche Medal (1956); Roger Adams Medal (1959); Davy Medal (1961); Nobel Prize, joint (1969); Award in Natural Product Chemistry Chemical Society (London) (1971); Royal Medal (1972); Longstaff Medal

(1972); BC Law Gold Medal (1972); Medal of the Society of Cosmetic Chemistry (1972); Order of the Rising Sun (2nd Class) by the Emperor of Japan (1972); Royal Medal (1972); Order of "Chevalier de la Legion d'Honneur" (1974); second centennial medal of the Priestley Chemistry Award (1974); Medal of the Union of Scientific Workers of Bulgaria (1978); Medal of the University of Sofia, Bulgaria (1978); Medal of the Academy of Sciences of Bulgaria (1978); Copley Medal (1980); Hanbury Medal (1981); Médaille d'Or de la Ville de Nancy (1984); Officier de la Legion dHonneur (1985); Cliff S. Hamilton Award in Organic Chemistry (1987); May and Baker Award (1987); American Chemical Society Award for Creativity in Organic Synthesis (1989); Kosolapoff Award, Auburn Univeristy (1991); Texas A&M University Award (1991); Oesper Award (1991); Chemical Pioneers Award from the American Institute of Chemistry (1993); Priestley Medal (1995); Lavoisier Medal from the French Chemical Society (1995).

Barton received many fellowships, memberships of learned societies around the world including Fellowship of the Royal Society (1954, Council: 1964–66) and the Royal Society of Edinburgh (1956). He was Honorary Fellow of Royal Society of Chemistry, London (1985) and of the Indian Chemical Society (1974). He was Foreign Fellow of the Indian National Science Academy (1977) and of the Hational Academy of Sciences of India (1990). He was Centennial Fellow of the American Chemical Society (1976). Barton was an Honorary Member of: the Sociedad Quimica de Mexica (1969); the Belgium Chemical Society (1970); the Chilean Chemical Society (1970); the Pharmaceutical Society of Japan (1970); the Royal Academy of Exact Sciences, Madrid (1971); the Acedemy of Pharmaceutical Sciences USA (1971); the Danish Academy of Sciences and Letters (1972); the National Academia of Exact, Physical and Natrual Sciences Buenos Aires (1973); the Hungarian Academy of Sciences (1976); the Societa Italiana per il Progresso delle Scienze (1977); the Chemical Society of Japan (1982); the Serbian Academy (1987); the Irish Academy of Science (1988); the Kisato Institute (1990); the Argentinean Society for Organic Chemistry (1993). He was a Foreign Honorary Member of the American Academy of Arts and Sciences (1960) and a Foreign Member of: the Acedemia das Ciencias da Lisboa (1971); the Accademia Nationale dei Lincei (1974); the Dutch Society of Sciences (1977). He was also a Member of the European Academy of Arts, Sciences and Humanities (1980), Membre Associétranger de l'Académie des Sciences (1978) and Membretranger de la Société de Chemie Théraputique (1981). Barton also held Membre d'Honneur: de l'Académie des Sciences (1983); de l'Université de Concepcion (1983); de l'Association de Chemie et de Pharmacie d'Uruguay (1983); de la Fondation de la Maisson de Chemie (1986). He was Corresponding Member of the Argentinean Chemical Society (1970), Foreign Associate of the National Academy of Sciences (1970) and Foreign Corresponding Member of the Slovenian Academy of Sciences (1985).

Barton also served on many Institutes including: President of the Organic Chemistry Division of the International Union of Pure and Applied Chemistry (1969–72); Fellow of Birkbeck College (1970); President of the Perkin Division of the Chemical Society (1971–73); President of the Chemical Society (1973–75); Member for the Council for Scientific Policy (1973); Directeur Honoraire de l'Institut de Chemie des Substances Naturelles (1986).

He was awarded the civic honours of Knight Bachelor (1972), Citizen d'Honneur de la ville de Gif-sur-Yvette (1986) and named Kentucky Colonel by the Governor of Kentucky (1989).

Barton was also Visiting Professor or named lecturer at many institutes around the world.

Barton's citation for election (date: 18/03/1954) as Fellow of the Royal Society reads [11]:

'Distinguished for his work in Organic Chemistry and particularly for his application of original theoretical concepts to the solution of structural problems among selected complex naturally occurring compounds. His work falls into four main divisions, (a) the development of relationships

between molecular rotations and structure, and their application to the solution of structural problems in the steroid field; (b) the stereochemistry of pyrolytic elimination reactions with special reference to the pyrolysis of chlorinated hydrocarbons; (c) structural studies, especially in the triterpenoid field; and (d) the development of the conception of conformational analysis and particularly its application to stereochemical problems in steroid, terpenoid and triterpenoid chemistry. A list of over 80 publications is appended..'

He was proposed by:

From personal knowledge. I. M. Heilbron; R. Robinson; Alex R. Todd; J. L. Simonsen; R. P. Linstead; E. L. Hirst; A. H. Cook; J. W. Cook; J. W. Cornforth.

Bibliography

1. Reproduced with permission.
2. From Nobel Lectures, Chemistry 1963–1970, Elsevier Publishing Company, Amsterdam, 1972.
3. Biographical Memoirs of Fellows of the Royal Society, 2002, vol. 48, pp. 1–23.
4. Barton Obituary by A. I. Scott - Tetrahedron 1998, 54, 8847–8917.
5. Conformation of the Steroid Nucleus Barton Experientia 1950, 6, 316.
6. J. Amer. Chem. Soc., 1960, 82, 2641.
7. Some Recollections of Gap Jumping - Autobiography - presented in the series Profiles, Pathways and Dreams ed. Jeffrey I. Seeman, pub. American Chemical Society, ISBN: 0-8412-1770-x. ISBN: 0-8412-1796-3(PBK).
8. Modern Conformational Analysis by Helena Dodziuk, VCH, 1995, ISBN: 1-56081-689-9.
9. Basic Organic Stereochemistry by Ernest L. Eliel, Samuel H. Wilen and Michael P. Doyle, Wiley Interscience, 2001, ISBN: 0-471-37499-7.
10. Stereochemistry by David G. Morris, RSC, 2001, ISBN: 0-85404-602-X.
11. Sackler database, Royal Society.

1969

© Nobel Foundation [1]

Odd Hassel (0.5)

"for their contributions to the development of the concept of conformation and its application in chemistry" [2]

During his presentation speech, Arne Fredga, then a member of the Nobel Committee for Chemistry of the Royal Swedish Academy of Sciences, commented of Hassel and his work [2]:

...Often a great number of conformations are possible, but some are more stable than others. These are statistically favoured. A ring of six carbon atoms can have two conformations, known as the chair and boat forms, which easily interchange. At room temperature, a molecule changes its conformation about a million times in a second. One of the conformations is, however, strongly predominant (about 99%). Professor Hassel has carried out fundamental investigations on this system and shown how heavy or bulky groups, attached to the carbon atoms, take up their positions relative to the ring and to each other.

The conformation is of great importance for the mode of reaction of the molecules. Reactive groups may be easily accessible, or they may also be more or less blocked by other groups. Knowledge of the conformation is therefore of great importance for explaining or predicting the mode of reaction of a certain molecule. It is always a good thing to know if an experiment has any chance of success...

...In the development of scientific ideas it is generally possible to trace contributions, elements of thought, from many sides. But often the decisive advances, the intellectual syntheses from different thoughts and suggestions can be attributed to one or two scientists, who stand out from the others. Professor Hassel's elegant work on six-membered rings, carried out with ever increasing precision, has laid a solid foundation for a dynamic chemistry in three dimensions...

...Professor Hassel. Not very long ago, organic chemists spoke of free rotation. Gradually they found that the rotation is restricted and that this fact is important. Many workers have contributed

to the development, but in our opinion your work on the cyclohexane system is of outstanding importance. In your work on decaline, you have also taken the important step to polycyclic systems and pointed the way for coming development. In recognition of your services to Chemical Science, the Royal Academy has decided to confer upon you the Nobel Prize. To me has been granted the privilege of conveying to you the most hearty congratulations of the Academy...

Odd Hassel was born in Kristiania (now Oslo) Norway on 17th May 1897, the son of Ernst August Hassel (a gynaecologist) and Mathilde, née Klaveness. Hassel died on 11th May 1981 [2,3].

The Early Years

Hassel enrolled in the University of Oslo (1915) where he studied chemistry with mathematics and physics as his subsidiary subjects. He graduated as a cand. real. in 1920. Hassel took a year touring around France and Italy prior to spending some time (six months) in Munich (1922), working in the laboratory of K. Fajans, where he worked on the sensibilisation of silver halides by organic dyes. Hassel then moved to the Kaiser Wilhelm Institute in Dahlem (near Berlin) with a Rockefeller Fellowship which was suggested by Haber. Hassel worked on X-ray crystallography, graduating with a Dr.Phil. from Berlin University (1924).

The Career

Year	*Post*	*Institute*
1925	Universitetsstipendiat	University of Oslo
1926	Dosent in physical and electro-chemistry	University of Oslo
1934 to 1964	Professor of Physical Chemistry	University of Oslo
1934	Head of the Department of Physical Chemistry	University of Oslo
1964	Retirement	

Hassel was arrested by Norwegian Nazis in 1943 and handed over to the German occupying forces shortly thereafter. He was not released until November 1944, by which time the University was all but deserted.

Hassel's early, pre-war interests concerned various aspects of inorganic chemistry. However, from *circa* 1930 onwards, he turned his attention towards the problems of molecular structure. Of particular interest to Hassel were studies of:

- The structure of six-membered rings, particularly cyclohexane and related compounds
- A study of charge-transfer compounds (from the early 1950s) i.e. pairs of electron-donor molecules (amines, ethers etc) with electron acceptors (e.g. halogen containing molecules etc.)

The work involved the adaptation of techniques not previously used in Norway for measuring:

- electron diffraction by vapours based on the rotating sector method
- electric dipole moments

Hassel compiled sufficient data that he was able to set up rules for the geometry (post war). He remained interested in the field of conformational analysis for the remainder of his career.

Hassel's Nobel Prize acceptance speech is dated 9th June 1970 with the title 'Structural aspects of interatomic charge-transfer bonding'. During his introductory remarks, Hassel commented [2]:

> *...During the later part of the 1940's the interest in complexes formed by donor and acceptor molecules was stimulated by quantitative spectroscopic work dealing first with solutions of iodine in henzene, later with a considerable number of liquid systems containing different combinations of donor and acceptor molecules dissolved in solvents regarded as "inert" with respect to the interacting species. From the spectroscopic data equilibrium constants and thermodynamic values associated with the formation of 1:1 complexes were evaluated. A quantum mechanical theory of the "complex resonance" was worked out by Mulliken which is of a very general nature and explains spectroscopical observations, but has not yet made possible reliable predictions regarding the atomic arrangements within the complexes.*
>
> *Direct interferometric structure determinations in the vapour phase are regarded virtually impossible in most cases because of the extremely low concentration of the complex which may be expected to be present. X-Ray crystallographic structure determinations have made it possible, however, to draw conclusions regarding atomic arrangements not only in solids, but even in isolated 1:1 complexes, conclusions which should be of value for theoretical workers trying to establish a more elaborate theory of charge-transfer interaction.*
>
> *Particular importance may be attributed to complexes in which direct bonding exists between one atom belonging to the donor molecule and another atom belonging to the acceptor molecule. Complexes of this kind are above all those formed by donor molecules containing atoms possessing "lone pair electrons" and halogen or halide molecules. A presentation and discussion of the general results obtained by X-ray analysis of solid adducts exhibiting charge-transfer bonding between such atoms might therefore be of some interest...*

Hassel went on to comment on some of the specific work undertaken by his group in Oslo:

> *...Early in the 1950's X-ray crystallographic investigations of halogen adducts were started in Oslo, beginning with the solid 1:1 1,4-dioxan-bromine compound. The most striking feature of the resulting crystal structure, the endless chains of alternating dioxan and bromine molecules depending on linear O-Br-Br-O arrangements running in a direction roughly equal to the "equatorial" direction in cyclohexane was rather unexpected. It proved that both atoms belonging to a particular halogen molecule may simultaneously be bonded to oxygen atoms, although probably not to the same oxygen atom. The existence of halogen molecule bridging between donor atoms contradicts previous assumptions according to which a charge-transfer bond formed by one of the atoms belonging to a particular halogen molecule creates a marked negative charge on the partner atom.*
>
> *The oxygen-bromine distance in the dioxan-bromine adduct is 2.7 Å and thus considerably larger than the sum of the covalent radii of oxygen and bromine, but at the same time definitively shorter than the sum of the corresponding van der Waals radii. The type of "polymerisation' of simple complexes into endless chains observed in the crystalline dioxin-bromine comhas also been observed in the crystalline 1:1 adducts of dioxan and chlorine, resp. iodine. A comparison between the oxygen-halogen separations and of the interhalogen bond lengths in the three 1,4- dioxan adducts*

leads to the conclusion that the former increases rather slowly from chlorine to iodine, indicating a certain degree of compensation of the effect of larger halogen radius by the increase in charge-transfer bond strength. On the other hand, the halogen-halogen bond length increases, although slowly, from chlorine to iodine compared with that observed in "free' halogen molecules, an observation which must also find its explanation in the stronger charge-transfer interaction between oxygen and halogen...

A detailed knowledge of the stereochemistry of all organic chemical, expecially those of biological importance, is mandatory to their synthesis. Many texts have been written to help the chemist appreciate the importance of conformational analysis, see for example [4–6].

Honours and Awards

Hassel was presented with honorary degrees from the Universities of Copenhagen (Dr.Phil. 1950) and Stockholm (Fil.Dr. 1960).

He received the Guldberg-Waage Medal (1964) from the Norwegian Chemical Society, the Gunnerus Medal from the Royal Norwegian Academy of Sciences (1964) and the Nobel Prize for Chemistry, joint (1969).

Hassel was an Honorary Fellow of the Chemical Society, London and of the Norwegian Chemical Society. He also held Fellowships with the Royal Danish Academy of Sciences, the Royal Norwegian Academy of Science and the Royal Swedish Academy of Sciences.

He was also appointed a Knight of the Order of St. Olav.

Bibliography

1. Reproduced with permission.
2. From Nobel Lectures, Chemistry 1963–1970, Elsevier Publishing Company, Amsterdam, 1972.
3. Who was who: 1981–1990.
4. Modern Conformational Analysis by Helena Dodziuk, VCH, 1995, ISBN: 1-56081-689-9.
5. Basic Organic Stereochemistry by Ernest L. Eliel, Samuel H. Wilen and Michael P. Doyle, Wiley Interscience, 2001, ISBN: 0-471-37499-7.
6. Stereochemistry by David G. Morris, RSC, 2001, ISBN: 0-85404-602-X.

1970

© Nobel Foundation [1]

Luis F. Leloir

"for his discovery of sugar nucleotides and their role in the biosynthesis of carbohydrates" [2]

Although the presentation speech for the 1970 Chemistry Nobel Prize was prepared by Karl Myrbäck, then a member of the Nobel Committee for Chemistry of the Royal Swedish Academy of Sciences, it was read by Professor Arne Tiselius [2]:

> *...In 1949 Leloir published the discovery which became the foundation for a remarkable development. He found that in a certain reaction, which results in the transformation of one sugar to another sugar, the participation of a so far unidentified substance was essential. He isolated the substance and determined its chemical nature. It turned out to be a compound of an unknown type, containing a sugar moiety bound to a nucleotide. Compounds of this type are now called sugar nucleotides. Leloir established that the transformation reaction does not occur in the sugars as such, but in the corresponding sugar nucleotides. To put it simply, one may say, that the linking with the nucleotide occasions an activation of the sugar moiety which makes the reaction possible.*
>
> *The remarkable aspect of the discovery was not the explanation of a single reaction, but Leloir's quick comprehension that he had found the key which would enable us to unravel an immense number of metabolic reactions. He ingeniously realized that a path had been opened to a field of research containing an accumulation of unsolved problems. In the twenty years that have elapsed since his initial discovery he has carried on his research in this field in an admirable manner...*
>
> *...Leloir soon found that besides the sugar nucleotide first isolated, several others of the same type occur in Nature, and many have also been found by other research workers. Today more than one hundred sugar nucleotides which are essential participants in various reactions are known and well characterized. Some of them have an action similar to that of the first isolated, namely in the transformations of simple sugars to other simple sugars or sugar derivatives.*

> *Still more important was Leloir's discovery that other sugar nucleotides have another action which occurs in the biological synthesis of compounds which are composed of or contain simple sugars or sugar derivatives. Leloir showed that all these syntheses are essentially transfer reactions. Sugar moieties from sugar nucleotides are transferred to accepting molecules which thereby increase in size. Probably the most sensational discovery made by Leloir was that the synthesis of the high-molecular polysaccharides also functions in this manner. The first example of the fundamental role of the sugar nucleotides in polysaccharide biosynthesis was found by Leloir in 1959 in the case of glycogen...*

Luis F. Leloir was born in Paris, of Argentine parents on 6th September 1906. His parents moved to Buenos Aires when Leloir was aged two. He married Amelia Zuberbuhler (1943) and they had a daughter, Amelia. Leloir died on 2nd (some record 3rd) December 1987 [2–4].

The Early Years

Leloir graduated with an M.D. degree in medicine from the University of Buenos Aires (1932), followed by some work under A. Houssay (Nobel Prize of Medicine, joint 1947) on the role of the adrenalin carbohydrate metabolism. This was followed by a period in the Biochemical Laboratory of Cambridge, England (from 1936) under Sir Frederick Gowland Hopkins, but working with Dixon, Edson and Green. Returning to Buenos Aires, he worked on the oxidation of fatty acids in liver with Muñoz, and on the formation of angiotensin with Menéndez, Fasciolo and Taquini.

The Career

Year	*Post*	*Institute*
1943	Resigned from post and moved to the United States[a]	
1944	Research Assistant in Dr. Carl F. Cori's laboratory and with D.E. Green in the College of Physicians and Surgeons	St. Louis, United States and then Columbia University, New York
1945	Director and later, as an additional post, Professor of Biochemistry in the Faculty of Sciences,	Instituto de Investigaciones Bioquímicas, Fundación Campomar, Buenos Aires and later (in addition) University of Buenos Aires
1962	Extraordinary Professor	Faculty of Exact Sciences, University of Buenos Aires

[a] Leloir's resignation from his post in 1943 was a protest of the sacking of Bernardo Houssay (Nobel Prize for Medicine (joint), 1947) by the Peron Government.

Leloir's lifelong interests lay in various aspects of biochemistry which is reflected in various investigations into:

- fatty acid oxidation
- the role of adenosine diphosphate glucose in that of starch
- the metabolism of galactose
- the discovery of glucose-1,6-diphosphate (G1,6DP)
- the role of a polyprenol, dolichol, in glucose transfer in animal tissues
- the isolation of uridine diphosphate glucose
- the isolation of guanosine diphosphate
- studies of the presence of microsomal acceptor lipid in animal tissues
- studies concerning glucose donors in the synthesis of trehalose (with Enrico Cabib, 1953)
- studies concerning glucose donors in the synthesis of sucrose (with Carlos E. Cardini and J.Chiriboga, 1955)

Leloir's Nobel Prize acceptance speech is dated 11th December 1970 with the title 'Two decades of research on the biosynthesis of saccharides'

> *...On returning to Buenos Aires in 1945 I started to work with R. Caputto and R. Trucco. Dr. Caputto had done some research on the mammary gland and had the idea that glycogen was transformed into lactose. At the time one had to rely on osazones for identification and we soon reached a dead end. On looking back I think that what we were observing was the degradation of glycogen by amylase.*
>
> *We then decided to study lactose breakdown by a yeast* ***Saccharomyces fragilis*** *with the idea that this would give us information on the mechanism of synthesis. In fact it did give us information but only after a long and tortuous process.*
>
> *First we studied the lactase, then the phosphorylation of galactose (Trucco, Caputto, Leloir and Mittehnan, 1948) and the transformation of galactose 1-phosphate...*
>
> *...The results of our experiments were very confusing because we did not know that we were dealing with two thermostable factors. Finally we realized what was happening and we concentrated on the purification of the factor involved in the second reaction. That is in the phosphoglucomutase reaction.*
>
> *We sent a letter to the editors of Archives of Biochemistry (Caputto, Leloir, Trucco, Cardini and Paladini, 1948) describing a new cofactor and mentioned that Kendall and Strickland (1938) had previously described as activation by fructose 1,6-diphosphate but that our cofactor was different.*
>
> *After we had sent the manuscript we happened to test again fructose 1,6-diphosphate and obtained a strong activation. Furthermore our purified preparations were loaded with fructose 1,6-phosphate. We had decided to ask the letter back but as a consequence of much worrying we struck on the idea that the activator might be glucose 1,6-diphosphate. Since the latter compound has the reducing group blocked we reasoned that it should be alkali-stable. Strangely enough everything turned out as expected. If it had not been for this mistake we might still be talking of the allosteric effect of fructose 1,6-di-phosphate on phosphoglucomutase.*

When we finished working with glucose 1,6-diphosphate we continued with the other cofactor. The concentrates were found to absorb light at 260 mμ and had a spectrum similar to that of adenosine but with some differences. At the time the only soluble nucleotides known to be present in tissues were the inosine and adenosine nucleotides. It was an exciting day when Caputto came in one morning with a Journal of Biological Chemistry which showed the absorption spectrum of uridine. It looked identical to that of our cofactor. After measuring glucose and phosphate content and doing a titration curve the structure shown (in Figure I) was proposed (Cardini, Paladini, Caputto and Leloir2, 1950; Caputto, Leloir, Cardini and Paladini', 1950). This first sugar nucleotide was named uridine diphosphate glucose: UDPG. Its structure was confirmed by synthesis some five years later by Todd and co-workers in Cambridge...

Many texts have been published concerning biosynthesis in general and of saccharides in particular since Leloir's pioneering studies, see for example [5–7].

Honours and Awards

Leloir received honorary degrees from the following Universities: Granada (Spain); La Plata (Argentina); Paris (France); Tucuman (Argentina).

He received awards and prizes from many Institutes including: the Argentine Scientific Society; the Benito Juarez (Mexico); the Bunge and Born Foundation (Argentina); the Gairdner Foundation (Canada); the Louisa Gross Horowitz (United States); the Helen Hay Whitney Foundation (United States); the Severo Vaccaro Foundation (Argentina); the Nobel Foundation who awarded the Nobel Prize for Chemistry for 1970.

Leloir held Fellowships/Memberships in a number of Academies including Member Academia Nacional de Medicina. Foreign Member of: the American Academy of Arts and Sciences; the National Academy of Sciences; the American Philosophical Society; the Pontificial Academy of Sciences; of the Royal Society. He was also a Honorary Member of the American Society of Biological Chemists and of the Biochemical Society (England).

He was also President of the Pan-American Association of Biochemical Societies and a Member of the Directorate of the National Research Council (1958–64).

Leloir was presented with the Légion d'Honneur.

Leloir's citation for election (date: 27/04/1972) as Foreign Member of the Royal Society reads [8]:

'Dr Luis Federico Leloir of Buenos Aires, Argentina distinguished for his work on carbohydrate metabolism and especially for his discovery of the sugar nucleotides.'

No proposers are listed in the Sackler database.

Bibliography

1. Reproduced with permission.
2. From Nobel Lectures, Chemistry 1963–1970, Elsevier Publishing Company, Amsterdam, 1972.
3. Biographical Memoirs of Fellows of the Royal Society, 1990, vol. 35, pp. 201–208.
4. Who was who: 1981–1990.

5. Bioorganic Chemistry: Deoxysugars, Polyketides and Related Classes: Synthesis, Biosynthesis, Enzymes (Springer Desktop Editions in Chemistry) ed. by J. Rohr, Springer 1st ed. 2000, ISBN: 354066971X.
6. Carbohydrate Chemistry (Oxford Chemistry Primers, 99), by B. G. Davis and A. J. Fairbanks, Oxford University Press, USA, 2002, ISBN: 0198558333.
7. Chemical Aspects of Biosynthesis (Oxford Chemistry Primers, No. 20) by John Mann, Oxford University Press, USA, 1995, ISBN: 0198556764.
8. Sackler database, Royal Society.

1971

Gerhard Herzberg

"for his contributions to the knowledge of electronic structure and geometry of molecules, particularly free radicals" [2]

During his presentation speech, Stig Claesson, then a member of the Royal Academy of Sciences, commented of Herzberg [2]:

...This year's Nobel Prize winner in Chemistry, Dr. Gerhard Herzberg, is generally considered to be the world's foremost molecular spectroscopist and his large Institute in Ottawa is the indisputed center for such research. It is quite exceptional, in the field of science, that a single individual, however distinguished, in this way can be the leader of a whole area of research of general importance. A noted English chemist has also said that the only institutions that have previously played such a role were the Cavendish laboratory in Cambridge and Bohr's Institute in Copenhagen...

...For a chemical reaction to occur the original molecules must in some way break up into fragments which rearrange to form the new molecules. These fragments, or intermediates, are called free radicals.

Free radicals are very difficult to study due to their short life-times - measured in millionths of a second. Herzberg therefore had ample opportunity to repeatedly demonstrate his exceptional experimental skill when the necessary spectroscopic technique was worked out.

Herzberg has so far performed extensive precision determinations of the properties of over thirty free radicals among which are to be found the radicals methyl and methylene - well known from organic chemistry. Among his exciting discoveries may be mentioned that radicals drastically change their shape with increasing energy. For example, methylene is linear in its ground state but bent in states of higher energy. Many of the most important results were only achieved after several years' work and some of the most exciting as late as at the end of the 1960's. One can therefore note that this year's prize is truly an award for contributions of great current interest...

Gerhard Herzberg was born in Osdorf, a suburb of Hamburg on 25th December 1904, the younger son of Albin and Ella Herzberg. He married Luise Herzberg, neé Oettinger (m 1929) and they had two children. Luise died in 1971 and Herzberg married Monika, née Tenthoff. Herzberg died on 3rd March 1999 [2,3].

The Early Years

Herzberg attended schools in his home town of Hamburg (including the Real Gymnasium of the Johanneum) prior to enrolling at the Darmstadt Institute of Technology. He obtained his Dr.Ing. degree under H. Rau (1928), followed by two years of postdoctoral work (1928 to 1930) under Max Born and James Franck at the University of Göttingen and at the University of Bristol. His Bristol work involved using a grating spectrograph and investigating the spectra of P_2, CP, N_2, C_2H_2 and CH_2O.

The Career

Year	*Post*	*Institute*
1930	Privatdozent (lecturer) and senior assistant – Physics Dept	Darmstadt Institute of Technology
1935	Forced to flee Germany took up a Guest Professorship. Appointed Research Professor of Physics a few months later	University of Saskatchewan (Saskatoon, Canada).
1945	Professor of Spectroscopy	Yerkes Observatory of the University of Chicago
1948	Principal Research Officer and shortly afterwards Director of the Division of Physics	National Research Council
1955	Director of the Division of Pure Physics	National Research Council
1969	Distinguished Research Scientist, Division of Physics	National Research Council

Herzberg's main contributions are to the field of spectroscopic techniques for studying free radicals, including:

- studies of atomic and molecular spectroscopy
- determining the structures of a large number of diatomic and polyatomic molecules
- the identification of certain molecules in planetary atmospheres, in comets, and in interstellar space
- a detailed study of methyl and methylene radiacals
- the structures of many free radicals

A biography of Herzberg was written by Stoicheff [5].

His Nobel Prize acceptance speech is dated 11th December 1971 with the title 'Spectroscopic Studies of Molecular Structure'. Herzberg's opening remarks concerned a detailed account of the spectra of radicals of small molecules [2] i.e.:

H_2, D_2, HD and H_2^+
O_2
CH and CH^+
C_2^+

He subsequently went on to discuss polyatomic molecules and radicals thus:

...In polyatomic molecules and radicals there are in general several geometrical parameters which are required to describe their structures (rather than one, r_e, for diatomic molecules). For a large number of stable molecules these parameters have been determined in their ground states by the techniques of infrared, Raman and microwave spectroscopy. In excited electronic states these parameters can only be obtained by a detailed study of electronic band systems in absorption or emission In many cases it has been found that the shape (point group) of the molecule in an excited state is different from its shape in the ground state. For example the molecules C_2H_2 and HCN, well-known to be linear in their ground states are found to be strongly bent in their first (singlet) excited states... It is clear that this behaviour throws an interesting light on the way in which the electronic structure determines molecular shape.

For polyatomic free radicals and ions one is dependent both for the ground states and the excited states on the study of electronic spectra to obtain the shapes and the geometrical parameters. Only for a few radicals have microwave and infrared spectra been obtained in the gaseous state. Not infrequently, just as for stable molecules, electronic spectra of radicals are diffuse or even continuous in which case no information about geometrical structure can be obtained. The BH_3 radical may be such a case. An additional difficulty in the interpretation of free radical spectra is the problem of identification, that is the problem to which particular radical an observed spectrum belongs. As an example of these difficulties I should like to discuss the history of the discovery of the spectra of CH_2 and C_3...

The ability to measure free radicals by a number of spectroscopic techniques has increased markedly since Herzberg pioneered his techniques, see for example [6–8].

Honours and Awards

Herzberg was presented with Honorary D.Sc. degrees from: McMaster University (1954); National University of Ireland (1956); Oxford University (1960); University of British Columbia (1964); Queen's University, Kingston (1965); University of New Brunswick (1966); Carleton University, Ottawa (1967); University of Chicago (1967); Memorial University, Newfoundland (1968); York University, Toronto (1969); University of Windsor (1970); Royal Military College of Canada (1971); Cambridge University (1972); Drexel University (1972); McGill University (1972); University of Montreal (1972); Université de Sherbrooke (1972); University of Manitoba (1973); Andhra University (1975); University of Bristol (1975); University of Delhi (1976); Osmania University (1976); University of Western Ontario (1976); Laval University (1979); St. Mary's University (1991). He was presented with Honorary LL.D. degrees from the University of: Saskatchewan (1953); Toronto (1958);

Dalhousie (1960); Alberta (1961); Simon Fraser (1972); St. Francis Xavier University (1972). He also holds the Honorary: Fil:Hed-Dr. from the University of Stockholm (1966); Dr.rer.nat. from the University of GIöttingen (1968) and the University of Hamburg (1974); the Honorary D.Phil. from the Weizmann Institute of Science (1976); a Dr.phil.nat. from the University of Frankfurt (1983); a Honorary Ph.D. from the University of Toledo (1984).

Herzberg was presented with: Mueller Alewyn Plakette Medal, Darmstadt (1927); Médaille de l'Université de Liège (1950); Henry Marshall Tory Medal, Royal Society of Canada (1953); Joy Kissen Mookerjee Gold Medal (1954); a Gold Medal for Achievement in Physics (1957); Medal of the Society for Applied Spectroscopy (1959); Médaille de l'Université de Liège (1960); Médaille de l'Université de Bruxelles (1960); Spectroscopy Award (1962); Frederic Ives Medal of the Optical Society of America (1964); Willard Gibbs Medal (1969); Gold Medal of the Professional Institute of the Public Service of Canada (1969); Faraday Medal (1970); Royal Medal of the Royal Society of London (1971); Linus Pauling Medal (1971); Nobel Prize in Chemistry (1971); Madison Marshall Award (1974); Earle K. Plyler Prize (1985); Jan Marcus Marci Memorial Medal (1987).

Herzberg was elected a Fellow of the Royal Society of Chemistry (1939), of the Royal Society of Canada (1939) and of the Royal Society of London (1951). He was a Member of the Pontifical Academy of Sciences, an Honorary Fellow of: the Indian Academy of Science (1954), the Indian Physical Society (1957) and the Chemical Society (1968). He was an Honorary Member of the Hungarian Academy of Sciences (1964), the Optical Society of America (1968), the Royal Irish Academy (1970, the Japanese Academy (1976) and the Chemistry Society of Japan (1978). He was also an Honorary Foreign Member of the Royal Swedish Academy of Sciences (1981). Herzberg was also a Foreign Associate of the National Academy of Sciences in Washington (1968).

He was a Companion of the Order of Canada (1968) and a Member of the Privy Council (1992).

Herzberg had been active as President or Vice President of several international commissions dealing with spectroscopy. He was also Vice President of the International Union of Pure and Applied Physics from 1957 to 1963. He held the offices of President of the Canadian Association of Physicists for the year 1956–57 and President of the Royal Society of Canada for the year 1966–67.

He was Bakerian Lecturer (1960) and Faraday Lecturer (1970).

Herzberg's citation for election (date: 15/03/1951) as Fellow of the Royal Society reads [9]:

> **'Dr. Herzberg is distinguished for his work on atomic and molecular spectra. He has contributed greatly to the determination of the structures of diatomic and polyatomic molecules. He has been prominent in the study of forbidden transitions, and has recently observed the quadrupole rotation vibration spectrum of hydrogen. He has identified CH as a constituent of the interstellar medium, and has produced the nuclear bands of comets in the laboratory. His three books on atomic and molecular spectra are internationally known as standard works on the subject.'**

He was proposed by:

From General Knowledge. C. J. Mackenzie; J. S. Foster; W. B. Lewis; E. W. Steacie; J. A. Gray; H. W. Thompson; G. B. B. M. Sutherland; S. Chandrasekhar; E. C. Bullard; E. A. Guggenheim.

Bibliography

1. Reproduced with permission.
2. From Les Prix Nobel en 1971, Editor Wilhelm Odelberg, [Nobel Foundation], Stockholm, 1972.

3. Biographical Memoirs of Fellows of the Royal Society, 2003, vol. 49, pp. 179–196.
4. Who's who: 1999.
5. Gerhard Herzberg - An Illustrious Life in Science - Boris Stoicheff FRS, McGill-Queen's University Press, ISBN: 0-660-18757-4.
6. Biomedical EPR - Part A: Free Radicals, Metals, Medicine and Physiology (Biological Magnetic Resonance) edited by Sandra S. Eaton, Gareth R. Eaton and Lawrence J. Berliner, Springer, 1st ed. 2004, ISBN: 0306485060.
7. EPR of Free Radicals in Solids: Trends in Methods and Applications (Progress in Theoretical Chemistry and Physics) edited by A. Lund and M. Shiotani, Springer, 1st ed. 2003, ISBN: 1402012497.
8. Infrared Vibration-Rotation Spectroscopy: From Free Radicals to the Infrared Sky by Geoffrey Duxbury, John Wiley & Sons, 2000, ISBN: 0471974196.
9. Sackler database, Royal Society.

1972

© Nobel Foundation [1]

Christian B. Anfinsen (0.5)

"for his work on ribonuclease, especially concerning the connection between the amino acid sequence and the biologically active conformation" [2]

During his presentation speech, Bo Malmström, of the Royal Academy of Sciences, commented of Anfinsen and his work [2]:

...From a chemical point of view enzymes are proteins. These are built up of 20 different amino acids which are linked together into long chains. Despite the fact that proteins have only 20 building blocks, there are thousands of enzymes, each with its specific properties. This large degree of variation becomes possible because the number and sequence of the amino acids in the chain can be varied. Ribonuclease was the first enzyme for which the complete amino acid sequence was determined thanks to contributions from Anfinsen and from Moore and Stein...

...Every living organism has its own characteristic pattern of enzymes. It can also produce a copy of itself, and this progeny has the same enzymes. An important question concerns the source of the information which has to be passed on from generation to generation for the enzyme pattern to be preserved. We know, thanks to contributions which have led to earlier Nobel Prize awards, that a specific molecule, called DNA, serves as the carrier of the traits of inheritance. These traits are expressed by DNA controlling the synthesis of enzymes. DNA accomplishes this by determining the sequence of the amino acids making up a particular protein molecule. An active enzyme does not, however, consist just of a long chain of amino acids linked together, but the chain is folded in space in a way which gives the molecule a globular form. What is the source of the information responsible for this specific folding of the peptide chain? It is this question in particular which has been the concern of Anfinsen's investigations. In a series of elegant experiments he showed that the necessary information is inherent in the linear sequence of amino acids in the peptide chain, so that no further genetic information than that found in DNA is necessary...

...Dr. Anfinsen, I have tried to explain your pioneering investigations showing that the linear sequence of amino acids in the enzyme ribonuclease determines the biologically active conformation of this enzyme. This finding has profound implications for our understanding of the way in which active enzyme molecules are formed in living cells...

Anfinsen was born in Monessen, Pennsylvania on 26th March 1916, the son of Christian Boehmer Anfinsen, Sr., a mechanical engineer, and Sophie, née Rasmussen. Both of Anfinsen's parents were Norwegian immigrants who taught their children the Norwegian language and heritage. Anfinsen married Florence Bernice Kenenger (1941) with whom he had three children. They were divorced in 1978. Anfinsen married Libby Esther Shulman the following year and converted to Orthodox Judaism. Anfinsen died of a heart attack in Randallstown, Maryland on 14th May 1995 [2,3].

The Early Years

Anfinsen was admitted to Swarthmore College on a scholarship (1933) from where he graduated with a B.A. degree (1937). He moved to the University of Pennsylvania where he graduated with an M.S. degree in organic chemistry (1939). He was a Visiting Investigator at the Carlsberg Laboratory in Copenhagen (1939–40), prior to studying biochemistry at Harvard Medical School from which he graduated with a Ph.D. (1943).

The Career

Year	*Post*	*Institute*
1943	Instructor and then Assistant Professor of Biological Chemistry	Harvard Medical School
1948	Assistant Professor of Biological Chemistry	Harvard Medical School
1950	Chief of the Laboratory of Cellular Physiology incorporating Metabolism from 1952	National Institute of Health
1962	Professor of Biological Chemistry	Harvard Medical School
1963	Chief, Laboratory of Chemical Biology	National Institute of Arthritis and Metabolic Diseases
1981	Retired from National Institute of Health	
1982–1995	Professor, Department of Biology	Johns Hopkins University

Although Anfinsen was based in the United States of America, he travelled extensively, sometimes with the help of grants. He was a Senior Fellow of the American Cancer Society working with Dr. Hugo Theorell at the Medical Nobel Institute (1947–48), a Rockefeller Fellow (1954), held a Guggenheim Fellowship (1958) in the Carlsberg Laboratory (Copenhagen) and was a Visiting Fellow of All Souls College, Oxford, England (1970).

Anfinsen remained active in biochemistry throughout his career. His investigations included work on:

- protein chemistry (from the early 1950s)
- the "thermodynamic principle" to describe protein folding in enzymes (1954–62)

- calculating the rate at which amino acids are polymerized
- studies of ribonuclease
- the discovery of a microsomal enzyme which catalyzes sulfhydryl-disulfide interchange
- an extracellular nuclease of Staphylococcus aureus
- the use of affinity chromatography techniques to identify amino acid sequence in enzymes (1966–68)
- interferon (1973)
- an investigation of the reversible denaturation of several proteins
- the relationship between structure and function in enzymes
- "hyperthermophilic bacteria" commences (1983)
- the non-uniform labelling in newly synthesized proteins (Steinberg)

Anfinsen was an editor of *Advances in Protein Chemistry*, served on the Editorial Board of the *Journal of Biological Chemistry and Proceedings of the National Academy of Sciences*. He wrote "*The Molecular Basis of Evolution*" (published in 1959).

Anfinsen's Nobel Prize acceptance speech is dated 11th December 1972 with the title 'Studies on the principles that govern the folding of protein chains'. During the course of his presentation, Anfinsen commented on the nucleation of folding thus:

> *...A chain of 149 amino acid residues with two rotatable bonds per residue, each bond probably having 2 or 3 permissible or favored orientations, would be able to assume on the order of 4^{149} to 9^{149} conformations in solution. The extreme rapidity of the refolding makes it essential that the process take place along a limited number of "pathways", even when the statistics are severely restricted by the kinds of stereochemical ground rules that are implicit in a so-called Ramachandran plot, It becomes necessary to postulate the existence of a limited number of allowable initiating events in the folding process. Such events, generally referred to as nucleations, are most likely to occur in parts of the polypeptide chain that can participate in conformational equilibria between random and cooperatively stabilized arrangements. The likelihood of a requirement for cooperative stabilization is high because, in aqueous solution, ionic or hydrogen bonded interactions would not be expected to compete effectively with interactions with solvent molecules and any thing less than a sizeable nucleus of interacting amino acid side chains would probably have a very short lifetime. Furthermore, it is important to stress that the amino acid sequences of polypeptide chains designed to be the fabric of protein molecules only make functional sense when they are in the three dimensional arrangement that characterizes them in the native protein structure. It seems reasonable to suggest that portions of a protein chain that can serve as nucleation sites for folding will be those that can "flicker" in and out of the conformation that they occupy in the final protein, and that they will form a relatively rigid structure, stabilized by a set of cooperative interactions. These nucleation centers, in what we have termed their "native format", might be expected to involve such potentially self-dependent substructures as helices, pleated sheets or beta-bends...*

Further examples of enzyme reactivity can be found in [4–5].

Honours and Awards

Anfinsen was presented with honorary degrees from many Universities including D. Sc. (hc) from: Swarthmore College (1965); Georgetown University (1967); New York Medical College (1969); the

University of Pennsylvania (1973); Gustavus Adolphus College (1975); Brandeis University (1977); Providence College, Rhode Island (1978); Yeshiva University, New York City, New York (1981); Adelphi University, New York (1987); University of Las Palmas, Canary Islands (Honorary) (1993). He was also presented with the Honorary degree of Doctor of Medicine, M.D. from the University of Naples, Italy (1981).

He was presented with many awards and prizes including: the Rockefeller Foundation Public Service Award (1954); the Hadassah Myrtle Leaf Award (1969); the Nobel Prize in Chemistry (1972); the Fritz Lippmann Life Sciences Medal, Paris, France (1979); the Case Institute of Technoogy, Centennial Medal (1980); the NIH Scientist Emeritus Award (1981); the Hebrew University of Jerusalem Medal (1985); the National Library of Medicine Medal, Bethesda, Maryland (1986); the Pioneer in Science Award, Bar Ilan University, Ramat Aviv, Israel (1990); the University of Texas Medical Branch, Centennial Symposium Medal, Galveston, Texas (1991); the Morris Brown College Research Symposium Award, Outstanding Nobelist (1991); the Department of Health, Education and Welfare, Distinguished Service Medal (1994); the Department of Health, Education and Welfare, Superior Service Medal (1994).

Anfinsen was elected Fellow or Member of numerous societies including the National Academy of Sciences (1963), the Royal Danish Academy (1964) and the Pontifical Academy of Science, The Vatican (1981). Anfinsen was a Member of the American Society of Biological Chemists, of the American Academy of Arts and Sciences, the National Academy of Sciences, the Washington Academy of Sceinces and the Federation of Scientists. He was an Honorary Fellow, Weizmann Institute of Science, Rehovot, Israel (1969).

He held many named lecturerships including: Leon Lecture, University of Pennsylvania (1969); EMBO Lecture for Sweden (1970); Jubilee Lecture, British Biochemical Society, England (1972); Mathers Lectures, Indiana University, Bloomington, Indiana (1974); Kempner Lectureship, University of Texas Medical Branch, Galveston, Texas (1975); Naff Lectures, University of Kentucky, Lexington, Kentucky (1977); Herbert A. Sober Memorial Lectureship (1978); Berson Memorial Lecture, Mount Sinai School of Medicine, New York City, New York (1979); Distinguished Scientist Lectures Series, Bard College, New York (1984); Theobald Smith Lecture Award, Albany College of Medicine, New York (1984); Genetic Engineering Lectures, Kon Kuk University, Seoul, Korea (1988).

Anfinsen was also: a Member of the Board of Governors of the Weizmann Institute of Science in Rehovot, Israel; President of the American Society of Biological Chemists for the Academic Year 1971–72; Participant, Encounter for the Universality of UNESCO, Paris (1975); Scientific Advisory Committee of Weizmann Institute Chairman (1974); a Consultant to ABC Cancer Foundation (1978). He was Visiting Professor at Harvard Medical School, Department of Biological Chemistry (1962–63) and also of Biochemistry, Weizmann Institute of Science (1981–82).

Many memorial lecturers or endowments were created to honour Anfinsen including: the Feinberg Graduate School, Christian B. Anfinsen Scholarship (established in 1974); Fogarty International Symposium in honor of Anfinsen on Contributions of Chemical Biology to Biomedical Science (1981); Protein Folding Symposium to Honor Christian B. Anfinsen, NIH, Bethesda, Maryland (1992); National Institutes of Health/Israel Alumni Association Lectureship established in honor of Christian B. Anfinsen, Rehovot, Israel: Memorial Biennial Lecture (1994); Johns Hopkins University Lectureship honoring Anfinsen (established in Baltimore, Maryland in 1994); National Institutes of Health Anfinsen Lectureship established (1994).

Bibliography

1. Reproduced with permission.
2. From Nobel Lectures, Chemistry 1971–1980, Editor-in-Charge Tore Frängsmyr, Editor Sture Forsén, World Scientific Publishing Co., Singapore, 1993.
3. Who was who: 1991–1995.
4. Enzymes: A Practical Introduction to Structure, Mechanism, and Data Analysis by Robert A. Copeland, Wiley-VCH, 2nd ed. 2000, ISBN: 0471359297.
5. Computational Approaches to Biochemical Reactivity (Understanding Chemical Reactivity) edited by Gábor Náray-Szabó and Arieh Warshel, Springer, 1st ed. 1997, ISBN: 0792345126.

1972

© Nobel Foundation [1]

Stanford Moore (0.25)

"for their contribution to the understanding of the connection between chemical structure and catalytic activity of the active centre of the ribonuclease molecule" [2]

During his presentation speech, Bo Malmström, of the Royal Academy of Sciences, commented of Anfinsen and his work [2]:

> *...The contributions of Moore and Stein concern another fundamental question regarding ribonuclease, namely the basis for its catalytic activity. The reacting substances, the substrates, are bound to an enzyme in what is generally called its active site. In the complex so formed there is an interaction between enzyme and substrate leading to a changed reactivity of the substrate. Knowledge about the structure of an enzyme is of little help in understanding this interaction if it is not possible to find the active site and to determine the chemical groups in it. Moore and Stein discovered as an important principle that the active site contains amino acids with an anomalously high reactivity compared to the same amino acids in free form. This high reactivity is of direct importance for the catalytic activity of the enzyme, but Moore and Stein also found it possible to utilize it to label two amino acids in the active site by chemical modification. In this way the position of these amino acids in the long peptide chain could be unambigously determined. Through these investigations Moore and Stein were able to provide a detailed picture of the active site of ribonuclease long before the three-dimensional structure of the enzyme had been determined...*

> *...Drs. Moore and Stein, I have attempted to summarize your fundamental contributions to our understanding of the relationship between chemical structure and catalytic activity in the enzyme ribonuclease. In particular, I have stressed your studies leading to the localization of two specific histidine residues in the active site of the enzyme. It is for these pioneering experiments that the Royal Academy of Sciences has decided to award this year's Nobel Prize in Chemistry to you together with Dr. Anfinsen...*

Stanford Moore was born on 4th September 1913 in Chicago, Illinois, the son of John Howard and Ruth Fowler Moore. He died in New York City on 23rd August 1982 [2,3].

The Early Years

Moore grew up in Nashville, Tennessee, spending part of his early years tutored from home. This was followed by his school years at a high school administered by the George Peabody College for Teachers in Nashville. Moore's interest in chemistry stemmed from one his teachers, R.O. Beauchamp. Moore entered Vanderbilt University, undecided between a career in chemistry or aeronautical engineering. A. Ingersoll, one of Vanderbilt's organic chemistry professors turned Moore's thoughts to the study of molecular architecture. Moore graduated with a B.S. *summa cum laude* (1935) where he majored in chemistry. He won a Wisconsin Alumni Research Foundation Fellowship for which he was recommended by the Vanderbilt staff. The Fellowship enabled Moore to enrol at the University of Wisconsin under Karl Paul Link, from where he graduated with his Ph.D. in biochemistry (1938), the subject for which was the characterization of carbohydrates as benzimidazole derivatives. Encouraged by Link, Moore went on to work under Max Bergmann, a former senior research associate to Emil Fischer, at the Rockefeller Institute for Medical Research. Stein, with whom Moore was to collaborate for many years, was also a member of the Bergmann group at this time.

The Career

Year	*Post*	*Institute*
1939	Assistant	Rockefeller Institute for Medical Research
1942	Associate	Rockefeller Institute for Medical Research
1949	Associate Member	Rockefeller Institute for Medical Research
1952	Professor and Member	Rockefeller Institute for Medical Research

Moore's military service was as a junior administrative officer in the Office of Scientific Research Development, where he worked on academic and industrial chemical projects. Initially based in Washington, Moore ended the War attached to the Headquarters of the United States Armed Forces in the Pacific Ocean Area, Hawaii.

Although based at the Rockefeller Institute for Medical Research for most of his career, Moore undertook a sabbatical year, six months of which were as the Francqui Chair at the University of Brussels (1950). He concentrated on amino acid analysis during this period, which was followed by six months work at the University of Cambridge, where he shared some laboratory space with Sanger (Nobel Laureate in Chemistry for 1958 and 1980).

Moore's interests covered aspects of biochemistry connected with:

- ribonuclease in general
- understanding of the connection between chemical structure and catalytic activity of the ribonuclease molecule
- the development of quantitative chromatographic methods for amino acid analysis
- automated amino acid analysis
- protein chemistry

Moore's Nobel Prize acceptance speech was combined with that of Stein. The combined presentation is dated 11th December 1972 with the title 'The chemical structures of pancreatic ribonuclease and deoxyribonuclease'. It is unusual for an acceptance speech to be combined in this way. For the purposes of this and Stein's entry, the author has arbitrarily chosen to record the purification and analysis of ribonuclease under this entry and the structure and special features under Stein's entry.

Moore and Stein comment on the Purification of Ribonuclease thus [2]:

...The first step in the study of the structure of ribonuclease was, of course, its purification. Ribonuclease was first described in 1920 by Jones, who showed that there was present in beef pancreas a relatively heat-stable enzyme capable of digesting yeast nucleic acid. Dubos and Thompson partially purified the enzyme some eighteen years later and in 1940 Kunitz described the isolation of bovine ribonuclease in crystalline form after fractionation by ammonium sulfate precipitation. In order to be as certain as possible that we were beginning the structural study with a single molecular species, we undertook to apply the potential resolving power of ion exchange chromatography to ribonuclease. While Werner Hirs, in our laboratory, was exploring the chromatographic purification of ribonuclease on the polymethacrylic resin Amberlite IRC-50, Paléus and Neilands, in Stockholm, were studying cytochrome C on the same exchanger. These two proteins were the first molecules of their size to be thus purified. The best resolution for ribonuclease is now obtained with an exchanger invented in Uppsala, a sulfoethyl cross-linked dextran, which was a development that grew from Porath and Flodin's experiments on gel filtration and drew upon Sober and Peterson's emphasis on the advantages of a carbohydrate matrix for the exchanger.

When pancreatic extracts were analyzed without prior fractionation, two peaks of enzymatic activity were observed by us by ion exchange chromatography and by Martin and Porter by partition chromatography. The major component, ribonuclease A, was selected for the first structural studies. (In later independent experiments, Plummer and Hirs isolated ribonuclease B in pure form from pancreatic juice and showed it to be the same as A but with the addition of a carbohydrate side chain attached to one asparagine residue.)

Amino Acid Analysis

The second step in the structural study of ribonuclease A was the determination of the empirical formula of the chromatographically homogeneous protein in terms of the constituent amino acids. Our appreciation of the importance of quantitative amino acid analysis began in the late 1930's when we had the special privilege of starting our postdoctoral studies in apprenticeship to Max Bergmann. In 1945 it was possible to take a new look at the subject in the light of the renaissance in chromatography stimulated by Martin and Synge in the early 1940's. In 1949, by combining a quantitative photometric ninhydrin method with elution of amino acids from starch columns by alcohol water eluents on an automatic fraction collector, we were able to analyze a protein hydrolysate in about two weeks by running three such chromatograms to resolve all overlaps. In the early 1950's, the process was speeded up to one week by turning to ion exchange chromatography on a sulfonated polystyrene resin. In 1958, in cooperation with Darrel Spackman the process was automated to give recorded curves and the speed was increased to give an overnight run. Shorter columns and faster flow rates permitted an analysis time of about 6 hours. Results from many academic and industrial laboratories have helped to make the procedures simpler and more rapid. Some recent users have adopted 2-hour systems and the ninhydrin reagent has been improved. In the 1970's, a number of industrially designed analyzers with increased automation have reduced the time for a complete analysis to about 1 hour and increased the sensitivity to the nanomole range. The sharing of knowledge among academic scientists and industrial designers of instruments and manufacturers of ion exchangers has played an important role in progress of biomedical research in this field.

A schematic diagram of the analyzer appears below.

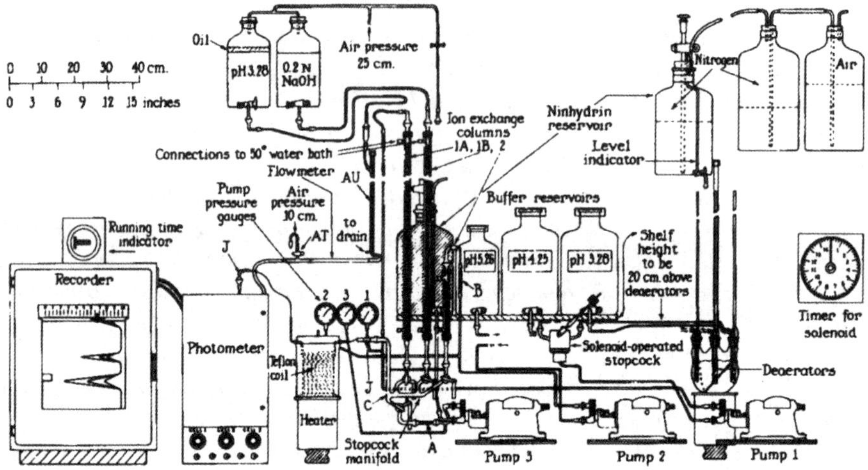

Reproduced from Moore and Stein's acceptance speech [2].

Further examples of enzyme reactivity can be found in [4–5].

Honours and Awards

Moore received Honorary Doctorates from the University of Brussels (1954) and from the University of Paris (1964), and an Honorary D.Sc. from the University of Wisconsin (1974).

He was presented with many awards and prizes including: the American Chemical Society Award in Chromatography and Electrophoresis, jointly with W. H. Stein (1964); the Richards Medal of the American Chemical Society (1972); the Linderstrom-Lang Medal, Copenhagen (1972); the Nobel Prize for Chemistry, jointly with W. H. Stein (1972).

Moore was a Fellow or Member of: the American Academy of Arts and Sciences, the American Society of Biological Chemists (Treasurer, 1956–59; Editorial Board, 1950–60; President, 1966); the American Chemical Society; the National Academy of Sciences. He was an Honorary Member of the Belgian Biochemical Society and a Foreign Correspondent of the American Academy of Arts and Sciences, the Belgian Royal Academy of Medicine, the Biochemical Society (Great Britain), the Harvey Society and the U.S. National Academy of Sciences (Chairman, Section of Biochemistry, 1970).

Moore was: Chairman of Panel on Proteins of the Committee on Growth of the National Research Council (1947-49); Secretary of the Commission on Proteins of the International Union of Pure and Applied Chemistry (1953–57); Chairman of the Organizing Committee for the Sixth International Congress of Biochemistry (1964); President of the Federation of American Societies for Experimental Biology (1970).

He was also a Visiting Professor at the University of Brussels (Francqui Chair, 1950–51) and of Health Sciences at the Vanderbilt University School of Medicine (1968).

Bibliography

1. Reproduced with permission.
2. From Nobel Lectures, Chemistry 1971–1980, Editor-in-Charge Tore Frängsmyr, Editor Sture Forsén, World Scientific Publishing Co., Singapore, 1993.
3. Who was who: 1981–1990.
4. Enzymes: A Practical Introduction to Structure, Mechanism, and Data Analysis by Robert A. Copeland, Wiley-VCH, 2nd ed. 2000, ISBN: 0471359297.
5. Computational Approaches to Biochemical Reactivity (Understanding Chemical Reactivity) edited by Gábor Náray-Szabó and Arieh Warshel, Springer, 1st ed. 1997, ISBN: 0792345126.

1972

© Nobel Foundation [1]

William H. Stein (0.25)

"for their contribution to the understanding of the connection between chemical structure and catalytic activity of the active centre of the ribonuclease molecule" [2]

During his presentation speech Bo Malmström, of the Royal Academy of Sciences, commented of Stein and his work [2]:

> *...The contributions of Moore and Stein concern another fundamental question regarding ribonuclease, namely the basis for its catalytic activity. The reacting substances, the substrates, are bound to an enzyme in what is generally called its active site. In the complex so formed there is an interaction between enzyme and substrate leading to a changed reactivity of the substrate. Knowledge about the structure of an enzyme is of little help in understanding this interaction if it is not possible to find the active site and to determine the chemical groups in it. Moore and Stein discovered as an important principle that the active site contains amino acids with an anomalously high reactivity compared to the same amino acids in free form. This high reactivity is of direct importance for the catalytic activity of the enzyme, but Moore and Stein also found it possible to utilize it to label two amino acids in the active site by chemical modification. In this way the position of these amino acids in the long peptide chain could be unambigously determined. Through these investigations Moore and Stein were able to provide a detailed picture of the active site of ribonuclease long before the three-dimensional structure of the enzyme had been determined...*

> *...Drs. Moore and Stein, I have attempted to summarize your fundamental contributions to our understanding of the relationship between chemical structure and catalytic activity in the enzyme ribonuclease. In particular, I have stressed your studies leading to the localization of two specific histidine residues in the active site of the enzyme. It is for these pioneering experiments that the Royal Academy of Sciences has decided to award this year's Nobel Prize in Chemistry to you together with Dr. Anfinsen...*

William H. Stein was born in New York City on the 25th June 1911, the second of three children, to Freed M. and Beatrice Borg Stein. He married Phoebe Hockstader (1936) and they had three sons William H. Jr. (1937), David F. (1939) and Robert J. (1944). Stein was diagnosed with Guillain-Barré syndrome (1969). Progressive paralysis followed, and despite a year in hospital, he remained a quadriplegic. Stein died suddenly of heart failure on 2nd February 1980 [2–4].

The Early Years

Stein's early education was at the Lincoln School of Teachers College of Columbia University in New York City, followed by the Phillips Exeter Academy in New England. Stein enrolled at Harvard for his undergraduate study where he majored in chemistry. Graduation (1933) was followed by a further year at Harvard after which he moved into biochemistry from organic chemistry. He went on to work under Clarke at the College of Physicians and Surgeons, Columbia University in New York for his doctorate, the subject of which was amino acid analysis of the protein elastin and graduated in 1937. Postdoctoral work followed without a break under Max Bergmann at the Rockefeller Institute, where he met Stanford Moore. Like everyone else at this time in the Bergmann group, Stein's work concerned protein research.

The Career

Year	*Post*	*Institute*
1939	Assistant	Rockefeller Institute for Medical Research
1943	Associate	Rockefeller Institute for Medical Research
1949	Associate Member	Rockefeller Institute for Medical Research
1952	Member	Rockefeller Institute for Medical Research
1955	Professor	Rockefeller Institute for Medical Research

Stein's contribution to the war effort is recorded by Moore thus [p 417, ref 3]:

> *...At this stage, the research on amino acid analysis was interrupted by the war years. The laboratory was engaged under contract to the Office of Scientific Research and Development to look for possible therapeutic agents for vesicant war gases through the study of the physiological mechanisms of mustard gas and nitrogen mustards. Stein was a coauthor of a series of fundamental papers concerned with the chemistry of the reactions of mustard gas and related compounds with the functional groups of amino acids and peptides...*

Stein's career interests were in:

- the determination of the amino acid composition of proteins
- column chromatography separation of amino acids (using potato starch)
- amino acid analysis
- the structure of proteins in general
- the structure of ribonuclease
- the development of an amino acid analyser

He was a Member of the Editorial Committee of the *Journal of Biological Chemistry* for six years, Chair for three (1958–61) and a Member of the Editorial Board of the *Journal of Biological Chemistry* (1962), Associate Editor (1964–1968), Editor (1968–1971).

Stein's Nobel Prize acceptance speech was combined with that of Moore. The combined presentation is dated 11th December 1972 with the title 'The chemical structures of pancreatic ribonuclease and deoxyribonuclease'. It is unusual for an acceptance speech to be combined in this way. For the purposes of this and Moore's entry, the author has arbitrarily chosen to record the structure and special features under this entry and the purification and analysis of ribonuclease under Moore's entry.

The structure of bovine ribonuclease A. Reproduced from Stein and Moore's Nobel Prize acceptance speech.

Further examples of enzyme reactivity can be found in [5,6].

Honours and Awards

Stein was presented with honorary degrees from Columbia University (1973) and from the Albert Einstein College of Medicine of Yeshiva University (1973).

He was presented with many awards and prizes including: the American Chemical Society Award in Chromatography and Electrophoresis, joint with S. Moore (1964); the Richards Medal of the American Chemical Society, joint with S. Moore (1972); the Kaj Linderstrøm-Lang Award, Copenhagen joint with S Moore (1972); the Nobel Prize for Chemistry, joint with S. Moore and C. B. Anfinsen (1972); the Award of Excellence Medal from the Columbia University Graduate Faculty and Alumni Association (1973).

Stein was a Member of many Academies including: the American Academy of Arts and Sciences (1960); the American Society of Biological Chemists; the American Chemical Society; the American Association for the Advancement of Science; the Biochemical Society of London; the Harvey Society of New York; the National Academy of Sciences (1960).

He also held positions as: Chairman of the Publications Committee for the sixth International Congress of Biochemistry (1964); Chairman of the US National Committee for Biochemistry (1968–69); Member of the Council of the Institute of Neurological Diseases and Blindness of the NIH (1961–66); Member of Medical Advisory Board, Hebrew University-Hadassah Medical School (1957–70); Trustee, Montefiore Hospital, New York (1947–74).

Stein was Philip Schaffer Lecturer at Washington University at St. Louis (1965), Harvey Lecturer (1956) and Phillips Lecturer at Haverford College (1962).

He was Visiting Professor at the University of Chicago (1961) and also Visiting Professor at Harvard University (1964).

Bibliography

1. Reproduced with permission.
2. From Nobel Lectures, Chemistry 1971–1980, Editor-in-Charge Tore Frängsmyr, Editor Sture Forsén, World Scientific Publishing Co., Singapore, 1993.
3. Biographical Memoir, written by Stanford Moore, The National Academy of Sciences 1987.
4. Who was who: 1971–1980.
5. Enzymes: A Practical Introduction to Structure, Mechanism, and Data Analysis by Robert A. Copeland, Wiley-VCH, 2nd ed. 2000, ISBN: 0471359297.
6. Computational Approaches to Biochemical Reactivity (Understanding Chemical Reactivity) edited by Gábor Náray-Szabó and Arieh Warshel, Springer, 1st ed. 1997, ISBN: 0792345126.

1973

Ernst Otto Fischer (0.5)

"for their pioneering work, performed independently, on the chemistry of the organometallic, so called sandwich compounds" [2]

During his presentation speech, Ingvar Lindqvist, of the Royal Academy of Sciences, commented of Fischer [2]:

> *...Now the science of chemistry involves, of course, more than flashes of visionary inspiration, and both Fischer and Wilkinson did not hesitate to confirm and develop the concept of the sandwich compound by an intensive experimental effort. This they did by the successful synthesis of a large number of compounds which were analogous to the initially discovered ferrocene (named by Woodward in analogy with benzene), but with other metals than iron, and with other flat molecules than the cyclopentadienyl group found in ferrocene. Amongst other things, Fischer managed to surprise chemists by preparing a sandwich of chromium between two benzene molecules. The culinary exploits were pursued further with the progress to open sandwiches, having a flat molecule on one side of the metal atom, and with only small molecules such as carbonyl, methyl or ethyl groups on the other side. Perhaps the most interesting development was when Wilkinson managed to prepare a sandwich compound with a direct chemical bond between the metal atom rhenium and hydrogen. This was a completely new type of chemical bond at the time.*
>
> *The discovery and confirmation of the new bonding and structural principles applying to sandwich compounds is the notable achievement of the two Nobel prize winners. It is a fact that every discovery within the realm of fundamental chemical research has potential applications. The very circumstance of the stability of sandwich compounds has led to involved theoretical discussions, which have also played their part in important advances in theoretical and applied research on catalysis. We cannot yet predict the practical importance that sandwich compounds will have in the future, nor whether they will finally be shown to occur in biological systems. One thing that is quite sure, however, is that all workers in the field of sandwich compounds in the future will be familiar with the names of Fischer and Wilkinson.*

Professor Fischer, Die Entdeckungen vollständig neuer Prinzipien der chemischen Bindung und Struktur sind immer grosse Augenblicke in der Geschichte der Chemie gewesen. Sie haben zu einer solchen Entdeckung in hervorragender Weise beigetragen. Ich überbringe Ihnen die wärmsten Glückwünsche der Königlichen Schwedischen Akademie der Wissenschaften...

Ernst Otto Fischer was born in Solln, near Munich, on 10th November 1918 as the third child of Karl T. Fischer (died 1953), and his wife, Valentine, *née* Danzer (died 1935) [2]. Fischer never married and currently lives in Munich.

The Early Years

After four years at elementary school, Fischer attended a grammar school (1929–1937) prior to a period of "work service" and two years' compulsory Military Service. The Second World War broke out towards the end of his two years and thus Fischer served in Poland, France and Russia. He started to study chemistry at the Technical College in Munich whilst on study leave (winter of 1941/2). His release by the Americans in the Autumn of 1945 allowed Fischer to resume his studies in Munich, once the Technical College was re-opened (1946). He graduated with a Dip. Chem. in 1949.

Fischer's doctorate studies, under Hieber at the Technische Hochschule München, were in inorganic chemistry. The title of his thesis was *"The Mechanisms of Carbon Monoxide Reactions of Nickel II Salts in the Presence of Dithionites and Sulfoxylates"*. Graduation in 1952 was followed by post-doctoral work at the College, also in collaboration with Hieber, the subject for which was transition metal and organo-metallic chemistry. Fischer's university teaching thesis was titled *"The Metal Complexes of Cyclopentadienes and Indenes"* which led to his first appointment.

The Career

Year	*Post*	*Institute*
1955	Lecturer	Technische Hochschule, München
1956	Study in the United States	
1957	Associate Professor	University of Munich
1959	Professor	University of Munich
1964	Professor and Director of Inorganic Chemistry	Technische Hochschule, München

Fischer's interests lay in the field of inorganic chemistry and included:

- carbyne complexes
- the preparation and characterization of dibenzenechromium
- metallic complexes of cyclopentadienes and indenes,
- metal-complexes of six-ringed aromatics, mono-, di- and oligo-olefins
- metalcarbonyl carbene and carbyne complexes
- the study of sandwich compounds
- the study of carbenes

Fischer's Nobel Prize acceptance speech is dated 11th December 1973 with the title 'On the road to carbene and carbyne complexes'. Fischer commented on the nature of the carbene bond and then on their possible reactions thus:

...The first X-ray structure determination, carried out by 0. S. Mills in collaboration with us, on pentacarbonyl [methoxy (phenyl) carbene] chromium (0) confirmed our notion about the bonds postulated earlier. Our notion was that the carbon atom of carbene is present in the sp3-hybridised form. It should therefore have a vacant p-orbital and should hence show an electron deficit.

This strong electron deficiency is compensated mainly through a pπ-pπ bond between one of the free electron pairs of the oxygen atom of the methoxygroup and the unoccupied p-state of the carbene carbon atom. A dπ-pπ reverse bonding takes place-to a smaller yet appreciable extent - from an occupied central metal orbital with a suitable symmetry to this vacant p-orbital of the carbene carbon. This can be deduced from the distances between the carbene carbon atom and the oxygen atom on the one hand, and between the carbene carbon atom and the central chromium atom on the other hand.

POSSIBLE REACTIONS OF CARBENE COMPLEXES

In what follows, I shall confine myself to carbene complexes of our type and I shall show, with some examples from recent times, the kinds of reactions that we could observe with these complexes.

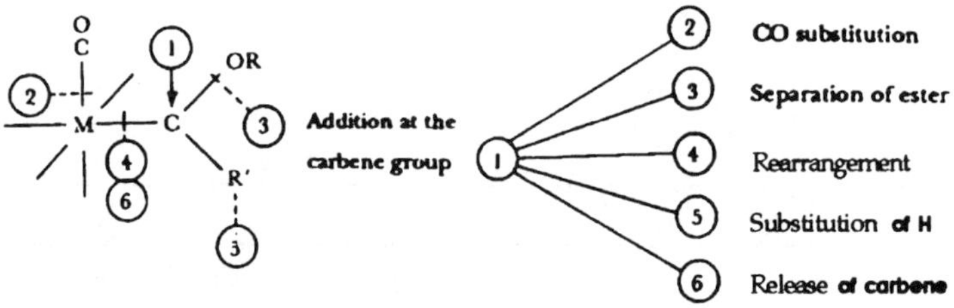

Reproduced from Fischer's Nobel acceptance speech.

We have already pointed out that the carbene carbon atom is an electrophilic centre and should therefore have a high nucleophilic reactivity. Hence, according to our present understanding, in most reactions a nucleophile would be added to the carbene carbon atom in the primary reaction stage. In some cases, for example in the case of some phosphines and tertiary amines, such addition products can be isolated in an analytically pure form under certain conditions. Another possible reaction is that the nucleophilic agent substitutes a carbon monoxide group in the complex while preserving the carbene ligand. The carbene complex can also be regarded in a very formal way from the point of view of an ester-like system

$$X=C\begin{matrix} / \, OCH_3 \\ \backslash \, R \end{matrix} \qquad \text{where } X = M(CO)_5 \text{ instead of } X = 0$$

since the oxygen atom as well as the metal atom in the $M(CO)_5$ radical have two electrons less than the number of electrons required for attaining the rare gas configuration. Therefore, it is not surprising that the OR radical can be substituted by amino, thio and seleno groups. This leads us to the synthesis of amino(organyl)carbene complexes, thio(organyl)carbene complexes and seleno (organyl)carbene complexes. The synthesis of the last two series of complexes requires some experimental skill.

We can also observe reactions which lead to a more stable arrangement of the entire system through a primary addition followed by a rearrangement. It can be noticed further that the hydrogen atoms of alkyl groups in the a position with respect to the carbene carbon atom are so acidic - because of the electron pull of the $M(CO)_5$ radical - that their acidity is similar to that of the hydrogen atoms in nitromethane. Finally the separation of the carbene ligand from the metal complex opens up new synthetic routes in organic chemistry...

Many sandwich compounds or metallocenes have been prepared and studied over the last twenty years or so, see for example those that appear in references [4,5,6].

Honours and Awards

Fischer was presented with Honorary Dr. rer. nat. by the Faculty of Chemistry and Pharmacy of the University of Munich (1972) and the Universities of Erlangen (1977) and Veszprem (1983). He was also awarded an Honorary D.Sc. from the University of Strathclyde (1975).

He was the holder of many prizes including: the Chemistry Prize by the Göttingen Academy of Sciences (1957); the Alfred Stock Memorial Prize (1959); the Nobel Prize for Chemistry, jointly with Geoffrey Wilkinson (1973).

Fischer was a Member of: the Mathematics/Natural Science section of the Bavarian Academy of Sciences (1964); of the German Academy of Scientists, Leopoldina (1969); the Austrian Academy of Science (1976); the Accademia dei Lincei (1976); Göttingen Akademie der Wissenschaften (1977); the Society of German Chemists. He was also a Centenary Foreign Fellow of the American Chemistry Society (1976).

He was Firestone Lecturer at the University of Wisconsin (1969) and the first Inorganic Chemistry Pacific West Coast Lecturer.

Fischer was: Visiting Professor at the University of Florida (1971); Visiting Professor at the University of Florida, Gainesville, USA, (1971); Arthur D. Little Visiting Professor at the Massachusetts Institute of Technology (1973); Visiting Distinguished Lecturer at the University of Rochester, Rochester, New York, USA. (1973).

Bibliography

1. Reproduced with permission.
2. From Nobel Lectures, Chemistry 1971–1980, Editor-in-Charge Tore Frängsmyr, Editor Sture Forsén, World Scientific Publishing Co., Singapore, 1993.
3. Who's who: 1996.
4. Ferrocenes ed Antonio Togni and Tamio Hayashi, VCH, 1995, ISBN: 3-527-29048-6.
5. Metallocenes vol. 1 and 2nd ed. Antonio Togni and Ronald L Halterman, Wiley-VCH, 1998, ISBN: 3-527-29539-9.
6. Metallocenes by Nicholas J. Long, Blackwell Science, 1998, ISBN: 0-632-04162-5.

1973

Geoffrey Wilkinson (0.5)

"for their pioneering work, performed independently, on the chemistry of the organometallic, so called sandwich compounds" [2]

During his presentation speech, Ingvar Lindqvist of the Royal Academy of Sciences, commented of Wilkinson and his work [2]:

> *...The facts were available for all the chemists of the world to see, but the right interpretation was lacking. Once the correct hypothesis was arrived at by fantasy or intuition, it readily lent itself to simple processes of logical deduction. I am of course referring to the way in which they together with the former Nobel laureate Woodward reached the conclusion that certain compounds could not be understood without the introduction of a new concept, namely that of the sandwich compound. This expression applies to the structure of these compounds, which can be thought of as a metal atom-the filling-sandwiched between two flat molecules. Now the science of chemistry involves, of course, more than flashes of visionary inspiration, and both Fischer and Wilkinson did not hesitate to confirm and develop the concept of the sandwich compound by an intensive experimental effort. This they did by the successful synthesis of a large number of compounds which were analogous to the initially discovered ferrocene (named by Woodward in analogy with benzene), but with other metals than iron, and with other flat molecules than the cyclopentadienyl group found in ferrocene. Amongst other things, Fischer managed to surprise chemists by preparing a sandwich of chromium between two benzene molecules. The culinary exploits were pursued further with the progress to open sandwiches, having a flat molecule on one side of the metal atom, and with only small molecules such as carbonyl, methyl or ethyl groups on the other side. Perhaps the most interesting development was when Wilkinson managed to prepare a sandwich compound with a direct chemical bond between the metal atom rhenium and hydrogen. This was a completely new type of chemical bond at the time...*

...Professor Wilkinson, The discovery of quite new types of chemical bonds and structures has always been considered as milestones in the history of chemistry. You have contributed to such a discovery in a decisive way. May I convey to you the warmest congratulations of the Royal Swedish Academy of Sciences...

Geoffrey Wilkinson was born in Springside, near Todmorden in the West Riding of Yorkshire, England, on 14th July 1921, the oldest of three children of Geoffrey Wilkinson (a painter) and Ruth née Crowther (a weaver). He married Lise Sølver (1951), the only daughter of Professor and Mrs. Svend Aage Schou, lately Rector of Denmark's Pharmaceutical High School. They had two daughters. He died on 26th September 1996 [2,3].

The Early Years

Wilkinson's interest in chemistry started at the age of 5 or 6 when he used to watch his father mixing his painting materials. He attended the local council primary school and Todmorden Secondary School, having won a County Scholarship in 1932. This small school also educated Sir John Cockroft (the 1951 Physics Nobel Laureate). He obtained a Royal Scholarship (1939) allowing him to enrol at Imperial College of Science and Technology from which he graduated in 1941 as the top student of all of the University of London colleges. His year also included Ralph Raphael (FPR 1962). D Barton (Chemistry Laureate 1969) was in the year above.

Wilkinson was directed to stay at Imperial College of Science and Technology to help with the war effort which he did under the supervision of H.V.A. Briscoe. Briscoe's mentor was Theodore Richards (Chemistry Laureate 1914). He later joined F.A. Paneth as part of the nuclear energy project and was sent to Montreal (later Chalk River) in Canada (Jan 1943), where he remained until 1946 (as a junior scientific officer). Wilkinson then moved to the University of California (as a Research Fellow) to work under Glenn T. Seaborg, where he worked on nuclear taxonomy, making many new neutron deficient isotopes using the cyclotrons, for the next four years. Wilkinson was the first non-American to be cleared by the US Atomic Energy Commission for work at the Lawrence Livermore Laboratory. He then returned to inorganic chemistry by accepting the post of Research Associate at the Massachusetts Institute of Technology (1950–51), where he worked on transition metal complexes such as carbonyls and olefin complexes.

The Career

Year	*Post*	*Institute*
1951	Assistant Professor	Harvard University
1956	Professor of Inorganic Chemistry, later renamed Sir Edward Frankland Chair of Inorganic Chemistry	Imperial College London
1988	Emeritus Professor and Senior Fellow	Imperial College London

Wilkinson was a Guggenheim Fellow in 1954 during which he studied with Jannik Bjerrum in Copenhagen.

He worked almost entirely on the complexes of transition metals, with other work concerning:

- coordination chemistry
- compounds of unsaturated hydrocarbons and with metal to hydrogen bonds
- excitation functions for protons on cobalt
- homogeneous catalytic reactions
- olefin complexes
- the chemistry of ruthenium, rhodium and rhenium in particular
- radiochemistry

His papers are deposited in the College Archives, Imperial College London.

Wilkinson's Nobel Prize acceptance speech is dated 11th December 1973 with the title 'The long search for stable transition metal alkyls'. During the course of his presentation, Wilkinson provides examples of 'sandwich' compounds that were made within his group at Imperial College:

...During the next few years a number of stable compounds in which alkyl or aryl groups were present were synthesised. The first was a phenyltitanium alkoxide $C_6H_5Ti(OC_3H_7)_3$ and others soon followed. However, in essentially all of the compounds a special type of ligand was present. These ligands were what are referred to as π-acid or π-bonding ligands. Examples are π-C_5H_5, CO. PEt_3, etc. Representative types of these alkyls are (VI–IX).

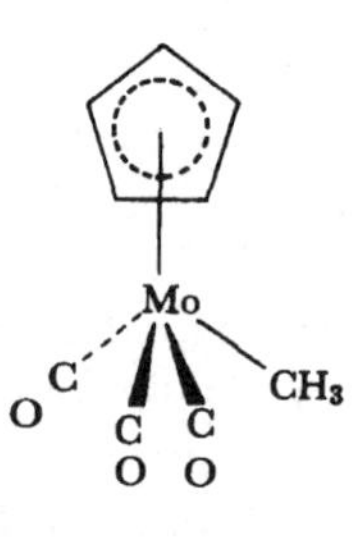

(VI)

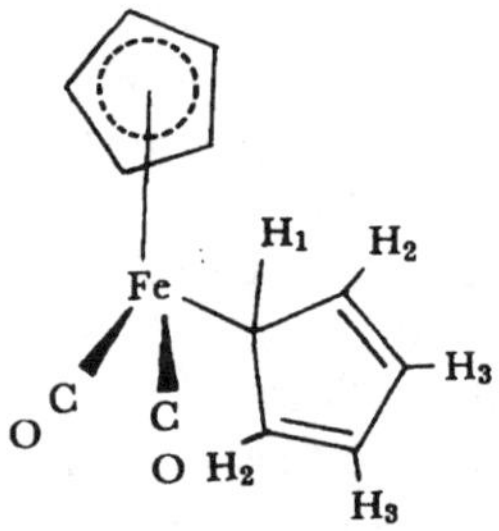

(VII)

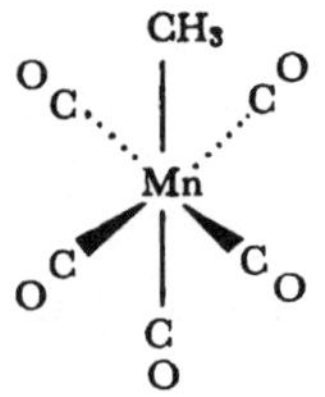

(VIII)

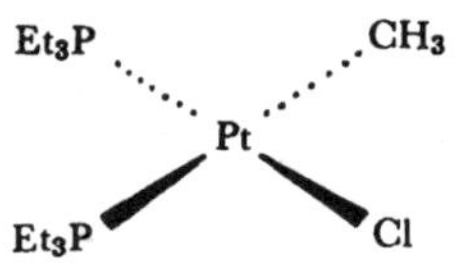

(IX)

Reproduced from Wilkinson's acceptance speech [2].

The compound (VII) turned out to he especially interesting. We observed only two signals in the proton magnetic resonance spectrum whereas the σ-C_5H_5 group alone should have had a complex spectrum; the infrared spectrum however was consistent with (VII). Because of the analogy with photography where, if one takes a picture of a moving wheel with a short exposure time (infrared), the picture is sharp, whereas if one uses a longer time (n.m.r.) it is blurred, I had to draw the conclusion that the σ-C_5H_5 was actually quite slowly rotating via a 1:2 shift. This was the first *recognition of what are now known as fluxional molecules, this particular type being called "ring whizzers".*

So the view developed that in order to have stable compounds with alkyl or aryl groups bound to a metal, some "stabilizing" group also had to be present. As recently as 1968, it was written "By any criterion, simple transition metal alkyls are unstable" and "In contrast to the simple alkyls, some metal complexes bearing other ligands in addition to alkyl or aryl groups are strikingly stable" [5].

However, it was not too well recognised that the presence of such "stabilizing" ligands, is no guarantee of stability because other factors are involved. Indeed, the activity of many metal complexes with π-honding ligands present in catalytic reactions depends on the lability of the metal of carbon bonds.

Despite all the intense study of metal-carbon bonds and the arguments about their stability, it is remarkable that only few bond energies are known. The available thermodynamic data show that M–C bonds are not exceptionally weak and the bond energies are quite comparable to those of non-transition metal to carbon bonds. It could be argued that since the compounds for which bond energies have been determined have "stabilizing" π-bonding ligands present the M–C bond energies are abnormally high, but this view cannot be sustained. There is no reason to assume this to be so, nor is there any reason to assume that carbon would differ appreciably in its capability to bond to transition metals compared to other first row elements such as oxygen and nitrogen or to the halogens. For both oxygen and nitrogen, metal compounds in high oxidation states are well known, e.g., the alkoxides such as $V(OR)_4$,and dialkylamides such as $W(NR_2)_6$ So, accepting that there is no thermodynamic reason for the instability of simple alkyls, the conclusion was that they are kinetically unstable...

Many sandwich compounds or metallocenes have been prepared and studied over the last twenty years or so, see for example those that appear in references [6–8].

Honours and Awards

Wilkinson was presented with Honorary D.Sc. degrees from Edinburgh University (1975); Granada University (1976); Columbia University (1978); Bath University (1980); Essex University (1989); Oxford University (1996).

He received many prizes and awards including: the American Chemistry Society Award in Inorganic chemistry (1965); the Lavoisier Medal (1968); the Chemical Society Award for Transition Metal Chemistry (1972); the Nobel Prize jointly with Fischer (1973); the Consejero de Honor (1974); the Hiroshima University Medal (1978); CASE Institute of Technology Centenary Medal (1980); the Royal Medal (1981); the Galileo Medal (1983); the Longstaff Medal (1987); the Polyhedron Prize (1989); the Medal of the University of Camerino (1989); the Messel Medal (1990); the Yovborg Jensen Award (1992); the Humphrey Davy Medal (1996).

He was elected: a Fellow of the Royal Society (1965); a Foreign Member of the Royal Danish Academy of Science and Arts (1968); a Member of the American Academy of Arts and Sciences (1970); a Foreign Associate of the National Academy of Sciences (1975); Centennial Foreign Fellow of the American Chemistry Society (1976). He was also an Honorary Fellow of Lady Margaret Hall, Oxford (1984) and an Honorary Citizen awarded by Todmorden Town Council (1990).

Wilkinson was knighted in 1976.

Wilkinson's citation for election (date: 18/03/1965) as Fellow of the Royal Society reads [9]:

Professor Geoffrey Wilkinson is one of the outstanding research workers in modern inorganic chemistry. He has published about one hundred research papers since 1946, a list of which is attached. The first twenty of these deal with nuclear chemistry to which subject he made noteworthy contributions during and after the 1939–45 war. The remainder are concerned chiefly with the chemistry of transition metal compounds in general and cyclopentadienyl compounds in particular. In regard to the volume of research work done, the originality displayed and the new types of substances which he has isolated, Wilkinson's contribution has not been surpassed in this field. Amongst the new types of compounds which he has prepared are substances containing unusual metal-hydrogen, metal-nitrosyl and metal-aromatic bonds. In all of his work there is a marked sense of purpose. His recognition of the unity of chemistry is reflected in the manner in which he has used the whole range of physical methods in the assignments of structure to the various metal-organic compounds which have now become closely associated with his name.

No proposers listed in the Sackler database.

Bibliography

1. Reproduced with permission.
2. From Nobel Lectures, Chemistry 1971–1980, Editor-in-Charge Tore Frängsmyr, Editor Sture Forsén, World Scientific Publishing Co., Singapore, 1993.
3. Biographical Memoirs of Fellows of the Royal Society, 2000, vol. 46, pp. 593–606.
4. Who's who: 1996.
5. G. W. Parshall and J. Kochi, *Organometal. Chem*, 1971, 29, 11.
6. Ferrocenes ed Antonio Togni and Tamio Hayashi, VCH, 1995, ISBN: 3-527-29048-6.
7. Metallocenes vol. 1 and 2nd ed. Antonio Togni and Ronald L Halterman, Wiley-VCH, 1998, ISBN: 3-527-29539-9.
8. Metallocenes by Nicholas J. Long, Blackwell Science, 1998, ISBN: 0-632-04162-5.
9. Sackler database, Royal Society.

1974

Paul J. Flory

"for his fundamental achievements, both theoretical and experimental, in the physical chemistry of the macromolecules" [2]

During his presentation speech, S. Claesson, then of the Royal Academy of Sciences, commented of Flory [2]:

...When chain molecules are dissolved in different solvents they become coiled to different degrees, depending on the interaction between repulsive and attractive forces in the solution. In a good solvent the chain molecules are extended. In a poor solvent, in contrast, the chain molecules assume a highly coiled form.

Professor Flory showed that if one takes a solution of extended chain molecules in a good solvent, and slowly cools the solution, then the molecules become progressively more coiled until they are no longer soluble.

Thus, there must be an intermediate temperature where the attractive and repulsive forces are balanced. At this temperature the molecules assume a kind of standard condition that can be used, generally, to characterize their properties.

This temperature Professor Flory named the theta temperature. A corresponding temperature exists for real gases at which they follow the ideal gas law. This temperature is called the Boyle temperature after Robert Boyle who discovered the gas laws. By analogy, the theta temperature for macromolecules is often referred to as the Flory temperature.

Professor Flory showed also that it was possible to define a constant for chain molecules, now called Flory's universal constant, which can be compared in significance to the gas constant...

...In recent years Professor Flory has investigated, both theoretically and experimentally, the relation between rotational characteristics of the chain links and the form of the chain molecules.

This is of fundamental significance for the understanding of both biological macromolecules and synthetic chain molecules...

Paul J. Flory was born on 19th June 1910, in Sterling, Illinois, the son of Ezra Flory, a clergyman-educator and Martha, née Brumbaugh a former schoolteacher. He married Emily Catherine, née Tabor, (1936) and they had three children Susan, Melinda and Paul John Flory, Jr.

Paul J. Flory died of a heart attack in Big Sur, California on 9th September 1985 [2,3].

The Early Years

Flory attended Manchester College, Indiana, which was at the time described by Flory as 'a liberal arts college'. One of his tutors, Carl W. Holl, Professor of Chemistry, inspired Flory to take up science. Graduating from Manchester in 1931, Flory enrolled in the Graduate School of the Ohio State University for postgraduate work under Herrick L. Johnston, concerning photochemistry and spectroscopy from which he graduated with a Ph.D. in 1934.

The Career

Year	*Post*	*Institute*
1934	Central Research Department	DuPont Company
1938	Basic Science Research Laboratory	University of Cincinnati
1940–1948	War work	
1948	George Fisher Baker Non-Resident Lecturer in Chemistry	Cornell University
1956–61	Executive Director	Mellon Institute, Pittsburgh
1961–75	J.G. Jackson - C.J. Wood Professor in Chemistry and Department Chairman (1969–71)	Stanford University
1976	Emeritus	

War work involved working on the development of synthetic rubber in:

- the Esso (now Exxon) Laboratories of the Standard Oil Development Company (1940–43)
- the Research Laboratory of the Goodyear Tyre and Rubber Company (1943–48)

Flory was invited to take up the George Fisher Baker Non-Resident Lecturership in Chemistry at the invitation of Peter J.W. Debye (Nobel Prize winner in Chemistry 1936) who was then Chairman of that Department.

Flory's research interests were all in the field of polymer chemistry and approximately the sequence in which the work was done include:

- basic polymer research (with Wallace Carothers who died in 1937 – the inventor of nylon and neoprene)
- theories of step-growth polymerization

- the science of nylon
- the theoretical basis of nylon science
- the behaviour of polymer solutions and blends
- the elastic behaviour of rubber
- liquid crystals
- the effect of excluded volume on the configuration of polymer chains
- the formulation of the hydrodynamic constant called theta
- the recognition of the Theta point at which excluded volume interactions are neutralized
- the spatial configuration of chain molecules
- the treatment of configuration-dependent properties by rigorous mathematical methods
- the thermodynamics of solutions

Flory published several books including:

- "*Statistical Mechanics Of Chain Molecules (Hardcover)*", Pub Butterworth Heinemann, 1969, ISBN: 0407026495
- "*Selected Works of Paul J. Flory*" (Hardcover) with Leo Mandelkern (Editor), Pub Stanford University Press, 1985, ISBN: 0804712778
- "*Principles of Polymer Chemistry (The George Fisher Baker non-resident lectureship in chemistry at Cornell University)*" Pub Cornell University Press (1953) ASIN: B0006ATLJA

Flory's Nobel Prize acceptance speech is dated 11th December 1974 with the title 'Spatial configuration of macromolecular chains'. Following some preliminary remarks concerning the history of macromolecules, Flory commented [2]:

> ...*Comprehension of the spatial relationships between the atoms of a molecule is a universal prerequisite for bridging the connection between the graphic formula and the properties of the substance so constituted. Structural chemistry has provided a wealth of information on bond lengths and bond angles. By means of this information the graphic formula, primarily a topological device, has been superseded by the structural formula and by the space model that affords a quantitative representation of the molecule in three dimensions. The stage was thus set for the consideration of rotations about chemical bonds, i.e., for conformational analysis of conventional organic compounds, especially cyclic ones. A proper account of bond rotations obviously is essential for a definitive analysis of the spatial geometry of a molecule whose structure permits such rotations*...

Flory then proceeded to provide a detailed account of some of his work, concluding:

> ...*The analysis of the spatial configurations of macromolecular chains presented above is addressed primarily to an isolated molecule as it exists, for example, in a dilute solution. On theoretical grounds, the results obtained should be equally applicable to the molecules as they occur in an amorphous polymer, even in total absence of a diluent. This assertion follows unambiguously from the statistical thermodynamics of mixing of polymer chains, including their mixtures with low molecular diluents. It has evoked much scepticism, however, and opinions to the contrary have been widespread. These opposing views stem primarily from qualitative arguments to the effect that difficulties inherent in the packing of long chains of consecutively connected segments to space-filling density can only be resolved either by alignment of the chains in bundle arrays, or by*

segregation of individual molecules in the form of compact globules. In either circumstance, the chain configuration would be altered drastically.

Whereas dense packing of polymer chains may appear to be a distressing task, a thorough examination of the problem leads to the firm conclusion that macromolecular chains whose structures offer sufficient flexibility are capable of meeting the challenge without departure or deviation from their intrinsic proclivities. In brief, the number of configurations the chains may assume is sufficiently great to guarantee numerous combinations of arrangements in which the condition of mutual exclusion of space is met throughout the system as a whole. Moreover, the task of packing chain molecules is not made easier by partial ordering of the chains or by segregating them. Any state of organization short of complete abandonment of disorder in favor of creation of a crystalline phase offers no advantage, in a statistical-thermodynamic sense.

Theoretical arguments aside, experimental evidence is compelling in showing the chains to occur in random configurations in amorphous polymers, and to further that these configurations correspond quantitatively with those of the unperturbed state discussed above. The evidence comes from a variety of sources: from investigations on rubber elasticity, chemical cyclization equilibria, thermodynamics of solutions, and, most recently, from neutron scatterring studies on protonated polymers in deuterated hosts (or vice versa). The investigations last mentioned go further, they confirm the prediction made twenty-five years ago that the excluded volume perturbation should be annulled in the bulk amorphous state. The excluded volume effect is therefore an aberration of the dilute solution, which, unfortunately, is the medium preferred for physicochemical characterization of macromolecules...

The physical chemistry of polymers is an ongoing topic, which is also applied to molecules of biological importance, see for example [5–7].

Honours and Awards

Flory received an Honorary DSc from Manchester University (1969), Honorary PhDs from the Weizmann Institute of Science (1976) and Clarkson College (1978) and Honorary ScDs from Manchester College (1950) and Ohio State University (1970).

He was presented with many awards and prizes including: the Nobel Prize for Chemistry (1974) and the Priestley Medal (1974); the US National Medal of Science (1974); the Perkin Medal (1977); the Carl-Dietrich-Harries Medal (1977); the Eringen Medal (1978).

Flory was a Fellow of the American Physics Society and a Member of the American Academy of Arts and Sciences, the American Chemistry Society, the American Philosophical Society and the National Academy of Sciences.

He was Chairman of the Committee on Macromolecular Chemistry (1955–59), Chairman of the Division of Chemistry and Chemical Technology (1966–68) and a Member of the National Research Council.

Bibliography

1. Reproduced with permission.
2. From Nobel Lectures, Chemistry 1971–1980, Editor-in-Charge Tore Frängsmyr, Editor Sture Forsén, World Scientific Publishing Co., Singapore, 1993.

3. Who was who: 1981–1990.
4. Physical Chemistry of Macromolecules: Basic Principles and Issues by S. F. Sun, Wiley-Interscience, 2nd ed. 2004, ISBN: 0471281387.
5. Physical Chemistry for the Biosciences by Raymond Chang, University Science, 2005, ISBN: 1891389335.
6. Chemistry and Physical Chemistry of Polymers by Yves Gnanou and Michel Fontanille, John Wiley & Sons Inc, Nov 2006, ISBN: 0471725439.

1975

John Warcup Cornforth (0.5)

"for his work on the stereochemistry of enzyme-catalyzed reactions" [2]

During his presentation speech, Arne Fredga, of the Royal Academy of Sciences, commented of Cornforth and his work [2]:

> *...Professor Cornforth has among other things studied the biological synthesis of the hydrocarbon squalene from six molecules of mevalonic acid. This hydrocarbon is necessary for the formation of steroids, which are of vital importance in many respects. The synthesis of squalene takes place in 14 steps and at each the enzyme must find the proper way. That means that there are just 214 = 16384 different routes and only one of them leads to squalene. If the enzyme should make a mistake in the first step (which it does not), the final result could be rubber or various other things but definitely not squalene. The problem at each step concerns which of two hydrogen atoms is to be eliminated, the right one or the left one. Professor Cornforth has shown which choice the enzyme makes at each of the 14 steps. For this purpose he has, with brilliant mastership, utilized the properties of the hydrogen isotopes: the ordinary hydrogen, the heavy hydrogen and the radioactive hydrogen. The lastmentioned isotope can only be used in tracer quantities, which means that only about one part per million of the participating molecules are radioactive. In a similar way, Professor Cornforth has studied several other biologically important reactions. All problems connected with the reaction mechanisms are not solved at that point, but the results constitute a very important step on the way...*

> *...Professor Cornforth. Enzymatic reactions have always had a certain air of magic, perhaps witchcraft. Of course this is due to our imperfect knowledge of what really happens. This air of magic is, however, gradually dispersing, and your contributions, utilizing the isotopes of hydrogen, imply most striking advances. The handling of compounds with chiral methyl groups is an achievement of the highest intellectual standard...*

...Let me also express our admiration for the skill and perseverance with which you have pursued your work in spite of a serious physical handicap. Perhaps it had not been possible without the never-failing help and support of Mrs. Cornforth. I think she should not be forgotten on this day...

John Warcup Cornforth was born on 7th September 1917 at Sydney in Australia, the elder son of four children of J. W. Cornforth and Hilda, née Eipper. Cornforth married Rita, née Harradence, also of Sydney and also an organic chemist, in 1941, and they have three children, one son and two daughters. Cornforth is an Emeritus Professor of University of Sussex and currently lives in Brighton [2,3,4].

The Early Years

Cornforth spent part of his childhood in Sydney and part in Armidale, New South Wales, Australia. He was about ten years old when the first signs of deafness (from otosclerosis) became noticeable. The transition to total deafness covered a period of ten years or so, during which Cornforth attended Sydney Boys' High School, where he was influenced by Leonard Basser in chemistry. He went on to attend Sydney University from the age of sixteen. He constructed a chemistry laboratory in his home (when he was fourteen), which gave him the practical acumen, as deafness prevented him for hearing the lectures at the university. Cornforth graduated with a University Medal and a B.Sc. (first-class) in 1937, followed by an M.Sc. in 1938. He won an 1851 Exhibition Scholarship (1939–42) which he used to work under Robert Robinson (1947 Nobel Chemistry Laureate) at Oxford University, from where he graduated with a D.Phil in 1941, the subject for which was steroid synthesis.

The Career

Year	*Post*	*Institute*
1946	Medical Staff	MRC Cambridge
1962	Co-Director of the Milstead Laboratory of Chemical Enzymology (with Popják)	Shell Research Ltd
1975	Royal Society Research Professor	University of Sussex
1982	Emeritus Professor	University of Sussex

Both Cornforth and his wife became part of the penicillin effort, which was a major part of the Robinson laboratory war work. Cornforth continued his collaboration with Robinson after joining the MRC.

Cornforth's main independent research work included:

- the first total synthesis of the non-aromatic steroids (simultaneously with Woodward)
- structure of cholesterol (with Popják)
- the study of the stereochemistry of enzymic reactions by means of asymmetry, artificially introduced by isotopic substitution

Cornforth helped to write *The Chemistry of Penicillin* (Princeton University Press, 1949).

His Nobel Prize acceptance speech is dated 12th December 1975 with the title 'Asymmetry and Enzyme Action'. Cornforth commented [2]:

> *...In 1960 we were checking on the formation of squalene from two molecules of farnesyl pyrophosphate. This reaction looks like a symmetrical coupling of two identical halves: in fact, we found that the process is attended by the exchange of one, and only one, hydrogen atom from one of the carbon atoms that become joined together in squalene. This non-symmetrical synthesis of a symmetrical molecule roused further my curiosity about the mechanism of the whole process, from mevalonic acid to squalene.*
>
> *Mevalonic acid has asymmetry of the ordinary kind, with Cabcd substitution at the central carbon atom, but this type of asymmetry is soon lost in the biochemical sequence. But mevalonic acid has three Caabc centres and all of these undergo changes in bonding on the way to squalene. Each of these six hydrogen atoms in these three groups is stereochemically distinct in the Ogston sense, and so it was possible in principle to follow the fate of a hydrogen atom from any one of these six positions, until it was either lost in the aqueous medium of reaction or came to a specific, and stereospecific, destination in a moleule of squalene. This fate is uniquely determined by, and therefore throws light on, the stereochemistry of the enzymic reactions in the biosynthetic sequence.*
>
> *Thus it became necessary to place a distinguishing mark on each of these six hydrogen atoms in turn, and this could not be done except by replacing normal hydrogen by one of its isotopes: the stable deuterium or the radioactive tritium. This problem was solved to a large extent by drawing on the vast store of organic chemical knowledge, especially that part which concerns the stereochemistry of reactions. In this way, the individual labelling of four out of the six hydrogen atoms was achieved by non-enzymic processes, and an organic reaction of known stereochemical preference was employed to define the sterochemistrv of an enzyme-mevaldate reductase which was used to generate a label on the fifth hydrogen: the sixth was also labelled, eventually, with the help of enzymes of known stereochemistry.*
>
> *These labelled mevalonic acids were introduced into enzyme preparations made from rat or pig liver. According to the nature of the preparation and to the presence or absence of co-factors or inhibitors, it was possible to execute the whole sequence from mevalonic acid to squalene or to stop at various intermediate stages; phosphomevalonic acid, isopentenyl pyrophosphate, or farnesyl pyrophosphate. Especially in experiments where the label was deuterium, unusually large amounts of product-typically, 50 milligrams - had to be accumulated from these enzymic incubations, and Popjak and I became familiar with the dialogue "How much can you make?" "How little do you need?"*
>
> *After the enzymic transformations, the products were examined to find out what had happened to the labelling isotope. When it was a question of whether the isotope had been lost from, or retained in, the product the procedure was relatively simple; one examined the product in a mass spectrometer for the presence or absence of deuterium, or one measured the radioactivity assignable to tritium. When the absolute stereochemistry at a labelled position was needed, it was necessary to use deuterium as the label and to degrade the product chemically, by reactions that either left the labelled centre undisturbed or altered it in a well-defined manner, to a substance suitable for examination by polarimetry or mass spectrometry. It was fortunate that this work coincided with the development of polarimeters sensitive enough to measure, in favourable cases, optical activity due solely to substitution of hydrogen by deuterium in specimens of a few milligrams. Our first measurements of this sort were, indeed, made on a prototype machine at the National Physical*

Laboratory. Since contamination by optically active material of the usual type could have been damaging, we had to develop a technique for re-crystallization in capillary tubes. This permitted the recrystallization of citric acid, for example, in milligram quantities from about two parts of water. Once the optical rotation was known it could be compared with the rotation of a sample into which a known absolute configuration had been built by enzymic and chemical synthesis.

The chemistry of enzyme-catalyzed reactions occupies the time of many chemists, some of the data from which is covered in references [5–7].

Honours and Awards

Cornforth was presented with Honorary degrees from many Universities and Institutes, including: ETH Zürich (1975); Dublin University (1976); Liverpool University (1976); Oxford University (1976); Warwick University (1976); Aberdeen University (1977); Hull University (1977); Sussex University (1977); Sydney University (1977).

He was presented with many awards and prizes including: the Corday Morgan Medal (1953); the Ciba Medal, joint with Popják (1965); the Flintoff Medal (1965); the Stouffer Prize (1967); the Davy Medal (1968); the Ernest Guenther Award (1968); the Prix Roussel (1972); the Nobel Prize for Chemistry, joint with Prelog (1975); the Copley Medal (1982).

Cornforth was elected: Fellow of the Royal Society (1953); Foreign Honorary Member of American Academy (1973); Corresponding Member of Australian Academy (1977); Foreign Associate of US National Academy of Science (1978); Foreign Member of Royal Netherlands Academy of Science (1978).

He received many civic honours including: CBE (1972); Kt (1977) and AC (1991).

Cornforth presented many named lectures including: Pedler Lecture (1968–69); Max Tishler, Harvard University (1970); Robert Robinson lecture (1971); the Sandin lectures, University of Alberta (1977).

He was also: Associate Professor in Molecular Science University of Warwick (1965–71); Visiting Professor of University of Sussex (1971–75); Honorary Professor of Beijing Medical University (1986–); Honorary Fellow of St Catherines' College (1976).

Bibliography

1. Reproduced with permission.
2. From Nobel Lectures, Chemistry 1971–1980, Editor-in-Charge Tore Frängsmyr, Editor Sture Forsén, World Scientific Publishing Co., Singapore, 1993.
3. Who's who entry: 2005.
4. Royal Society year book.
5. Organic Chemistry of Enzyme-Catalyzed Reactions, Revised Edition, Second Edition by Richard B. Silverman, Academic Press, 2nd ed. 2002, ISBN: 0126437319.
6. New Trends in Enzyme Catalysis and Biomimetic Chemical Reactions by Gertz I. Likhtenshtein, Springer, 1st ed. 2002, ISBN: 1402010060.
7. Enzyme Kinetics and Mechanism, Part F: Detection and Characterization of Enzyme Reaction Intermediates, Volume 354: Methods in Enzymology (Methods in Enzymology) by Daniel L. Purich, Academic Press, 2002, ISBN: 0121822575.

1975

© Nobel Foundation [1]

Vladimir Prelog (0.5)

"for his research into the stereochemistry of organic molecules and reactions" [2]

During his presentation speech, A. Fredga of the Royal Academy of Sciences, commented of Prelog [2]:

...Professor Prelog has worked in many fields of stereochemistry, and often the problems have been connected with the geometrical shapes of the molecules and their influence on the course of the reactions.

An impressive series of investigations deal with the "medium rings", i.e. molecules containing rings of 8 to 11 carbon atoms. Such rings are not rigid but rather limp. Parts of the ring which may seem rather distant may come into close contact with each other leading to unexpected reactions. Professor Prelog has been able to elucidate such reactions by utilizing the carbon isotopes. Many important investigations refer to reactions between chiral molecules. The term chiral is derived from a word in ancient Greek, meaning hand. The molecules are unsymmetrical and may exist in two forms differing in the same way as a right hand and a left hand. The molecules are so small that you can't see them, but one can gain much knowledge by studying the reactions between chiral molecules of different kind.

Professor Prelog has also made important contributions to enzyme chemistry. He has studied enzymatic reactions on small molecules and in particular oxidation or reduction processes. The experiments may be more or less successful depending on how the enzyme and the other molecule fit together geometrically. By systematic experiments with various small molecules of well-defined shapes, it was possible to construct a "map" of the active part of the enzyme molecule. The results have recently been confirmed in a special case by Swedish scientists using x-ray methods.

Professor Prelog has also with ingenuity and penetration discussed and analysed the fundamental concepts of stereochemistry, not least the conditions for chirality in large and complicated molecules...

Vladimir Prelog was born on 23rd July 1906 in Sarajevo, Bosnia (which at that time belonged to the Austrian-Hungarian Monarchy and later became part of Yugoslavia (1918)), the son of Milan Prelog, a high school teacher. He married Kamila, née Vitel, in Prague (1933). Their son, Jan, was born in Zürich in 1949. Prelog died on 6th (some say 7th) January 1998 [2–4].

The Early Years

As a young boy, Prelog stood in the procession of the Austrian Crown Prince Archduke Franz-Ferdinand and his wife Sophie (28th June 1914). The Archduke was assassinated shortly before the procession was due to reach Prelog. Throughout the remainder of his life, Prelog had an aversion to mass demonstrations. At the beginning of the First World War (1915), the family moved to Zagreb, the capital of Croatia, where Prelog attended the local gymnasium. He went on to study chemistry at the Institute of Technology in Prague, Czechoslovakia (1924 to 1929). His doctorate Supervisor was Emil Votocek, but he was greatly influenced by Rudolf Lukes. Prelog graduated with a Dr. Ing. in 1929.

The Career

Year	*Post*	*Institute*
1929	Chemist in the laboratory of G.J. Dríza	Prague
1935	Lecturer, then Associate Professor	University of Zagreb
1941	Privatdozent in Organic Chemistry Laboratory	Swiss Federal Institute of Technology (ETH) in Zurich
1947	Associate Professor	Swiss Federal Institute of Technology (ETH) in Zurich
1950	Professor	Swiss Federal Institute of Technology (ETH) in Zurich
1957–1976	Head, Laboratory of Organic Chemistry (succeeded Ruzicka)	Swiss Federal Institute of Technology (ETH) in Zurich

Prelog and his wife were able to escape to Switzerland in 1941, in part due to the invitations of Richard Kuhn (Nobel Chemistry Laureate 1938), and separately Leopold Ruzicka (Nobel Chemistry Laureate 1939). Ruzicka organized funding by CIBA Ltd, which allowed Prelog to start work in the Organic Chemistry Laboratory at the Swiss Federal Institute of Technology (ETH) in Zurich.

Prelog synthesized many natural products and worked on problems of stereochemistry, including:

- the synthesis of adamantine
- the synthesis of aialoids
- the synthesis of boromycin
- the synthesis of rifamycins
- stereochemical problems
- the development of a nomenclature for specifying stereoisomers

Prelog published some of his lectures in the series Profiles, Pathways, and Dreams in 1991 [5].

His Nobel Prize acceptance speech is dated 12th December 1975 with the title 'Chirality in Chemistry'. Prelog commented [2]:

Let us illustrate this by using as example an antibiotic isolated in our laboratories and named boromycin. This is a compound of medium complexity and has the molecular formula $C_{45}H_{74}O_{15}BN$. The van't Hoff - Le Bel model system allows an average student of chemistry to calculate that the constitutional formula of boromycin corresponds to 262,144 ($=2^{18}$) stereo-isomers. This is a rather large number, compared with the 2 million organic compounds which have hitherto been isolated or synthesized by thousands of hard-working chemists during almost two centuries, if a chemist were to set off to synthesize boromycin, he would not get very far from a knowledge of its constitutional formula alone. To approach his goal, he has to know what is the invariant part of it's spatial architecture. Moreover, he has to know processes, stereospecific reactions, which produce specifically the desired stereoisomers and not randomly all possible ones.

One problem in dealing with the multiplicity of stereoisomers is that of communication - how to transfer the information about their molecular architecture in space. This can be done, of course, by three-dimensional models (or their projections) constructed on the basis of coordinates obtainable by diffraction methods e.g. by X-ray crystal structure analysis. Such models that describe the complete molecular topography are invaluable for any detailed discussion of the molecule. However, they often include many structural details that are unnecessary for our purpose, i.e. specification of the particular stereoisomer. Indeed, some of these details may be dependent on the state (solid, liquid, vapour, solution) in which the molecule was observed. The very abundance of this information often makes it difficult to recognize, register and memorize that invariant aspect of the topography, the so-called primary structure, which is essential for specification and synthesis of the compound.

In 1954 I joined R S. Calm and Sir Christopher Ingold in their efforts to build up a system for specifying a particular stereoisomer by simple and unambiguous descriptors which could be easily assigned and deciphered. This system, which now carries our names, makes it possible to convey the essential information with the aid of a few conventions, letter symbols or numbers. In the rather complex model of boromycin which contains 136 atoms corresponding to 408 coordinates the primary structure is specified by 18 descriptors. They are:

2R,	3R,	4R,	7S,	9S,	15S,	16R,	22R,	2'R,	3'R,	4'R
0	0	0	1	1	I	0	0	0	0	0

7'S,	9'R,	13'R,	15'S,	16'R,	B R;	12,13 seq cis
1	0	0	1	0	0	(0)

The structure of boromycin is:

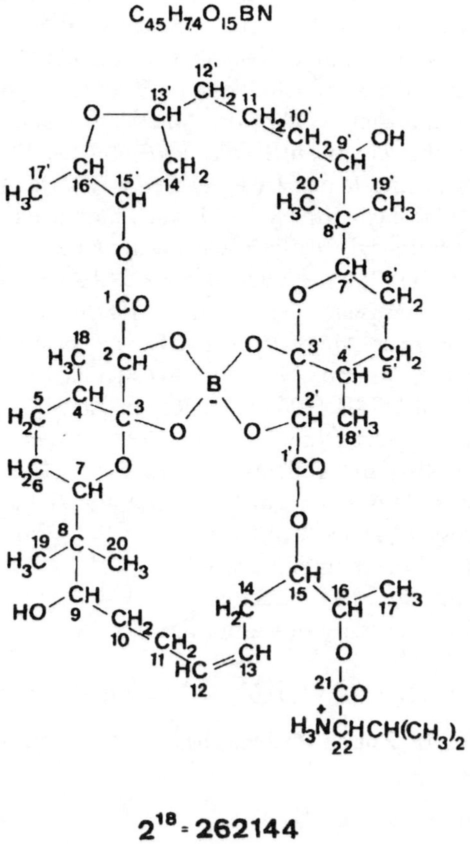

Diagram reproduced from Prelog's Nobel acceptance speech [2]

The issues of chirality and stereochemistry in organic molecules are still relevant, see for example [6–8].

Honours and Awards

Prelog received Honorary Doctorates from the: University of Zagreb (1954); University of Liverpool (1963); University of Paris (1963); University of Cambridge (1969); Université Libre de Bruxelles (1969); University of Manchester (1977); lnstituto Quimico del Sarriá Barcelona (1978); Weizmann Institute (1985); University of Ljubljana (1989); University of Osijek (1989); University of Prague (1992); University of Sarajevo, Bosnia (1996).

He was presented with many awards and prizes including the: Werner Medal and Award (1945); Stas Medal (1962); Medal of Honour (1962); Marcel Benoist Award (1965); Hanus Medal (1966); August-Wilhelm-von Hofmann Memorial Medal (1967); Davy Medal (1968); Roger Adams Award (1969); Paul Karrer Medal (1974); Nobel Prize for Chemistry, jointly with Cornforth (1975); Spiridon Brusina Medal (1975); Paracelsus Medal (1976); Emil Votocek Medal (1978); Hamilton Award (1983); Evans Award

(1983); Arun Guthikonda Memorial Award (1984); Adolfo Quilico Medal (1985); Bozo Tezak Medal of the Croatian Chemical Society (1989); Chirality Medal of the Chiral Discrimination Symposium (1992).

Prelog was a Fellow of the New York Academy of Sciences (1966), a Centennial Foreign Fellow of the American Chemical Society (1976) and a Member of: the Papal Academy of Science (1986). He was an Honorary Member of: the Société Chimique de Belgique (1962); the Czechoslovak Chemical Society and Croatian Chemical Society (1948); the American Academy of Arts and Sciences (1960); the Chemical Society (1960); the American Society of Biological Chemists (1965); the Academy of Pharmaceutical Sciences (1970); the Royal Irish Academy (1971); the Croatian Society for Nature Research (1975); the Serbian Chemical Society (1977); the Pharmaceutical Society of Japan (1977); the Institut Grand-Ducal de Luxembourg (1977); the Chemical Society of Japan (1978); La Asociacion de Quimicos del Instituto de Sarriá (1978); the Yugoslavenska Akademija Znanosti i Umjetnosti, Zagreb (1986); the Schweizerische Chemische Gesellschaft (1986); the Gesellschaft Oesterreichischer Chemiker (1986). He was a Foreign Member of: the Royal Society (1962); the lstituto Lombardo, Accademia di Scienze et Lettere, Milano (1964); the Accademia Nazionale dei Lincei, Rome (1965); the Academy of Sciences of the USSR (1966); the Royal Danish Academy of Sciences (1971); the American Philosophical Society (1976); the Académie des Sciences (1981); the Serbian Academy of Sciences and Arts (1981); the Russian Academy of Sciences (1994).

Prelog was presented with: the Order of Yugoslavian Star (1977); the Order of the Rising Sun, Japan (1977); the Order of the Yugoslav Banner with Golden Wreath (1986). He was an Honorary Citizen of the city of Zagreb, Croatia (1989) and an Honorary Citizen of the city of Sarajevo, Bosnia (1992).

Prelog was a Member of the Board of Directors of Ciba-Geigy Ltd (1960–78) and was a Visiting Professor of Chemistry at the University of Cambridge (1974).

Prelog's citation for election (date: 03/05/1962) as Foreign Member of the Royal Society reads [6]:

'Professor Vladimir Prelog of Zurich distinguished for his contribution to the development of modern stereochemistry'

No proposers are listed in the Sackler database.

Bibliography

1. Reproduced with permission.
2. From Nobel Lectures, Chemistry 1971–1980, Editor-in-Charge Tore Frängsmyr, Editor Sture Forsén, World Scientific Publishing Co., Singapore, 1993.
3. Biographical Memoirs of Fellows of the Royal Society, 2000, vol. 46, pp. 443–464.
4. Who's who: 1998.
5. My 132 Semesters of Chemistry Studies-presented in the series Profiles, Pathways and Dreams ed Jeffrey I Seeman, pub American Chemical Society, 1991, ISBN: 0-8412-1798.
6. Supramolecular Chirality (Topics in Current Chemistry) edited by Mercedes Crego-Calama and David N. Reinhoudt, Springer, 1st ed. June 2006, ISBN: 3540321519.
7. The Physical Chemistry of Chirality by Janice M. Hicks Publisher, An American Chemical Society Publication, 1st ed. 2001, ISBN: 0841237379.
8. When Topology Meets Chemistry: A Topological Look at Molecular Chirality (Outlooks) series editors Erica Flapan, Ronald Graham, John Barrow, Fan Chung, Ingrid Daubechies, Persi Diaconis and Don Zagier, Cambridge University Press, 2000, ISBN: 0521664829.
9. Sackler database, Royal Society.

1976

© Nobel Foundation [1]

William N. Lipscomb

"for his studies on the structure of boranes illuminating problems of chemical bonding" [2]

During his presentation speech, Gunnar Hägg of the Royal Academy of Sciences, commented of Lipscomb and his work [2]:

. . .The name borane is the collective term for compounds of hydrogen with boron, the latter element forming part of among other things boric acid and borax. A great number of boranes and related compounds are known, and it is this whole group of substances which has been studied by Lipscomb. In the eighties of the last century it was understood that such compounds exist in the gas mixture that is formed when alloys between boron and certain metals are decomposed by acids. But it was only from about 1912 that the German chemist Alfred Stock succeeded in producing some pure boranes.

The structures and bonding conditions of boranes remained, however, unknown until about 1950, and it is not without reason that they have been considered problematical. The experimental study of the boranes has been very difficult. They are in most cases unstable and chemically aggressive and must, therefore, as a rule be investigated at very low temperatures. But it was still more serious that their structure and bonding conditions were essentially different from what was known for other compounds. Stock had found borane molecules that, for example, consisted of in one case two boron and six hydrogen atoms and in another case ten boron and fourteen hydrogen atoms. But when the object was to determine how these atoms are bound to each other, i.e. the appearance of the molecule, and also the nature of the bonds which keep the atoms together within the molecule, one was left in the dark for many years. One might suppose that these bonds were similar to those between the atoms in the hydrogen compounds of carbon, for instance in the hydrocarbons in liquefied petroleum gas. In these, the bond between two neighbouring atoms usually involves two electrons, an electron pair. However, boron does not have as many bonding electrons as carbon and, therefore, all the bonds cannot be of this type. A new type of bond, where two electrons co-operate in binding three atoms together, and which thus can master this electron deficiency, was proposed in 1949 but it was not until the researches of Lipscomb from 1954 and onwards that the problems of borane chemistry could begin to be solved satisfactorily...

William N. Lipscomb was born in Cleveland, Ohio, USA, on 9th December 1919, the son of William Nunn Lipscomb Sr. and Edna Petterson, née Porter. He married Jean Craig Evans (1983) and they have one son and one daughter. Lipscomb has one daughter by a previous marriage. He can be contacted through Harvard University, Cambridge, MA. [2,3].

The Early Years

Lipscomb graduated from the University of Kentucky with his bachelors degree (1941) and from the California Institute of Technology with a Ph.D. (1946). Initially, Lipscomb entered Cal Tec to study physics, but was influenced by Linus Pauling (Nobel Laureate for Chemistry 1954) to switch to chemistry. Lipscomb worked as a Physical Chemist in the Office of Science Research and Development (1942–1946).

The Career

Year	*Post*	*Institute*
1946–50	Assistant Professor of Physical Chemistry	University of Minnesota
1950–54	Associate Professor	University of Minnesota
1952–54	Acting Chief of the Physical Chemistry Division	University of Minnesota
1959–71	Professor of Chemistry (Chair of Department 1962–65)	Harvard University
1971	Abbott and James Lawrence Professorship	Harvard University

Lipscomb's research covers many areas including:

- borane chemistry
- the relationship between three-dimensional structures of enzymes and how they catalyze reactions
- how enzymes are regulated by allosteric transformations
- the structure of the first leucine aminopeptidase
- allosteric enzymes
- aspartate transcarbamylase

- chorismate mutase
- fructose-1 6-bisphosphatase

Lipscomb has published his work in many scientific papers, made contributions of chapters to many books and written the following books:

- *Boron Hydrides*, W. A. Benjamin, Inc., 1963
- *Nuclear Magnetic Resonance Studies of Boron and Related Compounds*, with G. R. Eaton, New York, (1969)

His Nobel Prize acceptance speech is dated 11th December 1976 with the title 'The Boranes and their Relatives'. Lipscomb commented [2]:

> *...One of the rather simple rules which emerged was that nucleophilic attack occurs in a polyhedral carborane at a boron atom closest to carbon, while electrophilic attack was preferred at a boron furthest removed from carbon. Experimental studies of boranes and carboranes in which one or more hydrogens had been replaced by a halogen or by an amine group confirmed these predictions. The extended Hückel method became for a time the most widely used program for the study of molecular theory and reactions in complex organic and inorganic molecules. For example, the Woodward-Hoffmann rules, and related orbital concepts, were tested by their inventors with the use of this theory.*
>
> *I thought then that progress in structure determination, for new polyborane species and for substituted boranes and carboranes, would be greatly accelerated if the ^{11}B nuclear magnetic resonance spectra, rather than X-ray diffraction, could be used. One approach was empirical while the other was purely theoretical. This latter approach required the development of a theory for fairly reliable calculation of chemical shifts of ^{11}B from first principles of quantum mechanics. This theory, the coupled Hartree-Fock method for a molecule in a magnetic or electric field, yielded molecular constants good to a few to several percent for many diatomic molecules. A striking result was the prediction of the paramagnetic nature of diatomic BH. However, the application of this method to the complex boranes still lies in the future, even after 13 years of effort. While we understand some of the contributions to chemical shift, diamagnetic and large temperature-independent paramagnetic effects, the use of this method for structure determination of complex polyboranes is still somewhat limited.*
>
> *These programs yielded accurate self-consistent field molecular orbitals, which were explored in other areas of chemistry, as well as in the boranes and carboranes. One example is the first accurate calculation of the barrier to internal rotation in ethane. At this point, my research students and I set out to explore the gap between extended Hückel theory and self-consistent field theory. We achieved by stages a molecular orbital theory still being extended, which was applicable to large polyboranes and other molecules, and which had essentially the accuracy of self-consistent field methods. Molecular orbital methods which do not go beyond symmetry orbitals tend to make each molecule a separate case. Hence we began only a few years ago to explore the connections between these symmetry orbitals and the three-center bonds of the previous section...*

The boranes take on many shapes including:

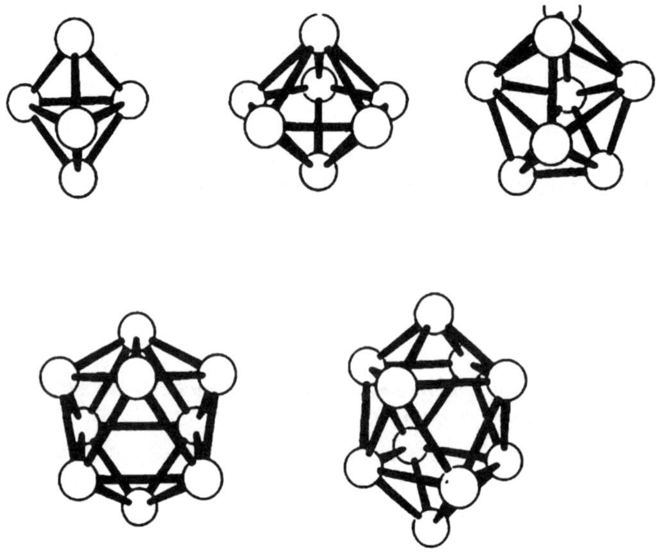

Reproduced from Lipscomb's Nobel Prize acceptance speech.

For further data concerning the structure and bonding of inorganic complexes, see for example [4,5].

Honours and Awards

Lipscomb has received Honorary degrees from many Universities including: Honorary M.A. Harvard University (1959); Honorary D.Sc. from Kentucky (1963); Long Island (1977); Rutgers (1979); Gustavos Adolphus (1980); Marietta (1981); Miami (1983). He was also conferred the Honorary degree of Doctor by the University of Munich (1976).

He is the holder of numerous awards including: the Bausch and Lomb Honorary Science Award (1937); the Sullivan Medallion, University of Kentucky (1941); the Distinguished Alumni Centennial Award (1965); the Award for Distinguished Service in the Advancement of Inorganic Chemistry; the Peter Debye Award in Physical Chemistry; the Harrison Howe Award; the Remsen Award; the George Ledlie Prize (1971).

Lipscomb is: a Member of the American Academy of Arts and Sciences (1959); a Member of the National Academy of Sciences USA (1961); a Foreign Member of the Royal Netherlands Academy of Sciences and Letters (1976); a Member of the International Association of Bioinorganic Scientists (1979); a Member of the International Academy of Quantum Molecular Science (1980); a Member of the Académie Européenne des Science, des Arts et des Lettres (1980) and a Member of the American Chemical Society.

He was also Chairman of the Minneapolis Section of the American Chemical Society (1949) and a Member of the Board of Directors, Dow Chemical Co. (1982–89). He is currently a Member of the Scientific Advisory Board, Robert Welch Foundation (1982–).

Bibliography

1. Reproduced with permission.
2. From Nobel Lectures, Chemistry 1971–1980, Editor-in-Charge Tore Frängsmyr, Editor Sture Forsén, World Scientific Publishing Co., Singapore, 1993.
3. Who's who entry: 2005.
4. Valency and Bonding: A Natural Bond Orbital Donor-Acceptor Perspective by Frank Weinhold and Clark R. Landis, Cambridge University Press, 2005, ISBN: 0521831288.
5. Principles and Applications of Density Functional Theory in Inorganic Chemistry II (Structure and Bonding) edited by N. Kaltsoyannis and J. E. McGrady, Springer, 1st ed. 2004, ISBN: 3540218610.

1977

Ilya Prigogine

"for his contributions to non-equilibrium thermodynamics, particularly the theory of dissipative structures" [2]

During his presentation speech, S. Claesson then of the Royal Academy of Sciences, commented of thermodynamics in general and Prigogine's work in particular [2]:

...Prigogine's great contribution lies in his successful development of a satisfactory theory of non-linear thermodynamics in states which are far removed from equilibrium. In doing so he has discovered phenomena and structures of completely new and completely unexpected types, with the result that this generalized, nonlinear and irreversible thermodynamics has already been given surprising applications in a wide variety of fields.

Prigogine has been particularly captivated by the problem of explaining how ordered structures - biological systems, for example - can develop from disorder. Even if Onsager's relations are utilized, the classical principles of equilibrium in thermodynamics still show that linear systems close to equilibrium always develop into states of disorder which are stable to perturbations and cannot explain the occurrence of ordered structures.

Prigogine and his assistants chose instead to study systems which follow non-linear kinetic laws and which, moreover, are in contact with their surroundings so that energy exchange can take place - open systems, in other words. If these systems are driven far from equilibrium, a completely different situation results. New systems can then be formed which display order in both time and space and which are stable to perturbations. Prigogine has called these systems dissipative systems, because they are formed and maintained by the dissipative processes which take place because of the exchange of energy between the system and its environment and because they disappear if that exchange ceases. They may be said to live in symbiosis with their environment.

The method which Prigogine has used to study the stability of the dissipative structures to perturbations is of very great general interest. It makes it possible to study the most varied problems, such as city traffic problems, the stability of insect communities, the development of ordered biological structures and the growth of cancer cells to mention but a few examples...

Ilya Prigogine was born in Moscow on 25th of January 1917, the younger son of Roman Prigogine (died 1974), a chemical engineer. Prigogine married Marina Prokopowicz on 25th February 1961. They had two children Yves Prigogine (5th July 1945) and Pascal Prigogine (6th February 1970). Prigogine died on 28th May 2003 [2,3].

The Early Years

According to his mother, Prigogine was able to read musical scores before he read printed words, however a difficult relationship with the Russian regime forced the family to leave Russia in 1921. The family lived in Germany for eight years (until 1929), prior to the move to Belgium. Prigogine attended secondary school in Brussels followed by a degree in chemistry at the Université Libre de Bruxelles. He graduated with the Licencié en Sciences Chimiques in 1939, the Licencié en Sciences Physiques in 1939, and Docteur en Sciences Chimiques in 1941. Prigogine became a Belgium citizen in 1949.

The Career

Year	*Post*	*Institute*
1947	Chargé de Cours	Université Libre de Bruxelles
1950	Professeur Extraordinaire	Université Libre de Bruxelles
1951–87	Professeur Ordinaire	Université Libre de Bruxelles
1967	Professor of Physics and Chemical Engineering, and Director	Ilya Prigogine Center for Studies in Statistical Mechanics and Complex Systems, University of Texas at Austin
1977	Regental Professor	University of Texas at Austin
1984	Ashbel Smith Professor	University of Texas at Austin

In addition to the above posts, Prigogine was Director, International Institutes of Physics and Chemistry founded by E. Solvay (from 1959) and Professeur Honoraire, Université Libre de Bruxelles (from 1987).

Prigogine's research interests included:

- various studies of thermodynamics
- theory of non-linear thermodynamics
- non-equilibrium thermodynamics
- theory of dissipative structures

He was co-author of *The Molecular Theory of Solutions*. (with V. Mathot and A. Bellemans)

Prigogine's Nobel Prize acceptance speech is dated 8th December 1977 with the title 'Time, Structure and Fluctuations'. Prigogine outlines his presentation thus [2]:

> *...It is the main thesis of this lecture that we are only at the beginning of a new development of theoretical chemistry and physics in which thermodynamic concepts will play an even more basic role. Because of the complexity of the subject we shall limit ourselves here mainly to conceptual problems. The conceptual problems have both macroscopic and microscopic aspects. For example, from the macroscopic point of view classical thermodynamics has largely clarified the concept of equilibrium structures such as crystals. Thermodynamic equilibrium may be characterized by the minimum of the Helmholtz free energy defined usually by*
>
> $$F = E - TS$$
>
> *Are most types of "organisations" around us of this nature? It is enough to ask such a question to see that the answer is negative. Obviously in a town, in a living system, we have a quite different type of functional order. To obtain a thermodynamic theory for this type of structure we have to show that that non-equilibrium may be a source of order. Irreversible processes may lead to a new type of dynamic states of matter which I have called "dissipative structures". Sections 2–4 are devoted to the thermodynamic theory of such structures.*
>
> *These structures are today of special interest in chemistry and biology. They manifest a coherent, supermolecular character which leads to new, quite spectacular manifestations; for example in biochemical cycles involving oscillatory enzymes.*
>
> *How do such coherent structures appear as the result of reactive collisions? this question is briefly discussed in Section 5. We emphasize that conventional chemical kinetics corresponds to a "mean field" theory very similar to the Van der Waal's theory of the equation of state or Weiss' theory of ferromagnetism. Exactly as in these cases the mean field theory breaks down near the instability where the new dissipative structures originate. Here (as in equilibrium theory) fluctuations play an essential role.*
>
> *In the last two sections we turn to the microscopic aspects. We briefly review the recent work done by our group in this direction. This work leads to a microscopic definition of irreversible processes. However this is only possible through a transformation theory which allows one to introduce new non-unitary equations of motion that explicitly display irreversibility and approach to thermodynamic equilibrium.*
>
> *The inclusion of thermodynamic elements leads to a reformulation of (classical or quantum) dynamics. This is a most surprising feature. Since the beginning of this century we were prepared to find new theoretical structures in the microworld of elementary particles or in the macroworld of cosmological dimensions. We see now that even for phenomena on our own level the incorporation of thermodynamic elements leads to new theoretical structures...*

Non-linear kinetics is studies today just as much in the field of solid state physics as it is in chemical and biological systems, see for [5–7].

Honours and Awards

Prigogine was presented with Honorary degrees from many Universities including: Newcastle (1966); Poitiers (1966); Chicago (1969); Bordeaux (1972); Liège (1977); Uppsala (1977); Aix-Marseille (1979); Georgetown (1980); Stevens Institute of Technology (1981); Krakow (1981); Rio de Janeiro (1981); Madrid (1985); Heriot-Watt University (1985); François Rabelais Université de Tours (1986); Nanjing (1986); Beijing (1986); Varanasi (1988); Buenos Aires (1989); Gustavus Adolphus College of the University of Minnesota (1990); Cagliari (1990); Siena (1990); Nice-Sophia-Antipolis (1991); Santiago (1991); Tucumán (1991); Philippines (1992); Moscow (1993); Romania (1993); Universidad Nacional de San Luis, Argentina (1994); Palermo, Buenos Aires (1994); Binghamton (1995); Valladolid (1995); Institut National Polytechnique de Lorraine, Nancy (1995); Brussels (1995); de "International Association of University Presidents" Séoul (1995); Kerala (1995); Urbino (1996); St. Petersburg (1996); Engineering Faculty, University of Xanthi (1996).

He received many awards and prizes including: Van Laar Prize of the Société Chimique de Belgique (1947); A. De Potter Prize of the Académie Royale de Belgique (1949); A.Wetrems Prize of the Académie Royale de Belgique (1950); Annual Prize of the Académie Royale de Belgique, Classe des Sciences (1952); Francqui Prize of the Francqui Foundation (1955); E.J. Solvay Prize of the Fonds National de la Recherche Scientifique (1965); Swante Arrenius Gold Medal (1969); Bourke Medal (1972); Cothenius Gold Medal (1975); Médaille de l'Association pour l'Avancement des Sciences (1975); Annual First Prize of the Southwest Science Forum (1976); Rumford Gold Medal (1976); Nobel Prize in Chemistry (1977); Umberto Biancamano Prize (1987); Karcher Medal (1978); Descartes medal (1979); Prix du Haut Comité de la Langue Française (1981); Honda Prize (1983); Award of the Gravity Research Foundation (1988); Médaille d'Or de l'Université de Padoue (Padua) (1988); Distinguished Service Medal, Austin (1989); International Foundation for Artificial Intelligence Award for Scientific Achievement (1990); Medal of the Imperial Order of the Rising Sun (gold and silver stars) (1991); Homer Smith Medal (1991); Summa Prize, Université Laval, Cite Universitaire (1993); Médaille d'Or de l'Université de Pavie (1994); Médaille 'Professeur Kampé de Fériet' (1994); Médaille de l'Ecole Normale Supérieure (1995); Medal of Piotr Kapitza décernée par l'"Academy of Natural Science," (1995).

Prigogine's civic honours included: Médaille de la Résistance; Commander of the Order of Léopold II (1961); Commander of the Order de la Couronne (1968); Médaille Civique de Première Classe (1972); Holder of the Grand Cross of the Order of Léopold II (1977); Commander of the National Order of Merit (1977); Citoyen d'Honneur de la Ville de Monpellier (1983); Honorary Citizen of the City of Dallas (1983); Commander of the Order of Arts and Letters (1984); Citoyen d'Honneur de la Commune d'Uccle (1984); Commandeur dans l'Ordre des Arts et des Lettres (1984); Médaille d'Or de la Ville de Pavie (1987); Médaille d'Or de la Ville d'Ostende (1987); Commander of French Legion of Honor (1988); Commander of the Legion of Honor (1989); Hereditary nobility and the personal title of Viscount, awarded by His Majesty, the King of Belgium (21st July 1989); Member of the Imperial Order of the Rising Sun, honored with the Gold and Silver Medals (1991); Médaille d'Or, Nice (1991); Médaille de la Ville de Milan (1993); Médaille de la Ville de Trieste (1993); Honorary Citizen of the City of Lubbock (1995); Honorary citizen of the town of Thessaloniki (1996); Honorary citizen and gold medal, Athens (1996); Medal of the town of Xanthi (1996); Medal of the Hellenic Parliament (1996).

He was: a Fellow, New York Academy of Sciences (1962); the World Academy of Art and Science (1986); American Chemical Society Centennial Fellow (1976); Foreign Fellow of the Indian National Science Academy (1979) and the Academia des Ciencias de Lisboa (1988). Prigogine was a Member,

Corresponding Member or Honorary Member of many societies in Belgium, Romania, Sweden, Germany, Austria, France, Italy, England, Argentina, United States of America, Poland, Tunisia, Russia and the former USSR.

Prigogine was an Honorary Professor of Benares Hindu University, Varanasi (1988)and Nanjing University, People's Republic of China (1986). He was Doyen d'honneur Honoris Causa de l'Institut Royal des Elites du Travail de Belgique Albert 1°, Brussels, Belgium (1995) and Visiting Professor, the University of Chicago, Department of Chemistry, Enrico Fermi Institute for Nuclear Studies and the Study of Metals (1961–1966).

Prigogine was a Member: Honorary Board of the International Society for Theoretical Chemical Physics, Nuremberg, Germany (1995); Advisory Board, Kothari Centre for Science, Ethics and Education of the University of Delhi (1995); Honorary Member, Korean Society of Chemistry (1996); Member, International Scientific Advisory Board (ISAB), UNESCO (1996); Founding Member, European Council of Research Libraries, Lausanne (1985); Member of the Editorial Board of the Ukrainian Physical Journal, URSS (1990); Member of the Consultative Council of the Scientific Section of the Fondazione dell'Istituto Bancario San Paolo di Torino per la cultura, la scienzia e l'arte, Turin, Italy (1988). Prigogine held many more appointments, the limitations of space prevent a further listing here.

Prigogine was also honoured by the creation of the Haute Ecole Libre Ilya Prigogine, Brussels (1996) and the 'Ilya Prigogine Chair of the Philosophy of Sciences' (School of Humanities and Social Sciences), Universidad de Palermo, Argentine (1996).

Bibliography

1. Reproduced with permission.
2. From Nobel Lectures, Chemistry 1971–1980, Editor-in-Charge Tore Frängsmyr, Editor Sture Forsén, World Scientific Publishing Co., Singapore, 1993.
3. Who's who entry: 2005.
4. Dissipative Structures and Chaos by Hazime Mori and Yoshiki Kuramoto, Springer 2001, ISBN: 3540627448.
5. Chemical Oscillations and Instabilities: Non-linear Chemical Kinetics (International Series of Monographs on Chemistry) by Peter Gray and Stephen K. Scott, Oxford University Press, USA, 2005, ISBN: 0198556462.
6. Introduction to Non-Linear Kinetics in Heterogeneous Catalysis by Yuri I. Pyantaanitsky, Mykola V. Pavlenko, Nina I. Ilchenko and Yuri I. Pyatnitsky, Nova Science Publishers, 2000, ISBN: 1560728191.

1978

© Nobel Foundation [1]

Peter D. Mitchell

"for his contribution to the understanding of biological energy transfer through the formulation of the chemiosmotic theory" [2]

During his presentation speech L. Ernster, then a member of the Royal Academy of Sciences, commented of Mitchell and his work [2]:

> *...The discoveries for which Peter Mitchell has been awarded this year's Nobel Prize for Chemistry relate to a field of biochemistry often referred to in recent years as bioenergetics, which is the study of those chemical processes responsible for supplying energy to living cells...*
>
> *...ATP serves as a universal energy currency for living cells. This compound is split by a variety of specific enzymes and the energy released is used for various energy-requiring processes. The regeneration of ATP by way of oxidative and photosynthetic phosphorylation thus plays a fundamental role in the energy supply of living cells...*
>
> *...At this stage, in 1961, Peter Mitchell put forward his chemiosmotic hypothesis. The basic idea of this hypothesis is that the enzymes of the electron-transport and ATP-synthesizing systems are localized in the membrane with a well-defined orientation and are functionally linked to a vectorial transfer of positively charged hydrogen ions, or protons, across the membrane. Thus, electron transport will give rise to an electrochemical proton gradient across the membrane which can serve as a driving force for ATP synthesis. A requisite for the establishment of a proton gradient is, of course, that the membrane itself is impermeable to protons, which explains the need for an intact membrane structure in oxidative and photosynthetic phosphorylation...*
>
> *...Dr. Mitchell, with ingenuity, courage and persistence you have innovated one of the classical fields of biochemistry. Your chemiosmotic theory has meant a breakthrough that has opened up new insights into the fundamental problems of bioenergetics. The details may need completion and adjustment; but the edifice you have raised will stand...*

Peter Dennis Mitchell was born in Mitcham, Surrey (England) on 29th September 1920, to Christopher Gibbs Mitchell O.B.E. and Kate Beatrice Dorothy, née Taplin. Mitchell's first wife was Eileen Rollo, by whom he had one son. His second marriage was to Helen French (1958) by whom he had two sons. Mitchell died on 10th April 1992 [2, 3].

The Early Years

Mitchell was educated at Queens College, Taunton prior to going up to Jesus College, Cambridge University. His main influence at Queens was the Headmaster, C.L. Wiseman, who was able to write a personal letter to the University authorities vouching for Mitchell following the results of Mitchell's 'dismally bad' scholarship examination. Mitchell studied biochemistry, chemistry, mathematics physics and physiology for part I, for which he obtained a class III result. He studied biochemistry for his part II, obtaining an upper second for his honours degree. His Ph.D. studies under J.F. Danielli in the Department of Biochemistry, Cambridge University commenced in 1942 but were protracted, because Mitchell undertook some research for the war effort which could not be used for his thesis. He received his doctorate for work on the mode of action of penicillin in 1951, a year after he started his post of University Demonstrator. Mitchell's first paper was published with Danielli in 1946 [5].

The Career

Year	*Post*	*Institute*
1950	Demonstrator	Department of Biochemistry, Cambridge University
1955	Head of Chemical Biology Unit (which he also set up)	Department of Zoology, Edinburgh University
1961	Senior Lectureship (additional post)	Department of Zoology, Edinburgh University
1962	Readership (additional post)	Department of Zoology, Edinburgh University
1963	Resigned due to acute gastric ulcers	
1963 to 1965	architect and master of works for the restoration of Glynn House	
1964–86	Founder and Director of Research	Glynn Research Laboratories
1987	Chairman and Honorary Director	Glynn Research Foundation

Mitchell founded a charitable trust with Jennifer Moyle, a former research colleague, known as Glynn Research Ltd. which financed the work of the Glynn Research Laboratories at Glynn House. The original endowment of about £250,000 was donated by Mitchell and his elder brother Christopher John Mitchell, who transferred their shares in Wimpey and Co. to Glynn Research Ltd.. Their uncle was Sir Godfrey Way Mitchell, was President of the firm of George Wimpey Ltd, and a notable public benefactor and industrialist.

The Memorandum and Articles of Association were signed on 1st April 1964 and incorporated on 21st April 1964. The aims were

> **'to promote and carry on scientific research, in particular in the field of fundamental biology, with a view to development and spread of scientific knowledge'**

The Glynn Research Laboratories undertook work concerning:

- chemiosmotic reactions
- reaction systems
- the mechanism of oxidative phosphorylation
- the communication between molecules

Mitchell's humanitarian interests concerned:

- the communication between individual people
- the use of money as an instrument of personal responsibility
- the monetary system

Mitchell wrote the two books:

Chemiosmotic Coupling in Oxidative and Photosynthetic Phosphorylation (1966)
Chemiosmotic Coupling and Energy Transduction (1968)

His Nobel Prize acceptance speech is dated 8th December 1978 with the title 'David Keilin's respiratory chain concept and its chemiosmotic consequences'. Mitchell's opening remarks were:

> *...It was obviously my hope that the chemiosmotic rationale of vectorial metabolism and biological energy transfer might one day come to be generally accepted, and I have done my best to argue in favour of that state of affairs for more than twenty years. But, it would have been much too presumptuous to have expected it to happen. Of course, I might have been wrong. and in any case, was it not the great Max Planck who remarked that a new scientific idea does not triumph by convincing its opponents, but rather because its opponents eventually die? The fact that what began as the chemiosmotic hypothesis has now been acclaimed as the chemiosmotic theory at the physiological level, even if not at the biochemical level - has therefore aroused me emotions of astonishment and delight in full and equal measure, which are all the more heartfelt because those who were formerly my most capable opponents are still in the prime of their scientific lives.*
>
> *I shall presently explain the difference between the physiological and the biochemical levels at which the chemiosmotic theory has helped to promote useful experimental research. But let me first say that my immediate and deepest impulse is to celebrate the fruition of the creative work and benevolent influence of the late David Keilin, one of the greatest of biochemists and - to me, at least - the kindest of men, whose marvelously simple studies of the cytochrome system in animals, plants and microorganisms, led to the original fundamental idea of aerobic energy metabolism: the concept of the respiratory chain. Perhaps the most fruitful (and surprising) outcome of the development of the notion of chemiosmotic reactions is the experimental stimulus and guidance it has provided in work designed to answer the following three elementary questions about respiratory*

chain systems and analogous photoredox chain systems:

- *What is it?*
- *What does it do?*
- *HOW does it do it?*

The mechanisms of bioenergetics and related processes are an ongoing issue, see for example [6–8].

Honours and Awards

Mitchell was awarded Honorary degrees by many Universities and Institutes, including: the Technische Universität Berlin (1976); Exeter University (1977); University of Chicago (1978); University of Liverpool (1979); University of Bristol (1980); University of Edinburgh (1980); University of Hull (1980); Jesus College, Cambridge (1980); University of East Anglia (1981); University of York (1982); University of Cambridge (1985); University of Manchester Institute of Science and Technology (1990); the University of Aberdeen (1990).

He was presented with many awards and prizes, including: CIBA Medal and Prize, British Biochemical Society (1973); the Warren Triennial Prize, jointly with Efraim Racker, USA Louis and Bert Freedman Foundation Award, New York Academy of Sciences (1974); the Wilhelm Feldberg (Anglo/German) Foundation Prize (1976); the Lewis S. Rosenstiel Award, USA (1977); the Nobel Prize (Chemistry) (1978); the Copley Medal (1981); the Gold Medal of Honour of Athens (1982).

Mitchell also elected Honorary (Foreign) Member (Associate, Fellow) of the Royal Society (1974); American Society of Biological Chemists (1975); the American Academy of Arts and Sciences (1975); the National Academy of Sciences, USA (1977); the Royal Society of Edinburgh (1979); the Japanese Biochemical Society (1984); the Society of General Microbiology, UK (1984); l'Académie de Sciences, France (1989); the Academy of Creators, USSR (1989); the Biochemical Society of USSR (1991).

Mitchell was Visiting Professor at KCL (1987–89).

In addition, Mitchell also gave a number of 'named' lectures including: the Fritz Lipmann Lecture of the Gesellschaft für Biologische Chemie (1978); Hans Krebs Lecturer (1978); Croonian Lecturer (1987); the Humphrey Davy Lecture at the Royal Institution of London (1980); the James Rennie Bequest Lecture of the University of Edinburgh (1980).

Mitchell's citation for election (date: 21/03/1974) as Fellow of the Royal Society reads [9]:

'Dr Mitchell's early work was mainly concerned with the relationships between surface structure, cell growth, osmotic properties and phosphate metabolism in bacterial and subcellular organelles. In 1958 he proposed the group-translocation theory of vectorial transport across membranes, and in 1961 the chemiosmotic hypothesis of oxidative phosphorylation. His novel and stimulating ideas have had a most important influence on research in the field of energy transduction.'

No proposers are listed in the Sackler database.

Bibliography

1. Reproduced with permission.
2. From Nobel Lectures, Chemistry 1971–1980, Editor-in-Charge Tore Frängsmyr, Editor Sture Forsén, World Scientific Publishing Co., Singapore, 1993.

3. Biographical Memoirs of Fellows of the Royal Society, 1994, vol. 40, pp. 281–305.
4. Who was who: 1991–1995.
5. Danielli, J.F., Danielli, M., Mitchell, P.D., Owen, L.N. & Shaw, G. (1946), Nature, Lond. 157, 217–218.
6. Bioenergetics 3 by David G. Nicholls and Stuart J. Ferguson, Academic Press, 3rd ed. 2002, ISBN: 0125181213.
7. Bioenergetics by Lars Garby and Poul S. Larsen, Cambridge University Press, 1995, ISBN: 0521451434.
8. Free Energy Transduction and Biochemical Cycle Kinetics by Terrell L. Hill, Dover Publications, 2004, ISBN: 0486441946.
9. Sackler database, Royal Society.

1979

Herbert C. Brown (0.5)

"for their development of the use of boron- and phosphorus-containing compounds, respectively, into important reagents in organic synthesis" [2]

During his presentation speech B. Lindberg, then of the Royal Academy of Sciences, commented of Brown [2]:

...One of the most important tasks for the present-day organic chemists, in basic as well as in applied research, is the synthesis of biologically active compounds. In order to achieve this, we need methods of joining carbon atoms to one another and to modify organic compounds in a variety of ways. A great many chemists devote their time to developing such methods. A few times in the history of chemistry have new synthetic methods been deemed so important that the originators have been awarded the Nobel Prize. This has once more happened, this year Brown and Wittig have been awarded the prize for chemistry for their development of boron and phosphorus compounds, respectively, into important reagents in organic synthesis.

Herbert C. Brown has systematically studied various boron compounds and their chemical reactions. He has shown how various specific reductions can be carried out using borohydrides. One of the simplest of these, sodium borohydride, has become one of the most used chemical reagents. The organoboranes, which he discovered, have become the most versatile reagents in organic synthesis. The exploitation of their chemistry has led to new methods for rearrangements, for addition to double bonds and for joining carbon atoms to one another...

...Professor Brown, Your discovery and exploration of borohydrides and organo boranes have given the chemists new and powerful tools for organic syntheses. May I convey to you the warmest congratulations of the Royal Swedish Academy of Sciences...

Herbert C. Brown was born in London, England on 22nd May 1912, the only son of four children (Ann 1909, Sophie 1916 and Riva 1918) of Charles Brovarnik and Pearl, née Gorinstein. He married Sarah, née Baylen, on 6th February 1937. Brown died on 20th December 2004 [2–4].

The Early Years

The family moved to Chicago in June 1914, the surname being anglicized to Brown sometime earlier. Brown attended the Haven School at Wabash and 16th Street, graduating at 12. He was offered the chance for advancement into an even higher class but did not want to be in the same class as his sister, Ann. Brown then attended Englewood High School on the South Side but left following the death of his father (1926), to work in the family store. He re-entered Englewood in February 1929 and graduated in 1930. During his time at Englewood, Brown ran the humour column of the school paper and won a national prize.

He entered Crane Junior College (1933) intending to study electrical engineering, but became interested in and then transferred to chemistry. The college closed shortly after Brown joined. He then went to night school at the Lewis Institute, and worked as a shoe clerk part-time, to finance his studies. He attended Wright Junior College (1934), graduated in 1935. He sat for a scholarship for the University of Chicago, which he entered in autumn of 1935. He received his B.Sc. degree in 1936 and his doctorate in 1938, and worked under M.S. Kharasch for his postdoctorate work, prior to taking up an Instructor's post with Schlesinger.

The Career

Year	*Post*	*Institute*
1939	Instructor	Wayne University
1943	Assistant Professor	Wayne University
1946	Associate Professor	Wayne University
1947	Professor of Inorganic Chemistry	Purdue University
1959	Wetherill Distinguished Professor	Purdue University
1960	Wetherill Research Professor	Purdue University
1978	Emeritus Professor	Purdue University

Brown's research interests included:

- The alkali metal hydride route to diborane
- The synthesis and use of sodium borohydride
- Hydroboration studies
- Organoboranes studies
- The boron route to asymmetric synthesis
- The study of steric effects
- Selective reductions
- Volatile compounds of uranium (early work)

Brown supervised 22 M.S. students, 141 Ph.D. students and many postdoctoral workers. He published many papers, the details for which can be found on the Purdue University web page devoted to H. C. Brown [3]. Brown also filed 63 patents, mainly with the US Patent Office.

Brown wrote many books including: *Organic Syntheses Via Boranes* Pub. John Wiley & Sons Inc. (1975) ISBN: 0471112801; *Boranes in organic chemistry*, (the George Fisher Baker non-resident lectureship in chemistry at Cornell University) Pub. Cornell University Press (1972) ISBN: 0801406811; *Organoboranes for Syntheses* (Acs Symposium Series) by P. V. Ramachandran (Editor), Herbert Charles Brown (Editor) Pub American Chemical Society (March, 2001) ISBN: 0841237085; *Hydrides of boron* ASIN: B0006AOUIC; *In Memory of H. L. Meerwein* (Topics in Current Chemistry 80) by Herbert Charles Brown, Hans Meerwein (Editor) Publisher: Springer-Verlag (August, 1979) ISBN: 0387093095; *Hydroboration* Pub W.A. Benjamin (1962) ASIN: B0006AXTTS; *The nonclassical ion problem* Pub Plenum Press (1976) ASIN: B0006Y3SXQ; Herbert C. Brown Pub. Chemical Heritage Foundation (1996) ASIN: B0006QNVPY; "*H.C. Brown--a celebration*": October 10–11, 1980; *to honor the award to Herbert C. Brown of the Nobel Prize in Chemistry for 1979* Pub. Dept. of Chemistry, Purdue University (1980) ASIN: B00071RI6G

Brown's Nobel Prize acceptance speech is dated 8th December 1979 with the title 'From little acorns to tall oaks – from boranes to organoboranes'. Brown's opening comment provides the background to his decision to enter boron research [2]:

> *...Why did I decide to undertake my doctorate research in the exotic field of boron hydrides? As it happened, my girl friend, Sarah Baylen, soon to become my wife, presented me with a graduation gift, Alfred Stock's book, The Hydrides of Boron and Silicon. I read this book and became interested in the subject. How did it happen that she selected this particular book? This was the time of the Depression. None of us had much money. It appears that she selected as her gift the most economical chemistry book ($2.06) available in the University of Chicago bookstore. Such are the developments that can shape a career!...*

The remainder of the lecture covered the following topics:

- The alkali metal hydride route to diborane and borohydrides
- Reductions with complex hydrides
- Hydroboration
- New hydroborating agents
- The versatile organoboranes
- Stitching and riveting
- Hydroboration of acetylenes
- Vinyl boranes
- Pheromones

The versatility of the boranes and hydroboranes is perhaps illustrated in the following manner.

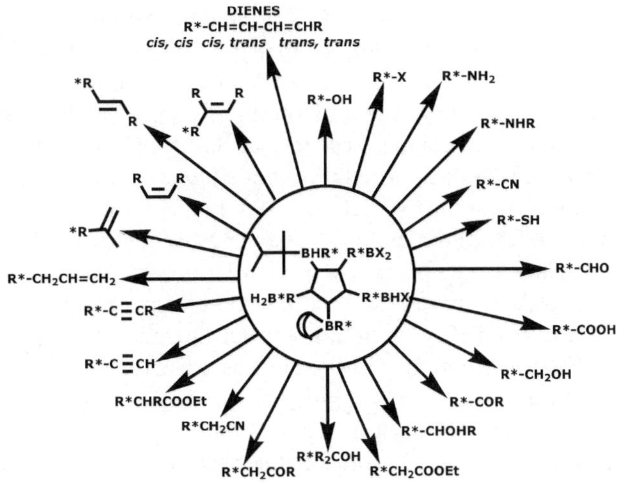

reproduced from the Purdue University website [3].

Brown published a text concerning hydroboration in 2000, which will provide the interested reader with an entry point into this subject [5].

Honours and Awards

Brown received many Honorary Doctorates including Honorary D.Sc. from the Universities of: Chicago (1968); Wayne State (1980); Lebanon Valley College (1980); Long Island (1980); The Hebrew University of Jerusalem (1980); Pontificia Univeristy (1980); Purdue (1981); Wales (1982); Butler (1982); Ball State (1985); Grado Academico Honorifico de Doctor Scientiae Honoris Caus, (1980). He was conferred with Honorary Doctor from the Universite de Paris-Sud (1982) and Doctorem Honoris Causa, Nicholas Copernicus University (1998).

Brown received many awards and prizes including: the Harrison Howe Award, Rochester Section of the American Chemical Society (1953); the William H. Nichols Medal, New York Section, American Chemical Society (1959); the Award for Creative Research in Organic Chemistry, American Chemical Society (1960); the S.O.C.M.A. Medal, Society of Organic Chemistry Manufacturing Association (1960); the Linus Pauling Medal, Oregon and Puget Sound Sections, American Chemical Society (1968); the National Medal of Science, Goverment of the United States (1969); the Roger Adams Medal, Organic Division, American Chemical Society (1971); the Charles Frederick Chandler Medal, Columbia University (1973); the Chemical Pioneer Award, American Institute of Chemists (1974); the Madison Marshall Award (1975); the City College Chemistry Alumni Award and Medal for Scientific Achievement, State University of New York, NYC (1976); the Elliot Cresson Medal, The Franklin Institute, Philadelphia (1978); the The Nobel Prize for Chemistry (1979); the Medal of Culture, Ministry of Education, Taiwan (1980); the Priestley Medal, American Chemical Society (1981); the Perkin Medal, American Section, Society of Chemical Industry (1982); the A.J. Beckman Memorial Medal, Colorado School of Mines (1983); the A.I.C. Gold Medal, American Institute of Chemists (1985); the Sixtieth Anniversary Commemorative Medal, Jewish Academy of Arts and Sciences (1986); the Sesfuicentennial Commemorative Medal, National Library of Medicine (1986); the National Academy of Sciences Award and Medal in Chemical Sciences, National Academy of

Sciences (1987); the G.M. Kosolapoff Award and Medal, Auburn Section, American Chemical Society (1987); the Medal of the Government of Japan: Order of the Rising Sun, Gold and Silver Star Emperor, Government of Japan (1989); the Oesper Award, Cincinnati Section, American Chemical Society (1990); the University Medal of Highest Honour, Kyung Hee University, Seoul, Korea (1995); the Distinguished Scholar Award, Indiana Academy of Science (1995); the Herbert C. Brown Medal and Award for Creative Research in Synthetic Methods (1998).

He was elected to National Academy of Science (1957) and to the American Academy of Arts and Sciences (1966), Fellowship of the Indiana National Science Academy (1977) and was made Honorary Fellow, Royal Society of Chemistry (1977). He was an Honorary Member of: the Chinese Chemical Society (1980); The Indiana Academy (1982); the Pharmaceutical Society of Japan (1982); Chemical Society of Japan (1982); the Gold Key National Honour Society (1983); of the Academy of Natural Sciences of the Russian Federal Republic (1991). He was a Corresponding Member of the Academia Mexicana de la Investigation Cientifica (1991), of the Academy of Arts and Science of Puerto Rico (1982) and Pontificia Universidad Catolica de Chile (1980).

Brown also received: the Sigma Xi Award, Purdue Section of Sigma Xi (1951); the Phi Lambda Upsilon Key for Honorary Membership, Phi Lambda Upsilon (1961); Honorary Member, Phi Lambda Upsilon (1962); Phi Beta Kappa (1972); Nobel Hall of Science, Museum of Science and Industry, Chicago, IL (1980); Honorary Old Master, Purdue University (1980); the H.C. Brown Laboratories of Chemistry were created (1987); Honorary Scholar, University of Wales, Swansea, Wales (1994); Honorary Professor, Laboratory of Organometallic Chemistry, Shanghai Institute of Organic Chemistry, Chinese Academy of Science (1994); Visitante Distinguido, BUSA V-MEX, Mexico (1996); Visitante Distinguido, la Universidad de Guanajuato, Mexico (1996); Order of the Griffin, Purdue University (1997). He was named one of the top 75 Distinguished Contributors to the Chemical Enterprise, Chemical & Engineering News (1998).

He also gave the Centenary Lectures and received the Medal of the Chemical Society, London, England (1955) and the C.K. Ingold Lectures and received the Medal, the Chemical Society, London (1978).

Bibliography

1. Reproduced with permission.
2. From Nobel Lectures, Chemistry 1971–1980, Editor-in-Charge Tore Frängsmyr, Editor Sture Forsén, World Scientific Publishing Co., Singapore, 1993.
3. From http://www.chem.purdue.edu/hcbrown/.
4. Who's who: 2004.
5. Hydroboration by H. C. Brown, Addison-Wesley Pub. Co., 2000, ISBN: 0805315004.

1979

Georg Wittig (0.5)

"for their development of the use of boron- and phosphorus-containing compounds, respectively, into important reagents in organic synthesis" [2]

During his presentation speech B. Lindberg, then a member of the Royal Academy of Sciences, commented on chemical reagents in general and Wittig's work in particular [2]:

...One of the most important tasks for the present-day organic chemists, in basic as well as in applied research, is the synthesis of biologically active compounds. In order to achieve this, we need methods of joining carbon atoms to one another and to modify organic compounds in a variety of ways. A great many chemists devote their time to developing such methods. A few times in the history of chemistry have new synthetic methods been deemed so important that the originators have been awarded the Nobel Prize. This has once more happened, this year Brown and Wittig have been awarded the prize for chemistry for their development of boron and phosphorus compounds, respectively, into important reagents in organic synthesis...

...Georg Wittig has provided many significant contributions in organic chemistry. The most important of these is the discovery of the synthetic method which bears his name, the Wittig reaction. In this, phosphorus ylides, a type of compound which he discovered, are allowed to react with carbonyl compounds. An exchange of groups takes place and the result is a compound in which two carbon atoms have been joined by a double bond. Since many natural products with biological activity contain such bonds, this elegant method has found widespread use, for example in the industrial synthesis of vitamin A...

Georg Wittig was born in Berlin on 16th June 1897, the son of Gustav Wittig and Martha, née Dombrowski. He married Waltraut Ernst (1930) and they had two daughters. Wittig died on 26th August 1987 [2, 3].

The Early Years

Wittig attended the Wilhelms Gymnasium in Kassel and the Faculty of Chemistry, Marburg University, from which he graduated with a doctorate and university teaching qualification.

The Career

Year	*Post*	*Institute*
1926	Lecturer	University of Marburg
1932	Head of Department	Technical College of Braunschweig
1937	Associate Professor	University of Freiburg
1944	Professor and Faculty Director	University of Tübingen
1956	Professor and Faculty Director	University of Heidelberg
1967	Professor Emeritus	

Wittig's research was varied, and included:

- The study of organic reactions of alkali metals
- The discovery of the carbonylolefins (1953)
- The discovery of the halogen-metal exchange reaction (simultaneously with H. Gilman)
- A study of the Stevens and Sommelet rearrangements
- The development of ylide chemistry including the discovery of phosphorous ylides
- The Wittig reaction

Wittig also published a textbook on stereochemistry (1930).

Wittig's Nobel Prize acceptance speech is dated 8th December 1979 with the title 'From diyles to ylides to my idyll'. Wittig began his talk by discussing his early work concerning carbon radicals [2]:

...Some 50 years ago I was fascinated by an idea which I investigated experimentally. The question was how ring strain acts on a ring if an accumulation of phenyl groups at two neighboring carbon atoms weakens the C–C linkage and predisposes to the formation of a diradical (for brevity called diyl) (Fig. 1). Among the many experimental results I choose the synthesis of the hydrocarbons 1 and 4, which we thought capable of diyl formation. Starting materials were appropriate dicarboxylic esters, which we transformed into the corresponding glycols. While these were obtained under the influence of phenylmagnesium halide only in modest yield, phenyllithium proved to be superior and was readily accessible by the method of K. Ziegler, using bromohenzene and lithium. The glycolates resulting from the reaction with potassium phenylisopropylide formed - on heating with methyl iodide - the corresponding dimethyl ethers, which supplied the equivalent hydrocarbons 1 and 4 by alkali metal splitting and demetalation with tetramethylethylene dibromide.

reproduced from Wittig's lecture [2]

The resulting tetraphenylbenzocyclobutane (1), however, rearranged to triphenyldihydroanthracene (3) (Fig. 2). In contrast, tetraphenyldihydrophenanthrene (4), which was prepared analogously,

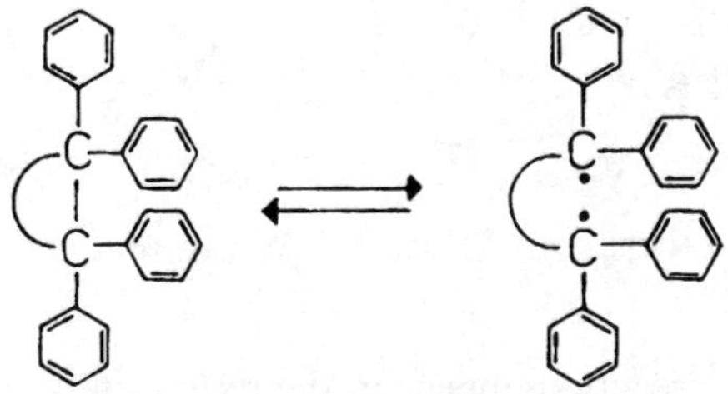

Figure 1 *Formation of diradical (2)*

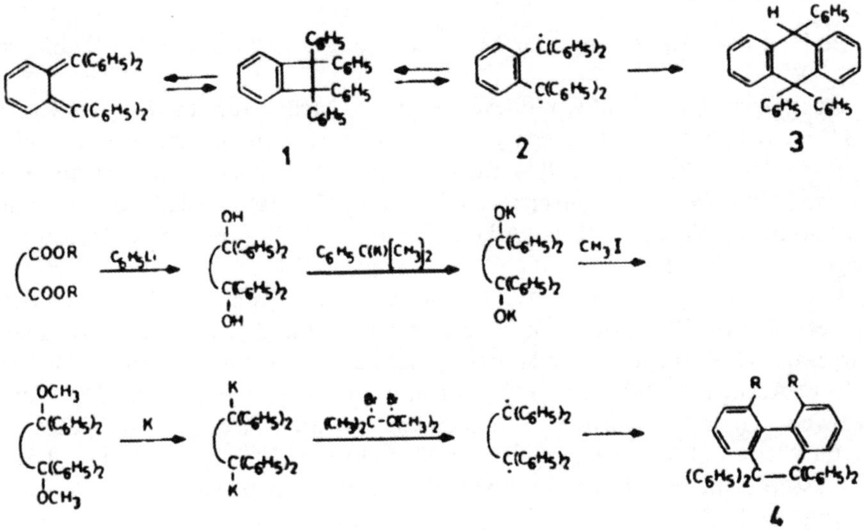

Figure 2 *Some reactions observed in the synthesis of 1 and 4 (3)*

proved to be a stable hydrocarbon, even when substituents R were introduced that forced the biphenyl system to twist. While 4 did not decompose at 340°C and was stable in solution against oxygen, its aryl-weakened C–C bond could be observed since it split with potassium into the ring-opened dipotassium derivate. The results of these investigations on formation of radicals and ring strain seem to indicate that ring closure is more likely to contribute to stabilization of the ethane bond...

reproduced from Wittig's lecture [2]

Wittig went on to discuss phosphorus ylids and then ultimately the basis for the Wittig reaction.

The Wittig Reaction

The reaction of an aldehyde or ketone with a triphenyl phosphonium ylide (often called a Wittig reagent) produces an alkene and triphenylphosphine oxide.

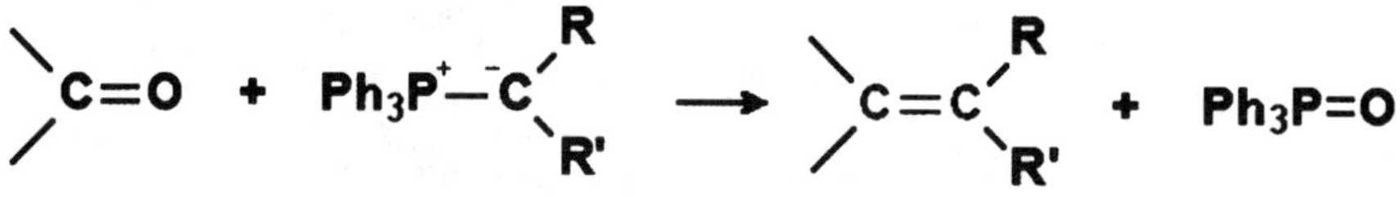

Many examples of complex syntheses which use the Wittig reaction are documented in the general literature.

Many examples of the Wittig reaction appear in the recent literature, see for example [4,5].

Honours and Awards

Wittig received Honorary degrees from the Universities of the Sorbonne (1957): Tübingen (1962) and Hamburg (1962).

He was presented with: the Adolf von Baeyer Memorial Medal from the German Chemical Society (1953); the Silver Medal from the University of Helsinki (1957); the Dannie Heinemann Award from the Göttingen Academy of Sciences (1965); the Otto Hahn Award for Chemistry and Physics (1967); the Paul Karrer Medal from the University of Zurich (1972); the Médaille de la Chaire Bruylants (University of Louvain) (1972); the Roger Adams Award from the American Chemical Society (1973); the Karl Ziegler Prize (1975).

Wittig was an Honorary Fellow of the Chemical Society (London) (1967); a Member of the Chemical Society of Peru (1965); the French Academy in 1971 and of the Society of Medical Sciences, Córdoba (Argentina) (1976). He was an Honorary Member of the Swiss Chemical Society (1963) and of the New York Academy of Sciences (1965). In addition to the above academies, Wittig was also a Member of several German academies including: the Bavarian Academy of Sciences; the German Academy of the Natural Scientist Leopoldina Halle and the Heidelberg Academy of Sciences.

Wittig was also presented with the Silver Medal from the City of Paris in 1969.

Bibliography

1. Reproduced with permission.
2. From Nobel Lectures, Chemistry 1971–1980, Editor-in-Charge Tore Frängsmyr, Editor Sture Forsén, World Scientific Publishing Co., Singapore, 1993.
3. Who's who entry: 1981–1990.
4. Name Reactions: A Collection of Detailed Reaction Mechanisms by Jie J. Li, Springer, 2nd ed. 2003, ISBN: 3540402039.
5. Strategic Applications of Named Reactions in Organic Synthesis by Laszlo Kurti, Barbara Czako, Academic Press 2005, ISBN: 0124297854.

1980

© Nobel Foundation [1]

Paul Berg (0.5)

"for his fundamental studies of the biochemistry of nucleic acids, with particular regard to recombinant-DNA" [2]

During his presentation speech, Bo G. Malmström, then a member of the Royal Academy of Sciences, commented of Berg and his work [2]:

> *...The machinery of life is made possible by a unique interplay between two groups of biological giant molecules, nucleic acids and the proteins, in the form of enzymes, the orchestra which plays the various expressions of life, motility, feeling, reproduction etc. DNA is the carrier of the genetic traits in the chromosomes of the cells, and it governs the chemical machinery by determining which enzymes a cell shall manufacture. We know through investigations which have earlier been awarded with Nobel Prizes, that the sequence of the building blocks in DNA, called nucleotides, determines the structure of a particular enzyme that the cell produces. The investigators who have been awarded this year's Nobel Prize in Chemistry, Paul Berg, Walter Gilbert and Frederick Sanger, have through their methodological contributions made it possible to penetrate into further depth in our understanding of the relationship between the chemical structure and biological function of the genetic material. Berg was the first one to construct a recombinant-DNA molecule, i.e. a molecule which contains part of DNA from different species, e.g. genes from a human being combined with part of a bacterial chromosome. Berg has also used his method to analyze the chromosome of a virus in considerable details...*

> *...The recombinant-DNA debate was started by the scientists themselves, after an initiative from Berg, warning for possible dangers with the new technique. Continued research has, however, shown that the concern for hypothetical risks has been unwarranted. The recombinant-DNA technique is instead, together with methods for the determination of nucleotide sequences, an extremely important tool to widen our understanding of the way in which the DNA molecule governs the chemical machinery of the cell...*

Paul Berg was born in New York on 30th June 1926. He married Mildred, née Levy, and they had one son. Berg is currently the Emeritus Cahill Professor in Cancer Research and Biochemistry and can be contacted through the Stanford University Medical Centre, Stanford, California 94305-5301. USA [2,3].

The Early Years

Berg's early years were spent in the gated community of Sea Gate, an area of Brooklyn. His ambition to be a scientist was formed by the time he was in junior high school. Berg was inspired in high school by Sophie Wolfe. He graduated from high school (1943) and enrolled at New York's City College to study chemical engineering. Berg enrolled in the Navy to be a flyer during the War, and studied biochemistry at Pennsylvania State University whilst waiting to be called to pre-flight training. He also served on a submarine chaser towards the end of the War, prior to returning to University in 1946, and graduated two years later with a B.Sc. degree. Berg then enrolled at Western Reserve University where he worked on C1 metabolism, specifically the conversion of formic acid, formaldehyde and methanol to the fully reduced state of methyl groups in methionine. Berg graduated from Western Reserve University with a Ph.D. in 1952, followed by two periods of postdoctoral work (1952–54). The first was with Herman M. Kalckar, Institute of Cytophysiology, Copenhagen, followed by a period with A. Kornberg, Washington University School of Medicine, St. Louis, M. He entered the American Cancer Society, Department of Microbiology, Washington University School of Medicine in 1954 as a Scholar in Cancer Research [4].

The Career

Year	*Post*	*Institute*
1955	Assistant to Associate Professor of Microbiology	Washington University School of Medicine
1959	Associate Professor of Biochemistry	Stanford University School of Medicine
1960	Professor, Department of Biochemistry	Stanford University School of Medicine
1969	Chairman, Department of Biochemistry	Stanford University School of Medicine
1970	Sam, Lulu and Jack Willson Professor of Biochemistry	Stanford University School of Medicine
1973–83	Non-Resident Fellow	Salk Institute
1985–2000	Director	Beckman Center for Molecular and Genetic Medicine
1994	Vivian K. and Robert W. Cahill Professor in Biochemistry and Cancer Research	Stanford University School of Medicine
2000	Emeritus	Stanford University School of Medicine

Berg has studied many aspects of biochemistry including:

- studies with mammalian cells
- the introduction of new genes into mammalian cells

- developing a general way to join two DNA molecules together in vitro
- the recombinant DNA technology
- the analysis of genetic recombination in eukaryotic cells; work on the process of gene conversion
- exploring the signal transduction pathway from the T cell receptor as part of the world wide work concerning HIV
- the mechanism of recombinational repair of double-strand breaks in DNA

Public policy issues (more recent interests)

- recombinant DNA
- embryonic stem cells

Berg has written several books including:

- *Glossary of terms frequently used in protein chemistry*, Pub: American Institute of Physics (1962) ASIN: B0007FC51A
- *Genes and Genomes: A Changing Perspective* (with Maxine Singer), Pub: University Science Books (1991), ISBN: 0935702172
- *Dealing With Genes: The Language of Heredity* (Hardcover) (with Maxine Singer) Pub: University Science Books (1992) ISBN: 0935702695
- *Exploring Genetic Mechanisms* (with Maxine Singer) Pub: University Science Books (1997) ISBN: 0935702709
- *George Beadle, an Uncommon Farmer: The Emergence of Genetics in the 20th Century* (New England Monographs in Geography) (with Maxine Singer) Pub: Cold Spring Harbor Laboratory Press (2003) ISBN: 0879696885.

Berg's Nobel Prize lecture is dated 8th December 1980 with the title 'Dissections and reconstitutions of Genes and Chromosomes'. During his lecture, Berg comments on his early experiements [2]:

> *...This background led me to consider, soon after beginning work with the tumor viruses, whether SV40 could be used to transduce new genes into mammalian cells. Initially, I had serious reservations about the success of such a venture because of the predictably low probability of generating specific recombinants between virus and cell DNA and the limited capability for selecting or screening animal cells that had acquired specific genetic properties. But it seemed that one possible way out of this difficulty, at least one worth trying, was to produce the desired SV4O transducing genomes synthetically. Consequently, in about 1970, I began to plan the construction,* ***in vitro****, of recombinant DNA molecules with SV4O and selected non-viral DNA segments. The goal was to propagate such recombinant genomes in suitable animal cells, either as autonomously replicating or integrated DNA molecules. At the time there were few if any animal genes available for recombination with SV40 DNA but I anticipated that a variety of suitable genes would eventually be isolated. Therefore the first task was to devise a general way to join together,* ***in vitro****, any two different DNA molecules...*

> *...The resulting hybrid DNA was approximately three times the size of SV40 DNA and, therefore, could not be propagated as an encapsidated virus But we intended to test whether the E. coli galactose genes would be expressed after introduction into the chromosomes of cultured animal cells. Moreover, since the λdv gal plasmid could replicate autonomously in E. coli, we also planned*

to determine if SV4O DNA would be propagated in E. coli cells and if any SV4O genes would be expressed in the bacterial host. Although the SV4O-λdv gal recombinant DNA could not have replicated in E. coli a gene needed for replication of the plasmid DNA in E. coli had been inactivated by the insertion of the SV4O DNA-a relatively simple modification of the procedure - the use of hdv gal dimeric DNA as acceptor for the SV4O DNA insert could have circumvented this difficulty. Nevertheless, because many colleagues expressed concern about the potential risks of disseminating E. coli containing SW40 oncogenes. the experiments with this recombinant DNA were discontinued.

...Since that time there has been an explosive growth in the application of recombinant DNA methods for a number of novel purposes and challenging problems...

See, for example, further texts concerning recombinant DNA from the many works that have been published since Berg was awarded the Nobel Prize [5–7].

Honours and Awards

Berg has been presented with the Honorary Doctor of Science degrees from the following Universities: Rochester (1978); Yale (1978); Washington (1986); St. Louis (1986); Oregon State University (1989); Pennsylvania State (1995).

He is the recipient of the following awards and medals: the Eli Lilly Award in Biochemistry (1959); the Henry J. Kaiser Award for Excellence in Teaching at Stanford University School of Medicine (1969, 1972); the Distinguished Alumnus Award, Pennsylvania State University (1972); the V.D. Mattia Prize of the Roche Institute for Molecular Biology (1972); the Sarasota Medical Awards for Achievement and Excellence (1979); the Annual Award of the Gairdner Foundation (1980); the Nobel Award in Chemistry, joint with Gilbert and Sanger (1980); the Albert Lasker Basic Medical Research Award (1980); the New York Academy of Sciences Award (1980); the American Association for the Advancement of Science Scientific Freedom and Responsibility Award (1982); the National Medal of Science (1983); the National Library of Medicine Medal (1986); the Special Achievement Award, Odyssey Biomedical Corporation (1989).

Berg is a Fellow of American Association for the Advancement of Science (1991) and of the American Academy of Microbiology (1992). He is a Member of: the National Academy of Sciences (1966); the American Academy of Arts and Sciences (1966); the Institute of Medicine, National Academy of Science (1974); the American Philosophical Society (1983); the American Academy of Achievement (1988) and of the Pontifical Academy of Sciences (1996). He is: a Foreign Member of the French Academy of Sciences (1981) and of the Royal Society, London (1992); Honorary Member of the Academy of Natural Sciences of the Russian Federal Republic (1991); the Alpha Omega Alpha Honour Medical Society (1992) and of the AMBO/AMBL (1994). Berg is also an Associate Member of EMBO (1984). He was California Scientist of the Year (1963), President, American Society of Biological Chemists (1975) and Sustained Leadership at National Level, Research! America (2003).

Berg has served on many public committees including: Member, Board of Scientific Advisors of Jane Coffin Childs Foundation for Medical Research; Member, Advisory Boards to National Institutes of Health, American Cancer Society, National Science Foundation, Massachusetts Institute of Technology, and Harvard University; elected to the Council of National Academy of Science; Member and Chairman, International Advisory Board of the Basel Institute of Immunology; Chairman, Whitehead Institute Board of Advisory Scientists; Chairman, National Advisory Committee, Human Genome Project; Trustee, Rockefeller University; Chairman, Board of Directors,

National Foundation for Biomedical Research; Chairman, Public Policy Committee, American Society for Cell Biology; Advisory Editorial Board, Molecular Medicine Today; Advisory Panel, Human Genome Education Program; Chairman, Scientific Advisory Board, Beckman Foundation; the NAS-CSIS Roundtable on Biotechnology and Bioterrorism.

Bibliography

1. Reproduced with permission.
2. From Nobel Lectures, Chemistry 1971–1980, Editor-in-Charge Tore Frängsmyr, Editor Sture Forsén, World Scientific Publishing Co., Singapore, 1993.
3. Who's who entry: 1981–1990.
4. http://med.stanford.edu/profiles/Paul_Berg/.
5. Manipulation and Expression of Recombinant DNA, Second Edition by Sue Carson and Dominique Robertson, Academic Press 2nd ed. 2005, ISBN: 0120884186.
6. Molecular Biotechnology: Principles and Applications of Recombinant DNA by Bernard R. Glick and Jack J. Pasternak, ASM Press, 3rd ed. 2003, ISBN: 1555812694.
7. Recombinant DNA Principles Methodologies by James Greene, CRC, Illustrated edition 1998, ISBN: 0824799895.

1980

© Nobel Foundation [1]

Walter Gilbert (0.25)

"for their contributions concerning the determination of base sequences in nucleic acids" [2]

During his presentation speech Bo G. Malmström, then of the Royal Academy of Sciences, commented of Gilbert [2]:

> *...The machinery of life is made possible by a unique interplay between two groups of biological giant molecules, nucleic acids and the proteins, in the form of enzymes, the orchestra which plays the various expressions of life, motility, feeling, reproduction etc. DNA is the carrier of the genetic traits in the chromosomes of the cells, and it governs the chemical machinery by determining which enzymes a cell shall manufacture. We know through investigations which have earlier been awarded with Nobel Prizes, that the sequence of the building blocks in DNA, called nucleotides, determines the structure of a particular enzyme that the cell produces. The investigators who have been awarded this year's Nobel Prize in Chemistry, Paul Berg, Walter Gilbert and Frederick Sanger, have through their methodological contributions made it possible to penetrate into further depth in our understanding of the relationship between the chemical structure and biological function of the genetic material. Berg was the first one to construct a recombinant-DNA molecule, i.e. a molecule which contains part of DNA from different species, e.g. genes from a human being combined with part of a bacterial chromosome. Berg has also used his method to analyze the chromosome of a virus in considerable details.*

Gilbert and Sanger have independently developed separate methods for the determination of the exact sequence of the building blocks in DNA. Gilbert has in addition studied those parts of DNA in a bacterial chromosome which control the transcription of the genetic message in the cell. Sanger has, for example, determined the complete nucleotide sequence for a small virus, whose DNA still consists of not less than 5,375 building blocks...

Walter Gilbert was born on 21st March 1932 in Boston, Massachusetts, the only son of Richard V. Gilbert and Emma, née Cohen. Gilber has one younger sister. He married Celia Stone in 1953 and they have two children, John Richard and Kate. Gilbert can be contacted through the Dept. of Molecular and Cellular Biology, Harvard University [2,3].

The Early Years

Gilbert was educated at home for the first few years by his mother, who was a child psychologist. The family moved to Washington D.C. (1939) and thereafter Gilbert was educated in public schools, and later at the Sidwell Friends high school. He was a member of a minerological society and an astronomical society as a child, and became interested in inorganic chemistry whilst still at high school. He graduated with his bachelors degree from Harvard University, where he majored in chemistry and physics, and his doctorate from University of Cambridge. His Ph.D. thesis, under Abdus Salam, concerned '*dispersion relations for elementary particle scattering: an effort to use a notion of causality, formulated as a mathematical property of analyticity of the scattering amplitude*'. He graduated with a Ph.D in mathematics in 1957 and returned to Harvard University spending a postdoctoral year as J. Schwinger's Assistant (1957–58), prior to becoming an Assistant Professor of Physics.

The Career

Year	*Post*	*Institute*
1959	Assistant Professor of Physics	Harvard University
1964	Associate Professor of Biophysics	Harvard University
1968	Professor of Biochemistry	Harvard University
1972–81	American Cancer Professor of Molecular Biology	Harvard University
1978	Joint Founder and Director	Biogen NV
1981–84	Chairman and Principle Executive Officer	Biogen NV
1985–	Carl M. Loeb University Professor, Dept. of Molecular and Cellular Biology	Harvard University
1992–	Vice Chairman	Myriad Genetics Inc

Gilbert was a Guggenheim Fellow in 1968–69, which he spent in Paris.

His early career concerned his interests in physics, particular problems of theory, which later developed into experimental physics.

His interests in biochemistry started in the summer of 1960 and have involved (in approximate date order):

- work on messenger RNA
- a study how proteins are synthesized
- the mechanism of protein synthesis
- the isolation of the lactose repressor (with Benno Müller-Hill)
- the invention of the rolling circle model, which describes one of the two ways DNA molecules duplicate themselves (David Dressler in the late 1960s)

- the development of a rapid chemical DNA sequencing technique (with Allan Maxam in the mid 1970s)
- the development of some of the recombinant DNA techniques
- work on bacterial strains that expressed the mammalian gene product, insulin (with Lydia Villa Komaroff and Argiris Efstratiadis in the late 1970s)
- the natural synthesis of useful proteins in bacteria
- the structure of genes and the evolution of DNA sequences.

Gilbert's Nobel Prize lecture is dated 8th December 1980 with the title 'DNA sequencing and gene structure'. During the course of his presentation, Gilbert commented [2]:

> *...During our conversation (with another colleague) I had an idea for an experiment, which ultimately underlies our sequencing method. We knew we could obtain a defined DNA fragment, 55 base-pairs long, which carried near its center the region to which the lac repressor bound. This fragment was made by cutting the DNA sequentially with two different restriction engymes, each defining one end of the fragment. Secondly, I knew that at every base along the DNA at which methylation occurred, that base could be removed by heat. Furthermore, once that had happened, only a sugar would be left holding the DNA chain together, and that sugar could be hydrolysed, in principle, in alkali to break the DNA chain. I put these ideas together by conjecturing that if we labelled one end of one strand of the DNA fragment with radioactive phosphate, we might determine the point of methylation by measuring the distance between the labelled end and the point of breakage. We could get such labelled DNA by isolating a DNA fragment (by length by electrophoresis through polyacrylamide gels) made by cutting with one restriction enzyme, labelling both ends of that fragment and then cutting it again with a second restriction enzyme to release two separable double-stranded fragments, each having a label at one end but not the other. Using polynucleotide kinase this procedure would introduce a radioactive label into the 5' end of one of the DNA strands of the fragment bearing the operator while leaving the other unlabelled. If we then modified that DNA with dimethyl sulfate so that only an occasional adenine or guanine would be methylated, heated, and cleaved the DNA with alkali at the point of depurination, we would release among other fragments a labelled fragment extending from the unique point of labelling to the first point of breakage. Any fragments from the other strand would be unlabelled, as would any fragments arising beyond the first point of breakage. If we could separate these fragments by size, as we could in principle by electrophoresis on a polyacrylamide gel, we might be able to associate the labelled fragments back to the known sequence and thus identify each guanine and adenine in the operator that had been modified by dimethyl sulfate...*

Gilbert also commented on his gene work:

> *...The first genes to be sequenced, those in bacteria, yielded an expected struca contiguous series of codons lying upon the DNA between an initiation signal and one of the terminator signals. Before the position at which the RNA copy will start, there lies a site for the RNA polymerase, interacting with the Pribnow box, a region of sequence homology lying one turn of the helix before the initial base of the messenger RNA, and also with another region of homolthirty-five bases before the start. Thus one could understand the bacterial gene in terms of a binding site for the RNA polymerase, and further binding sites for repressors and activator proteins around and under the polymerase. Alternatively, the control on transcription could be exercised by a control of the termination*

function: new proteins or an elegant translation control could determine whether or not the polymerase would read past a stop signal into a new gene...

He concluded:

...At the same moment that the rapid sequencing methods and the molecular cloning gave us the promise of being able to work out the structure of any gene, the ability to achieve a complete understanding of the genetic material, Nature revealed herself to be more complex than we had imagined. We cannot read the gene product directly from the chromosome by DNA sequencing alone. We must appeal to the sequence of the actual protein, or at least the sequence of the mature messenger RNA, to learn the intron-exon structure of the gene. Nontheless the hope exists, that as we look down on the sequence of DNA in the chromosome, we will not learn simply the primary structure of the gene products, but we will learn aspects of the functional structure of the proteins put together over evolutionary time as exons linked through introns...

Understanding DNA sequencing has been the focus of very many scientists over the last twenty years and more, see for example [4–6].

Honours and Awards

Gilbert was presented with Honorary D.Sc. degrees from many Universities including: Chicago (1978); Columbia (1978); Rochester (1979); Yeshiva (1981).

Among his many awards and prizes is the Nobel Prize for Chemistry, jointly with Sanger (1980).

Gilbert is a Member of: the American Academy of Arts and Sciences (1968); the National Academy of Sciences (1976); the American Physics Society and the American Society of Biochemists. He is a Foreign Member of the Royal Society (1987). Gilbert is the current Chairman of the Harvard Society of Fellows.

Bibliography

1. Reproduced with permission.
2. From Nobel Lectures, Chemistry 1971–1980, Editor-in-Charge Tore Frängsmyr, Editor Sture Forsén, World Scientific Publishing Co., Singapore, 1993.
3. Who's who entry: 1981–1990.
4. Genetics: Analysis of Genes and Genomes by Daniel L. Hartl and Elizabeth W. Jones, Jones & Bartlett Publishers, 6th ed. 2004, ISBN: 0763715115.
5. Computational Genome Analysis: An Introduction by Richard C. Deonier, Simon Tavaré and Michael S. Waterman, Springer, 1st ed. 2005, ISBN: 0387987851.
6. Analysis of Genes and Genomes by Richard J. Reece, John Wiley & Sons, 2004, ISBN: 0470843802.

1980

© Nobel Foundation [1]

Frederick Sanger (0.25)

"for their contributions concerning the determination of base sequences in nucleic acids" [2]

Sanger is the only chemist to be awarded the Nobel Prize for Chemistry twice (see 1958). His earlier work has been documented in his 1958 entry, this entry concentrates on his later accomplishments.

During his presentation speech Bo G. Malmström, then of the Royal Academy of Sciences, commented of Sanger and his work [2]:

> *...The machinery of life is made possible by a unique interplay between two groups of biological giant molecules, nucleic acids and the proteins, in the form of enzymes, the orchestra which plays the various expressions of life, motility, feeling, reproduction etc. DNA is the carrier of the genetic traits in the chromosomes of the cells, and it governs the chemical machinery by determining which enzymes a cell shall manufacture. We know through investigations which have earlier been awarded with Nobel Prizes, that the sequence of the building blocks in DNA, called nucleotides, determines the structure of a particular enzyme that the cell produces. The investigators who have been awarded this year's Nobel Prize in chemistry, Paul Berg, Walter Gilbert and Frederick Sanger, have through their methodological contributions made it possible to penetrate into further depth in our understanding of the relationship between the chemical structure and biological function of the genetic material. Berg was the first one to construct a recombinant-DNA molecule, i.e. a molecule which contains part of DNA from different species, e.g. genes from a human being combined with part of a bacterial chromosome. Berg has also used his method to analyze the chromosome of a virus in considerable details.*
>
> *Gilbert and Sanger have independently developed separate methods for the determination of the exact sequence of the building blocks in DNA. Gilbert has in addition studied those parts of DNA in a bacterial chromosome which control the transcription of the genetic message in the cell. Sanger has, for example, determined the complete nucleotide sequence for a small virus, whose DNA still consists of not less than 5,375 building blocks...*

Frederick Sanger was born on 13th August 1918 in Rendcombe, Gloucestershire, the second son of Frederick Sanger M.D., and his wife Cicely. Sanger married Margaret Joan Howe (1940). They have two sons and one daughter, Robin (1943), Peter (1946) and Sally Joan (1960). He is retired and lives in Cambridgeshire, England [2,3].

The Early Years

Sanger was educated at Bryanston School and at St. John's College, Cambridge, graduating with a B.A. degree in natural sciences (1939). He worked under Dr. A. Neuberger on the metabolism of the amino acid lysine (1940–1943) and graduated with a Ph.D. degree in 1943. Sanger has carried out research in the Department of Biochemistry at Cambridge University since 1940.

The Career

Year	*Post*	*Institute*
1944	Beit Memorial Fellowship for Medical Research	Department of Biochemistry at Cambridge
1951	Member of the External Staff	Medical Research Council
1961–83	Head of the Division of Protein Chemistry	MRC Laboratory for Molecular Biology at Cambridge

Sanger's post insulin work was concerned with:

- nucleic acids
- "sequencing" both nucleic acids and proteins

Sanger published some of his papers in book form in 1996 [4], and a biography of Sanger was published in 1969 [5].

His 1980 Nobel Prize lecture is dated 8th December 1980 with the title 'Determination of nucleotide sequences in DNA'. During his lecture, Sanger commented on the various techniques that he and his co-workers developed during the course of this work [2]:

> *...Following my work on amino acid sequences in proteins I turned my attention to RNA and, with G.G. Brownlee and B.G. Barrell, developed a relatively rapid small-scale method for the fractionation of* ^{32}*P-labelled oligonucleotides. This became the basis for most subsequent studies of RNA sequences. The general approach used in these studies, and in those on proteins, depended on the principle of partial degradation. The large molecules were broken down, usually by suitable enzymes, to give smaller products which were then separated from each other and their sequence determined. When sufficient results had been obtained they were fitted together by a process of deduction to give the complete sequence. This approach was necessarily rather slow and tedious, often involving successive digestions and fractionations, and it was not easy to apply it to the larger DNA molecules. When we first studied DNA some significant sequences of about 50 nucleotides in*

length were obtained with this method, but it seemed that to be able to sequence genetic material a new approach was desirable and we turned our attention to the use of copying procedures...

COPYING PROCEDURES

...In the RNA field these procedures had been pioneered by C. Weissmann and his colleagues in their studies on the RNA sequence of the bacteriophage. Qβ.Qβ contains a replicase that will synthesize a complementary copy of the single-stranded RNA chain, starting from its 3′ end. These workers devised elegant procedures involving pulse-labelling with radioactively labelled nucleotides, from which sequences could be deduced.

For DNA sequences we have used the enzyme DNA polymerase, which copies single-stranded DNA. The enzyme requires a primer, which is a single-stranded oligonucleotide having a sequence that is complementary to, and therefore able to hybridize with, a region on the DNA being sequenced (the template). Mononucleotide residues are added sequentially to the 3′ end of the primer from the corresponding deoxynucleoside triphosphates, making a complementary copy of the template DNA. By using triphosphates containing ^{32}P in the α position the newly synthesized DNA can be labelled. In the early experiments synthetic oligonucleotides were used as primers, but after the discovery of restriction enzymes it was more convenient to use fragments resulting from their action as they were much more easily obtained.

The copying procedure was used initially to prepare a short specific region of labelled DNA which could then be subjected to partial digestion procedures. One of the difficulties of sequencing DNA was to find specific methods for breaking it down into small fragments. No suitable enzymes were known that would recognise only one nucelotide. However, Berg, Fancher & Chamberlin had shown earlier that under certain conditions it was possible to incorporibonucleotides, in place of the normal deoxyribonucleotides, into DNA chains with DNA polymerase. Thus, for instance, if copying were carried out using riboCTP and the other three deoxynucleoside triphosphates, a chain could be built up in which the C residues were in the ribo form. Bonds involving ribonucleotides could be broken by alkali under conditions where those involving the deoxynucleotides were not, so that a specific splitting at C residues could be obtained. Using this method we were able to extend our sequencing studies to some extent. However extensive fractionations and analyses were still required...

THE 'DIDEOXY' METHOD

...More recently we have developed another similar method which uses specific chain-terminating analogues of the normal deoxynucleoside triphosphates. This method is both quicker and more accurate than the plus and minus technique. It was used to complete the sequence of φX174, to determine the sequence of a related bacteriophage, G4, and has now been applied to mammalian mitochondrial DNA.

The analogues most widely used are the dideoxynucleoside triphosphates. They are the same as the normal deoxynucleoside triphosphates but lack the 3′ hydroxyl group. They can be incorporated into a growing DNA chain by DNA polymerase but act as terminators because, once they are incorporated, the chain contains no 3′ hydroxyl group and so no other nucelocan be added.

Primer and template are denatured to separate the two strands of the primer, which is usually a restriction enzyme fragment, and then annealed to form the primer-template complex. The mixture is then divided into four samples. One (the T sample) is incubated with DNA polymerase in the presence of a mixture of ddTTP (dideoxy thymidine triphosphate) and a low concentration of TTP, together with the other three deoxynucleoside triphosphates...

Understanding DNA sequencing has been the focus of very many scientists over the last twenty years and more, see for example [6–8]. A recent article concerning Sanger and his work appeared in Chemistry World in 2005 [9].

Honours and Awards

Sanger holds Honorary D.Sc. degrees from the Universities of: Leicester (1968); Oxford (1970); Strasbourg (1970); Cambridge (1983).

He has been presented with many awards and prizes including: the Corday-Morgan Medal (1951); the Nobel Prize for Chemistry (1958); the Alfred Benson Prize (1966); the Royal Medal (1969); the Sir Frederick Gowland Hopkins Memorial Medal (1971); the Gairdner Foundation Annual Award (1971), 1979); the William Bate Hardy Prize, Cambridge Philisophical Society (1976); the Hanbury Memorial Medal (1976); the Copley Medal (1977); the Albert Lasker Award (1979); the Horwitz Prize (1979); the Biochem Analysis Prize (1980); the Nobel Prize for Chemistry, joint the Berg and Gilbert (1980); Gold Medal RSM (1983).

Sanger is: a Fellow of the Royal Society (1954); a Foreign Member of the American Academy of Arts and Sciences (1958); an Honorary Member of the American Society of Biological Chemists (1961); a Foreign Member of the Accociation of the National Academy of Sciences (1967); a Fellow King's College Cambridge University (1954, Honorary Fellow since 1983).

Sanger holds: the Order of Merit (1986); a Companion of Honour (1981); a Commander of the British Empire (1963).

He was Croonian lecturer in 1975.

Bibliography

1. Reproduced with permission.
2. From Nobel Lectures, Chemistry 1971–1980, Editor-in-Charge Tore Frängsmyr, Editor Sture Forsén, World Scientific Publishing Co., Singapore, 1993.
3. Who's who entry: 1981–1990.
4. *Selected Papers of Frederick Sanger: (With Commentaries)* (Series in 20th Century Biology) (Hardcover) by Frederick Sanger (Ed.), Margaret Dowding (Ed.), Pub: World Scientific Publishing Company (1996) ISBN: 9810224303.
5. Frederick Sanger: The man who mapped out a chemical of life, (Great men of science) by Alvin Silverstein, Pub: John Day Co., (1969) ASIN: B0006CZ76E.

6. Genetics: Analysis of Genes and Genomes by Daniel L. Hartl and Elizabeth W. Jones, Jones & Bartlett Publishers, 6th ed. 2004, ISBN: 0763715115.
7. Computational Genome Analysis: An Introduction by Richard C. Deonier, Simon Tavaré and Michael S. Waterman, Springer, 1st ed. 2005, ISBN: 0387987851.
8. Analysis of Genes and Genomes by Richard J. Reece, John Wiley & Sons, 2004, ISBN: 0470843802.
9. Chemistry World 2005, 2. No. 12, p. 44.

1981

Kenichi Fukui (0.5)

"for their theories, developed independently, concerning the course of chemical reactions" [2]

During his presentation speech, Inga Fischer-Hjalmars, then of the Royal Academy of Sciences, commented of molecular orbital theory in general and Fukui's work in particular [2]:

> *...What is it then that decides the sequence of events during the collision? One governing factor is the energy. The new molecules are found at a lower energy level than the original ones. Sometimes, the change will take place without difficulty. The reaction complex is simply sliding down an energy slope. But in general some hindrance must be overcome. It is necessary first to go upwards before starting the downhill ride. Then, the problem is to find the lowest passage over the height barrier. Frequently, rather much is known about the starting material and the final product, about the energy valley of the starting point and the valley of the final destination. But about the character of the ridge in between very little has been known. This year's laureates in chemistry have helped us to foresee the obstacles and to, find the best way to the final goal. The barriers depend on the fact that the electronic orbitals must be transformed. Now, there is a large number of electrons in every molecule, each with its orbital. A drastic simplification of the barrier problem was achieved in the beginning of the 1950's when Kenichi Fukui discovered that only a few orbitals, those with highest energy, are dominating the frontier of the reaction. He therefore called them frontier orbitals. By use of his theory Fukui found the laws for many groups of organic-chemical reactions. An example: Naphthalene is an important initial material in dye-stuff industry, among others. For a long time the puzzling fact had been known that hydrogen atoms at different positions in the naphthalene molecule are reacting most unequally. The explanation came first through Fukui's theory...*

> *...Professor Fukui, Professor Hoffmann. Each of you have independently developed important theories of chemical reactivity. The concepts of frontier orbitals and conservation of orbital symmetry have revealed completely new aspects of the interaction between molecules in collision. Through drastic simplifications you have been able to make beautiful generalizations. From your*

theoretical work new tools have emerged of the greatest importance for the design of chemical experiments. In recognition of your outstanding work the Royal Academy of Sciences has decided to award this year's Nobel Prize for Chemistry to you...

Kenichi Fukui was born in Nara, Japan on 4th October 1918, the the eldest of three sons of Ryokichi Fukui, a trade merchant and factory manager, and Chie Fukui, née Sugisawa. He married Tomoe Horie (1947) and they had two children Tetsuya (son) and Miyako (daughter). Fukui died in Kyoto, Japan on 9th January 1998 having undergone surgery for cancer of the stomach in the winter of 1997. His tomb is on the hillside of Nigashiyama [2–4].

The Early Years

Fukui entered Tamade Daini Primary School (1925) thereafter moved to Imamiya Middle School in 1931, followed by Osaka High School (1935–38). He was not particularly interested in chemistry whilst at high school. His father asked the advice of Professor Gen-itsu Kita of Kyoto Imperial University for a possible career path for Fukui, the younger. Prof. Kita suggested that Ryokichi, one of his juniors from the same native province, should send him to the Department of Industrial Chemistry with which he was then affiliated. He entered the Department of Industrial Chemistry, Kyoto Imperial University (1937) from which he graduated in March 1941. His graduate training was in the Department of Fuel Chemistry, also at Kyoto Imperial University.

The Career

Year	*Post*	*Institute*
1941	Synthetic fuel chemist	Army Fuel Laboratory
1943	Lecturer	Fuel Chemistry Department of Kyoto Imperial University
1945	Assistant professor	Kyoto Imperial University
1951–1982	Professor	Kyoto Imperial University
April 1971–March 1973	Dean, Faculty of Engineering	Kyoto Imperial University
1982	Professor Emeritus	Kyoto Imperial University
1982–1988	President	Kyoto Institute of Technology
1988	Professor Emeritus	Kyoto Institute of Technology
1988–1998	Director	Institute for Fundamental Chemistry

Fukui was a National Science Foundation Senior Foreign Scientist Fellow (Feb.–July, 1970). Fukui's research interests involved many aspects of chemistry including:

- catalytic engineering
- the theory of chemical reactions
- reaction engineering
- the statistical theory of gellation

- organic synthesis by inorganic salts
- polymerization kinetics and catalysts

Papers concerning his work on frontier orbitals started to appear in the literature in the early 1950's. Since that time Fukui has published work concerning:

- a correlation between the frontier electron density and the chemical reactivity in aromatic hydrocarbons
- chemical reactivity theory
- the frontier orbital approach to chemical reactions
- the geometrical shape of reacting molecules

Fukui has published papers in Japanese and English journals as well as two books concerning molecular orbital theory:

- *Theory of Orientation and Stereoselection (Reactivity and Structure Concepts in Organic Chemistry, Vol. 2)* by Kenichi Fukui Pub: Springer-Verlag (1975) ISBN: 0387074260
- *Frontier Orbitals and Reaction Paths: Selected Papers of Kenichi Fukui* (World Scientific Series in 20th Century Chemistry, Vol. 7) by Kenichi Fukui (Ed.), Hiroshi Fujimoto (Ed.) Pub: World Scientific Publishing Company (1997) ISBN: 9810222416.

Kukui's Nobel Prize lecture is dated 8th December 1981 with the title 'The role of frontier orbitals in chemical reactions'. During the course of his lecture Fukui comments on the differences between his approach to molecular orbital theory and that of Hoffmann [2]:

> *...By considering the HOMO-LUMO interactions between the fragments of a conjugated chain divided into parts, the frontier orbital theory can yield selection rules which are absolutely equivalent to those obtained from the principle called "the conservation of orbital symmetry" by Woodward and Hoffmann. One point that I may stress here is, as was pointed out by Fujimoto, Inagaki and myself, that the electron delocalization between the particular orbitals interprets definitely in terms of orbital symmetries the formation and breaking of chemical bonds which, I believe, should be a key for perceiving chemical reaction processes.*
>
> *In the cycloaddition of butadiene and ethylene, both the interaction between the HOMO of diene and the LUMO of dienophile and that between the LUMO of diene and the HOMO of dienophile stabilize the interacting system. If one is interested in the local property of interaction, however, one may recognize the clear distinction between the roles of the two types of orbital interactions. The HOMO of ethylene and the LUMO of butadiene are both symmetric with regards to the symmetry plane retained throughout the course of cycloaddition. This signifies that each of the carbon atoms of ethylene are bound to both of the terminal carbons of butadiene. The chemical bonding between the diene and dienophile thus generated may be something like the one in a loosely bound complex, e.g., protonation to an olelinic double bond. On the contrary, the HOMO of butadiene and the LUMO of ethylene are antisymmetric. The interaction between these orbitals leads, therefore, to two separated chemical bonds, each of which combines a carbon atom of ethylene and a terminal carbon atom of butadiene. Needless to say it is the interaction*

between the HOMO of diene and the LUMO of dienophile that is of importance for the occurrence of concerted cycloaddition.

In this way, it turned out in the course of time that the electron delocalization between HOMO and LUMO generally became the principal factor deterthe easiness of a chemical reaction and the stereoselective path, irreof intra- and intermolecular processes. . .

The fundamental principles of molecular orbital theory and practice have been well documented over the years, see for eample [6–8].

Honours and Awards

Fukui was presented with a Honorary D.Sc. from the University of Wales (1992).

Fukui received many awards and prizes including: the Japan Academy Medal (1962); the Order of Culture (1981); the Nobel Prize for Chemistry, jointly with R. Hoffmann (1981); the Person of Cultural Merits (1981); the Imperial Honour of the Grand Cordon of the Order of the Rising Sun (1988).

He was: a Foreign Associate of the National Academy of Sciences (1981); a Foreign Honorary Member of the American Academy of Arts and Sciences (1983); a Member of the Japan Academy (1983) and of the Pontifical Academy of Sciences (1985); a Honorary Member of the Royal Institution of Great Britain (1989); a Member of the European Academy of Arts, Sciences and Humanities (1981); a Foreign Member of the Royal Society (1989).

Additionally, Fukui was: a Member of the International Academy of Quantum Molecular Science (France) (from July 1970); a Councillor of Kyoto University (1970–1973); a member of the US-Japan Eminent Scientist Exchange Program Chemist (Sept. 1973); a Counsellor of the Institute for Molecular Science (from Jan. 1976); Vice-President of the Chemical Society of Japan, March 1978–Feb. 1979); Chairman of the Executive Committee for the 3rd International Congress of Quantum Chemistry, Kyoto, (Oct.–Nov. 1979); President of the Chemical Society of Japan (1983–1984).

Fukui's citation for election (date: 29/06/1989) as Foreign Member of the Royal Society reads [9]:

'Co-recipient (with Roald Hoffmann) of the 1981 Nobel Prize for Chemistry. Foreign Associate, U.S. National Academy of Sciences, Member of the Pontifical Academy of Sciences, 1985. Professor Fukui is renowned for the successful explanation and prediction of the course of various (especially organic) chemical reactions in a unified fashion using quantum mechanics. In his seminal studies of the stereoselectivity of organic compounds, he recognized the crucial role of electron delocalisation within the reactants and was the first to emphasize the importance of the so-called frontier orbitals which encompass the LUMO (lowest unoccupied) and HOMO (highest unoccupied) states, now an integral part of theoretical and other branches of chemistry. The detailed course of cyclic additions and intramolecular arrangements, for example, is interpretable in terms of the symmetry of frontier orbitals. The concepts which he has introduced have, both in his own hands and this of investigators throughout the world, proved enormously productive and clarifying. In recent years he has pioneered new wave-mechanical techniques for evaluating absolute rates of chemical reactions on potential energy surfaces, and new non-empirical methods of visualizing chemical reactions by so-called orbital picture graphics.'

No proposers are listed in the Sackler database.

Bibliography

1. Reproduced with permission.
2. From Nobel Lectures, Chemistry 1981–1990, Editor-in-Charge Tore Frängsmyr, Editor Bo G. Malmström, World Scientific Publishing Co., Singapore, 1992.
3. Biographical Memoirs of Fellows of the Royal Society, 2001, vol. 47, pp. 223–237.
4. Who's who: 1998.
5. Frontier Orbitals and Organic Chemical Reactions by Ian Fleming, John Wiley and Sons, 1976, ISBN: 0471-01819-8.
6. Frontier Orbitals and Organic Chemical Reactions by Ian Fleming, John Wiley and Sons, 1976, ISBN: 0471-01819-8.
7. An Introduction to Molecular Orbitals by Yves Jean, Francois Volatron, Jeremy Burdett, Oxford University Press, USA, 1993, ISBN: 0195069188.
8. Elements of Molecular Symmetry by Yngve Öhrn, Wiley-Interscience, 2000, ISBN: 0471363235.
9. Sackler database, Royal Society.

1981

Roald Hoffmann (0.5)

"for their theories, developed independently, concerning the course of chemical reactions" [2]

During his presentation speech, Inga Fischer-Hjalmars, then of the Royal Academy of Sciences, commented of Hoffmann and his work [2]:

...What is it then that decides the sequence of events during the collision? One governing factor is the energy. The new molecules are found at a lower energy level than the original ones. Sometimes, the change will take place without difficulty. The reaction complex is simply sliding down an energy slope. But in general some hindrance must be overcome. It is necessary first to go upwards before starting the downhill ride. Then, the problem is to find the lowest passage over the height barrier. Frequently, rather much is known about the starting material and the final product, about the energy valley of the starting point and the valley of the final destination. But about the character of the ridge in between very little has been known. This year's laureates in chemistry have helped us to foresee the obstacles and to, find the best way to the final goal. The barriers depend on the fact that the electronic orbitals must be transformed. Now, there is a large number of electrons in every molecule, each with its orbital...

...Many molecules have no stereo-symmetry. Then, the molecule and its mirror image have very different effects, as the right-hand and left-hand are functioning differently. The finger-tips of a right-hand can easily match those of a left-hand, but not those of another right-hand. Usually, chemical reactions will give rise to a mixture of right-and left-molecules, but our bodies are only producing one kind. For the preparation of vitamins and drugs that are to react in the body it is therefore often necessary to use methods giving only right- or only left-molecules. Such methods have been found through an exquisite combination of theory and experiment. The problem to prepare vitamin B12 was attacked among others by the brilliant molecule builder Woodward at Harvard. There was also the theoretician Roald Hoffmann. Together, Hoffmann and Woodward discovered that not only the energy of the orbital but also its symmetry is decisive for the reactivity. Thus, the Woodward-Eschenmoser synthesis of vitamin B12 could be accomplished.

Hoffmann continued to develop the theory of orbital symmetry to an exceedingly practical instrument for synthetic work of widely different character. At the same time, Fukui showed that the frontier orbital theory leads to another powerful method of solving the intricate problems of stereochemistry. In this way, the theoreticians Fukui and Hoffmann have radically changed the conditions for the design of chemical experiments...

Roald Hoffmann was born in Zloczów, Poland, the son of Hillel Safran and Clara, née Rosen, on 18th July 1937. Safran was killed during the war and his mother married Paul Hoffmann. Together they had a daughter Elinor (1954). Hoffmann married Eva Börjesson (1960) and they have a son, Hillel Jan and a daughter Ingrid Helena. Hoffmann is the current Frank H.T. Rhodes Professor of Humane Letters at Cornell University [2,3,4].

The Early Years

As a young Polish Jew, Hoffmann went to a ghetto (1941), then a labour camp. His father smuggled both his mother and Hoffmann out of the camp (1943), and for the remainder of the war they hid in the attic of a school house, in a nearby village in the Ukraine. Most of his family were killed by the Nazis, including his father. The remaining family members were freed by the Red Army (June 1944) and moved to Przemysl, and then to Krakow, where Hoffmann went to school.

Prolonged moves through various European displacement camps were brought to an end when the family arrived in the United States (1949). Hoffmann attended schools in Brooklyn and Stuyvesant High School prior to attending Columbia College as a premedical student (1955). He graduated from Columbia College with a B.A. in 1958. Hoffmann began graduate work initially under W.E. Moffitt (who died within a year of Hoffamnn's entry), followed by M.P. Gouterman. Hoffmann obtained a scholarship from P.O. Lowdin's Quantum Chemistry Group at Uppsala to attend a Summer School (1959), where he met his future wife. His return to Harvard was shortlived as he and Eva went to the Soviet Union as part of the US-USSR graduate student exchange. Hoffmann worked for nine months at Moscow University with A.S. Davydov on excitor theory. Hoffmann returned to Harvard to complete his doctorate, switching to work under Lipscomb (Nobel Chemistry Laureate 1976) on extended Hückel methods, obtaining his Ph.D. in 1962, having previously been awarded an M.A. from Harvard in 1960.

The Career

Year	*Post*	*Institute*
1962	Junior Fellow	Society of Fellows, Harvard University
1965	Associate Professor, then Professor	Cornell University
1974	John A. Newman Professor of Physical Science	Cornell University
1996–	Frank H.T. Rhodes Professor of Humane Letters and Professor of Chemistry	Cornell University

Hoffman commented of his relationship with R.B. Woodward [2]:

. . .in the Spring of 1964 R.B. Woodward asked me some questions about what subsequently came to be called electrocyclic reactions. That last year at Harvard was exciting. I was learning organic chemistry at a great pace, and I had gained access to a superior mind. R.B. Woodward possessed clarity of thought, powers of concentration, encyclopedic knowledge of chemistry, and an aesthetic sense unparalleled in modern chemistry. He taught me, and I have taught others. . .

Hoffmann's many interests has resulted in a diverse range of work including:

- the conservation of orbital symmetry
- the electronic structure of extended systems in one-, two-, and three dimensions
- frontier orbital arguments
- the COOP curve (the solid state analogue of an overlap population)
- studies of the platinocyanides, Chevrel phases, transition metal carbides
- displacive transitions in NiAs, MnP and NiP
- new metallic forms of carbon
- the interaction of CH_4, acetylene and CO with specific metal faces
- the development of the extended Hückel method
- the structure and reactivity of inorganic and organometallic molecules
- the analysis of cluster bonding and the equilibrium geometries and conformational preferences of olefin and polyene metal carbonyl complexes
- the geometries, polytopal rearrangement and substitution site preferences of five, six, seven and eight coordination
- the electronic structure of extended systems in one-, two-, and three dimensions.

Hoffmann participated in the production of a television course in introductory chemistry titled "*The World of Chemistry*" (1986–88). This series of 26 half-hour programmes was developed at the University of Maryland and produced by Richard Thomas. Hoffmann was the presenter or narrator for the series.

He has published more than 450 scientific articles and papers.

Hoffmann has also written three chemistry books: *Building bridges between inorganic and organic chemistry*, Pub: Cornell University (1981) ASIN: B00070OKMC; *Solids and Surfaces: A Chemist's View of Bonding in Extended Structures*, Pub: Wiley-VCH (1989) ISBN: 0471187100; *Computational Chemistry of Solid State Materials: A Guide for Materials Scientists, Chemists, Physicists and others*, with Richard V. Dronskowski Pub: John Wiley & Sons (2006) ISBN: 3527314105. He has also written numerous books of poetry including: *The Metamict State* (Contemporary Poetry Series) (Hardcover) Pub: Univ Pr of Florida (1987) ISBN: 0813008697; *Gaps and Verges: Poems (Contemporary Poetry Series)* Pub: University Press of Florida (1990) ISBN: 081300943X; *Chemistry Imagined: Reflections on Science*, with Vivian Torrence Pub: Smithsonian Inst Pr; Reissue edition (1995) ISBN: 1560985399; *The Same and Not the Same*, Pub: Columbia University Press (1997) ISBN: 0231101392; *Old Wine New Flasks: Reflections on Science and Jewish Tradition*, with Shira Leibowitz Schmidt Pub: W.H. Freeman & Company (1997) ISBN: 0716728990; *Lo Mismo y Lo No Mismo*, Pub: Fondo de Cultura Economica USA (1998) in Spanish ISBN: 9681654455; *Memory Effects*, Pub: Columbia College, Dept. of Art and Design (1999) ISBN: 0932026435; *Oxygen*, with Carl Djerassi Publ: Wiley-VCH; 1st ed. (2001) ISBN: 3527304134; *Chemical Sciences in the 20th Century: Bridging Boundaries*, with Carsten Reinhardt Pub: Wiley-VCH (2001) ISBN: 3527302719; *Soliton: Poems* (New Odyssey Series) Pub: Truman State University Press (2002) ISBN: 1931112193.

Hoffmann's Nobel Prize lecture is dated 8th December 1981 with the title 'Building bridges between inorganic and organic chemistry'. Hoffman's opening remarks concerned R.B. Woodward [2]:

> *...R. B, Woodward, a supreme patterner of chaos, was one of my teachers. I dedicate this lecture to him, for it is our collaboration on orbital symmetry conservation, the electronic factors which govern the course of chemical reactions, which is recognized by half of the 1981 Nobel Prize in Chemistry. From Woodward I learned much: the significance of the experimental stimulus to theory, the trait of constructing explanations, the importance of aesthetics in science. I will try to show you how these characteristics of chemical theory may be applied to the construction of conceptual bridges between inorganic and organic chemistry...*

He defines the isolobal approach as:

...A word is needed to describe the resemblance of the two fragments, CH_3 and d^7 ML_5. They are certainly not isostructural, nor are they isoelectronic. However, both possess a frontier orbital which looks approximately the same for the two fragments. We will call two fragments isolobal if the number, symmetry properties, approximate energy and shape of the frontier orbitals and the number of electrons in them are similar-not identical, but similar. Thus CH_3 is isolobal with $Mn(CO)_5$...

He concluded:

...The isolobal analogy is a model. It is the duty of our scientific craft to push it to its extremes, and being only a model it is certain to fail somewhere. For any model, as ingenious a construction as it might be, is bound to abstract only a piece of reality. The reader has seen just how far the model can be pushed and he or she has seen where it breaks down.

The pleasing aspect of this particular model is that it brings together different subfields of our central science. We are separated, split asunder - organic, inorganic, physical, biological, analytical chemists - by the very largesse of our creation. The variety of molecules we create, and the methods we use to study them breed jargon and specialization. Yet underneath the seeming complexity there must be a deep unity. I think this approach would have pleased R. B. W...

The fundamental principles of molecular orbital theory and practice have been well documented over the years, see for example [5–7].

Honours and Awards

Hofmann has received over twenty five Honorary D.Sc. degrees including those from the Universities of: Yale (1980); Columbia (1982); Hartford (1982); City University of NY (1983); Puerto Rico (1983); La Plata (1984); Uruguay (1984); Colgate (1985); State University of NY at Binghamton (1985); Rennes (1986); Ben Gurion (1989); Carleton (1989); Lehigh University (1989); Maryland (1990); Athens (1991); Bar Ilan (1991); St. Petersburg (1991); Thessalonika (1991); Barcelona (1992); Ohio State (1993). He was also presented with the degree of Hon. D.Tech. from the Royal Institute of Technology, Stockholm (1977).

Hofmann is the only person ever to have received the American Chemical Society's awards in three different specific subfields of chemistry, namely: the Pimentel Award (Chemical Education); the Award Priestley Medal (Inorganic Chemistry); the A.C. Cope Award, jointly with R.B. Woodward (Organic Chemistry). He has also been presented with two other ASC awards. He was also awarded the Nobel Prize in Chemistry, jointly with Kenichi Fukui (1981), the Monsanto Award and the National Medal of Science.

Hoffman holds many Fellowships/Memberships in many of the worldwide learned societies and academies, including: the American Academy of Arts and Sciences; the American Philosophical Society; the Finnish Society of Sciences and Letters; the Indian National Science Academy; the National Academy of Sciences; the Nordrhein Westfällische Academy of Sciences; the Royal Society; the Royal Swedish Academy of Sciences and the Russian Academy of Sciences.

Bibliography

1. Reproduced with permission.
2. From Nobel Lectures, Chemistry 1981–1990, Editor-in-Charge Tore Frängsmyr, Editor Bo G. Malmström, World Scientific Publishing Co., Singapore, 1992.
3. Who's who: 2005.
4. Hoffman home page http://hamiltonian.chem.cornell.edu/.
5. Frontier Orbitals and Organic Chemical Reactions by Ian Fleming, John Wiley and Sons, 1976, ISBN: 0471-01819-8.
6. An Introduction to Molecular Orbitals by Yves Jean, Francois Volatron, Jeremy Burdett, Oxford University Press, USA, 1993, ISBN: 0195069188.
7. Elements of Molecular Symmetry by Yngve Öhrn, Wiley-Interscience, 2000, ISBN: 0471363235.

1982

Aaron Klug

"for his development of crystallographic electron microscopy and his structural elucidation of biologically important nucleic acid-protein complexes" [2]

During his presentation speech, Bo G. Malmström, then a member of the Royal Academy of Sciences, commented of Klug [2]:

> *...Giant molecules have a tendency to aggregate, and biological function is often associated with complicated molecular aggregates. In the chromosomes of the cell nucleus, for example, the hereditary material is present in the form of chromatin, a giant aggregate between DNA and thousands of protein molecules. In viruses, which represent the border between living and dead matter, there are simpler aggregates between nucleic acids and proteins. A virus can be said to be genetic material without a cell of its own, and the structure of viruses can provide clues to the more complicated organisation of the hereditary material in higher organisms.*
>
> *Large molecular aggregates can seldom be obtained in a form which allows structural determination by X-ray diffraction. The investigator who has been awarded with this year's Nobel Prize in chemistry, Aaron Klug, has developed a method to study the structure of molecular aggregates from biological systems. His technique is based on an ingenious combination of electron microscopy with principles taken from diffraction methods. Electron microscopy has long been used to depict the structural components of the cell, but its power of resolution is limited by a lack of contrast in the picture. Klug has shown that even picture; seemingly lacking in contrast may contain a large amount of structural information, which can be made available by a mathematical manipulation of the picture.*
>
> *With this technique, in combination with other methods of structural chemistry, Klug has inter alia investigated viruses and chromatin of the cell nucleus. His virus studies have illuminated an important biochemical principle, according to which the complicated molecular aggregates in the*

cell are formed spontaneously from their components. The chromatin investigations have provided clues to the structural control of the reading of the genetic message in DNA. In a long-term perspective they will undoubtedly be of crucial importance for our understanding of the nature of cancer, in which the control of the growth and division of cells by the genetic material no longer functions...

Aaron Klug was born on 11th August 1926, the son of Lazar and Bella (née Silin) Klug in Zelvas, Lithuania. Klug married Liebe Bobrow. They had two sons; Adam (1954) and David (1963), of whom one is now deceased. Klug can be contacted through the Medical Research Council, Cambridge [2,3].

The Early Years

Although Klug was born in Zelvas, he was taken to South Africa as a child of two, and grew up in Durban. Klug attended Durban High School where he read a book called *Microbe Hunters* by Paul de Kruif, which influenced him to study medicine at University as a way into microbiology. He enrolled at the University of Witwatersrand in Johannesburg, from which he graduated with a B.Sc. in science followed by an M.Sc. in physics from the University of Cape Town (1947–48). Whilst studying for his Masters degree, Klug was influenced by R.W. James, an X-ray crystallographer, which led Klug into a strong interest in how matter is organized at the molecular level. Supported by an 1851 Exhibition Scholarship and also by a Rouse-Ball Research Studentship to Trinity College, Klug went to Cambridge (1949–52) under D.R. Hartree, where he studied the cooling of steel through the austenite-pearlite transition for his Ph.D., which he gained in physics.

Klug worked for a year on a postdoctorate with F.J.W. Roughton, Colloid Science Dept., University of Cambridge (1953), on simultaneous diffusion and chemical reaction, such as occurs when oxygen enters a red blood cell. He then obtained a Nuffield Fellowship (1953) and used it to work in J.D. Bernal's department in Birkbeck College in London where he worked alongside Rosalind Franklin but on a different research topic to Franklin, namely the protein ribonuclease. Klug was awarded the Sc.D. from the University of Cambridge.

The Career

Year	*Post*	*Institute*
1958	Head, Virus Structure Research Group	Birkbeck College
1962	Member of staff	MRC Laboratory of Molecular Biology, University of Cambridge
1978	Joint Head, Division of Structural Studies	MRC Laboratory of Molecular Biology, University of Cambridge
1986–96	Director	MRC Laboratory of Molecular Biology, University of Cambridge
1996–	Member of staff	MRC Laboratory of Molecular Biology, University of Cambridge

Klug was also Director of Studies in Natural Science, Peterhouse College.

Klug's main interests are with:

- the interactions of proteins with nucleic acids
- the elucidation of the structures of large biological molecules
- the of method of 3-D image reconstruction in electron microscopy
- the discovery of the zinc finger family of transcription factors
- the design of artificial factors to switch genes on and off.

Klug wrote '*Towards Sustainable Consumption: a European Perspective*' with Brian Heap and Jennifer Kent, Pub: The Royal Society (2000) ISBN: 0854035370.

His Nobel Prize lecture is dated 8th December 1982 with the title 'From Macromolecules to biological assemblies'. Klug commented on his work [2]:

...The general method of reconstruction which we developed is based on the projection theorem, which states that the two-dimensional Fourier transform of a plane projection of a three-dimensional density distribution is identical to the corresponding central section of the three-dimensional transform normal to the direction of view. The three-dimensional transform can therefore be built up section by section using transforms of different views of the object, and the three-dimensional reconstruction then produced by Fourier inversion. The important feature of the method is that it tells one how many different views are needed for a required resolution and how these are to be recombined into a three-dimensional map of the object. The process is both quantitative and free from arbitrary assumptions. The approach is similar to conventional X-ray crystallography, except that the phases of the X-ray diffraction pattern cannot be measured directly, whereas here they can be computed from a digitised image. Were it not for radiation damage, the different views could be collected from a single particle by using a tilting stage in the microscope, but more realistically one must use several particles in different but identifiable orientations. In general, it is desirable to combine data from different particles so that imperfections can be averaged out...

Phase contrast microscopy

...Electron microscopy, combined with some method of image analysis, when applied to negatively stained specimens, has proved ideal for determining the arrangement and shape of small protein subunits within natural or artificial arrays, including two dimensional crystals and macromolecular assemblies such as viruses and microtubules. The structural information obtainable has proved to be highly reliable with respect to detail down to about the 20 Å or 15 Å level. It became clear, however, that the degree of detail revealed was limited by the granularity of the negative stain and the fidelity with which it follows the surface of the specimen. To obtain much higher resolution information, better than about 10 Å, one should dispense with the stain and view the protein itself. At high resolution, there is a second problem: irradiation damage. This can be reduced by cutting down the illuminating beam, but the statistical noise is then increased, and the raw image becomes less and less reliable. However, this difficulty can be overcome satisfactorily by imaging ordered arrays of molecules, so that the information from the different molecules can be averaged, as described above, to give a statistically significant picture. The first problem of replacing the negative stain, yet avoiding dehydration. can be solved in two ways. One, now being intensively studied, is to use frozen hydrated specimens. The second, tried method is that of Unwin and Henderson, who, in their radical

approach to determining the structure of unstained biological specimens by electron microscopy used a dried-down solution of glucose to preserve the material.

The question then arose as to how this unstained specimen, effectively transparent to electrons, is to be visualized. In the light microscopy of transparent specimens the well-known Zernike phase contrast method is used. Here the phase of the scattered beams relative to the unscattered beam are shifted by means of a phase plate and then the scattered and unscattered beams are allowed to interfere in the image plane to produce an image. A successful electrostatic phase contrast device for electron microscopy, quite analogous to the phase plate used in light microscopy, was constructed by Unwin, but it is not easy to make or use. A practical way of producing phase contrast in the electron microscope is simply to record the image, with the objective lens underfocussed, and this was the method used by Unwin and Henderson...

Honours and Awards

Klug has received many Honorary degrees including: Hon D.Sc. Chicago (1978); Columbia University (1978); Witwatersrand (1984); Hull (1985); St. Andrews (1987); Western Ontario (1991); Warwick (1994); Cape Town (1997); Weizmann Institute (1997); Stirling (1998); London (2000) and Oxford (2001). Hon Doctor Strasbourg (1978); Hon Ph.D. Jerusalem (1984); Hon. Dr.Fil. Stockholm (1980); Hon. Litt.D. Cantab (1998).

He has been presented with many awards and prizes including: the Heineken Prize, Royal Netherlands Academy of Sciences (1979); the Louisa Gross Horowitz Prize, Columbia University (1981); the Nobel Prize in Chemistry (1982); the Gold Medal of Merit, University of Cape Town (1983); the Copley Medal (1985); the Harden Medal, Biochemical Society (1985); the Baly Medal (1987); the William Bate Hardy Prize, Cambridge Philosophical Society (1996).

He is: a Fellow of the Royal Society 1969; a Foreign Associate National Academy of Sciences, USA (1984); a Foreign Member of the Max Planck Society (1984) and of the Japanese Academy (1996); a Foreign Honorary Member American Academy of Arts and Sciences (1969); an Honorary FRCP (1987); a Foreign Associate Academy des Sciences Paris (1989); a Honorary Fellow of the Royal College of Pathology (1991); a Fellow of the American Philosophical Society (1996); a Foreign Associate of the French Academy of Sciences; a Honorary Fellow of Trinity College, Cambridge University (1983); a Honorary Professor, University of Cambridge (1989); a Honorary Fellow of Peterhouse and Trinity College, Cambridge (1993, Fellow 1962–93); a Founder F.Med.Sci. (1998).

Klug has presented the following lectures: Carter-Wallace, Princeton (1972); Leeuwenhoek (1973); Dunham Medical School (1975); Harvey NY (1979); Lane, Stanford University (1983); Silliman, Yale University (1985); Nishma Meml. Tokyo (1986); Pauli, ETH Zürich (1986); Cetus, University of California (1987); Konrad Bloch, Harvard University (1988); Steenbock, University of Wisconsin (1989); National, US Biophysical Society, Washington (1993); William and Mary, Leiden (1996); Blackett Memorial (1996).

He was knighted in 1988; became a Member of the Order of Merit (1995) and a Member of the Council for Science and Technology (1993–2000). The Gold Order of Mapungubwe was bestowed on Klug for his contribution to medicine in 2005.

He served on the Council of the Royal Society (1988–90, 1995–2000, Vice President for the period (1989–90) and President 1995–2000). Klug was Chair of the International Board of Overseers for the Kreitman Foundation.

Bibliography

1. Reproduced with permission.
2. From Nobel Lectures, Chemistry 1981–1990, Editor-in-Charge Tore Frängsmyr, Editor Bo G. Malmström, World Scientific Publishing Co., Singapore, 1992.
3. Who's who: 2005.
4. Principles of Protein X-ray Crystallography (Springer Advanced Texts in Chemistry), 2nd ed. by Jan Drenth, Springer, 2002, ISBN: 0387985875.
5. A Practical Guide to X-Ray Crystallography of Biomacromolecules by Albrecht Messerschmidt, Wiley-VCH for publication Jan 2007, ISBN: 3527313966.
6. Protein Crystallography by T.L. Blundell and L.N. Johnson, Academic Press, 1976, ISBN: 0-12-108350-0.

1983

© Nobel Foundation [1]

Henry Taube

"for his work on the mechanisms of electron transfer reactions, especially in metal complexes" [2]

During his presentation speech, Ingvar Lindqvist, then of the Royal Academy of Sciences, commented of inorganic chemistry in general and Taube's work in particular [2]:

> *...Henry Taube has been awarded the 1983 Nobel prize in chemistry for his studies of the mechanisms of electron transfer-reactions, particularly of metal complexes. I will not, during these few minutes, try to give a survey of the rich scientific production of Taube. Instead, I will choose one chemical reaction and attempt to show how our way of looking at this and similar reactions has changed drastically thanks to Taube...*
>
> *...A trivalent cobolt ion surrounded by five ammonia molecules and one chloride ion is reacting with a bivalent chromium ion which probably is surrounded by six water molecules. During the reaction there are formed a bivalent cobalt ion surrounded by five ammonia molecules and one water molecule and a trivalent chromium ion surrounded by five water molecules and one chloride ion. The electron transfer is thus connected with a chloride ion transfer from cobolt to chromium. This description is certainly much more complete and gives a clear picture of the conditions before and after the reaction. It does not tell anything, however, about how the reaction has taken place.*
>
> *Taube has now proceeded one step further and has shown us exactly how the reaction occurs. The first step is the formation of a larger complex where the chromium ion is attached to the cobolt ion via the chloride ion which thus functions as a bridge between the two metal ions. This requires that a water molecule leaves the chromium ion to give place for the bridging chloride ion. The chromium ion will thus be surrounded by five water molecules and the bridging chloride ion. It is only when this bridge has formed that the electron transfer can take place making the cobalt ion bivalent and the chromium ion trivalent. Finally the bridge is opened and the chloride ion follows the trivalent chromium ion while the cobolt ion must take up a molecule of water to replace the chloride ion. In an*

impressive series of investigations Taube has developed and refined this idea and it will in the future be a natural part of the set of paradigms of chemistry, which thus in a decisive way has been enriched by the contributions of Henry Taube...

Henry Taube was born in Neudorf, Saskatchewan, Canada on 30th November 1915, the son of Samuel Taube and Albertina, née Tiledetski. He became a naturalized US citizen in 1942. Taube married Mary Alice Wesche and they had two sons and two daughters. Taube was Professor Emeritus at Stanford University [2,3] and died in November 2005.

The Early Years

Taube attended high school at Luther College in Regina, Saskatchewan, followed by studies at University of Saskatchewan, where he received his B.Sc degree (1935) and M.Sc (1937) degree, under J.W.T. Spinks. He graduated from the University of California, Berkeley with his Ph.D under W.C. Bray (1940).

The Career

Year	*Post*	*Institute*
1940	Instructor	University of California, Berkeley
1941	Instructor and Assistant Professor	Cornell University
1946	Assistant Professor, Associate Professor, Professor (Chairman, Department of Chemistry 1956–59)	University of Chicago
1962	Professor (Chairman, Department of Chemistry 1972–74 & 1978–79)	Stanford University
1986	Professor Emeritus	

Taube was a Guggenheim Fellow, 1949 and 1955.

He continued to teach for several years following the award of the Nobel Prize. He once commented that there was:

"*considerable improvement in the attentiveness of my students, though little in the message or the way it was transmitted had changed.*" [4].

Taube's interests were many and included:

- basic chemistry of osmium and ruthenium
- charge transfer as affecting properties including the reactivity of ligands
- effects arising from back-bonding
- mixed valence molecules
- mechanisms of "atom" and electron transfer reactions
- metal complexes
- the reactivity of co-ordination compounds

He has published over 350 scientific articles

Taube's Nobel Prize lecture is dated 8th December 1983 with the title 'Electron transfer between metal complexes – retrospective'. Taube commented [2]:

> *...Chemical reactions are commonly classified into two categories: substitution or oxidation-reduction. The latter can always be viewed as involving electron transfer, though it is agreed that when we consider the mechanisms in solution, electron transfer is not as simple as it is in the $Ne + Ne^+$ case. Rearrangement of atoms always attend the changes in electron count at each center, and these must be allowed for. I will, however, simplify the subject by considering only processes of simple chemistry: those in which electron transfer leaves each of the reaction partners in a stable oxidation state. While substitution reactions can be discussed without concern for oxidation-reduction reactions, the reverse is not true. The changes that take place at each center when the electron count is changed is an essential part of the "electron transfer" process, and may be the dominating influence in fixing the rate of the reaction. Moreover, most of the early definitive experiments have depended on exploiting the substitution characteristics of the reactants, and of the products. Thus, the attention which will be devoted to the substitution properties of the metal ions is not a digression but is an integral part of the subject.*
>
> *An appropriate place to begin this account is with the advent of artificial radioactivity. This enormously increased the scope of isotopic tracer methods applied to chemistry, and made it possible to measure the rates of a large number of oxidation reduction reactions such as:*

$$*Fe^{2+}(aq) + Fe^{3+}(aq) = *Fe^{3+}(aq) + Fe^{2+}(aq)$$

ELECRONIC STRUCTURE AND MECHANISM

> *...A major theme in my own research, on returning from leave, was to try to understand the large differences in rate, noted qualitatively in my early work, and later made more quantitative, for the reactions $Cr^{2+}(aq)$ with carboxylate complexes of $(NH_3)_5Co(III)$. Many of the ligands were dicarboxylic acids, and to explain the observation that when a conjugated bond system connects the two carboxylates, reaction is usually more rapid than it is for the saturated analog, it seemed reasonable then to assume that in the case of conjugated ligands, the reducing agent attacks the exo carboxyl (remote attack), the conjugated bond system serving as a "conduit" for electron transfer. In adopting this view the tacit assumption was made that the reactions are non-adiabatic so that the extent of electronic coupling would be reflected in the rate. In retrospect, this assumption is naive, because the effect of conjugation would be exerted even if the reducing agent attacked at the endo carboxyl...*

CONCLUDING REMARKS

> *...In this paper I have focussed rather narrowly on electron transfer reactions between metal complexes. The separation of this subclass from other possible ones which can be assembled from the reactant categories: metal complexes, organic molecules, molecules derived from other non-metallic elements, any of the above in excited states, electrodes, is not totally arbitrary as it might be were it dictated solely by limitations of space. Admittedly, all the possible electron transfer processes are governed by the same principles, at least when these are stated in a general enough*

way. But as these principles manifest themselves in the different subclasses, the desriptive chemistry can be quite different, and these differences are the fabric of chemistry. The subclass which has been treated has perhaps been the most thoroughly studied, yet, as the article by Sutin shows, our understanding at a basic level is far from complete. Even for the much studied Fe^{3+}/Fe^{2+} self exchange reaction, which served to introduce this subject, the important question of whether the reaction is adiabatic or not has not been settled to everyone's satisfaction. Still a great deal of progress has been made. The descriptive matter has increased enormously since 1940, and our understanding of it, both in scope and depth, has more than kept pace with observations. A great deal of progress has also been made in many of the other subclasses-for example in the study of electrode processes, and in "atom transfer" reactions (as a specific case the use of transition metal species to carry the oxidizing capacity of O_2 to a substrate such as an organic molecule).

Both are of the greatest importance in industrial applications, and the latter also in reaching an understanding of the chemistry of living cells. Because the subclasses are interrelated, progress in one leads to progress in another...

Electron transfer processes, perhaps more so for molecules of biological importance, continue to be investigated, see for example [5–7].

Honours and Awards

Taube was presented with a Honorary D.Sc. degree from the Universities of: Chicago (1983); the Polytechnic Institute, New York (1984); State University of New York (1985); Guelph (1987); Seton Hall (1988); Lajos Kossuth (1988); Northwestern (1990). He was also presented with a Honorary Doctorate (LL.D.) from the University of Saskatchewan (1973) and a Honorary Ph.D. from the Hebrew University of Jerusalem (1979).

Taube is holder of the following awards and prizes: the American Chemical Society Award for Nuclear Applications in Chemistry (1955); the Harrison Howe Award, Rochester Section, ACS (1960); the Chandler Medal, Columbia University (1964); the John Gamble Kirkwood Award, New Haven Section, ACS (1966); the ACS Award for Distinguished Service in the Advancement of Inorganic Chemistry (1967); Nichols Medal, New York, ACS (1971); Willard Gibbs Medal, Chicago Section, ACS (1971); the F.P. Dwyer Medal, University of New South Wales, Australia (1973); the National Medal of Science, Washington, D.C. (1977); the Allied Chemical Award for Excellence in Graduate Teaching & Innovative Science (1979); T.W. Richards Medal of the Northeastern Section, ACS (1980); the ACS Award in Inorganic Chemistry of the Monsanto Company (1981); the Linus Pauling Award, Puget Sound Section, ACS (1981); the National Academy of Sciences Award in Chemical Sciences (1983); the Bailar Medal, University of Illinois (1983); the Robert A. Welch Foundation Award in Chemistry (1983); the Nobel Prize in Chemistry (1983); the Priestley Medal, ACS (1985); the Distinguished Achievement Award, International Precious Metals Institute (1986); the Oesper Award, The Cincinnati Section of the American Chemical Society (1986); the G. M. Kosolapoff Award, Auburn Section, ACS (1990).

Taube is a Member of: the American Academy of Arts & Sciences; the American Chemical Society; American Philosophical Society; the National Academy of Sciences; the Royal Danish Academy of Sciences & Letters; of the Royal Physiographical Society of Lund. He is a Honorary Fellow of the Royal Society of Chemistry (1989) and of the Indian Chemical Society (1989). He is a Honorary Member of: the College of Chemists of Catalonia and Beleares (1984); the Canadian Society for

Chemistry (1986); the Hungarian Academy of Sciences (1988). He is a Foreign Member of the Finnish Academy of Science & Letters and of the Royal Society. Taube is a Corresponding Member of: the Brazilian Academy of Sciences; the Australian Academy of Science; the Academy of Arts and Science of Puerto Rico (1985). He is a Foreign Associate of the Engineering Academy of Japan. Taube is also a Member of: Phi Beta Kappa; Sigma Xi; and Phi Lamda Upsilon (Honorary).

He has also consulted for Catalytica Associates, Inc., Mountain View, California.

Bibliography

1. Reproduced with permission.
2. From Nobel Lectures, Chemistry 1981–1990, Editor-in-Charge Tore Frängsmyr, Editor Bo G. Malmström, World Scientific Publishing Co., Singapore, 1992.
3. Who's who: 2005.
4. From Stanford University web site http://news-service.stanford.edu/news/2001/october3/taube-103.html.
5. Electron Transfer Reactions: Inorganic, Organometallic, and Biological Applications (Advances in Chemistry Series) by Stephan S. Isied, An American Chemical Society Publication, 1997, ISBN: 0841234566.
6. Electron Transfer and Radical Processes in Transition-Metal Chemistry by Didier Astruc, Wiley-VCH, 1995, ISBN: 0471185884.
7. Electron Transfer in Chemistry and Biology: An Introduction to the Theory (Wiley Series in Theoretical Chemistry) by Alexander M. Kuznetsov and Jens Ulstrup, John Wiley & Sons, 1999, ISBN: 0471967491.

1984

© Nobel Foundation [1]

Robert Bruce Merrifield

"for his development of methodology for chemical synthesis on a solid matrix" [2]

During his presentation speech, Bengt Lindberg, then of the Royal Academy of Sciences commented on Merrifield's work [2]:

...The chemical reactions which take place in living organisms are not spontaneous, but require the involvement of catalysts. These catalysts are called proteins and are composed of chains of amino acids called peptides. A number of hormones and other substances which regulate different life processes are also peptides. There are about 20 naturally occurring amino acids which are found in such peptides and since the chains can be very long, the number of possible variations is virtually unlimited...

...When Vincent du Vigneaud synthesized a peptide hormone, oxytocin, which is a nonapeptide, for the first time, this represented a great step forward which was rewarded with the Nobel prize for 1955. To use a similar approach for synthesizing a peptide containing 100 or more amino acid residues is truly a heroic task, requiring a very large amount of work and chemicals. This task can be compared to climbing a high mountain peak in the Himalayas, which begins with a large expedition carrying much equipment and ends, if all goes well, with a few lightly equipped alpinists reaching the top.

Therefore, Merrifield's development during the 1960's of a method for carrying out peptide synthesis on a solid matrix revolutionized the field. He attached the first amino acid to an insoluble polymer, a plastic material in the form of small spheres. Subsequently, the other amino acids were added one after one and only after the entire peptide chain had been synthesized was it released from the polymer. The advantages of this method are considerable. The complicated purification of the product after each synthetic step is replaced by simply washing the polymer to which the peptide is attached, so that loss of product is avoided completely. At the same time, the yield for each individual step is increased to 99.5% or better, a goal which cannot be achieved with conventional methods, but which is extremely important in syntheses involving a large number of steps. Finally, this method can be automated and automatic peptide synthesizers are now commercially available.

Thousands of different peptides of different sizes, as well as proteins, peptide hormones and analogues of these compounds have now been synthesized using this method. One milestone in this respect was the synthesis of an active enzyme, ribonuclease, containing 124 amino acid residues…

Robert Bruce Merrifield was born in Fort Worth, Texas, on 15th July 1921, the only son of George E. Merrifield and Lorene, née Lucas. He married Elizabeth Furlong (20th June 1949) and they had six children, one son and five daughters. Merrifield was an Emeritus Professor of the Rockefeller Institute for Medical Research [2, 3, 4] until his death at his Bergen County home in Cresskill, N.J. on 14th May 2006 after a long illness.

The Early Years

Merrifield attended nine grade schools and two high schools before graduating from Montebello High School (1939). He attended the Pasadena Junior College, followed by several years at the University of California at Los Angeles during which he obtained a B.A. (1943) in chemistry. Merrifield worked as a chemist for Philip R. Park Research Foundation (1943–44) prior to re-joining UCLA for his Ph.D. under M.S. Dunn in biochemistry (1949), on work concerning microbiological methods for the quantitation of the pyrimidines.

The Career

Year	*Post*	*Institute*
1948	Research Assistant	UCLA Medical School
1949	Assistant to Associate Professor	Rockefeller Institute for Medical Research
1966	Professor	Rockefeller Institute for Medical Research
1984	John D. Rockefeller Jr Emeritus Professor	Rockefeller Institute for Medical Research

Merrifield was a Nobel Guest Professor in Uppsala in 1968.

Merrifield has investigated, studied and written about a range of issues concerning peptides and their synthesis, including:

- the mechanism of antibacterial activity
- work on a dinucleotide growth factor (which he discovered in graduate school)
- a study of glucagons (a hyperglycemic peptide of the pancreas)
- work on peptide growth factors (which Woolley had discovered earlier)
- studies on peptides including antibiotics, antigens, enzymes, growth factors, hormones, ion carriers, neuropeptides and toxins
- the development and application of solid phase peptide synthesis (started in 1959)
- peptide hormones (using total chemical synthesis of analogs)

Merrifield's method of polypeptide synthesis has greatly facilitated the manufacture of insulin, interferon and other drugs.

He wrote an autobiography in 2001 [5], and was an associate editor of the International Journal of Peptide and Protein Research.

Merrifield's Nobel Prize lecture is dated 8th December 1984 with the title 'Solid Phase Synthesis'. A synopsis of this work was also published in [6]. During his lecture [2], Merrifield commented:

SYNTHESIS ON A SOLID MATRIX

...One day I had an idea about how the goal of a more efficient synthesis might be achieved. The plan was to assemble a peptide chain in a stepwise manner while it was attached at one end to a solid support. With the growing chain covalently anchored to an insoluble matrix at all stages of the synthesis the peptide would also be completely insoluble and, furthermore, would be in a suitable physical form to permit rapid filtration and washing. Therefore, after completion of each of the synthetic reactions the mixture could be filtered and thoroughly washed to remove excess reactants and by-products. The intermediate peptides in the synthesis would thus be purified by a very simple, rapid procedure rather than by the usual tedious crystallization methods. When a multistep process, such as the preparation of a long polypeptide or protein is contemplated, the saving in time and effort and materials could be very large. The fact that all of the steps just described are heterogeneous reactions between a soluble reagent in the liquid phase and the growing peptide chain in the insoluble solid phase led to the introduction of the name "solid phase peptide synthesis"...

SOLID PHASE PEPTIDE SYNTHESIS

...A detailed scheme for the synthesis of peptides is shown in Figure 3. Each of the steps has been modified in many ways, but the chemistry shown here has served well and has been applied to the synthesis of large numbers of peptides (9). The carboxyterminal amino acid is blocked at the amino end by a tertbutyloxycarbonyl (Boc) group and is covalently attached to the resin support as a benzyl ester via the chloromethyl group. Side chain functional groups must also be blocked, usually with benzyl-based derivatives. The synthesis depends on the differential sensitivity of these two classes of protecting groups to acid, which is greater than l000:1. The Boc group is completely removed with 50% trifuoroacetic acid in dichloromethane, with minimal loss of the anchoring bond or of the other protecting groups. The resulting α amine salt is neutral with a tertiary amine such as diisopropylethylamine and the free amine of the resin-bound amino acid is then ready to couple with a second Boc-amino acid. The latter must be activated for the reaction to occur. The simplest and most often used procedure is activation with dicyclohexylcarbodiimide, but active esters, anhydrides, and many other activated derivatives have been successfully applied. All of these reactions are carried out under non-aqueous conditions in organic solvents that swell the resin and accelerate the rates. Dichloromethane and dimethylformamide are the solvents of choice.

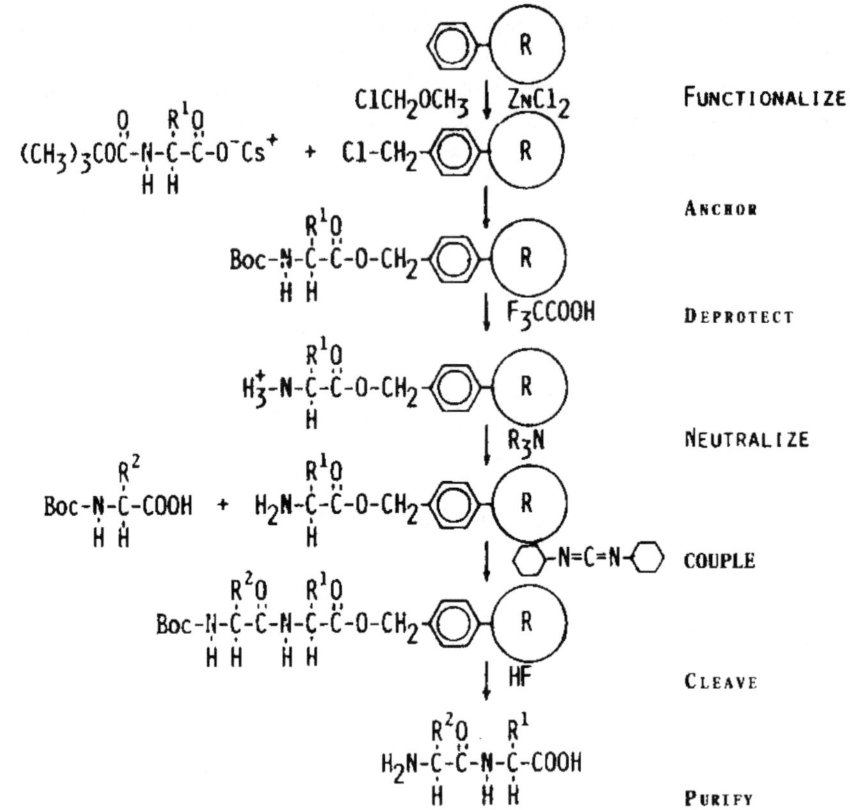

Fig. 3. A scheme for solid phase peptide synthesis.

Reproduced from Merrifield's Nobel Prize lecture [2].

To extend the peptide chain the deprotection, neutralization and coupling steps are repeated for each of the succeeding amino acids until the desired sequence has been assembled. Finally, the completed peptide is deprotected and cleaved from the solid support. With the chemistry described here, this is accomplished by treatment with a strong anhydrous acid such as HF. The free peptide is then purified by suitable procedures.

It is very important that the repetitive steps proceed rapidly, in high yields, and with minimal side reactions in order to prevent the accumulation of excessive amounts of by-products. Much of our effort has been directed toward developing and evaluating these requirements...

The technique of solid-phase peptide synthesis has matured considerably over the last twenty years or so, and is now the standard approach for the routine production of peptides. Many texts have been published outlining advancements and modifications, see for example [7–9].

Honours and Awards

Merrifield received many honorary degrees from various universities including: the University of Colorado (1969); Uppsala University (1970); Yale University (1971); the Medical College of Ohio (1972); Newark College of Engineering (1972); Colgate University (1977); Boston College (1984).

He received many awards and prizes including: the Lasker Award for Basic Medical Research (1969); the Gairdner Award (1970); the Intra-Science Award (1970); the American Chemical Society Award for Creative Work in Synthetic Organic Chemistry (1972); the Nichols Medal (1973); the Instrument Specialties Company Award of the University of Nebraska (1977); the 2nd Alan E. Pierce Award of the American Peptide Symposium (1979); the Nobel Prize for Chemistry (1984); the Hirschmann Award in Peptide Chemistry (1990); the Josef Rudinger Award from the European Peptide Society (1990). He was also presented with the Order of San Carlos (1984).

Merrifield was a Member of: the American Chemistry Society; the American Society of Biological Chemists; the American Institute of Chemists; of the National Academy of Sciences.

He was also a member of the Editorial Board of Analytical Biochemistry.

Bibliography

1. Reproduced with permission.
2. From Nobel Lectures, Chemistry 1981–1990, Editor-in-Charge Tore Frängsmyr, Editor Bo G. Malmström, World Scientific Publishing Co., Singapore, 1992.
3. Who's who: 2005.
4. http://www.rockefeller.edu/emeriti/merrifield/merrifield-lab.html.
5. Merrifield Bruce, Life During a Golden Age of Peptide Chemistry - The Concept and Development of Solid-Phase Peptide Synthesis. Oxford University Press: Oxford, 2001.
6. The Concept and Development of Solid-Phase Peptide Synthesis - presented in the series Profiles, Pathways and Dreams, ed Jeffrey I Seeman, pub American Chemical Society.
7. Fmoc Solid Phase Peptide Synthesis: A Practical Approach (Practical Approach Series) by W. C. Chan and Peter D. White, Oxford University Press, 2000, ISBN: 0199637245.
8. Solid-Phase Synthesis: A Practical Guide by Fernando Albericio, CRC, 2000, ISBN: 0824703596.
9. Optimization of Solid-Phase Combinatorial Synthesis by Bing Yan, CRC, 1st ed. 2002, ISBN: 0824706544.

1985

© Nobel Foundation [1]

Herbert A. Hauptman (0.5)

"for their outstanding achievements in the development of direct methods for the determination of crystal structures" [2]

During his presentation speech, Ingvar Lindqvist of the Royal Academy of Sciences commented of Hauptman and his work [2]:

...It was not, however, possible to determine crystal structures without some assumptions or guesses, because the phase differences between the different scattered X-rays were not known. The crystallographers had to use a trial and error method.

Several methodological improvements have since taken place, but it has for a long time been considered a great scientific achievement to determine the molecular structures of organic molecules as large as penicillin or vitamin B12. As late as 1964 Dorothy Hodgkins was awarded the Nobel Prize in chemistry for such structure determinations.

It therefore was met with great interest and much opposition and discussion when Herbert Hauptman and Jerome Karle during the years 1950–56 published a series of papers in which they claimed to have found a general method, a "direct" method for solving the phase problem, thus opening the possibility to determine the structure directly from the experimental results without any further assumptions. Hauptman and Karle built their method on two established facts. One was that the electron density in a molecule can never be negative - there are electrons or there are not. The other fact was that the number of experimental results is large enough to permit application of statistical methods. Recent developments have shown that they were right and the production of modern computers has strongly contributed to the rapidity and efficiency of their methods. These methods are now so efficient that structure determinations for which the Nobel Prize was awarded in 1964 can today be made by a clever beginner...

> *...Herbert Hauptman and Jerome Karle, Your basic development of the direct methods for X-ray crystallographic structure determination has given the chemists an efficient tool for faster and more detailed studies of the structures of molecules and therefore also for the study of chemical reactions...*

Herbert A. Hauptman was born in New York City on 14th February 1917, the oldest of three sons (the others are Manuel and Robert) of Israel Hauptman and Leah, née Rosenfeld. He married Edith Citrynell on 10th November 1940 and they have two daughters Barbara (1947) and Carol (1950). Hauptman can be contacted via the Hauptman-Woodward Medical Research Inst., Buffalo, NY 14203 [2, 3].

The Early Years

Hauptman has been reading about most areas of science and mathematics since he learned to read. He obtained his B.S. degree in mathematics from the City College of New York (1937) and his M.A. degree in mathematics from Columbia University (1939). Hauptman's career-long collaboration with Jerome Karle (joint winner of the Nobel Chemistry Prize for 1985 – see next entry) started in 1947 at the Naval Research Laboratory in Washington. D.C. Hauptman also decided to enrol at the University of Maryland for a Ph.D. at this time (completed in 1954).

The Career

Year	*Post*	*Institute*
1940	Statistician	Census Bureau
1943, 1946–47	Electronics Instructor	Air Force
1943	Weather Officer	Military Service
1947–70	Physicist-Mathematician	Naval Research Laboratory
1956–70	Representative of University of Maryland Mathematics Dept.	
1965–67	Head, Mathematical Physics Branch	
1967–68	Acting Superintendent, Mathematics & Information Sciences Division	
1968–69	Head, Applied Mathematics Branch, Mathematics Division	
1969–70	Head, Mathematics Staff, Optical Sciences Division	
1956–70	Part-time Professor, Mathematics	University of Maryland
1970–72	Head, Mathematical Biophysics Lab	Hauptman-Woodward Institute
1972	Deputy Research Director	Hauptman-Woodward Institute
1972	Executive Vice President and Research Director	Hauptman-Woodward Institute
1986	President and Research Director	Hauptman-Woodward Institute
1988–92	Member of the General Research Support Review Committee, Division of Research Resources	National Institutes of Health

Hauptman has also had appointments at the State University of New York at Buffalo in the Biophysical Sciences. He was appointed Research Professor (1970) in the Dept. of Computer Science & Engineering and Adjunct Professor (1991) in the Department of Structural Biology and appointed Distinguished Professor (2001) [4,5].

Hauptman's interests have included:

- molecular structures of crystallized materials
- the introduction of the concepts of structure invariants and seminvariants
- developing special linear combinations of the phases
- developing non-centrosymmetric space groups
- providing the link between the observed diffraction intensities with the needed phases of the structure factors
- work on 'the phase problem'
- formulating the neighborhood principle and extension concept
- developing probabilistic theories of the higher order structure invariants and seminvariants
- work on isomorphous replacement and anomalous dispersion
- formulating the phase problem of X-ray crystallography as a minimal principle.

Hauptman has written over 170 publications and three books. They are:

- *Solution of the phase problem. 1. The centrosymmetric crystal* with Jerome Karle, (American Crystallographic Association. ACA monograph), Pub: American Crystallographic Association (1953) ASIN: B0007IU1T0
- *Solution of the phase problem. I. The centrosystemmetric crystal* (ACA monograph), Pub: American Crystallographic Association, (1953), ASIN: B0007G0OSA
- *101 Great Ideas for Introducing Key Concepts in Mathematics: A Resource for Secondary School Teachers* (Paperback) with Alfred S. Posamentier, Pub: Corwin Press, (2000), ISBN: 0761975136

His Nobel Prize lecture is dated 9th December 1985 with the title 'Direct Methods and Anomalous Dispersion'. Hauptman's detailed lecture concluded with:

> *...In this article the goal of integrating the techniques of direct methods with anomalous dispersion is realized. Specifically, the conditional probability distribution of the three-phase structure invariant, assuming as known the six magnitudes in its first neighborhood, is obtained. In the favorable case that the variance of the distribution happens to be small, the distribution yields a reliable estimate of the invariant (the neighborhood principle). It is particularly noteworthy that, in strong contrast to all previous work, the estimate is unique in the whole interval $(-\pi,\pi)$ and that any estimate in this range is possible (even, for example, in the vicinity of $\pm\pi/2$ or π). The first applications of this work using error-free diffraction data have been made, and these show that in a typical case some tens of thousands of three-phase structure invariants may be estimated with unprecedented accuracy, even for a macromolecular crystal structure. Some preliminary calculations on a number of structures, not detailed here, show that the accuracy of the estimates depends in some complicated way on the complexity of the crystal structure, the number of anomalous scatterers, the strength of the anomalous signal, and the range of $\sin\theta/\lambda$. With smaller*

structures the accuracy may be greatly increased, average errors of only three or four degrees for thousands of invariants being not uncommon.

It should be stated in conclusion that the availability of reliable estimates for large numbers of the three-phase structure invariants implies that the traditional machinery of direct methods, in particular the tangent formula, suitably modified to accommodate the non-zero estimates of the invariants, may be carried over without essential change to estimate the values of the individual phases and thus to facilitate structure determination via anomalous dispersion. . .it seems likely that, in time, even macromolecules will prove to be solvable in this way. It is clear, too, that, owing to the ability to estimate both the sine and cosine invariants, that is to say both the signs and magnitudes of the invariants, the unique enantiomorph determined by the observed intensities is automatically obtained. In fact the first application of these techniques using experimental diffraction data has already facilitated the solution of the unknown macromolecular structure Cd, Zn Metallothionein. . .

Many developments concerning the methods for X-ray crystallographic structure determination have been documented over the last 20 years, see for example [6–8].

Honours and Awards

Hauptman has received many honorary degrees from universities including: the University of Maryland (1985); City College of New York (1986); D'Youville College, Buffalo, NY (1989); the University of Parma, Italy (1989); Bar-Ilan Univ. Israel (1990); Columbia University, New York, NY (1990); Technical University of Lodz, Poland (1992); Queen's University, Kingston, Canada (1993); Niagara University, New York (1996); the University of Toledo, Ohio (1999).

He has been presented with many awards and prizes including: the Belden Prize (Gold Medal), Mathematics (1935); the Scientific Res. Society of America, Pure Science Award, Naval Research Laboratory (1959); the Patterson Award, jointly with Jerome Karle (1984); the Nobel Prize in Chemistry, jointly with Jerome Karle (1985); the Gold Plate Award, American Academy of Achievement, Salute to Excellence Weekend, Washington, D.C. (1986); the National Library of Medicine Medal (1986); the Norton Medal, SUNY (1986); the Schoellkopf Award, American Chemical Society (1986); the Townsend Harris Medal for 1986, City College of New York (1986); the Western New York Man of the Year Award, Buffalo Chamber of Commerce (1986); the Citizen of the Year Award, Buffalo Evening News (1986); the Recipient of Medal from Jewish Academy of Arts and Sciences (1986); the Inducted into Nobel Hall of Science, Museum of Science and Industry, Chicago, Illinois (1986); the Western New York Man of the Year Award - Buffalo Chamber of Commerce (1986); the Establishment of the Eccles-Hauptman Student Award - SUNY at Buffalo (1987); the Cooke Award - State University of New York (SUNY) at Buffalo (1987); the 1987 Honoree, Western New Yorker of the Year (1987); Elected Member of Townsend Harris Hall of Fame (1989); the Dirac Medal for the Advancement of Theoretical Physics, University of New South Wales, Australia (1991); the President's Award - Annual Health Care Industries Association Award for Excellence in Western New York Health Care (1994); the City of Buffalo - Citizen of Distinction (1995); the Health Care Heroes Award for Lifetime Achievement - Business First (2000).

Hauptman is a Fellow of the Washington Academy of Sciences and of the American Association for the Advancement of Science. He is a Member of: the National Academy of Sciences; the Polish National Academy of Sciences; the New York Academy of Sciences; the American Crystallographic Association; the American Mathematical Society; the American Physical Society Member, Endocrine

Society; the Mathematical Association of America; the Cosmos Club; the Sigma Xi, Buffalo, NY Chapter.

He was inducted as a Fellow of the Jewish Academy of Arts and Sciences, was President of the Philosophical Society of Washington (1969–70) and of the Association of Independent Research Institutes (1979–80). He is an Honorary Member Phi Beta Kappa (1987). He was appointed Chairman of the Board of Directors, NYS Institute on Superconductivity (1988) and a consultant at Prometheus, Inc. (1990).

Bibliography

1. Reproduced with permission.
2. From Nobel Lectures, Chemistry 1981–1990, Editor-in-Charge Tore Frängsmyr, Editor Bo G. Malmström, World Scientific Publishing Co., Singapore, 1992.
3. Who's who: 2005.
4. http://www.hwi.buffalo.edu/.
5. http://www.hwi.buffalo.edu/personnel/Scientist/Hauptman/Hauptman.html.
6. Crystal Structure Determination by Werner Massa and R.O. Gould (Translator), Springer, 2nd ed. 2004, ISBN: 3540206442.
7. Structure Determination by X-Ray Crystallography by Mark F.C. Ladd and Rex A. Palmer, Springer, 4th ed. 2003, ISBN: 0306474549.
8. Current Methods and Optimization Algorithms for the Refinement of X-Ray Crystal Structures by J. URIA, J.F. Van der Maelen Uria and M. Moore, Taylor & Francis, 1999, ISBN: 9056992511.

1985

© Nobel Foundation [1]

Jerome Karle (0.5)

"for their outstanding achievements in the development of direct methods for the determination of crystal structures" [2]

During his presentation speech, Ingvar Lindqvist then of the Royal Academy of Sciences, commented of x-ray chrystalography in general and Karle's work in particular [2]:

...It therefore was met with great interest and much opposition and discussion when Herbert Hauptman and Jerome Karle during the years 1950–56 published a series of papers in which they claimed to have found a general method, a "direct" method for solving the phase problem, thus opening the possibility to determine the structure directly from the experimental results without any further assumptions. Hauptman and Karle built their method on two established facts. One was that the electron density in a molecule can never be negative-there are electrons or there are not. The other fact was that the number of experimental results is large enough to permit application of statistical methods. Recent developments have shown that they were right and the production of modern computers has strongly contributed to the rapidity and efficiency of their methods. These methods are now so efficient that structure determinations for which the Nobel Prize was awarded in 1964 can today be made by a clever beginner.

At the same time it has been more and more important for the chemists to know the exact structures of molecules which take part in important chemical and biochemical reactions. One could without exaggeration say that it is only in the last ten years that chemistry has developed into a truly molecular era. Molecules with desired structures and properties can be produced and the molecular mechanism is known for increasingly more reactions.

It is this importance to chemistry which has motivated a Nobel Prize in chemistry to the mathematician Herbert Hauptman and the physicist Jerome Karle. Another way to express it is, that the imagination and ingenuity of the laureates have made it unnecessary to exercise these qualities in normal structure determinations. On the other hand they have increased the possibilities for the chemists to use their imagination and their ingenuity.

Herbert Hauptman and Jerome Karle, Your basic development of the direct methods for X-ray crystallographic structure determination has given the chemists an efficient tool for faster and more detailed studies of the structures of molecules and therefore also for the study of chemical reactions...

Jerome Karle was born in New York City on 18th June 1918, the son of Louis Karfunkle and Sadie Helen née Kun. He married Isabella Lugoski in 1942 and they had three daughters Louise (1946), Jean (1950) and Madeleine (1955). He can be contacted through the Naval Research Laboratory, Washington [2,3].

The Early Years

Karle attended schools in the New York City public school system, his high school being Abraham Lincoln. He entered the City College of New York (1933), graduating with a BS degree in 1937. Karle moved to Harvard University where he studied biology, receiving his master's degree in 1938. He then went to work with the New York State Health Department in Albany prior to entering the Chemistry Department of the University of Michigan (1940) under Lawrence O. Brockway (one of the first graduate students of Linus Pauling), where he worked on the molecular structure of gaseous molecules for his MS (1942) and Ph.D., graduating in 1944. The practical work for his thesis was completed by the summer of 1943, and so Karle spent sometime, from the summer of 1943, working on the Manhattan Project at the University of Chicago where he was joined by his wife, Isabella, who also worked on the Manhattan Project. They returned to the University of Michigan (1944), where Karle undertook project work for the Naval Research Laboratory concerning structure of monolayers of long-chain hydrocarbon films involved in the boundary lubrication of metallic surfaces. Both husband and wife moved to Washington where they both started to work for the Naval Research Laboratory (1946).

The Career

Year	*Post*	*Institute*
1946–58	Head, Electron Diffraction Section	Naval Research Laboratory
1958–68	Head of Diffraction Branch	Naval Research Laboratory
1951–70	Professor (part time)	University of Maryland
1968	Chief Scientist, Laboratory for the Structure of Matter	Naval Research Laboratory

Karle's research interests have involved:

- diffraction theory and its application
- new techniques for structure determination
- developing the quantitative aspects of gas electron diffraction analysis
- the development of the direct methods for crystal structure analysis
- structure determinations for centrosymmetric crystals
- developing a procedure for crystal structure determination

- the symbolic addition procedure
- direct phase determinations of crystal structures (published in 1964)
- the derivation of a variance formula
- gas electron diffraction
- theoretical studies of internal rotation
- coherent diffraction associated with excitation processes
- theoretical work in crystal structure analysis that included the derivation of a "tangent formula" for phase determination
- derived heuristic joint probability distributions
- developing procedures for determining atomic arrangements in amorphous materials (with John Konnert and Peter D'Antonio)
- the development of an exact, linear algebraic theory that includes any number and type of anomalous scatterer and any number of wavelengths

Karle published the following book with Herbert A. Hauptman in 1953: *Solution of the phase problem: 1. The centrosymmetric crystal*, (American Crystallographic Association. ACA monograph), Pub: American Crystallographic Association, (1953), ASIN: B0007IU1T0.

Karle's Nobel Prize lecture is dated 9th December 1985 with the title 'Recovering phase information from intensity data'. Karle introduced his presentation thus [2]:

> *...The concept of a crystal is that of a solid body in which the atomic or molecular units are so arranged as to form an array having three dimensional periodicity. Because of the periodicity, it is possible to describe the arrangements of the atomic composition by means of Fourier series. The type of Fourier series that is used in crystal structure analysis represents the electron density distribution in a crystal This is indeed equivalent to representing the structure of a crystal since the atomic locations are represented by the regions of highest electron density in the electron distributions.*
>
> *The experimental technique used for examining the structure of crystals is called diffraction. In a diffraction experiment, rays are made to impinge on a crystalline substance of interest and, given the proper geometric conditions, the rays are scattered as if they were bouncing off large numbers of different planes imagined to be cutting through the crystal. The collected intensities of scatter(often 5000-10000 in number) are called a scattering pattern or diffraction pattern and comprise the experimental data from which the structure of the crystal of interest is to be elucidated. The most commonly used rays are Roentgen rays or X-rays, as they are usually called. Other rays, composed of neutrons or electrons, also have their purposes...*
>
> *...As this article proceeds, it will elaborate on a number of the items discussed in this introductory part, describe some interesting applications and discuss briefly some research paths and opportunities for the future...*

Karle broke down his presentation into the following topics:

- Electron Density Distribution
  - Overdeterminacy
  - Some attempts, some successes
  - Non-negativity and general inequalities

- Formulas for phase determination
  - Noncentrosymmetric drystals
- Practical Phase Determination
- Applications
  - Early applications of the symbolic addition procedure
  - Identification and stereoconfiguration
- Rearrangements
- Conformation
- Analysis of macromolecular structure: Recent Interests

Karle concluded:

> *...The development of optimal strategies for the use of triplet phase invariants awaits practical experience. If the heavy-atom (anomalously scattering) substructure is known, there are a number of techniques for obtaining values for individual phases... In that case, it is quite possible that the triplet phase invariants would be used mainly for phase extension and refinement. In those circumstances when the heavy-atom (anomalously scattering) substructure is not known and resists analysis, the triplet phase invariants may be able to play a primary role in reaching a solution to the structure since the triplet phase invariants can be evaluated without knowledge of the substructure.*
>
> *The enhanced opportunities to carry out multiple wavelength anomalous dispersion experiments with high intensity and tunable synchrotron x-ray sources should facilitate the development of applications of the new theoretical formulations. Coupled with the major advances in computing, the new theories have the potential to improve the facility with which macromolecular structure research is performed...*

Many developments concerning the methods for x-ray crystallographic structure determination have been documented over the last 20 years, see for example [4–6].

Honours and Awards

Karle's awards include the Nobel Prize in Chemistry, jointly with Herbert A. Hauptman (1985).

He is a Fellow of American Physical Society, a Member of the National Academy of Sciences and of the American Philosophical Society. Karle is a member of numerous other professional societies as well.

Karle has served as President of the International Union of Crystallography and as Chairman of the Chemistry Section of the National Academy of Sciences (1973–75).

He was a Professorial Lecturer in the University College of the University of Maryland and a Visiting Professor at the University of Kiel in Germany.

Bibliography

1. Reproduced with permission.
2. From Nobel Lectures, Chemistry 1981–1990, Editor-in-Charge Tore Frängsmyr, Editor Bo G. Malmström, World Scientific Publishing Co., Singapore, 1992.
3. Who's who: 2005.
4. Crystal Structure Determination by Werner Massa and R.O. Gould (Translator), Springer, 2nd ed. 2004, ISBN: 3540206442.
5. Structure Determination by X-Ray Crystallography by Mark F.C. Ladd and Rex A. Palmer, Springer, 4th ed. 2003, ISBN: 0306474549.
6. Current Methods and Optimization Algorithms for the Refinement of X-Ray Crystal Structures by J. URIA, J.F. Van der Maelen Uria and M. Moore, Taylor & Francis, 1999, ISBN: 9056992511.

1986

© Nobel Foundation [1]

Dudley R. Herschbach (0.33)

"for their contributions concerning the dynamics of chemical elementary processes" [2]

During his presentation speech, Sture Forsén, then of the Royal Academy of Sciences, commented of Herschbach and his coworkers [2]:

> *...Let us assume that we choose for our study only one of the many chemical reactions that take place in a flame. Superficially this reaction may appear very simple and easy to understand. At a closer look, however, we find that Nature is elusive and complexity still prevails. Many difficulties await him who would like to study chemical reactions in their most intimate molecular details. The reaction event proper is a molecular drama that takes place under an exceedingly short time span - of the order of a millionth of a millionth of a second-times scientists refer to as a "picosecond". How is it at all possible to obtain detailed information of what goes on under such short time? Most of our knowledge has been gained through a deliberate simplification of the reacting system and through a strict control of the conditions of the reaction. Furthermore our knowledge is to a large extent indirect and based upon a detailed analysis of the initial conditions as well as of the results of the reaction event...*
>
> *...This year's Nobel prize winners of chemistry have through their brilliant work in a decisive way enlarged our knowledge of the detailed events in chemical reactions. Reactions between molecules have been studied at low pressures by letting beams of molecules and/or atoms meet at one point in space. The energy of the reacting molecules or atoms has been controlled and the properties of the products formed - their chemical composition, their angular distribution from the place of collision, their speed and their rotational and vibrational energy-have been studied. Through experiments of this kind the prizewinners of this year have been able to paint a very detailed picture of the molecular drama occurring between the opening scene and the finale...*

...Professor Herschbach, Professor Lee and Professor Polanyi. Your brilliant research into the finer molecular details of chemical reactions has greatly advanced our knowledge in this central area of chemistry. Your approaches to the problem have differed in detail but your goals have been the same. You have, in a way that is truly admirable, combined extraordinary experimental skill with deep theoretical insight. In recognition of your services to chemistry and to natural science as a whole the Royal Academy of Sciences has decided to confer upon you this year's Nobel Prize for Chemistry...

Dudley R. Herschbach was born in San Jose, California on 18th June 1932, the first of six children of Robert and Dorothy Herschbach. He married Georgene Botyos (1964). They have two daughters, Lisa and Brenda. Herschbach is the current Frank B. Baird, Jr. Professor of Science at Harvard University and can be contacted through the Department of Chemistry at 12 Oxford Street, Cambridge, MA 02138 [2,3,4].

The Early Years

Herschbach's boyhood was spent a few miles outside San Jose. He attended Campbell High School and took all the science and mathematics courses offered. He entered Stanford University (1950) where Harold Johnston inspired the young Herschbach with his passion for chemical kinetics. He received his B.S. in mathematics in 1954 and the M.S. in chemistry in 1955 under Johnston with a thesis titled: "*Theoretical Pre-exponential Factors for Bimolecular Reactions.*" He continued his graduate studies at Harvard, receiving the A.M. in Physics (1956) and his Ph.D. in Chemical Physics (1958). His doctoral thesis, under E. Bright Wilson, Jr., was titled: "*Internal Rotation and Microwave Spectroscopy*".

The Career

Year	*Post*	*Institute*
1957	Junior Fellow, Society of Fellows	Harvard University
1959	Assistant Professor	University of California
1961	Associate Professor	University of California
1963	Professor of Chemistry	Harvard University
1976	Frank B. Baird, Jr. Professor of Science	Harvard University

Herschbach was a Guggenheim Fellow at Freiburg University (1968) and a Sherman Fairchild Scholar at the California Institute of Technology (1976). He has been active in various departmental posts during his career including: Chairman of the Chemical Physics program (1964–1977); Chemistry Department (1977–1980); Member of the Faculty Council (1980–1983); Co Master with his wife of Currier House (1981–1986). He serves as Chair of the Board of Trustees of Science Service, which publishes Science News and conducts the Intel Science Talent Search, and the Intel International Science and Engineering Fair at Harvard University.

Herschbach's research has covered topics such as:

- the reactions of alkali atoms with alkyl iodides
- prototype modes of reaction dynamics which could be correlated with the electronic structure of the target molecule
- the construction of a "supermachine" designed to extend the scope of crossed-beam experiments (with Lee)
- the simultaneous measurement of three or four vector properties of reactive collisions
- liquid-phase reactions by solvating reactant molecules during a supersonic expansion
- the determination of solute-solvent intermolecular forces
- theoretical studies of intermolecular forces
- rotor assisted molecular cooling
- strongly-correlated many-particle systems.

Herschbach has served as an Associate Editor of the Journal of Physical Chemistry (since 1980). He has published over 400 papers, and written the following books:

- *Tables for the Internal Rotation Problem*, Pub: Harvard University, (1957), ASIN: B0007FE-XME
- *Molecular Symmetry, Structure and Spectra*, Pub: World Scientific Publishing Company, 1st edition (2003), ISBN: 9810217749
- *Molecular Collisions Chem Physics*, Pub: World Scientific Publishing Company, 1st edition, (2004), ISBN: 9810217978.

Herschbach's Nobel Prize lecture is dated 8th December 1986 with the title 'Molecular dynamics of elementary chemical reactions'. Herschbach commented on the emergence of chemical dynamics [2]:

> *...Over the past 30 years a new level has been attained by study of the intimate molecular dynamics of individual reactive collisions. One of the chief experimental approaches is molecular beam scattering. This involves forming the reagent molecules into two collimated beams, each so dilute that collisions within them are negligible. The two beams intersect in a vacuum and the direction and velocity of the product molecules emitted from the collision zone are measured. A host of reactions can now be studied readily in this way - by virtue of an extremely sensitive mass spectrometric detector and the use of supersonic nozzles which generate beams with greatly enhanced intensity and with collision energies much higher than provided by ordinary thermal sources. The coupling of spectroscopic techniques with molecular beams has provided further advances in selectivity and sensitivity, particularly by use of intense laser light sources.*
>
> *The reaction properties now accessible include the disposal of energy among translation, rotation, and vibration of the product molecules: angles of product emission; angular momentum and its orientation in space: and variation of reaction yield and other attributes with impact energy, closeness of collision, rotational orientation or vibrational excitation of the target molecule. From the earliest stages these experiments have stimulated and responded to a vigorous outburst of theoretical developments. Especially helpful are computer simulations in which dynamical properties are predicted for various postulated forces by Monte Carlo sampling of huge numbers of calculated collisions trajectories. Both the laboratory experiments and computer simulations have prompted a variety of insightful mechanical models as well as very useful diagnostic procedures based on information theory. Ab initio electronic structure calculations have also begun to*

contribute significantly to the exploration of reaction dynamics, although (except for $H + H_2$) satisfying overall accuracy has yet to be achieved for potential energy surfaces…

Pursuit of Vector Correlations

…In addition to the product angular distribution, many other directional or vector properties of reactions are now accessible in beam experiments, especially by means of laser-induced fluorescence techniques developed by Zare which exploit the orientation dependence of the dipole selection rule for electronic excitation. These vector properties offer much information not provided by energetic or other scalar properties, and their study is now emerging as a vigorous field referred to as dynamical stereochemistry or stereodynamics…using laser-induced fluorescence, especially since the great sensitivity of the method may allow the alignment of individual rotation-vibration states to be measured as a function of the scattering angle…

The concepts of reaction dynamics have been applied to many systems, see for example [5,6,7].

Honours and Awards

Herschbach has been presented with honorary degrees form many universities and institutes including: the University of Toronto (1977); Dartmouth College (1992); Charles University, Prague (1993); Wheaton College (1995); Franklin and Marshall College (1998).

He has received many awards and prizes including: the Pure Chemistry Prize of the American Chemical Society (1965); the Linus Pauling Medal (1978); the Michael Polanyi Medal (1981); the Irving Langmuir Prize of the Americal Physical Society (1983); the Nobel Prize in Chemistry, jointly with Yuan T. Lee and John C. Polanyi (1986); the National Medal of Science (1991); the Jaroslav Heyrovsky Medal (1992); the Sierra Nevada Distinguished Chemist Award (1993); the Kosolapoff Award of the ACS (1994); the William Walker Prize (1994); the Council of Scientific Society President's Award for Support of Science (1999).

Herschbach is a Fellow of: the American Academy of Arts and Sciences (1967); the National Academy of Sciences (1967); the American Philosophical Society; the Royal Chemical Society of Great Britain.

He was Visiting Professor at Göttingen University (1963); Visiting Fellow of the Joins Institute of Laboratory Astrophysics (1969). He was named by Chemical Engineering News among 75 leading contributors to the chemical enterprise in the past 75 years (1998).

Herschbach served as a consultant to Aerodyne Corporation, the Fluorocarbon Research Panel, and Los Alamos National Laboratory. He was appointed an Exxon Faculty Fellow in 1981 and regularly visited the Corporate Research Laboratory in New Jersey to participate in projects.

Bibliography

1. Reproduced with permission.
2. From Nobel Lectures, Chemistry 1981–1990, Editor-in-Charge Tore Frängsmyr, Editor Bo G. Malmström, World Scientific Publishing Co., Singapore, 1992.
3. Who's who: 2005.
4. http://www.chem.harvard.edu/herschbach/dudley.php.

5. Chemical Kinetics and Reaction Dynamics by Paul L. Houston, McGraw-Hill Science/Engineering/Math, 2001, ISBN: 0072435372.
6. Molecular Reaction Dynamics by Raphael D. Levine, Cambridge University Press, 2005, ISBN: 052184276X.
7. Modern Trends in Chemical Reaction Dynamics: Experiment and Theory (Advanced Series in Physical Chemistry-Vol. 14) edited by Xueming Yang and Kopin Liu, World Scientific Publishing Company, 2004, ISBN: 9812385681.

1986

© Nobel Foundation [1]

Yuan T. Lee (0.33)

"for their contributions concerning the dynamics of chemical elementary processes" [2]

During his presentation speech, Sture Forsén, then of the Royal Academy of Sciences, commented of Lee and his coworkers [2]:

> *...Let us assume that we choose for our study only one of the many chemical reactions that take place in a flame. Superficially this reaction may appear very simple and easy to understand. At a closer look, however, we find that Nature is elusive and complexity still prevails. Many difficulties await him who would like to study chemical reactions in their most intimate molecular details. The reaction event proper is a molecular drama that takes place under an exceedingly short time span - of the order of a millionth of a millionth of a second-times scientists refer to as a "picosecond". How is it at all possible to obtain detailed information of what goes on under such short time? Most of our knowledge has been gained through a deliberate simplification of the reacting system and through a strict control of the conditions of the reaction. Furthermore our knowledge is to a large extent indirect and based upon a detailed analysis of the initial conditions as well as of the results of the reaction event...*

> *...This year's Nobel prize winners of chemistry have through their brilliant work in a decisive way enlarged our knowledge of the detailed events in chemical reactions. Reactions between molecules have been studied at low pressures by letting beams of molecules and/or atoms meet at one point in space. The energy of the reacting molecules or atoms has been controlled and the properties of the products formed - their chemical composition, their angular distribution from the place of collision, their speed and their rotational and vibrational energy - have been studied. Through experiments of this kind the prizewinners of this year have been able to paint a very detailed picture of the molecular drama occurring between the opening scene and the finale...*

...Professor Herschbach, Professor Lee and Professor Polanyi. Your brilliant research into the finer molecular details of chemical reactions has greatly advanced our knowledge in this central area of chemistry. Your approaches to the problem have differed in detail but your goals have been the same. You have, in a way that is truly admirable, combined extraordinary experimental skill with deep theoretical insight. In recognition of your services to chemistry and to natural science as a whole the Royal Academy of Sciences has decided to confer upon you this years Nobel Prize for Chemistry...

Yuan Tseh Lee was born on 19th November 1936 in Hsinchu, Taiwan, the son of T'se Fan Lee (an artist) and Pei Tsai (a schoolteacher). Lee met his wife, Bernice Chinli Wu, in primary school. They have two sons, Ted (1963) and Sidney (1966) and one daughter Charlotte (1969). Lee is currently the President of the Academia Sinica in Taiwan and can be contacted through that organization [2,3,4].

The Early Years

Lee read a biography of Madame Marie Curie at an early age which started his interest in science. His primary and secondary education in Hsinchu, which was under Japanese occupation, was frequently disrupted during world war bombing raids, forcing the people to relocate to the mountains. Nevertheless, he completed high school with extraordinarily high marks and was able to enter the National Taiwan University (1955) without taking the usual entrance exam. He graduated with a B.Sc. degree (1959) and enrolled at the National Tsinghua University (1959) for his master's degree concerning the natural radioisotopes contained in Hukutolite, a mineral of hot spring sediment under Professor H. Hamaguchi. Graduation in 1961 with an M.Sc. was followed by military service which disrupted his education (1961–1962). Following military service, he enrolled at the University of California at Berkeley (1962) under B. Mahan, concerning chemiionization processes of electronically excited alkali atoms. He stayed in Mahan's group after his graduation with a Ph.D. (1965), during which time he worked on ion molecule reactive scattering experiments, using ion beam techniques measuring energy and angular distributions (with Ron Gentry).

The Career

Year	*Post*	*Institute*
1968	Assistant Professor of Chemistry	James Frank Institute of the University of Chicago
1971	Associate Professor	James Frank Institute of the University of Chicago
1973	Professor of Chemistry	James Frank Institute of the University of Chicago
1974	Professor of Chemistry	University of California, Berkeley
1991	University Professor of Chemistry	University of California, Berkeley
1994–	President	Academia Sinica, Taiwan

Lee was also Principal Investigator of the Materials and Molecular Research Division, Laurence Berkeley Laboratory (1974–94). He was Alfred P. Sloan Fellow (1969–1971), Camille and Henry Dreyfus Foundation Teacher Scholar Grant Receipient (1971–1974) and also a John Simon Guggenheim Fellow (1976–1977).

Lee's academic interests lay in the fields of:

- the construction of a new generation state-of-the-art crossed molecular beams apparatus
- the study of reaction dynamics
- photochemical processes
- molecular spectroscopy

Current interests:

- various political and academic issues
- furthering the cause of democracy
- improving the education system

Lee's Nobel Price lecture is dated 8th December 1986 with the title 'Molecular beam studies of elementary chemical processes'. Lee commented on crossed molecular beam experiments and in particular, the measurements of angular and velocity distributions of products thus [2]:

...If the motion of individual atoms were observable during reactive collisions between molecules, it would be possible to understand exactly how a chemical reaction takes place by just following the motion of these atoms. Unfortunately, despite recent advances in microscope technology that allow us to observe the static arrangement of atoms in a solid, we are still far from being able to follow the motion of atoms in the gas phase in real time. The idea of crossed molecular beams experiments is in a sense to "visualize" the details of a chemical reaction by tracing the trajectories of the reaction products. This is done by first defining the velocities, approach angle, and other initial conditions of the reactants, and then measuring the velocity and angular distribution of the products...

...There are two sources of background molecules in the detector that one has to deal with in a crossed molecular beams experiment: the inherent background in the detector chamber and the background caused by the effusion of molecules from the collision chamber into the detector when the beams are on. The former is mainly due to outgassing from the materials used for the construction of the chamber and to limitations imposed by the performance of the ultrahigh vacuum pumping equipment. Reduction of the latter requires many stages of differential pumping using buffer chamber...

...Fortunately there is a way to reduce this background without substantially increasing the pumping speed in the collision chamber. Recognizing that at a pressure of 10 torr the mean free path between molecular collisions in the collision chamber is more than 100 meters, which is two orders of magnitude larger than the size of a typical scattering apparatus, one realizes that almost all of the "straight through" background will come from those molecules that bounce off the surface which is in the line-of-sight of the detector, and not from gas phase collisions that occur in the viewing window of the detector. Installing a small liquid helium cooled surface opposite the detector and behind the collision region, such that the detector line-of-sight always faces a cold surface will help to eliminate this background since the surface will trap essentially all condensable molecules that impinge on it...

...In the investigation of reaction dynamics, lasers have become increasingly important. Not only are they used extensively for the preparation of reagents and quantum state specific detection of

products, but they have also become indispensable for the investigation of the dynamics and mechanisms of photo-chemical processes.

One of the more exciting application of lasers in crossed molecular beams experiments is the control of the alignment and orientation of electronically excited orbitals before a reactive encounter. For example, in the reaction of Na with 0_2, *if linear polarized dye lasers are used to sequentially excite Na atoms from the 3S to 3P to 4D states, the electronically excited 4D orbital can be aligned along the polarization direction of the electric field vector of the lasers. Consequently the effect of the alignment of the excited orbital on chemical reactivity can be studied in detail by simply rotating the polarization of the lasers with respect to the relative velocity vector.*

For many atom-molecule reactions that proceed directly without forming long-lived complexes, for example,

$$K+CH_3I, \qquad F+H_2 \text{ and } D_2 \qquad Na(4D)$$

the dependence of chemical reactivity on the molecular orientation can be obtained from measurements of product angular distributions. For symmetric top moleculars, control of molecular orientation in the laboratory frame is possible, and careful investigations of the orientation dependence of chemical reactivity have been carried out for many systems. The combination of laser induced alignment of excited atomic orbitals and measurements of product angular distributions provide the first opportunity for the detailed experimental probing of the correlation between the alignment of the excited atomic orbital and the orientation of the molecule in a reactive encounter between an atom and a molecule...

The concepts of reaction dynamics have been applied to many systems, see for example [5,6,7].

Honours and Awards

Lee has been awarded a Honorary D.Sc. from the University of Waterloo (1986).

He has been presented with many awards and prizes including: the Ernest O. Lawrence Award U.S. Department of Energy (1981); the Harrison Howe Award, Rochester Section, American Chemical Society (1983); the Sherman Fairchild Distinguished Scholar, California Institute of Technology (1983); the National Medal of Science (1986); the Nobel Prize for Chemistry, jointly with Dudley R. Herschbach and John C. Polanyi (1986); the Peter Debye Award of Physical Chemistry, American Chemical Society (1986).

He is a Fellow of the American Academy of Arts and Science (1975) and of the American Physical Society (1976). He is also a Member of: the National Academy of Sciences (1979); the Academia Sinica, Taiwan, China (1980); of the National Academy of Sciences (1979).

Lee was Honorary Professor of the Institute of Chemistry, Chinese Academy of Science, Beijing, China (1980), and also of Fudan University, Shanghai, China (1980). He held the Miller Professorship at the University of California, Berkeley, California (1981–1982). Lee was also Fairchild Distinguished Scholar (1983) and Honorary Professor of the Chinese University of Science and Technology, Hofei, Anhuei, China (1986).

Lee became a naturalized U.S. Citizen (1974), which he later renounced (1994), on his return to Taiwan.

Bibliography

1. Reproduced with permission.
2. From Nobel Lectures, Chemistry 1981–1990, Editor-in-Charge Tore Frängsmyr, Editor Bo G. Malmström, World Scientific Publishing Co., Singapore, 1992.
3. Who's who: 2005.
4. http://chem.berkeley.edu/people/emeriti/lee.html.
5. Chemical Kinetics and Reaction Dynamics by Paul L. Houston, McGraw-Hill Science/Engineering/Math, 2001, ISBN: 0072435372.
6. Molecular Reaction Dynamics by Raphael D. Levine, Cambridge University Press, 2005, ISBN: 052184276X.
7. Modern Trends in Chemical Reaction Dynamics: Experiment and Theory (Advanced Series in Physical Chemistry - Vol. 14) edited by Xueming Yang and Kopin Liu, World Scientific Publishing Company, 2004, ISBN: 9812385681.

1986

John C. Polanyi (0.33)

"for their contributions concerning the dynamics of chemical elementary process" [2]:

During his presentation speech, Sture Forsén, then of the Royal Academy of Sciences, commented of Polanyi and his coworkers [2]:

...Let us assume that we choose for our study only one of the many chemical reactions that take place in a flame. Superficially this reaction may appear very simple and easy to understand. At a closer look, however, we find that Nature is elusive and complexity still prevails. Many difficulties await him who would like to study chemical reactions in their most intimate molecular details. The reaction event proper is a molecular drama that takes place under an exceedingly short time span - of the order of a millionth of a millionth of a second-times scientists refer to as a "picosecond". How is it at all possible to obtain detailed information of what goes on under such short time? Most of our knowledge has been gained through a deliberate simplification of the reacting system and through a strict control of the conditions of the reaction. Furthermore our knowledge is to a large extent indirect and based upon a detailed analysis of the initial conditions as well as of the results of the reaction event...

...This year's Nobel prize winners of chemistry have through their brilliant work in a decisive way enlarged our knowledge of the detailed events in chemical reactions. Reactions between molecules have been studied at low pressures by letting beams of molecules and/or atoms meet at one point in space. The energy of the reacting molecules or atoms has been controlled and the properties of the products formed - their chemical composition, their angular distribution from the place of collision, their speed and their rotational and vibrational energy-have been studied. Through experiments of this kind the prizewinners of this year have been able to paint a very detailed picture of the molecular drama occurring between the opening scene and the finale...

...Professor Herschbach, Professor Lee and Professor Polanyi. Your brilliant research into the finer molecular details of chemical reactions has greatly advanced our knowledge in this central area of chemistry. Your approaches to the problem have differed in detail but your goals have been the same. You have, in a way that is truly admirable, combined extraordinary experimental skill with deep theoretical insight. In recognition of your services to chemistry and to natural science as a whole the Royal Academy of Sciences has decided to confer upon you this years Nobel Prize for Chemistry...

John Charles Polanyi was born on 23rd January 1929 in Berlin, Germany of Hungarian parents, Michael and Magda Elizabeth Polanyi. The family moved to England in 1933 where he received his education. He married Anne (Sue) Ferrar Davidson in 1958. They have two children, Margaret Alexandra (1961), and Michael Ferrar (1963). The Honourable Charles Polanyi is the University Professor and Professor of Chemistry at the University of Toronto and can be contacted through the Department of Chemistry in St. George Street, Toronto [2, 3, 4].

The Early Years

Polanyi attended Manchester Grammar School where his interests were in history, politics and science. He decided to study chemistry at Manchester University obtaining his B.Sc. in 1949, and his Ph.D. in 1952. He was a Postdoctoral Fellow at the National Research Council Laboratories in Ottawa, Canada (1952–1954) and a Research Associate at Princeton University (1954–1956). He was awarded a D.Sc. in 1964.

The Career

Year	*Post*	*Institute*
1954–1956	Research Associate	Princeton University
1956–1957	Lecturer	University of Toronto
1957–1960	Assistant Professor	University of Toronto
1960–1962	Associate Professor	University of Toronto
1962	Professor	University of Toronto
January 1974–present	University Professor	University of Toronto

Polanyi was a Sloan Foundation Fellow (1959–1963) and a Guggenheim Memorial Fellow (1979–1980).

His research interests have involved:

- the motion of hydrogen chloride molecules, newly born in the reaction of atomic hydrogen with molecular chlorine
- the establishment of a new field called 'reaction dynamics' - the study of the motions of molecules in the course of the molecular collisions
- gaseous reaction dynamics by measurement of state-specified rates
- 'transition state spectroscopy'

- surface aligned photochemistry
- the molecular motions in chemical reactions in gases and at surfaces

His non-research interests, for which he has published over 100 articles include:

- science policy
- the control of armaments
- the impact of science on society

He has published 245 papers (up to 2005) and produced a film titled '*Concepts in Reaction Dynamics*' (1970) and is co-author of *The Dangers of Nuclear War: A Pugwash Symposium* with Franklyn Griffiths Pub: Univ of Toronto Pr (1979) ISBN: 0802063896.

Polanyi's Nobel Prize lecture is dated 8th December 1986 with the title 'Some concepts in reactions dynamics'. During the course of his presentation, Polanyi described some of the methods that have been used during the course of his work. The result of these techniques are further summarized in his lecture [2]:

> *...Our principal experimental method has been the study of infrared chemiluminescence. The most valuable theoretical tool has been, and remains, the computer-integration of the classical equations of motion. Though both of these methods belong to the modern period of reaction dynamics, both have clear antecedents in earlier times.*

> *...Infrared chemiluminescence derives from the presence of vibrationally excited molecules in the products of reaction. Indirect evidence for the existence of such species was obtained by M. Polanyi and his co-workers in 1928 in the course of studies of the reactions of alkali metal atoms and halogens. Following a suggestion by Bates and Nicolet, and by Herzberg, McKinley, Garvin, and Boudart looked for and found visible emission from vibrationally excited hydroxyl radicals formed in the reaction of atomic hydrogen with ozone. This finding was followed soon after by the identification, by absorption spectroscopy of vibrational excitation in the products of the reactions of atomic oxygen with* NO_2 *and* ClO_2*...*

> *...In closing I mention two further approaches which could assist materially in the quest for understanding of the choreography of chemical reaction. In the first, attempts are being made to observe the molecular partners while they are, so to speak, on the stage, rather than immediately prior to and following the reactive dance. This we term "transition state spectroscopy" (TSS); it is a young but burgeoning field. In the second novel approach, the intention, stated a little grandiosely, is to have a hand in writing the script according to which the dynamics occurs; the reagents are aligned by the forces at a crystal surface and held in a fixed arrangement immediately prior to the initiation of the reaction by light, thus restricting the subsequent pattern of motion. This field of "surface aligned photochemistry" (SAP) has only very recently made its appearance. SAP may initiate reaction part way across the pes and hence is related to TSS since the adsorbed reactants can be held together in configurations that locate them at the outset part way up the reactive barrier. In the SAP experiments performed till now the forces holding the reactants to the surface have been physisorptive...*

The concepts of reaction dynamics have been applied to many systems, see for example [5, 6, 7].

Honours and Awards

Polanyi has been presented with many honorary degrees including Hon. D.Sc. from: the University of Waterloo (1970); the Memorial University of Newfoundland (1976); McMaster University (1977); Carleton University (1981); Harvard University (1982); Brock University (1984); Rensselaer Polytechnic Institute (1984); Université Laval (1987); University of Lethbridge (1987); University of Ottawa (1987); Université de Sherbrooke (1987); University of Victoria (1987); Manchester University (1988); York University (1988); Acadia University (1989); University of Montreal (1989); Weizmann Institute of Science (1989); University of Bari, Italy (1990); University of British Columbia (1990); McGill University (1990); Queen's University (1992); Free University Berlin (1993); Laurentian University (1995); Liverpool University (1995); University of Toronto (1995). He holds the Hon. LL.D.: from Trent University (1977); Dalhousie University (1983); St. Francis Xavier University (1984); Concordia University (1990); University of Calgary (1994).

He has been awarded many prizes including: the Marlow Medal of Faraday Society (1962); the Centenary Medal and Lectureship Chemical Society of Great Britain (1965); the Steacie Prize for the Natural Sciences, joint with N. Bartlett (1965); the British Chemical Society Award (1971); the Chemical Institute of Canada Medal (1976); the Henry Marshall Tory Medal of the Royal Society of Canada (1977); the Sherman Fairchild Distinguished Scholar of the California Institute of Technology, (1982); the Wolf Prize in Chemistry jointly with G. Pimentel (1982); the Nobel Prize in Chemistry jointly with D.R. Herschbach and Y.T. Lee (1986); the Izaak Walton Killam Memorial Prize (1988); the Michael Polanyi Medal and Lecture, Royal Society of Chemistry (1989); the Royal Medal, Royal Society of London (1989); the Bakerian Prize Lecture, the Royal Society of London (1994).

Among the many affiliations to learned societies, Polanyi is a Fellow of the: Royal Society of Canada (1966); Royal Society of London (1971); Royal Society of Edinburgh (1988). He is an Honorary Fellow of the: Royal Society of Chemistry (1991); Chemical Institute of Canada (1991); National Institute of Chemistry, Singapore (1997). He is a Member of the Pontifical Academy of Sciences, Rome (1986) and of the Croatian Academy of Sciences and Arts (1992). He is also an Honorary Foreign Member of the American Academy of Arts and Science (1976), a Foreign Associate National Academy of Sciences (1978), a Distinguished Anniversary Fellow of The Australian National University (1996) and a Foreign Member of the Russian Academy of Sciences (1999). He is an Honorary Fellow of Trinity College, University of Toronto (1995).

Polanyi is a Companion Order of Canada (1979), a Knight Grand Cross Sovereign Military Order of St. John of Jerusalem "Knights of Malta," (1987), a Member of the Queen's Privy Council for Canada (1992) and a Member Board of the Premier's Council (1994).

He was a Founding Member and Chairman of the Canadian Pugwash Committee (1960) and a Senior Fellow of Massey College, University of Toronto (1963).

Polanyi has served on numerous committees including: Member of the Science Advisory Board, Max Planck Institute for Quantum Optics (1982–1992); Member of the National Advisory Board on Science and Technology (1987–1989); Member of the Board of the Ontario Laser and Lightwave Research Centre (1988–1992); President and Founding Member of the Canadian Committee of Scientists and Scholars; Founding Member of the Royal Society of Canada Committee on Scholarly Freedom; a Member of the American Academy of Arts and Science Committee on International Security Studies; a Member of the Board Canadian Centre for Global Security; a Board Member of the Steacie Institute for Molecular Sciences (1991–present); Co-Chair International Consultative Group on Improving United Nation's Rapid Reaction Capability (1995). Polanyi also served as an Honorary Consultant for the Institute for Molecular Science, Okazaki, Japan (1989–1992).

Polanyi has lectured in institutes around the world, many of which carried awards. The first of these was the Noranda Award and Lectureship of the Chemical Institute of Canada (1967). There are a further 40–50 named lecture series that he has given, concluding with the Llewellyn-Thomas Lecture, University of Toronto (1999). Details of the rest of these may be obtained from his official website [4].

The Government of the Province of Ontario has established a fund to provide up to five prizes (annually) to outstanding young researchers in the early stages of their career who are currently working at an Ontario university in the honour of John Charles Polanyi. Each award has a value of fifteen thousand dollars and are available in the areas of Physics, Chemistry, Physiology or Medicine, Literature, and Economic Science.

Bibliography

1. Reproduced with permission.
2. From Nobel Lectures, Chemistry 1981–1990, Editor-in-Charge Tore Frängsmyr, Editor Bo G. Malmström, World Scientific Publishing Co., Singapore, 1992.
3. Who's who: 2005.
4. http://www.utoronto.ca/jpolanyi/.
5. Chemical Kinetics and Reaction Dynamics by Paul L. Houston, McGraw-Hill Science/Engineering/Math, 2001, ISBN: 0072435372.
6. Molecular Reaction Dynamics by Raphael D. Levine, Cambridge University Press, 2005, ISBN: 052184276X.
7. Modern Trends in Chemical Reaction Dynamics: Experiment and Theory (Advanced Series in Physical Chemistry - Vol. 14) edited by Xueming Yang and Kopin Liu, World Scientific Publishing Company, 2004, ISBN: 9812385681.

1987

© Nobel Foundation [1]

Donald J. Cram (0.33)

"for their development and use of molecules with structure-specific interactions of high selectivity" [2]

During his presentation speech, Salo Gronowitz, then of the Royal Academy of Sciences, commented of Cram and his work [2]:

> *...Organic chemists have been wondering for a long time, how much of the very large biomolecule is really necessary for achieving the desired result. How much of the molecule is really needed to form the hole or cleft in which the key has to fit for the reaction to take place. Organic chemists have dreamt about preparing molecules that mimic biomolecules in the laboratory, molecules that for instance have the same catalytic effects as enzymes, which chemical evolution has taken millions of years to develop...*

> *...The explosive development of the art of organic synthesis during the last decades, which is reflected in the Nobel Prizes in Chemistry to Robert Woodward in 1965, to Herbert Brown and Georg Wittig in 1979 and to Bruce Merrifield in 1984, has made it possible for Donald Cram and Jean-Marie Lehn to very skilfully design more complex molecules containing holes and clefts which bind inorganic as well as organic positive ions even more selectively. Other synthetic molecules can bind various negative ions as well as neutral molecules. Even molecules which can differentiate between mirror image forms of a molecule have been prepared. It has also been possible to synthesize molecules which enzymes in their mode of action and strongly accelerate various types of chemical reactions. Other organic compounds have been prepared which transport ions through biological membranes. By means of their penetrating investigations the laureates have also elucidated the factors that govern complex formation, and the changes that occur in both partners' chemical and physical properties, and how this can be used in practice for different purposes...*

...Professor Cram...in these few minutes I have tried to explain your fundamental work in the field of molecular recognition and the great importance and consequences your results have in many branches of chemistry, such as analytical chemistry, organic synthesis, coordination chemistry and bioinorganic and bioorganic chemistry...

Donald J. Cram was born in Chester, Vermont, USA on the 22nd April 1919, the fourth child and only son, of William Moffet and Joanna Shelley Cram. He married his first wife, Jean Turner Cram in 1940. They were divorced in 1968. He married Jane L. Maxwell in 1969. Cram died on 17th June 2001 of cancer in his home in Palm Desert, California [2, 3, 4].

The Early Years

Cram was four years of age when his father died of pneumonia (aged 53). His older sister commented that Cram was 'precocious, curious, and constantly in, or causing, trouble' [2]. Cram reported that 'determined to walk at seven months, I pulled a pan of fresh eggs down from a table onto my head'. Cram further commented 'I usually was marked "A" in attitude and accomplishment, "B" in effort, and "C" in obedience' on school reports [2]. Cram left home at sixteen, working for a number of employers until he enrolled himself at Winwood, a little private school on Long Island, New York. There he studied chemistry, taught himself geometry from a book and won a $6,000 four-year National Rollins College Honor Scholarship. He was offered three posts for teaching assistantships (out of seventeen) and accepted the University of Nebraska, from which he graduated in 1942 with an MS degree (for work conducted under N. O. Cromwell). He worked for Merck & Co. during the war years (1942–45), completing his time there on their penicillin project. Tishler arranged for Cram to attend Harvard University (in 1945) from which he graduated with a Ph.D. (1947).

The Career

Year	*Post*	*Institute*
1942	Chemist	Merck & Co
1945	PhD	Harvard University
1947	Instructor and ACS Fellow	University of California
1948	Assistant Professor	University of California
1951	Associate Professor	University of California
1956	Professor of Chemistry	University of California
1985–95	S. Winstein Professor of Chemistry	University of California

Cram was a Guggenheim Fellow (1955) and a University Professor (1988–90).
He studied many aspects of host-guest chemistry including:

- complexation chemistry
- correlations between structure and binding, structure and catalytic properties, structure and physical properties, and structure and separation
- the correlation of crystal structures with solutions structures (using NMR spectra)

- reactions of guest molecules imprisoned in the interior of rigidly hollow spherical host molecules (carceplexes)
- potential uses of carceplexes include drug and radiation delivery systems, light switches, or molecular information storage and retrieval display systems (MISARDS)
- the synthesis and study of benzyne in the inner phase of a carcerand.
- structural recognition (including chiral recognition) in binding

Cram published more than 400 research papers and several books, including his autobiography [5]. His best-known book was *Organic Chemistry* with George S. Hammond of the California Institute of Technology (pub in 1959). This was the first organic chemistry text book to be organized by reactions and not classes of chemicals. He published three books that are no longer available, namely: *Elements of Organic Chemistry* (with D.H. Richards and G. S. Hammond, three translations); *Essence of Organic Chemistry* (with J.M. Cram, one translation); plus the monograph, *Fundamentals of Carbanion Chemistry*" (one translation). Some of his other books were:

- *Study Guide and Solutions Manual to Accompany the Essence of Organic Chemistry*, with Jane M. Cram, Pub: Addison-Wesley, (1978), ISBN: 0201010321
- *Essence of Organic Chemistry* (Addison-Wesley Series in Chemistry) with Jane M. Cram, Pub: Addison Wesley Publishing Company, (1978), ISBN: 0201010313
- *Container Molecules and Their Guests: Monographs in Supramolecular Chemistry* (Monographs in Supramolecular Chemistry) with Jane M. Cram, Pub: Royal Society of Chemistry, (1994), ISBN: 0851869726

His Nobel Prize lecture was dated 8th December 1987 with the title 'The design of molecular hosts, guests, and their complexes'. Cram set the scene for his own entry into the field of guest-host crown ethers [2]:

> *...In 1967, Pedersen's first papers appeared which reported that alkali metal ions bind crown ethers to form highly structured complexes. We immediately recognized this work as an entree into a general field. The 1969 papers on the design, synthesis, and binding properties of the cryptands by J.-M. Lehn. J.-P. Sauvage. and B. Dietrich further demonstrated the attractions and opportunities of complexation chemistry. Although we tried to interest graduate students in synthesizing chiral crown ethers from 1968 on, the efforts were unsuccessful. In 1970 we insisted that several postdoctoral co-workers enter the field. During 1973, we published five Communications on the subject. In 1974 with Jane M Cram, we published a general article entitled "Host-Guest Chemistry", which defined our approach to this research...*

Some time later, Cram and his group were involved in the synthesis of large, complex nonpolar guests [2].

> *...Absent among the millions of organic compounds hitherto reported are closed-surface hosts with enforced interiors large enough to imprison behind covalent bars, guests the size of ordinary solvent molecules. After much thought and molecular model examination, we chose 48 as the target for synthesis of the first molecular cell. The term carcerand was applied to this class of compound. The synthesis involved treating Cs_2CO_3, with a solution in $(CH_3)_2NCHO(CH_2)_4O$ of equal molar amounts of cavitands 46 and 47 under an atmosphere of argon. The first question to be answered was: what guest compounds would be trapped inside during the shell closure? This question is akin to*

asking whether two soup bowls closed rim-to-rim under the surface of a kettle of stew would net any stew. The answer was that 48 "contained" essentially every kind of component of the medium present during ring closure...

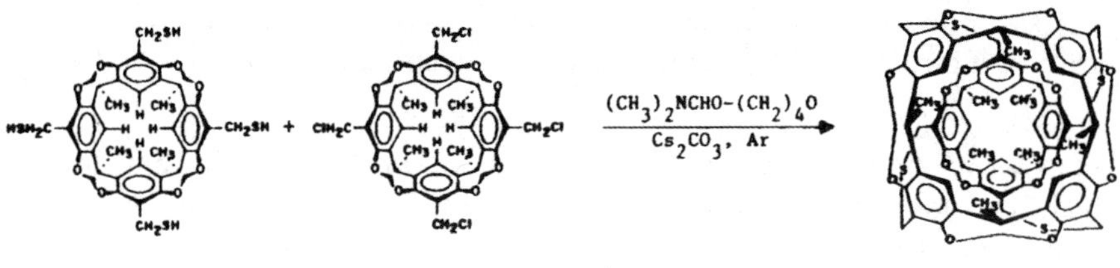

46 47 48

Reproduced from Cram's Nobel Prize lecture [2].

For further information concerning crown ethers, see for example [6–8].

Honours and Awards

Cram was presented with Honorary Doctor degrees from Uppsala University (1977) and Rollins College (1988). He was also presented with Honorary Doctor of Science degrees from: the University of Southern California (1983); the University of Nebraska (1989); the University of Western Ontario (1990); the University of Sheffield (1991).

He has been presented with many prizes and awards including: ACS award, for Creative Work in Synthetic Organic Chemistry (1987); the Arthur C. Cope Award for Distinguished Achievement in Organic Chemistry (1974); the Roger Adams Award in Organic Chemistry (1985); the Willard Gibbs medal (1985); the Tolman Medal (1985); the Nobel Prize for Chemistry, jointly with Lehn and Pedersen (1987); the Armand Hammer Award (1992); the President's National Medal of Science (1993); the Glenn T. Seaborg Award; the National Academy of Sciences Award in Chemical Sciences.

Cram was elected to Membership in the National Academy of Sciences (1961) and the American Academy of Arts and Sciences (1967). He was elected a Honorary Fellow of the Royal Society of Chemistry (1989).

He was California Scientist of the Year (1974), President of the National Medical Sciences and Honorary Member of Sigma Xi, Phi, Beta Kappa.

Cram was Chemistry Lecturer and Medalist of the Royal Institute of Chemistry (UK) (1976).

He consulted for: Upjohn Co (1952–88); Eastman Kodak (1982–91); Technicon Co (1984–91); Istituto Guido Donegani, Milan (1988–91).

Bibliography

1. Reproduced with permission.
2. From Nobel Lectures, Chemistry 1981–1990, Editor-in-Charge Tore Frängsmyr, Editor Bo G. Malmström, World Scientific Publishing Co., Singapore, 1992.
3. Who's who: 2005.
4. http://www.chem.ucla.edu/dept/Faculty/cram.html.

5. Cram, Donald J., *From Design to Discovery*, Oxford University Press, Oxford, 2001.
6. Crown Ethers and Analogous Compounds (Studies in Organic Chemistry, Vol. 45) ed by Michio Hiraoka, Elsevier Publishing Company, 1992, ISBN: 0444881913.
7. Inclusion Compounds: Volume 4: Key Organic Host Systems, by J. L. Atwood, J. E. D. Davies, D. D. MacNicol, Oxford University Press, USA, 2005, ISBN: 0198552920.
8. The Pedersen Memorial Issue (Advances in Inclusion Science) by R.M. Izatt (Editor), J.S. Bradshaw (Editor), Springer, 1st ed. 1999, ISBN: 0792317238.

1987

Jean-Marie Lehn (0.33)

"for their development and use of molecules with structure-specific interactions of high selectivity" [2]

During his presentation speech, Salo Gronowitz, then of the Royal Academy of Sciences, commented of Lehn [2]:

...Organic chemists have dreamt about preparing molecules that mimic biomolecules in the laboratory, molecules that for instance have the same catalytic effects as enzymes, which chemical evolution has taken millions of years to develop...

..The explosive development of the art of organic synthesis during the last decades, which is reflected in the Nobel Prizes in Chemistry to Robert Woodward in 1965, to Herbert Brown and Georg Wittig in 1979 and to Bruce Merrifield in 1984, has made it possible for Donald Cram and Jean-Marie Lehn to very skilfully design more complex molecules containing holes and clefts which bind inorganic as well as organic positive ions even more selectively. Other synthetic molecules can bind various negative ions as well as neutral molecules. Even molecules which can differentiate between mirror image forms of a molecule have been prepared. It has also been possible to synthesize molecules which enzymes in their mode of action and strongly accelerate various types of chemical reactions. Other organic compounds have been prepared which transport ions through biological membranes. By means of their penetrating investigations the laureates have also elucidated the factors that govern complex formation, and the changes that occur in both partners' chemical and physical properties, and how this can be used in practice for different purposes.

Through their work the laureates laid the foundation to what today is one of the most expansive chemical research areas, for which Cram has coined the term host-guest chemistry, while Lehn calls it supramolecular chemistry. Their research has been of enormous importance for the development of coordination chemistry, organic synthesis, analytical chemistry, bioinorganic and bioorganic chemistry, it is no longer science fiction to prepare supermolecules which are better and more versatile catalysts than the highly specialized enzymes. The dream may soon become reality. Through their work Cram, Lehn and Pedersen have shown the way...

Jean-Marie Lehn was born in Rosheim, Alsace on 30th September 1939, the eldest of four sons of Pierre and Marie, née Salomon. He married Sylvie Lederer (1965) and they have two sons, David (1966) and Mathias (1969). Lehn is currently Directeur and Professor at Collège de France of the Laboratoire de Chimie Supramoléculaire, Institut de Science et d'Ingénierie Supramoléculaires, Strasbourg [2, 3, 4].

The Early Years

Lehn grew up in Rosheim during the Second World War, attending primary school after the war had finished. He attended a nearby high school, the Collège Freppel (in Obernai), from 1950 to 1957, where he obtained the baccalaureat in philosophy (July 1957) and in Experimental Sciences (Sep 1957). He entered the University of Strasbourg thereafter studying physical, chemical and natural sciences, during which he was influenced by G. Ourisson towards research in organic chemistry. Graduating with the degree of Licencié-ès-Sciences (Bachelor), he entered Ourisson's laboratory (Oct 1960) in the *CNRS* for his Ph.D. degree, where he worked on issues relating to conformational and physico-chemical properties of triterpenes (completing his Ph.D. in 1963).

Lehn's postdoctoral years were spent with R.B. Woodward at Harvard University, on the total synthesis of Vitamin B_{12}. He took a course in quantum mechanics and performed his first computations with Roald Hoffmann.

The Career

Year	*Post*	*Institute*
1960	Scientist	CNRS
1966	Assistant Professor	University of Strasbourg
1970	Associate Professor	University Louis Pasteur, Strasbourg
1970 – to date	Professor of Chemistry	University Louis Pasteur, Strasbourg
1979 – to date	Professor of Chemistry (additional responsibility)	Collège de France

Lehn's work over the last twenty years has involved about 150 collaborators from over twenty countries, during which he has studied:

- the relationship between the structure of a molecule and its properties (1965–1970)
- NMR studies of conformational rate processes
- nitrogen inversion
- quadrupolar relaxation
- molecular motions
- liquid structure
- ab initio quantum chemical computations of inversion barriers
- the design of cation cryptates, later termed "supramolecular chemistry" (started Oct 1967)

- artificial photosynthesis
- the storage and chemical conversion of solar energy
- the photolysis of water
- the photoreduction of carbon dioxide
- the design and properties of "programmed" supramolecular systems. Photoactive and electroactive cryptates
- Molecular recognition, molecular receptors and coreceptors
- Anion coordination chemistry: anion cryptates
- Supramolecular catalysis
- Transport processes
- Selfassembly and selforganisation
- Supramolecular materials
- Chemionics
- Semiochemistry
- Photochemistry and energy storage

Lehn has written about 400 publications and review papers, including a book titled *Supramolecular Chemistry: Concepts and Perspectives* Pub: Wiley-VCH (1995) ISBN: 3527293116.

Lehn's Nobel Prize lecture is dated 8th December 1987 with the title 'Supramolecular chemistry – scope and perspectives, molecules – supermolecules – molecular devices'. Somewhat unusually, Lehn provided an abstract of his presentation which fully details the scope of his lecture.

Abstract

...Supramolecular chemistry is the chemistry of the intermolecular bond, covering the structures and functions of the entities formed by association of two or more chemical species. Molecular recognition in the supermolecules formed by receptor-substrate binding rests on the principles of molecular complementarity - as found in spherical and tetrahedral recognition, linear recognition by coreceptors, metalloreceptors, amphiphilic receptors, anion coordination. Supramolecular catalysis by receptors bearing reactive groups effects bond cleavage reactions as well as synthetic, bond formation via cocatalysis. Lipophilic receptor molecules act as selective carriers for various substrates and allow to set up coupled transport processes linked to electron and proton gradients or to light. Whereas endo-receptors bind substrates in molecular cavities by convergent interactions, exo-receptors rely on interactions between the surfaces of the receptor and the substrate; thus new types of receptors such as the metallonucleates may be designed. In combination with polymolecular assemblies, recepcarriers and catalysts may lead to molecular and supramolecular devices, defined as structurally organized and functionally integrated chemical systems built on supramolecular architectures. Their recognition, transfer and transformation features are analyzed specifically from the point of view of molecular devices that would operate via photons, electrons or ions, thus defining fields of molecular photonics, electronics and ionics. Introduction of photosensitive groups yields photoactive receptors for the design of light conversion and charge separation centers. Redox active polyolefinic chains represent molecular wires for electron transfer through membranes. Tubular mesophases formed by stacking of suitable macrocyclic receptors may lead to ion channels. Molecular self assembling occurs with acyclic ligands that form complexes of double helical structure. Such developments in molecular and supramolecular design and engineering open

perspectives towards the realization of molecular photonic, electronic and ionic devices, that would perform highly selective recognition, reaction and transfer operations for signal and information processing at the molecular level...

Lehn concluded...

...The present text was aimed at presenting the scope, providing illustration and exploring perspectives of supramolecular chemistry. Its conceptual framework has been progressively laid down and the very active research on molecular recognition, catalysis and transport, together with extension to molecular surfaces and polymolecular assemblies, is building up a vast body of knowledge on molecular behaviour at the supramolecular level. It is clear that much basic chemistry remains to be done on the design and realization of numerous other systems and processes that await to be imagined. The results obtained may also be analyzed in the view of developing components for molecular devices that would perform highly selective functions of recognition, transformation, transfer, regulation and communication, and allow signal and information processing. This implies operation via intermolecular interactions and incorporation of the time dimension into recognition events. One may note that such functions have analogies with features of expert systems, thus linking processes of artificial intelligence and molecular behaviour.

These developments in molecular and supramolecular science and engineeroffer exciting perspectives at the frontiers of chemistry with physics and biology. Of course, even with past achievements and present activities, exand predictions can only be tentative; yet, on such an occasion for the celebration of chemistry, it appears justified to try looking out into the future for "He (or she) who sits at the bottom of a well to contemplate the sky, will find it small" (Han Yu, 768–824)...

For further information concerning crown ethers see for example [5–7].

Honours and Awards

Lehn has been presented with many honorary degrees in Europe and the United States of America.

He was presented with many prizes and awards including: the Nobel Prize for Chemistry, jointly with Cram and Pedersen, (1987); the Davy medal from the Royal Society (1997); the Orden pour le mérite für Wissenchaften und Künste, FRG (1990).

Among the many affiliations to learned institutes and academies, he is an Honorary Fellow of the Royal Society of Chemistry (1987) and a Foreign Member of the Royal Society (1993).

His civic honours include Officier, Ordre National du Mérite, 1993 (Chevalier 1976) and Commandeur, Légion d'Honneur, 1996 (Chevalier 1983, Officier 1988).

Lehn has been a Visiting Professor at many institutions including Harvard University, the E.T.H. in Zürich and the Universities of Cambridge, Barcelona, Frankfurt.

He is a Member of the Board of Directors of Ciba Speciality Chemicals, a Member of the Scientific Board of Aventis and a Member of the Scientific Board of the Novartis Venture Fund (Basel).

Bibliography

1. Reproduced with permission.
2. From Nobel Lectures, Chemistry 1981–1990, Editor-in-Charge Tore Frängsmyr, Editor Bo G. Malmström, World Scientific Publishing Co., Singapore, 1992.

3. Who's who: 2005.
4. http://www-isis.u-strasbg.fr/supra/.
5. Crown Ethers and Analogous Compounds (Studies in Organic Chemistry, Vol. 45) ed. by Michio Hiraoka, Elsevier Publishing Company, 1992, ISBN: 0444881913.
6. Inclusion Compounds: Volume 4: Key Organic Host Systems, by J. L. Atwood, J. E. D. Davies, D. D. MacNicol, Oxford University Press, USA, 2005, ISBN: 0198552920.
7. The Pedersen Memorial Issue (Advances in Inclusion Science) by R.M. Izatt (Editor), J.S. Bradshaw (Editor), Springer, 1st ed. 1999, ISBN: 0792317238.

1987

© Nobel Foundation [1]

Charles J. Pedersen (0.33)

"for their development and use of molecules with structure-specific interactions of high selectivity" [2]

During his presentation speech, Salo Gronowitz, then of the Royal Academy of Sciences, commented of Pedersen [2]:

...Organic chemists have been wondering for a long time, how much of the very large biomolecule is really necessary for achieving the desired result. How much of the molecule is really needed to form the hole or cleft in which the key has to fit for the reaction to take place. Organic chemists have dreamt about preparing molecules that mimic biomolecules in the laboratory, molecules that for instance have the same catalytic effects as enzymes, which chemical evolution has taken millions of years to develop.

A breakthrough towards this goal occurred when Charles Pedersen in the 1960's prepared cyclic compounds that contained carbon and oxygen atoms in rings of 18 up to 40 atoms. As their form was similar to a royal crown, he called these new compounds crown ethers. He showed that these compounds had many peculiar and unexpected properties. For example, they formed strong complexes with the alkali metal ions, lithium, sodium, potassium, rubidium and cesium, which previously had been difficult to bind. Of these spherical ions, lithium is the smallest and cesium the largest. Depending upon the size of the synthesized crown, potassium, for example, could fit inside, while the cesium ion was too large to be trapped...

...Mister Pedersen, in these few minutes I have tried to explain your fundamental work in the field of molecular recognition and the great importance and consequences your results have in many branches of chemistry, such as analytical chemistry, organic synthesis, coordination chemistry and bioinorganic and bioorganic chemistry...

Charles J. Pedersen was born in Pusan, Korea on 3rd October 1904, the son of Brede Pedersen, a Norwegian marine engineer, and Takino Yasui. He had a sister, Astrid (1899) and an elder brother, who died in childhood prior to Pedersen's birth. Pedersen married Susan Ault (1947) and they had two daughters. He died of cancer in Salem, New Jersey on 26th October 1989 [2,3].

The Early Years

Pedersen was sent to Japan when aged eight, where he attended a convent school in Nagasaki for two years followed by studies at St. Joseph College, a preparatory school run by a Roman Catholic religious order of priests and brothers called the Society of Mary, in Yokohama, Japan. His father encouraged him to attend an American university for his degree, so Pedersen duly enrolled in the University of Dayton, Ohio, where he studied for a BS Chemical Engineering, graduating in 1926. This was followed by a master's degree in organic chemistry at the Massachusetts Institute of Technology, graduating in 1927. Pedersen chose not to stay on for a Ph.D.

The Career

Year	*Post*	*Institute*
1927–1969	Chemist	DuPont Company in Wilmington, Delaware
1947	Research Associate	DuPont Company in Wilmington, Delaware
1969	Retirement	

Pedersen's 42 year career at DuPont covered many aspects of chemistry including (in approximately date order):

- the search for oil-solvable precipitants for copper
- the preparation of the first good metal deactivator for petroleum products
- a study of the effects of various ligands on the catalytic properties of copper and the transition elements, the oxidative degradation of the substrates of petroleum products and rubber
- antioxidants
- the photochemistry of some new phthalocyanine adducts and of quinoneimine dioxides
- the development of polymerization initiators
- the preparation of some novel polymers
- the study the effects of bi- and multidentate phenolic ligands on the catalytic properties of the vanadyl group, VO. In the course of these investigations one of his experiments yielded an unexpected small quantity of unknown white crystals which he eventually identified as dibenzo18-crown-6, first crown ether
- the study of crown ethers.

Much of the work was either personally undertaken by Pedersen or with the help of an assistant.

The Nobel Committee, in announcing the award to Pedersen (and two others, for unrelated work), said this [4]:

"In 1967, Charles J. Pedersen published two works, which have now become classics, describing methods of synthesising cyclic polyethers, which he named crown ethers. Pedersen showed that these compounds have remarkable and unexpected properties and that they can even bind the alkali metal ions of lithium, sodium, potassium, rubidium and caesium into complexes in which the lithium ion is the smallest and the caesium ion the largest. He also found that, depending on the structure of the crown ether, potassium could for instance be bound before caesium. Simply expressed, the selectivity is determined by the fact that different crown ethers include 'holes'.

The original paper appeared in the Journal of the American Chemical Society in 1967, and took five years to undertake the work and write the 20 or so closely typed pages [5].

One of Pedersen's retirement hobbies was to write poetry. A volume of his works was published posthumously [6].

Pedersen's Nobel Prize lecture is dated 8th December 1987 with the title 'The discovery of crown ethers'. Pedersen commented:

...I will divide my lecture into three parts.

First, because every discovery takes place in more than a scientific context, I would like to touch on my life and background. In the weeks since it was announced that I would share this year's prize in chemistry, people have expressed as much interest in my early life as they have in my later work. So I think it appropriate to express myself on the matter. It may also be that details of my past have more than casual bearing on my work.

Second, I would like to describe for you my research program and some of the specific events that led to the discovery of the first crown ether. Since I am the only one who knows at firsthand the excitement and pleasure of the discovery, I will devote a portion of my time to sharing this experience with you.

And third, I would like to discuss the properties and preparation of crown ethers. In doing so, I hope I will convey to you that I was always a "hands-on" chemist; I took satisfaction from what I did in the laboratory. Also, I was very much an industrial chemist and was always interested in the potential application of my work, In fact, when I submitted my first major paper on the discovery of the crown ethers, the editor of the Journal of the American Chemical Society, Marshall Gates, remarked that my descriptions were replete with industrial jargon. Fortunately he published the paper anyway...

He continued:

...In 1960 when I returned to investigations in coordination chemistry, I decided to study the effects of bi- and multidentate phenolic ligands on the catalytic properties of the vanadyl group, VO. The multidendate ligand I selected is the bis [2-(o-hydroxyphenoxy)ethyl] ether. As I proceeded, I knew that the partially protected catechol was contaminated with about 10 percent unreacted catechol. But I decided to use this mixture for the second step anyway since purification would be required at the end. The reactions were carried out as outlined and gave a product mixture in the form of an unattractive goo. Initial attempts at purification gave a small quantity (about 0.4 percent yield) of white crystals which drew attention by their silky, fibrous structure and apparent insolubility in hydroxylic solvents.

The appearance of the small quantity of the unknown should have put me in a quandary. It probably was not the target compound because that would be obtained in a higher yield. My objective was to prepare and test a particular compound for a particular purpose. Had I followed this line, I would have doomed the crown ethers to oblivion until such a time as another investigator would retrace my steps and make the better choice at the critical moment. Crown ethers, however, were in no danger, because of my natural curiosity. Without hesitation, I began study of the unknown...

For further information concerning crown ethers see for example [7–9].

Honours and Awards

In 2001, DuPont issued a series of press releases on the occasion of the 200th anniversary of its founding. One press release recited:

"Late DuPont scientist Charles J. Pedersen is the only winner of the Nobel Prize in Chemistry never to possess a earned or honorary Ph.D. Several offers to bestow honorary degrees were extended to Charlie, who retired from DuPont in 1969 and died in 1989. However, he declined all of them, believing his achievement was evidence that an individual didn't need a doctorate to do excellent scientific work."

He was awarded the Nobel Prize for Chemistry, jointly with Cram and Lehn (1987) and a Medal for Excellence by the DuPont Research Fellows (1987).

Bibliography

1. Reproduced with permission.
2. From Nobel Lectures, Chemistry 1981–1990, Editor-in-Charge Tore Frängsmyr, Editor Bo G. Malmström, World Scientific Publishing Co., Singapore, 1992.
3. Who's who: 2005.
4. www.nobelprize.org.
5. "*Macrocyclic Polyethers for Complexing Metals*," J. Am. Chem. Soc. 1967, 89, 7017.
6. *The poetry of Charles J. Pedersen*, Pub: B. Cleveland and S. Evans, (1990), ASIN: B0006EYN26.
7. Crown Ethers and Analogous Compounds (Studies in Organic Chemistry, Vol. 45), ed. by Michio Hiraoka, Elsevier Publishing Company, 1992, ISBN: 0444881913.
8. Inclusion Compounds: Volume 4: Key Organic Host Systems, by J. L. Atwood, J. E. D. Davies, D. D. MacNicol, Oxford University Press, USA, 2005, ISBN: 0198552920.
9. The Pedersen Memorial Issue (Advances in Inclusion Science) by R.M. Izatt (Editor), J.S. Bradshaw (Editor) Springer, 1st ed. 1999, ISBN: 0792317238.

1988

© Nobel Foundation [1]

Johann Deisenhofer (0.33)

"for the determination of the three-dimensional structure of a photosynthetic reaction centre" [2]

During his presentation speech, Bo G. Malmström of the Royal Swedish Academy of Sciences, commented of Deisenhofer and his work [2]:

...For a long time it has been impossible to prepare membrane-bound proteins in a form allowing the determination of the detailed structure in three dimensions. Before 1984, there were only rather fuzzy structural pictures available for a few membrane proteins. These had been derived with the aid of an electron microscopic method developed by the Englishman Aaron Klug, who was awarded the Nobel Prize in chemistry in 1982 for this achievement. But the situation had actually drastically changed in 1982, when Hartmut Michel thanks to systematic experiments succeeded in preparing highly ordered crystals of a photosynthetic reaction center from a bacterium. With these crystals he could in the period 1982–1985, in collaboration with Johann Deisenhofer and Robert Huber, determine the structure of the reaction center in atomic detail.

The structural determination awarded has led to a giant leap in our understanding of fundamental reactions in photosynthesis, the most important chemical reaction in the biosphere of our earth. But it has also consequences far outside the field of photosynthesis research. Not only photosynthesis and respiration are associated with membrane-bound proteins but also many other central biological functions, e.g. the transport of nutrients into cells, hormone action or nerve impulses. Proteins participating in these processes must span biological membranes, and the structure of the reaction center has delineated the structural principles for such proteins. Michel's methodological contribution has, in addition, the consequence that there is now hope that we can determine detailed structures also for many other membrane proteins. Not least important is the fact that the reaction center structure has given theoretical chemists an indispensable tool in their efforts to understand how biologic electron transfer over very large distances on a molecular scale can occur as rapidly as in one billionth

(American English, trillionth) of a second. In a longer perspective it is possible that such research can lead to important energy technology in the form of artificial photosynthesis...

Johann Deisenhofer was born on 30th September 1943 in Zusamaltheim, Bavaria, the first child and only son, of Thekla and Johann Deisenhofer. He married Kirsten Fischer in 1989. He holds the Virginia and Edward Linthicum Distinguished Chair in Biomolecular Science Regental Professor at the UT Southwestern Medical Center at Dallas. He is also an Investigator at the Howard Hughes Medical Institute and can be contacted through either organization [2, 3, 4, 5].

The Early Years

Deisenhofer attended primary school at Zusamaltheim (1949–1956), followed by periods in three further schools: Knabenmittelschule HI. Kreuz, Donanwoerth (1956), Staatliche Realschule Wertingen (1957–1959); Holbein Gymnasium, Augsburg (1959–1963). Deisenhofer passed the "Abitur" examination (1963), which allowed him to enter university using a "Stipendium für besonders Begabte" of the "Bayerische Staatsministerium für Unterricht und Kultus". Military service (Autumn 1965 - spring 1967) in the German Bundeswehr interrupted his education, he nevertheless began to study physics at the Technische Universität München after his military service. Initially interested in solid state physics, Deisenhofer's interests slowly changed to that of biophysical problems, following experimental work under Karl-Friedrich Renk in Dransfeld's lab. This interest was mainly sparked by Klaus Dransfeld. Deisenhofer graduated with a Diploma in Physics in 1971.

Deisenhofer's Ph.D. studies (1971–74) were at the newly named Max-Planck-Institut für Biochemie under Robert Huber, on the crystallographic refinement of the structure of bovine pancreatic trypsin inhibitor. Huber offered Deisenhofer a two year postdoctoral fellowship (1974–76), which Deisenhofer accepted.

The Career

Year	*Post*	*Institute*
1976–1988	Staff Scientist	Max-Planck-Institut für Biochemie
1988–Present	Investigator	Howard Hughes Medical Institute
1988–Present	Virginia and Edward Linthicum Distinguished Chair in Biomolecular Science Regental Professor	UT Southwestern Medical Center at Dallas

In his autobiographical sketch reported in the Nobel web archive [4], Deisenhofer commented:

...In 1982, Hartmut Michel, who had come to Martinsried together with Dieter Oesterhelt, reported in one of Huber's group seminars about his spectacular success with the crystallization of the photosynthetic reaction center from Rhodopseudomonas viridis. After discussions between Hartmut and myself, and after Robert Huber had given his agreement, I joined the reaction center project in order to determine the three-dimensional structure of this molecule...

...In a surprisingly short time, at the end of 1983, we came to a point where the success of the project was at the horizon. It still took almost two years until we had worked out the complete structure, and two more years to refine the model at 2.3Å resolution...

Deisenhofer's work includes [5]:

- chaperonins
- determining the three-dimensional structures of biological macromolecules by X-ray crystallography
- integral membrane proteins
- membrane-associated proteins
- proteins involved in cholesterol metabolism
- proteins involved in energy or electron transfer
- proteins that utilize light as an energy source
- proteins with repeating sequence motifs.

Deisenhofer wrote a book, the title of which is:

The Photosynthetic Reaction Center with James R. Norris (Editor), Pub: Academic Press, (1993), ISBN: 0122086627.

Deisenhofer's Nobel Prize lecture was presented with Hartmut Michel and is dated 8th December 1988 with the title 'The photosynthetic reaction center from the purple bacterium *Rhodopsedomonas Viridis*'. Michel's task was the crystal preparation whilst Deisenhofer concentrated on producing the x-ray reflection intensity data. Deisenhofer commented [2]:

Determination of the Structure

...In spring 1982, I joined H.M. in order to determine the three-dimensional structure of the reaction centre. The tetragonal crystals have unit cell dimensions of $a = b = 223.5$Å $c = 113.6$Å, *and the symmetry of space group* $P4_32_12$ *(Michel, 1982; Deisenhofer et al., 1984). As it turned out, there is one reaction centre with a "molecular weight" of 145000 Daltons in the asymmetric unit...*

Collection of X-ray Reflection Intensity Data

...For data collection we used the rotation method with a rotating anode X-ray generator as the source, and photographic film as the detector (Deisenhofer et al., 1984). The large unit cell of the reaction centre crystals, in combination with the resolution limit of the diffraction pattern at 2.9Å limited the rotation interval per film exposure to 0.5°, so that more than two thirds of the reflections on any given film were partially recorded. However, the long lifetime of the crystals in the X-ray beam at about 0°C, and their positional stability allowed to add up partially recorded reflections from successive exposures, so that their treatment did not present a serious problem. Nevertheless, it took about three to four months to collect a complete data set. Data collection for the heavy atom derivatives was speeded up by choosing a rotation interval of 0.6° per exposure....

Model Building

Map interpretation and model building were done in three stages: At first the prosthetic groups in the reaction centre were identified. We found the four heme groups, the four bacteriochlorophyll-b, the two bacteriopheophytin-b, and one quinone (Deisenhofer et al., 1984). Next, the polypeptide chains were built with polyalanine sequence, except in the amino terminal regions of the subunits, L, M, and cytochrome where partial amino acid sequences were known (Michel et al., 1983), and could be used to distinguish between the subunits. At that stage, some use was made of the local symmetry of the subunits L and M. Finally, as the gene sequences of the reaction centre subunits were determined (Michel et al., 1985; Michel et al., 1986a; Weyer et al., 1986a), the model of the protein subunits was completed. The sequence information led to an overall verification but also to a number of minor corrections of the polypeptide backbone model, since in the previous model building stage the electron density was not always clear enough to allow determination of the correct number of amino acids.

Our tools for model building were interactive graphics display systems: a black and white Vector General 3400 system, and later a color Evans & Sutherland PS 300. On both systems we used Aiwyn Jones' program package FRODO (Jones, 1978). The model library of this package was extended to include bacteriochlorophyll-b, bacteriopheophytin-b, menaquinone-7, and ubiquinone-1. Frequent use was made of the real-space-refinement facility in FRODO which allowed to correctly place long stretches of helical structure into the electron density.

Structure Overview

...It is a complex of four protein subunits, and of 14 cofactors. The protein subunits are called H (heavy), M (medium), L (light), and cytochrome; the names H, M, and L were chosen according to the apparent molecular weights of the subunits, as determined by electrophoresis. The core of the complex is formed by the subunits L and M, and their associated cofactors: four bacteriochlorophyll-b (BChl-b), two bacteriopheophytin-b (BPh-b), one nonheme iron, two quinones, one carotenoid. Structural properties, e.g. the hydrophobic nature of the protein surface, and functional considerations strongly indicate that the subunits L and M span the bacterial membrane...Each of the subunits L and M contains five membrane spanning polypeptide segments, folded into long helices. The polypeptide segments connecting the transmembrane helices form flat surfaces parallel to the membrane surfaces...

Other examples of the study of photosynthetic reactions centers include [6, 7].

Honours and Awards

Deisenhofer is holder of: the Biological Physics Prize of the American Physical Society, jointly with Hartmut Michel (1986); the Otto Bayer Prize, jointly with Hartmut Michel (1988); the Nobel Prize for Chemistry, jointly with Robert Huber and Hartmut Michel (1988).

He is a Fellow of the American Association for the Advancement of Science, and a Member of the Academia Europaea, the National Academy of Sciences, the German Academy of Natural Scientists Leopoldina, the American Crystallographic Association and Sigma Xi.

He was presented with Honorary Citizenships of Zusamaltheim and of Dallas, and is a Member of the Order of the Federal Republic of Germany.

Bibliography

1. Reproduced with permission.
2. From Nobel Lectures, Chemistry 1981–1990, Editor-in-Charge Tore Frängsmyr, Editor Bo G. Malmström, World Scientific Publishing Co., Singapore, 1992.
3. Who's who: 2005.
4. www.nobelprize.org.
5. http://www8.utsouthwestern.edu/utsw/cda/dept108256/files/158233.html.
6. Electrogenic Reactions in Photosynthetic Reactions Centres of Purple Bacteria/Eucaryotic Features of Thermoacidophilic Archaebacteria (Soviet Scientific Reviews Series, Section D), by PRANGISHVILI, D. A. Prangishvili and G. D. Muskhelishvili, Taylor & Francis, 1991,ISBN: 3718651149.
7. Investigation of the oxygenic photosynthetic reaction centre photosystem II with specific isotope labelling: Synthesis and incorporation of stable-isotope labelled (S)-phenylalanine and (S)-tyrosine by Saskia Apollonia Maria Nieuwenhuis, S.A.M. Nieuwenhuis, 1998, ASIN: B0006FCPDO.

1988

© Nobel Foundation [1]

Robert Huber (0.33)

"for the determination of the three-dimensional structure of a photosynthetic reaction centre" [2]

During his presentation speech, Bo G. Malmström, of the Royal Swedish Academy of Sciences, commented of Huber and his work [2]:

> *...For a long time it has been impossible to prepare membrane-bound proteins in a form allowing the determination of the detailed structure in three dimensions. Before 1984, there were only rather fuzzy structural pictures available for a few membrane proteins. These had been derived with the aid of an electron microscopic method developed by the Englishman Aaron Klug, who was awarded the Nobel Prize in Chemistry in 1982 for this achievement. But the situation had actually drastically changed in 1982, when Hartmut Michel thanks to systematic experiments succeeded in preparing highly ordered crystals of a photosynthetic reaction center from a bacterium. With these crystals he could in the period 1982–1985, in collaboration with Johann Deisenhofer and Robert Huber, determine the structure of the reaction center in atomic detail.*
>
> *The structural determination awarded has led to a giant leap in our understanding of fundamental reactions in photosynthesis, the most important chemical reaction in the biosphere of our earth. But it has also consequences far outside the field of photosynthesis research. Not only photosynthesis and respiration are associated with membrane-bound proteins but also many other central biological functions, e.g. the transport of nutrients into cells, hormone action or nerve impulses. Proteins participating in these processes must span biological membranes, and the structure of the reaction center has delineated the structural principles for such proteins. Michel's methodological contribution has, in addition, the consequence that there is now hope that we can determine detailed structures also for many other membrane proteins. Not least important is the fact that the reaction center structure has given theoretical chemists an indispensable tool in their efforts to understand how biologic electron transfer over very large distances on a molecular scale can occur as rapidly as in one billionth (American English, trillionth) of a second. In a longer perspective it is*

possible that such research can lead to important energy technology in the form of artificial photosynthesis...

Robert Huber was born on 20th February 1937 in Munich, the first son of two children of Sebastian (a bank cashier) and Helene (a housewife) Huber. He married Christa Essig (1960) and they have four children, two daughters (1961 and 1976) and two sons (1963 and 1966). Huber is the Director of the Max-Planck-Institut für Biochemie and Professor at the Technische Universität München and can be contacted through either organization [2,3].

The Early Years

Huber attended the Humanistische Karls-Gymnasium in Munich (1946–56), prior to study at the Technische Universität in München, through a stipend of the Bayerisches Ministerium für Erziehung und Kultur and later, of the Studienstiftung des Deutschen Volkes. He graduated with a BS in chemistry (1960), having completed a project on crystallographic studies of the insect metamorphosis hormone ecdysone under W. Hoppe. Part of these studies were made in Karlson's laboratory at the Physiologisch-Chemisches Institut der Universität München. He continued at the Technical University graduating with a Dr. rer. Nat (1963), the thesis work for which concerned the crystal structure of a diazo compound.

The Career

Year	*Post*	*Institute*
1968	Lecturer	Technical University, Munich
1972 – to date	Director	Max Plank Institute for Biochemistry
1976 – to date	Associate Professor (additional post)	Technical University, Munich

Huber has been a Member of the Max Plank Society since 1972.

Huber's group has made many advances and developed many methods for protein crystallography, leading to the development of refinement in protein crystallography (with Steigemann, Deisenhofer, Remington). The group has also developed the Patterson search methods (with Bartels and Fehlhammer) and computer programmes for:

- area detector data collection (MADNES, with Pflugrath, Messerschmidt)
- computer graphics and electron density interpretation and refinement (FRODO, with Jones)
- intensity data evaluation and absorption correction (FILME, with Bartels, Bennett, Schwager)
- protein crystallographic computing (PROTEIN, with Steigemann)

The above techniques have been used to study:

- crystallographic work on the insect protein erythrocruorin (1976, with Formanek)
- inhibitors of cysteine proteases
- creatine hydrolase (with Hoeffken)

- the structures of citrate synthase in different states of ligation (with Remington, Wiegand)
- proteolytic enzymes and their natural inhibitors
- a number of structure determinations of organic compounds
- the basic pancreatic trypsin inhibitor (1970)
- a-amylase (with Pflugrath, Wiegand)
- immunoglobulins and their fragments (early 1970s)
- the first glycoprotein to be analysed in atomic detail (with Colman, Deisenhofer, Epp, Marquart, Matsushima)
- the elucidation of the structure and the chemical nature of the selenium moiety in glutathione peroxidase (with Ladenstein, Epp)
- proteins involved in excitation energy and electron transfer
- heavy riboflavin synthase (with Ladenstein).bilin-binding protein
- light-harvesting proteins (with Schirmer, Bode)
- ascorbate oxidase (with Messerschmidt, Ladenstein)
- the trypsinogen system (with Bode)

Huber is editor of the Journal of Molecular Biology (1976 – to date) and has written *The Handbook of Metalloproteins*: 2 Volume Set, with Albrecht Messerschmidt (Editor), Thomas Poulos (Editor) and Karl Wieghardt (Editor), Pub: John Wiley & Sons; (2001), ISBN: 0471627437

Huber's Nobel Prize lecture is dated 8th December 1988 with the title 'A Structural basis of light energy and electron transfer in biology'. During his *abstract*, Huber commented [2]:

> *...Aspects of inramolecular light energy and electron transfer will be discussed for three protein cofactor complexes, whose three-dimensional structures have been elucidated by X-ray crystallography: Components of light harvesting cyanobacterial phycobilisomes, the purple bacterial reaction centre, and the blue multi-copper oxidases. A wealth of functional data is available for these systems which allow specific correlations between structure and function and general conclusions about light energy and electron transfer in biological materials to be made...*

He went on to discuss

> *...will focus in my lecture on Boltzmann's 'improbable transition forms', namely, excited electronic states and charge transfer states in modern terminology, the structures of biological materials involved and the interplay of cofactors (pigments and metals) and proteins. I will discuss some aspects of the photosynthetic centre of Rps. viriilis...and of functionally related systems, whose structures have been studied in my laboratory: Light harvesting cyanobacterial phycobilisomes and blue oxidases. A wealth of structural and functional data is available for these three systems, which make them uniquely appropriate examples from which to derive general principles of light energy and electron transfer in biological materials. Indeed, there are very few systems known in sufficient detail for such purposes.*

We strive to understand the underlying physical principles of light and electron conduction in biological materials with considerable hope for success as these processes appear to be more tractable than other biological reactions, which involve diffusive motions of substrates and products and intramolecular motions. Large-scale motions have been identified in many proteins and shown to be essential for many functions (Bennett & Huber, 1983; Huber, 1988). Theoretical treatments of these reactions have to take flexibility and solvent into account and become theoretically tractable only by applying the rather severe approximations of molecular dynamics (Karplus & McCammon, 1981; Burkert & Allinger, 1982) or by limiting the system to a few active site residues, which can then be treated by quantum mechanical methods.

Light and electron transfer processes seem to be amenable to a more quantitative theoretical treatment. The substrates are immaterial or very small and the transfer processes on which I focus are intramolecular and far removed from solvent. Molecular motions seem to be unimportant, as shown by generally small temperature dependences. The components active in energy and electron transport are cofactors, which, in a first approximation suffice for a theoretical analysis, simplifying calculations considerably...

Other examples of the study of photosynthetic reactions centers include [4,5].

Honours and Awards

Huber has received Honorary D.Sc. degrees from the: Catholic University of Louvain (1987); University of Ljubljana (1989); University of Tor Vergata (1991); University of Barcelona (2000); University of Lisbon (2000).

He has received many awards and prizes including: the E.K. Frey Medal Gesellschaft für Chirurgie (1972); the Otto-Warburg Medal (1977); the Emil von Behring Medal (1982); the Keilin Medal (1987); the Richard Kuhn Medal (1987); the Nobel Prize for Chemistry, jointly with Deisenhofer and Michel (1988); the E.K. Frey-E. Werle Gedächtnismedaille (1989); the Kone Award (1990); the Sir Hans Krebs Medal (1992); the Linus Pauling Medal (1993–94); the Distinguished Service Award (1995); the Max Bergmann Medal (1997); the Max Tishler Prize (1997); the Bayerischer Maximiliansorden für Wissenschaft und Kunst (1993); the Grosse Verdienstkreuz mit Stern und Schulterband (1997).

Huber is a Member of: the EMBO (also Member of Council) (1976); the Deutsche Chemische Gesellschaft (1988); the Ges für Biologische Chemie (1988); the Bavarian Academy of Science (1988). He is an Associate Fellow of the Third World Academy of Sciences (1995), a Foreign Associate of the National Academy of Sciences USA (1995), a Fellow of the American Academy of Microbiology (1996) and a Foreign Member of the Royal Society (1999). He is a Corresponding Member of the Croatian Academy of Sciences and Arts, and a Honorary Member of the American Society of Biological Chemists. He is a Member of the: Deutsche Akademie der Naturforscher Leopoldina; Accademia Nazionale dei Lincei, Rome; the Orden Pour le Mérite für Wissenschaften und Künste; the Swedish Society of Biophysics; the Japanese Biochemical Society.

Bibliography

1. Reproduced with permission.
2. From Nobel Lectures, Chemistry 1981–1990, Editor-in-Charge Tore Frängsmyr, Editor Bo G. Malmström, World Scientific Publishing Co., Singapore, 1992.
3. Who's who: 2005.
4. Electrogenic Reactions in Photosynthetic Reactions Centres of Purple Bacteria/Eucaryotic Features of Thermoacidophilic Archaebacteria, (Soviet Scientific Reviews Series, Section D), by D. A. Prangishvili and G. D. Muskhelishvili, Taylor & Francis, 1991, ISBN: 3718651149.
5. Investigation of the oxygenic photosynthetic reaction centre photosystem II with specific isotope labelling: Synthesis and incorporation of stable-isotope labelled (S)-phenylalanine and (S)-tyrosine, by Saskia Apollonia Maria Nieuwenhuis, S.A.M. Nieuwenhuis, 1998, ASIN: B0006FCPDO.

1988

© Nobel Foundation [1]

Hartmut Michel (0.33)

"for the determination of the three-dimensional structure of a photosynthetic reaction centre" [2]

During his presentation speech, Bo G. Malmström, of the Royal Swedish Academy of Sciences, commented of Michel and his work [2]:

> *...For a long time it has been impossible to prepare membrane-bound proteins in a form allowing the determination of the detailed structure in three dimensions. Before 1984, there were only rather fuzzy structural pictures available for a few membrane proteins. These had been derived with the aid of an electron microscopic method developed by the Englishman Aaron Klug, who was awarded the Nobel Prize in Chemistry in 1982 for this achievement. But the situation had actually drastically changed in 1982, when Hartmut Michel thanks to systematic experiments succeeded in preparing highly ordered crystals of a photosynthetic reaction center from a bacterium. With these crystals he could in the period 1982–1985, in collaboration with Johann Deisenhofer and Robert Huber, determine the structure of the reaction center in atomic detail.*
>
> *The structural determination awarded has led to a giant leap in our understanding of fundamental reactions in photosynthesis, the most important chemical reaction in the biosphere of our earth. But it has also consequences far outside the field of photosynthesis research. Not only photosynthesis and respiration are associated with membrane-bound proteins but also many other central biological functions, e.g. the transport of nutrients into cells, hormone action or nerve impulses. Proteins participating in these processes must span biological membranes, and the structure of the reaction center has delineated the structural principles for such proteins. Michel's methodological contribution has, in addition, the consequence that there is now hope that we can determine detailed structures also for many other membrane proteins. Not least important is the fact that the reaction center structure has given theoretical chemists an indispensable tool in their efforts to understand how biologic electron transfer over very large distances on a molecular scale can occur as rapidly as in one billionth*

(American English, trillionth) of a second. In a longer perspective it is possible that such research can lead to important energy technology in the form of artificial photosynthesis...

Hartmut Michel was born in Ludwigsburg, Württemberg on the 18th July 1948, the elder son of Karl (a factory joiner) and Frieda (a dressmaker) Michel. He married Ilona Leger and they have two children Andrea (1981) and Robert Joachim (1984). Michel is the Director of the Department of Molecular Membrane Biology at the Max-Planck-Institute of Biochemistry and Adjunct Professor (Biochemistry) at the University of Frankfurt [2, 3].

The Early Years

Michel was an avid reader of books concerning archaeology, ethnology, geography and zoology as a child (aged eleven). His favourite school subjects were biology, chemistry, history and physics. He entered the University of Tübingen (1969–74) (following military service in the army), during which he spent a year in various biochemistry labs at the University of Munich and the Max-Planck-Institute of Biochemistry. He completed his Dr. rer. nat. under D. Oesterhelt at the University of Würzburg (1977), during which he correlated the intracellular levels of adenosine di- and triphosphate with the electrochemical proton gradient across the halobacterial cell membrane. He remained at the University of Würzburg (1977–1979) for his postdoctoral studies, during which he tried to fuse delipidated bacteriorbodopsin with bacterial vesicles, in order to achieve light-driven amino acid uptake.

The Career

Year	*Post*	*Institute*
1979–1987	Group Leader (Membrane Biochemistry)	Max-Planck-Institute of Biochemistry
1987 to date	Director, Department of Molecular Membrane Biology	Max-Planck-Institute of Biochemistry
1988 to date	Adjunct Professor (Biochemistry)	University of Frankfurt

Michel spent a year of ‘Habilitation’ (1986) at the University of Munich.

Michel’s interests in biochemistry have included:

- obtaining the first crystals of the photosynthetic reaction centre from the purple bacterium *Rhodopseudomonas viridis* with Johann Deisenhofer (1981)
- understanding the function and mechanism of membrane proteins based on accurately known structures
- the crystallization of membrane proteins and the subsequent X-ray crystallographic analysis.
- molecular bioenergetics
- understanding function and mechanism
- investigations into the development of methods to overproduce receptors, especially G-protein coupled receptors
- structural membrane genomics

- a study of the enzymes of the methane biosynthesis pathway
- investigations of the mechanism of action of Cytochrome c Oxidase
- a study of membrane bound transport proteins from the hyperthermophilic eubacterium Aquifex aeolicus

Michel edited the book *Crystallization of Membrane Proteins* Pub: Crc Pr I Llc (1990) ISBN: 0849348161.

Michel's Nobel Prize lecture was co-authored with Deisenhofer and is dated 8th December 1988 with the title 'The photosynthetic reaction center from the purple bacterium *Rhodopseudomonas Viridis*'. At the time, Michel's expertise lay in sample preparation which he covered in the presentation [2]:

1.1 The Background

> *...As in many instances of new scientific developments and technical inventions accidental observation caused the beginning of the experiments, which ultimately resulted in the elucidation of the three dimensional structure of a photosynthetic reaction centre. This initiating observation was the formation of solid, most likely glass-like aggregates in August 1978, when bacteriorhodopsin delipidated according to Happe and Overath (1976) was stored in the freezer. From there on I was convinced that it should be possible not only to obtain these solid bodies but also to produce three dimensional crystals. The availability of well ordered three-dimensional crystals is the prerequisite for a high resolution X-ray crystallographic analysis, which - despite the progress made by Henderson and Unwin (1975) with electron microscopy and electron diffraction on bacteriorhodopsin - was and still is the only way to obtain detailed structural knowledge of large biological macromolecules...*

1.2 A More Systematic Approach

> *...Based on the properties of membrane proteins, a new strategy was developed (Michel, 1983). Membrane proteins are embedded into the electrically insulating lipid bilayers. The difficulties in handling membrane proteins reside in the amphipathic nature of their surface. They possess a hydrophobic surface where in the membrane they are in contact with the alkane chains of the lipids and they have a polar surface where they are in contact with the aqueous phases on both sides of the membrane and the polar head-groups of the lipids. As a result, membrane proteins are not soluble in aqueous buffers or in organic solvents of low dielectric constant. In order to solubilize membrane proteins one has to add detergents. Detergents are amphiphilic molecules which form micelles above a certain concentration, the so-called critical micellar concentration. The detergent micelles take up the membrane proteins and shield the hydrophobic surface parts of the membrane protein from contact with water...*

1.3 The Turn to Classical Photosynthesis

> *...Frustrated from the lack of the final breakthrough with bacteriorhodopsin, which is partly due to the absence of large extramembranous domains in this protein, I looked for more promising membrane proteins to be crystallized. My choices were the photosynthetic reaction centres from the purple bacteria Rhodospirillum rubrum and Rhodopseudomonas viridis and the light- harvesting*

chlorophyll a/b protein from spinach. It was influenced by the fact that these proteins (or protein complexes) were said to be part of a two-dimensional crystalline array already in their native environment. As additional benefits, they were available in large quantities, could easily be isolated, were colored and denaturation of the proteins was indicated by color changes.

I learnt about the Rhodopseudomonas viridis system when E. Wehrli from the ETH Zurich presented the results of electron microscopical studies during a workshop at Burg Gemen, Germany, in June 1979 (Baumeister and Vogell, 1980). Initially, I received some isolated photosynthetic membranes from him in December 1980. At that time I had moved with D. Oesterhelt to the Max-Planck-Institut für Biochemie at Martinsried near Munich and I was just back from a stay at the MRC in Cambridge where we had done Xray diffraction experiments on the bacteriorhodopsin crystals. I isolated the reaction centres using hydroxyapatite chromatography according to a published procedure (Clayton and Clayton, 1978) and tried to crystallize it, without success. I developed a new isolation procedure using only molecular sieve chromatography, tried it again and met immediate success (Michel, 1982b). The conditions were nearly identical to those found to be optimal for bacteriorhodopsin. The exception was that I could use N,N-dimethyldodecytamine-N-oxide as detergent instead of octylglucopyranoside...

Other examples of the study of photosynthetic reactions centers include [4,5].

Honours and Awards

Michel has been awarded the: Biophysics Prize of the American Physical Society, jointly with J. Deisenhofer (1986); Otto Klung-Preis for chemistry (1986); Leibniz-Preis of the Deutsche Forschungsgemeinschaft; Otto-Bayer-Preis, jointly with J. Deisenhofer (1988); Nobel Prize, jointly with J. Deisenhofer and R. Huber (1988).

Bibliography

1. Reproduced with permission.
2. From Nobel Lectures, Chemistry 1981–1990, Editor-in-Charge Tore Frängsmyr, Editor Bo G. Malmström, World Scientific Publishing Co., Singapore, 1992.
3. Who's who: 2005.
4. Electrogenic Reactions in Photosynthetic Reactions Centres of Purple Bacteria/Eucaryotic Features of Thermoacidophilic Archaebacteria, (Soviet Scientific Reviews Series, Section D), by D. A. Prangishvili and G. D. Muskhelishvili, Taylor & Francis, 1991, ISBN: 3718651149.
5. Investigation of the oxygenic photosynthetic reaction centre photosystem II with specific isotope labelling: Synthesis and incorporation of stable-isotope labelled (S)-phenylalanine and (S)-tyrosine by Saskia Apollonia Maria Nieuwenhuis, S.A.M. Nieuwenhuis, 1998, ASIN: B0006FCPDO.

1989

© Nobel Foundation [1]

Sidney Altman (0.5)

"for their discovery of catalytic properties of RNA" [2]

During his presentation speech on the 10th December 1989, Bertil Andersson, of the Royal Swedish Academy of Sciences, commented of Altman and his work [2]:

> *...As recently as the early 1980s, the generally accepted view among scientists was that enzymes were proteins. The idea of proteins having a monopole of biocatalytic capacity has been deeply rooted, and created a fundamental dogma of biochemistry. This is the very basic perspective in which we have to regard the discovery today being rewarded with the Nobel Prize for Chemistry. When Sidney Altman showed that the enzyme denoted RNaseP only needed RNA in order to function, and when Thomas Cech discovered self-catalytic splicing of a nucleic acid fragment from an immature RNA molecule, this dogma was well and truly holed below the waterline. They had shown that RNA can have catalytic capacity and can function as an enzyme. The discovery of catalytic RNA came as a great surprise and was indeed met with a certain amount of scepticism. Who could ever have suspected that scientists, as recently as in our own decade, were missing such a fundamental component in their understanding of the molecular prerequisites of life? Altman's and Cech's discoveries not only mean that the introductory chapters of our chemistry and biology textbooks will have to be rewritten, they also herald a new way of thinking and are a call to new biochemical research.*
>
> *The discovery of catalytic properties in RNA also gives us a new insight into the way in which biological processes once began on this earth, billions of years ago. Researchers have wondered which were the first biological molecules. How could life begin if the DNA molecules of the genetic code can only be reproduced and deciphered with the aid of protein enzymes, and proteins can only be produced by means of genetic information from DNA? Which came first, the chicken or the egg? Altman and Cech have now found the missing link. Probably it was the RNA molecule that came first. This molecule has the properties needed by an original biomolecule, because it is capable of being both genetic code and enzyme at one and the same time...*

Sidney Altman was born Montreal, Canada, on 7th May 1939, the second son of Victor Altman (the owner of a grocery store) and Ray, née Arlin, who worked in a textile mill. Altman married Ann Korner (1972) and they have two children Daniel (1974) and Leah (1977). Altman is the current Sterling Professor of Molecular, Cellular & Developmental Biology at Yale University. Altman holds dual citizenship, both in Canada and in the United States [2–4].

The Early Years

Altman grew up in the Notre Dame de Grace suburb of Montreal, and as an avid book reader. He was given a copy of *Explaining The Atom* by Selig Hecht when twelve years old, which started the lifelong interest in science. At that time, Einstein's public profile influenced Altman towards physics, which he studied for his BSc at the Massachusetts Institute of Technology (MIT), graduating in 1960. During his final year at MIT, Altman took a course, presented by Cyrus Levinthal, in molecular biology. Eighteen months graduate work in physics at Columbia University was followed by PhD studies in molecular biology at the University of Colorado medical school, under L. Lerman, studying the effects of acridines on the replication of bacteriophage T4 DNA. He graduated with his PhD (1967), and followed this by two years of postdoctoral work in M. Meselson's laboratory at Harvard University, where he studied DNA endonuclease involved in the replication and recombination of T4 DNA.

With Lerman's help, Altman then moved to the Medical Research Council Laboratory of Molecular Biology in Cambridge University, in Crick and Brenner's laboratory, working on RNase P and the enzymatic properties of the RNA subunit of that enzyme.

The Career

Year	*Post*	*Institute*
1971	Assistant Professor	Yale University
1980	Professor	Yale University

Altman has been Chairman of the Biology Department (1983–1985) and Dean of Yale College (1985–1989) and is the current Sterling Professor of Biology at Yale University.

His research interests have involved:

- the discovery of the first radiochemically pure precursor to a tRNA molecule
- post-transcriptional RNA processing as a means of gene regulation
- genetic and biochemical analysis of events in tRNA biosynthesis (in E. coli, yeast and human tissue)
- the action of specific RNases
- an investigation of the novel enzyme M1 RNA (the RNA subunit of RNase P from E. coli)
- the nature of the catalytic site
- enzyme-substrate interactions
- a feasibility study the goal of which is the use of both human and E. coli RNase P as a means to inactivate any targeted RNA (in vivo or in vitro)

- a study of drug resistance in E. coli and influenza virus infection
- the function and structure of ribonuclease P in both bacteria and human cells
- exploring the use of RNase P and external guide sequences to activate various genes in bacteria and mammalian cells.

Altman has published many articles in learned journals. He has also been a Member of the international committee which reviewed fundamental research in Sweden, the results from which were published as:

International review of Swedish research in fundamental chemistry: A report to the Research Council by an international review committee with the following members, Sidney Altman... [et al.] Pub: Swedish Natural Science Research Council (1995) ISBN: 9154603420.

He was also the editor of *Transfer RNA* (Publication of the Joint Center for Urban Studies of the Mas), Pub: The MIT Press, (1978), ISBN: 0262010569.

Altman's Nobel Prize lecture is dated 8th December 1989 with the title 'Enzymatic cleavage of RNA by RNA'.

> *...The transfer of genetic information from nucleic acid to protein inside cells can be represented as shown in the figure. This simple scheme reflects accurately the fact that the information contained in the linear arrangement of the subunits of DNA is copied accurately into the linear arrangement of subunits of RNA which, in turn, is translated by machinery inside the cell into proteins, the macromolecules responsible for governing many of the important biochemical processes* ***in vivo****. The function of the straightforward transfer of information is carried out by a class of molecules called messenger RNAs (mRNAs). The diagram shown does not elaborate on the properties of other RNA molecules that are transcribed from DNA, namely transfer RNA (tRNA) and ribosomal RNA (rRNA) and many other minor species of RNA found in vivo that had no identifiable function prior to 1976, nor does it indicate that the information in DNA and RNA can be replicated as daughter DNA and RNA molecules, respectively (see Crick, 1970, for further discussion).*

> *Ribosomes are complexes, which in* ***Escherichia coli****, are made of about 50 proteins and three RNA molecules. It is on these particles that mRNA directs the synthesis of protein from free amino acids. LRNA molecules perform an adaptor function in the sense that they match particular amino acids to a group of three specific nucleotides in the mRNA to be translated and ensure that the growing polypeptide (protein) chain contains the right linear sequence of amino acid subunits. Thus, rRNA and tRNA participate in the process of information transfer inside cells but they clearly do so in a comparatively complex manner...*

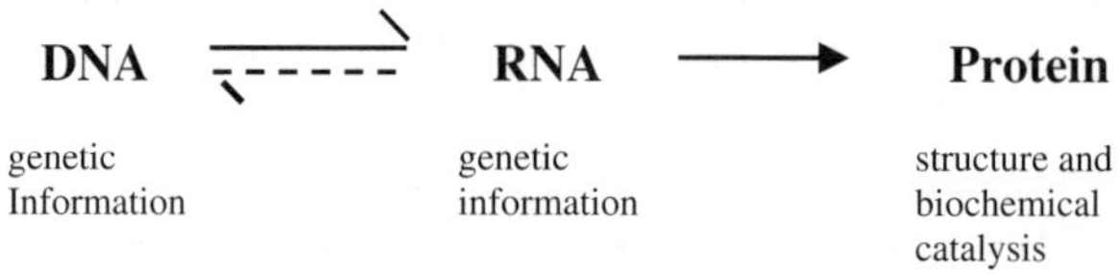

> *...My work on RNA began as a study of certain mutants that disrupted the ability of tRNA molecules to function normally during translation (Altman, 1971). This research, in turn, led to the identification of another stable RNA molecule that had, unexpectedly, all the properties of an enzyme (Guerrier-Takada et at., 1983). Aside from its intrinsic interests to students of catalysis*

and enzymology, our finding of enzymatic activity associated with RNA has stimulated reconsideration of the role of RNA in biochemical systems today (see Cech, 1987, and Altman, 1989, for reviews) as illustrated in the figure, and of the nature of complex biochemical systems eons ago (Darnell and Doolittle, 1986; Westheimer, 1986; Weiner and Maizels, 1987; Joyce, 1989).

As was first pointed out over twenty years ago by Woese (1967), Crick (1968) and Orgel (1968), if RNA can act as a catalyst the origin of the genetic code plays a much less critical role in the early stages of evolution of the first biochemical systems that were capable of replicating themselves. Indeed, the variety of biochemical reactions now known to be governed by RNA... allows one to consider the possibility that a large number of diverse enzymatic reactions took place in the absence of protein. To add further substance to these ideas about life on earth over a billion years ago, it is important to understand exactly how catalytic RNA, as we know it, works and what role it plays ***in vivo*** *today...*

All aspects of RNA have continued to be studied throughout the world, some of the data for which appears in books, see for example the following texts [5–7].

Honours and Awards

Altman has received the: National Institutes of Health Merit Award (1989); the Nobel Prize in Chemistry, jointly with Cech (1989); the Rosenstiel Award for Basic Biomedical Research (1989); the Yale Science and Engineering Association Award (1990).

Altman has also been a Member of the Council on Foreign Relations – a private organization in U.S. foreign policy.

Bibliography

1. Reproduced with permission.
2. From Nobel Lectures, Chemistry 1981–1990, Editor-in-Charge Tore Frängsmyr, Editor Bo G. Malmström, World Scientific Publishing Co., Singapore, 1992.
3. http://www.science.ca/scientists/scientistprofile.php?pID=3.
4. http://www.biology.yale.edu/facultystaff/altman.html.
5. Non-Coding RNAs: Molecular Biology and Molecular Medicine (Molecular Biology Intelligence Unit (Unnumbered).), Springer, 2003, ISBN: 0306478358.
6. Catalytic Rna (Nucleic Acids and Molecular Biology, Vol 10) edited by Fritz Eckstein and David M. J. Lilley, Springer-Verlag Telos, 1996, ISBN: 3540607951.
7. RNA Interference Technology: From Basic Science to Drug Development edited by Krishnarao Appasani, Cambridge University Press, 2005, ISBN: 0521836778.

1989

© Nobel Foundation [1]

Thomas R. Cech (0.5)

"for their discovery of catalytic properties of RNA" [2]

During his presentation speech on 10th December 1989, Bertil Andersson, of the Royal Swedish Academy of Sciences, commented of Cech and his work [2]:

...As recently as the early 1980s, the generally accepted view among scientists was that enzymes were proteins. The idea of proteins having a monopole of biocatalytic capacity has been deeply rooted, and created a fundamental dogma of biochemistry. This is the very basic perspective in which we have to regard the discovery today being rewarded with the Nobel Prize for Chemistry. When Sidney Altman showed that the enzyme denoted RNaseP only needed RNA in order to function, and when Thomas Cech discovered self-catalytic splicing of a nucleic acid fragment from an immature RNA molecule, this dogma was well and truly holed below the waterline. They had shown that RNA can have catalytic capacity and can function as an enzyme. The discovery of catalytic RNA came as a great surprise and was indeed met with a certain amount of scepticism. Who could ever have suspected that scientists, as recently as in our own decade, were missing such a fundamental component in their understanding of the molecular prerequisites of life? Altman's and Cech's discoveries not only mean that the introductory chapters of our chemistry and biology textbooks will have to be rewritten, they also herald a new way of thinking and are a call to new biochemical research.

The discovery of catalytic properties in RNA also gives us a new insight into the way in which biological processes once began on this earth, billions of years ago. Researchers have wondered which were the first biological molecules. How could life begin if the DNA molecules of the genetic code can only be reproduced and deciphered with the aid of protein enzymes, and proteins can only be produced by means of genetic information from DNA? Which came first, the chicken or the egg? Altman and Cech have now found the missing link. Probably it was the RNA molecule that came first. This molecule has the properties needed by an original biomolecule, because it is capable of being both genetic code and enzyme at one and the same time...

Thomas Cech was born in Chicago, USA, on 8th December 1947, the son of Robert Franklin Cech (a physician) and Annette Marie, née Cerveny (homemaker), and brother to Barbara and Richard. Cech married Carol Lynn Martison (also a biochemist) in 1970. They have two daughters Allison (1982) and Jennifer (1986). Cech is President of the Howard Hughes Medical Institute and Distinguished Professor at the University of Colorado [2,3].

The Early Years

Arguably, Cech's earliest scientific influence was his father, who was a medical practitioner with a strong interest in physics. In the fourth grade, Cech recollects knocking on the '*Geology professors' doors at the University of Iowa, asking to see models of crystal structures and to discuss meteorites and fossils*' [2]. He entered Grinnell College (1966), during which his interests developed towards biological chemistry, although he graduated with a chemistry degree (B.A.). He met his future wife whilst at Grinnell College. They both entered the University of California at Berkeley (1970), where Cech's thesis, under J. Hearst, concerned chromosome structure and function. Graduation for both Carol and Cech, with their respective PhD's in 1975, was followed by postdoctoral positions in Massachusetts. Cech went to the Massachusetts Institute of Technology, where he worked in Mary-Lou Pardue's laboratory, and Carol to Harvard University.

The Career

Year	*Post*	*Institute*
1978–83	Assistant then Associate Professor	University of Colorado
1983–to date	Professor of Chemistry and Biochemistry and of Molecular, Cellular and Devt Biology	University of Colorado
1990–to date	Distinguished Professor	University of Colorado
1988–99	Investigator	Howard Hughes Medical Institute
2000–to date	President	Howard Hughes Medical Institute

Cech was a National Science Foundation Fellow (1970–75), a Public Health Service Research Fellow, National Cancer Institute (1975–77) and a Guggenheim Fellow (1985–86). He is an American Cancer Society Lifetime Research Professor (1987) and a Professor of Biochemistry, Biophysics and Genetics at the University of Colorado Health Sciences Center in Denver.

Cech has been active in many areas of biochemistry including [4,5,6]:

- studies concerning the structure and function of catalytic RNA molecules
- the chemical and the biological mechanisms of RNA catalysis
- the discovery of the catalytic protein subunit of telomerase and cloned and sequenced genes
- the activity and regulation of telomerase
- encoding telomerase clones into Euplotes, S. pombe, human, Tetrahymena and Oxytricha
- the structure of ribozyme 247 nt at 5 Å resolution

- the structural biology of ribozymes
- the structures of active group I introns
- investigating protein sequence specificity of ssDNA

He has been written a book called *Genes We Share With Yeast, Flies, Worms and Mice: New Clues to Human Health and Disease* by Thomas R. Cech (Foreword), Maya Pines (Introduction), Pub: Diane Pub Co (J, 2003) ISBN: 0756720192.

Cech is a Deputy Editor of *Science* and a Member of the Editorial Board of *Genes and Development*.

His Nobel Prize lecture is dated 8th December 1989 with the title 'Self-splicing and enzymatic activity of an intervening sequence RNA from *Tetrahymena*'. During the course of his presentation, Cech commented [2]:

> *...Unlike most nuclear genes, which are embedded in giant chromosomes, the genes for rRNA in* ***Tetrahymena*** *are located on small DNA molecules in the nucleoli; they are extrachromosomal (Engberg et al., 1974; Gall, 1974). Furthermore, in the transcriptionally active macronucleus the gene is amplified to a level of ~10,000 copies (Yao et al., 1974). These properties made it possible to purify a significant amount of the rDNA. The ability to purify the gene was not in itself a major attraction, because by this time the availability of recombinant DNA techniques ensured that no gene would long escape isolation and sequence analysis. Rather the attractive feature was the prospect of being able to isolate the gene complete with its associated structural proteins and proteins that regulated transcription (the synthesis of an RNA copy by RNA polymerase)...*
>
> *...RNA Splicing in vitro*
>
> *The first step towards biochemical dissection of the transcriptional process was to see if rRNA synthesis would proceed in a crude cell-free system. We isolated nuclei from* ***T. thermophila*** *(the nuclei provided both the RNA polymerase and the ribosomal chromatin templates) and incubated them with the nucleoside triphosphates and salts necessary for transcription. We also included α-amanitin, a mushroom toxin known to inhibit the polymerases that transcribe mRNA, tRNA and other small RNAs, which enabled us to focus on synthesis of the large rRNA...*
>
> *...self-splicing unrecognized*
>
> *Our strategy for purifying the splicing enzyme was conventional. We would find conditions in which pre-rRNA transcription occurred but splicing was inhibited, and purify the accumulated pre-rRNA to use as a substrate. We would then treat the pre-rRNA with extracts of* ***Tetrahymena*** *nuclei, which we already knew contained splicing activity, and would use gel electrophoresis to monitor the splicing reaction. Finally, we would obtain ever purer subfractions of the nuclear extract that retained activity, eventually isolating the splicing enzyme.*
>
> *Isolation of unspliced pre-rRNA substrate proved to be straightforward (Cech et al., 1981). Art Zaug purified this RNA by standard SDS-phenol extraction and used it in a series of RNA splicing reactions. One set of test tubes contained substrate RNA and nuclear extract dissolved in the same solution of simple salts and nucleotides that had been conducive to RNA transcription and splicing in our earlier experiments with intact nuclei. Another tube, containing the same components except with the nuclear extract omitted, was to serve as the "splicing minus" control.*

The very first attempt was successful. The RNA in the tubes containing the nuclear extract gave rise to the small hand of RNA characteristic of the IVS. Surprisingly, however, the control RNA incubated only in salts arid nucleotides produced the same amount of IVS. "Well, Art, this looks very encouraging, except you must have made some mistake making up the control sample." Yet several careful repetitions of the experiment gave the same result: release of the IVS occurred independent of the addition of nuclear extract, and therefore apparently independent of any enzyme. I became concerned that we weren't observing RNA splicing at all; perhaps the RNA we had been calling "precursor" had already been spliced in vivo…

All aspects of RNA have continued to be studied throughout the world, some of the data for which appears in books, see for example the following texts [7–9].

Honours and Awards

Cech has received Honorary DSc degrees from: Grinnell College (1987); University of Chicago (1991); Drury College (1994); Colorado College (1999); University of Maryland (2000); Williams College (2000); Charles University, Prague (2002).

He has been presented with many awards and prizes including: the Medal of the American Institute of Chemists (1970); the USPHS Research Career Development Award (1980); the Harrison Howe Award (1984); the Young Scientist Award from the Passano Foundation (1984); the Pfizer Award in Enzyme Chemistry (1985); the U. S. Steel Award in Molecular Biology (1987); the V.D. Mattia Award (1987); the Gairdner Foundation Award (1988); the Heineken Prize (1988); the Lasker Award (1988); the Nobel Prize in Chemistry, jointly with S Altman (1989); the Rosenstiel Award (1989); the Warren Triennial Prize (1989); the National Medal of Science (1995); the Gregor Mendel Medal (2002). He has also been presented with an Award in Molecular Biology from the U.S. National Academy of Sciences.

Cech is a member of the U. S. National Academy of Sciences (1987) and of the American Academy of Arts and Sciences (1988).

Bibliography

1. Reproduced with permission.
2. From Nobel Lectures, Chemistry 1981–1990, Editor-in-Charge Tore Frängsmyr, Editor Bo G. Malmström, World Scientific Publishing Co., Singapore, 1992.
3. The International Who's who: 2005. ISBN: 1-85743-256-8.
4. http://www.hhmi.org/news/prez.html.
5. http://www.hhmi.org/research/investigators/cech_bio.html.
6. http://cechlab.colorado.edu/welcome.html.
7. Non-Coding RNAs: Molecular Biology and Molecular Medicine (Molecular Biology Intelligence Unit (Unnumbered).), Springer, 2003, ISBN: 0306478358.
8. Catalytic Rna (Nucleic Acids and Molecular Biology, Vol 10) edited by Fritz Eckstein and David M. J. Lilley, Springer-Verlag Telos, 1996, ISBN: 3540607951.
9. RNA Interference Technology: From Basic Science to Drug Development edited by Krishnarao Appasani, Cambridge University Press, 2005, ISBN: 0521836778.

1990

© Nobel Foundation [1]

Elias James Corey

"for his development of the theory and methodology of organic synthesis" [2]

During his presentation speech, Salo Gronowitz, of the Royal Swedish Academy of Sciences, commented of Corey and his work [2]:

> *...The process of synthetic planning has been compared to a game of three-dimensional chess using 40 pieces on each side. But the problem of synthesis may be even harder than this. Over 35,000 usable methods of synthesis are described in the chemical literature, each with its possibilities and its limitations. During synthesis, moreover, new methods appear which can modify the strategy. It is like allowing new moves during a game of chess.*
>
> *Beginning in the 1960s Professor E.J. Corey coined the term and developed the concept of retrosynthetic analysis. Starting from the structure of the molecule he was to produce, the target molecule, he established rules for how it should be dissected into smaller parts, and what strategic bonds should be broken. In this way, less complicated building blocks were obtained, which could later be assembled in the process of synthesis. These building blocks were then analyzed in the same way until simple compounds had been reached, whose synthesis was already described in the literature or which were commercially available. Corey showed that strict logical retrosynthetic analysis was amenable to computer programming. He is the leader in this rapidly developing field...*
>
> *...Corey's most important syntheses are concerned with prostaglandins and related compounds. These often very unstable compounds are responsible for multifarious and vital regulatory functions of significance in reproduction, blood coagulation and normal and pathological processess in the immune system...*
>
> *...To perform these total syntheses successfully, Corey was also obliged to develop some fifty entirely new or considerably improved synthesis reactions. His systematic use of different types of organometallic reagents has revolutionized recent techniques of synthesis in many respects. In*

recent years, he has also introduced a number of very effective enzyme-like catalysts, which yield only one mirror isomer of the target product in certain types of synthetically important reactions...

Elias James Corey was born on 12th July 1928 in Methuen, Massachusetts, one of two sons and two daughters of Elias (a successful business man) and Fatina, née Hasham. His birth name was William, which was changed by his mother on the death of Corey's father when Corey was eighteen months old. Corey married Claire Higham (Sept 1961) and they have two sons David Reid and John, and one daughter Susan. Corey is the Sheldon Emery Research Professor of Chemistry at Harvard University [2,3,4].

The Early Years

Corey attended the Saint Laurence O'Toole Elementary School in Lawrence, where he was taught by sisters of the Catholic order of Notre Dame de Namour (1933–40). He graduated from Lawrence Public High School (1945) and after a few weeks entered the Massachusetts Institute of Technology (MIT), where he studied chemistry, mathematics and physics. Corey graduated with a BS degree (1948) and stayed at MIT, where he worked under J. C. Sheehan on synthetic penicillin. His PhD thesis has the title '*The synthesis of N,N-diacylamino acids and analogs of penicillin*' and was presented in 1951 and he started his career as an Instructor the same year.

The Career

Year	*Post*	*Institute*
1951	Instructor	University of Illinois, Urbana-Champaign
1953	Assistant Professor	University of Illinois, Urbana-Champaign
1955	Professor of Chemistry	University of Illinois, Urbana-Champaign
1959	Professor of Chemistry	Harvard University
1968	Sheldon Emery Professor	Harvard University

Corey was a Guggenheim Fellow (1957), dividing his time between Harvard, to which he had been invited by Prof. Robert B. Woodward, and Europe.

Corey and his group have synthesized over 100 naturally occurring compounds (including gibberellic acid, longifolene and gingkolid B). He also worked in numerous other areas including, in approximately chronological order:

- the application of molecular orbital theory to the understanding of the transition states for various reactions in three dimensional (*i.e.* stereochemical) detail
- stereoelectronic concepts
- enantioselective synthesis
- metal complexes

- new reactions for synthesis
- enzyme chemistry
- the first chemical syntheses of prostaglandins
- eicosanoids
- retrosynthetic analysis
- the logic of chemical synthesis
- new methods of synthesis
- molecular catalysts and robots
- theoretical organic chemistry and reaction mechanisms
- organometallic chemistry
- bioorganic and enzyme chemistry
- the application of computers to organic chemical problems

Perhaps his most well-known book is *The Logic of Chemical Synthesis* written with Xue-Min Cheng Pub: John Wiley & Sons Inc; 1 edition, (1989), ISBN: 0471509795. Corey has also published *Collected papers of E.J. Corey*, ASIN: B0007KCQ82

Corey's Nobel Prize lecture is dated 8th December 1990 with the title 'The logic of chemical synthesis: Multistep synthesis of complex carbogenic molecules'. Corey defined a TGT as a synthetic target, and a retron as a keying structural subunit [2]. He then commented:

> *...The major types of strategies which are of value in retrosynthetic analysis may be summarized briefly as follows.*
>
> (i) *Transform-based strategies - long range search or look-ahead to apply a powerfully simplifying transform (or a tactical combination of simplifying transforms) to a TGT with certain appropriate keying features. The retron required for application of a powerful transform may not be present in a complex TGT and a number of antithetic steps (subgoals) may be needed to establish it.*
>
> (ii) *Structure-goal strategies-directed at the structure of a potential intermediate or potential starting material. Such a goal greatly narrows a retrosynthetic search and allows the application of bidirectional search techniques.*
>
> (iii) *Topological strategies - the identification of one or more individual bond disconnections or correlated bond-pair disconnections as strategic. Topological strategies may also lead to the recognition of a key substructure for disassembly or to the use of rearrangement transforms.*
>
> (iv) *Stereochemical strategies - general strategies which clear, i.e. remove, stereocenters and stereorelationships under stereocontrol. Such stereocontrol can arise from transform-mechanism control or substrate-structure control. In the case of the former the retron for a particular transform contains critical stereochemical information (absolute or relative) on one or more stereocenters. Stereochemical strategies may also dictate the retention of certain stereocenter during retrosynthetic processing or the joining of atoms in three-dimensional proximity. A major function of stereochemical strategies is the achievement of an experimentally valid clearance of stereo-centers, including clearance of molecular chirality.*
>
> (v) *Functional group-based strategies. The retrosynthetic reduction of molecular complexity involving functional groups (FG's) takes various forms. Single FG's or pairs of FG's (and the interconnecting atom path) can key directly the disconnection of a TGT skeleton to form simpler molecules or signal the application of transforms which replace functional groups by hydrogen. Functional group interchange (FGI) is a commonly used tactic for generating the retrons of simplifying transforms*

from a TGT. FG's may key transforms which stereo-selectively remove stereocenters, break topologically strategic bonds or join proximate atoms to form rings.

(vi) *"Other" types of strategies. The recognition of substructural units within a TGT which represent major obstacles to synthesis often provides major strategic input. Certain other strategies result from the requirements of a particular problem, for example a requirement that several related target structures be synthesized from a common intermediate. A TGT which resists retrosynthetic simplification may require the invention of new chemical methodology. The recognition of obstacles to synthesis provides a stimulus for the discovery of such novel processes. The application of a chain of hypotheses to guide the search for an effective line of retrosynthetic analysis is important.*

Other strategies deal with optimization of a synthetic design after a set of pathways has been generated antithetically, specifically for the ordering of synthetic steps, the use of protection or activation steps, or the determination of alternate paths...

Many examples of the use of retro-synthetic analysis have been documented in the literature, some of which have been included in the following texts [5–7].

Honours and Awards

Corey has received an Honorary AM degree from Harvard (1959), Honorary DSc degrees from the Universities of: Chicago (1968); Hofstra (1974); Oxford (1982); Liège (1985); Illinois (1985); an honorary ScD degree from the University of Cambridge (2000).

He has received numerous awards and prizes from the US, Europe and Asia including the US National Medal of Science (1988) and the Nobel Prize for Chemistry (1990).

Corey is a Foreign Member of the Royal Society (1998).

He was a Member of the Board of Directors, physical sciences, Alfred P. Sloan Foundation and a Member of the Advisory Board of the Robert A. Welch Foundation.

Bibliography

1. Reproduced with permission.
2. From Nobel Lectures, Chemistry 1981–1990, Editor-in-Charge Tore Frängsmyr, Editor Bo G. Malmström, World Scientific Publishing Co., Singapore, 1992.
3. Who's who: 2005.
4. http://www.chem.harvard.edu/faculty/corey.html.
5. Modern Methods of Organic Synthesis by W. Carruthers and Iain Coldham, Cambridge University Press, 4th ed. 2004, ISBN: 0521778301.
6. Organic Synthesis (Oxford Chemistry Primers, 31) by Christine L. Willis and Martin Wills, Oxford University Press, 1996, ISBN: 0198557914.
7. Classics in Total Synthesis: Targets, Strategies, Methods by K. C. Nicolaou and E. J. Sorensen, Wiley-VCH, 1996, ISBN: 3527292314.

1991

© Nobel Foundation [1]

Richard R. Ernst

"for his contributions to the development of the methodology of high resolution nuclear magnetic resonance (NMR) spectroscopy" [2]

During his presentation speech, Sture Forsén, of the Royal Swedish Academy of Sciences, commented of Ernst and his work [2]:

...When Fourier transform NMR was introduced around 1970 it had a tremendous impact on the applicability of the NMR technique to chemistry. It now became feasible to study very weak signals from small amounts of material or from important elements with magnetic nuclei that are rare in nature, for example ^{13}C and ^{15}N. The Achilles heel of the NMR technique had hitherto been its poor sensitivity, but now this obstacle was largely removed.

A later revolutionary development in NMR, in which this year's Chemistry Laureate played a leading role, was the introduction of more than one frequency dimension, 2, 3 or higher. In 2D NMR, the chemical "piano" is hit with pulses of varying lengths and intervals. This allows chemists to extract many parameters of interest from the NMR spectra with great ease. Hopelessly muddled and hard-to-interpret spectra can be spread out and thereby simplify the analysis - much as a two-dimensional map of a landscape would be superior to a mere silhouette. 2D NMR makes it possible to find out which specific atoms in a molecule are closely linked by chemical bonds or which atoms are near each other in space, or which atoms take part in chemical exchange reactions, and much more. A whole new range of experiments have become possible and multidimensional NMR has substantially increased the range of applications to chemistry. The new method has been a prerequisite for the very important applications of NMR in structural biology that have taken place during the past decade...

...Professor Ernst, You have played a leading role in several of the most significant methodological developments that have taken place in the field of NMR spectroscopy over the past two decades; developments that have had a lasting impact on the way modern chemistry is conducted. You have, in

an admirable way, combined excellent experimental know-how with extraordinary theoretical insight. In recognition of your services to chemistry, and to natural science as a whole, the Royal Swedish Academy of Sciences has decided to confer upon you this year's Nobel Prize for Chemistry...

Richard Robert Ernst was born in Winterthur, Switzerland on 14th August 1933, the only son of the three children of Robert Ernst (an architect) and Irma, née Brunner. He married Magdalena Kielholz (1963) and they have three children Anna Magdalena, Katharina Elisabeth and Hans-Martin Walter. Ernst can be contacted through the Laboratorium für Physikalische Chemie, ETH, Hönggerberg [2–4].

The Early Years

Ernst found a case filled with chemicals in the loft of his house when he was 13 years old, the former property of a deceased uncle who was a metallurgical engineer who was interested in chemistry and photography. Ernst experimented with these chemicals, causing both explosions and unusual and toxic gases in his house. He has been a committed chemist from those times, reading all of the chemistry books that became available. He enrolled in the Swiss Federal Institute of Technology in Zurich (ETH-Z) to study chemistry, following his early education at nearby Winterthur schools.

Ernst entered ETH-Z with an above average knowledge in chemistry and turned to books such as *Theoretical Chemistry* by S. Glasstone, from which he learned quantum mechanics, spectroscopy, statistical mechanics and statistical thermodynamics. Graduation with a DipChemEng (1956) was followed by a period of military service (1956–57), after which he enrolled for a PhD under Günthard at ETH-Z. Ernst worked on designing and building advanced electronic equipment for improved NMR spectrometers, in particular the construction of high sensitivity radio frequency preamplifiers and high sensitivity probe assemblies. At that time, 25MHz was considered standard. Ernst graduated with a PhD in 1962 and chose to start his career in industry.

The Career

Year	*Post*	*Institute*
1963–68	Scientist	Varian Associates, California
1968–70	Lecturer in Physical Chemistry and Group Leader in Nuclear Magnetic Resonance Spectroscopy	Swiss Federal Institute of Technology
1970	Assistant Professor	Swiss Federal Institute of Technology
1972–76	Associate Professor	Swiss Federal Institute of Technology
1976	Professor (and for part of the time Director) of Physical Chemistry	Swiss Federal Institute of Technology
1998	Retired	

Whilst Ernst has been involved in various aspects of nuclear magnetic resonance (NMR) spectroscopy for most of his adult life, there are two phases, industrial and academic, in approximately date order Ernst has been involved with:

Varian work

- some of the experiments that led to Fourier transform (FT) NMR
- the development in carbon-13 spectroscopy using 'noise decoupling'
- numerous computer applications in spectroscopy
- studies concerning automated experiments and improved data processing.

Academic work

- liquid state NMR
- solid state NMR
- two-dimensional NMR
- novel pulse techniques
- tomography (medical magnetic resonance)
- the structure determination of biopolymers in solution (with Kurt Wüthrich)
- the study of intramolecular dynamics
- time-domain NMR with repetitive pulse experiments
- Fourier double resonance
- the first pulsed time-domain chemically-induced dynamic nuclear polarization (CIDNP) experiments
- the development of multi-dimensional spectroscopy
- the nmr spectra of elements other than C and H (ie heteronuclear).

Throughout his career, Ernst has been involved with scientific publishing, having been on the editorial board of 10 scientific journals, and written two books:
Principles of Nuclear Magnetic Resonance in One and Two Dimensions (International Series of Monographs on Chemistry), with Geoffrey Bodenhausen and Alexander Wokaun, Pub: Oxford University Press; Reprint edition (Oct 1990) ISBN: 0198556470 and *Structural Identification of Organic Compounds with Spectroscopic Techniques* with Yong-Cheng Ning, Pub: John Wiley & Sons, (April 2005), ISBN: 3527312404.

Ernst's Nobel Prize lecture is dated 9th December 1992 with the title 'Nuclear Magnetic Resonance Fourier Transform Spectroscopy'. Ernst cited 131 references throughout his lecture which covered many aspects of his work. His comments of two-dimensional Fourier transform spectroscopy were [2]:

> *...As long as purely spectroscopic measurements are made, determining the eigenfrequencies or normal modes of a system, one-dimensional spectroscopy is fully adequate. In NMR, this applies to the measurement of the chemical shifts that characterize the local chemical environment of the different nuclei. However, no information can be obtained in this manner on the spatial and topological relations between the observed nuclei.*
>
> *There are two important pair interactions in nuclear spin systems, the scalar through-bond electron-mediated spin-spin interaction, the so-called J coupling, and the through-space magnetic dipolar interaction...*

Ernst commented on heteronuclear experiments thus:

> *...In addition to the homonuclear 2D experiments discussed so far, at least an equal number of heteronuclear experiments has been proposed and introduced in the routine spectroscopy*

laboratory. Of greatest practical importance are heteronuclear shift correlation spectra that correlate the chemical shifts of directly bonded or remotely connected heteronuclei...

And on three three-dimensional fourier spectroscopy:

...No new principles are required to develop 3D spectroscopy that is just a logical extension of 2D spectroscopy. Instead of a single mixing process which relates two frequency variables, two sequential mixing processes relate three frequencies: the origin frequency ω_1 the relay frequency ω_2, and the detection frequency ω_3. In this sense, a 3D experiment can be considered as the combination of two 2D experiments. Obviously, a very large number of possible 3D experiments can be conHowever, only few of them have so far proved to be indispensable....

NMR is a mandatory analytical tool in organic chemistry which has seen many developments, see for example [5–7].

Honours and Awards

Ernst has received many Honorary degrees including: the University Antwerpen; the Babes-Bolyai University; the University of Cluj-Napoca; the University Montpellier; the ETH Lausanne; the Technische Universität Munich; the Universität Zurich.

He has received many awards and prizes including the: Marcel Benoist Prize (1986); Horwitz Prize (1991); Nobel Prize for Chemistry (1991); Wolf Prize for Chemistry, jointly with Alexander Pines (1991).

Ernst is a Fellow of the American Physical Society and a Foreign Member of the Royal Society (1998). He is a member of: the Deutsche Akademie Leopoldina; the Korean Academy of Science and Technology; the US National Academy of Sciences; the Royal Academy of Sciences, London; the Russian Academy of Sciences. He is also an Honorary Member of many further societies.

Ernst is or has been: President of the Research Council (1990); a Member of the Board of Spectrospin AG (Vice President 1989); President of the Reseach Council of ETH Zurich; a Member of the COST Committee; a Member of the Hochschulrat; a Member of the Foundation Marcel Benoist; a Member of the Swiss Science Council; a Member of the Technische Universität Munich; Vice President of the Board of Bruker AG, Fällanden.

Bibliography

1. Reproduced with permission.
2. From Nobel Lectures, Chemistry 1991–1995, Editor Bo G. Malmström, World Scientific Publishing Co., Singapore, 1997.
3. Who's who: 2005.
4. http://www.verw.ethz.ch/cgi-win/whoShow.exe/ws7?ID=1009&lang=engl.
5. Multidimensional NMR in Liquids – Basic Principles and Experimental Methods by Frank J.M. van de Ven, VCH, 1995, ISBN: 1-56081-665-1.
6. NMR Spectroscopy Techniques 2nd ed, by Martha D. Bruch, Marcel Dekker Inc., 1996, ISBN: 0-8247-9450-8.
7. NMR in Chemistry, 4th ed, by J.W. Akitt and B.E. Mann, Stanley Thornes, 2000, ISBN: 0-7487-4344-8.

1992

Rudolph A. Marcus

"for his contributions to the theory of electron transfer reactions in chemical systems" [2]

During his presentation speech, Lennart Eberson, of the Royal Swedish Academy of Sciences, commented of Marcus and his work [2]:

...This problem led to lively discussion around 1950. Marcus became interested when he happened to read through some papers from a symposium on electron transfer reactions, where the American chemist Willard Libby had suggested that a well-known spectroscopic principle known as the Franck-Condon principle might apply to the movement of an electron between two molecules. Marcus realized that this ought to create an energy barrier, which might explain the slow electron transfer between bivalent and trivalent iron in an aqueous solution. To enable the two iron ions to exchange an electron, a number of water molecules in their surroundings must be rearranged. This increases the energy of the system temporarily, and at some point the electron can jump without violating the restrictions of the Franck-Condon principle.

In 1956, Marcus published a mathematical model for this type of reaction, based on classic theories of physical chemistry. He was able to calculate the size of the energy barrier, using simple quantities such as ionic radii and ionic charges. He later extended the theory to cover electron transfer between different kinds of molecules and derived simple mathematical expressions known as "the quadratic equation" and "the cross-equation." These could be tested empirically and led to new experimental programs in all branches of chemistry. The Marcus theory greatly contributed to our understanding of such widely varying phenomena as the capture of light energy in green plants, electron transfer in biological systems, inorganic and organic oxidation and reduction processes and photochemical electron transfer.

The quadratic equation predicts that electron transfer reactions will occur more slowly the larger the driving force of the reaction is. This phenomenon received its own name, "the inverted region." To a chemist, the phenomenon is just as unexpected as when a skier finds himself gliding more slowly down a slope the steeper it is. In 1965, Marcus himself suggested that certain chemiluminescent reactions ("cold light") might serve as an example of the inverted region. Only after 1985, however, could further examples of such reactions be demonstrated. The most improbable prediction in his theory had thereby been verified...

Rudolph A. Marcus was born on 21st July 1923 in Montreal, Canada, the son of Meyer Marcus and Esther, née Cohen. Marcus married Laura Hearne in 1949 (deceased 2003). They had three sons, Alan, Kenneth, and Raymond. Marcus became a US citizen in 1958 and is currently the Arthur Amos Noyes Professor of Chemistry at the California Institute of Technology in Pasadena, USA [2–4].

The Early Years

Marcus attended Baron Byng High School, during the early years of which he became interested in chemistry, having previously been interested in mathematics. From there he attended McGill University, from which he graduated with a B.Sc. in chemistry (1943) and a Ph.D. (1946) under C. A. Winkler, a former graduate student of Hinshelwood (Chemistry Nobel Laureate 1956). Marcus undertook post-doctoral work at the National Research Council of Canada in Ottawa in E.W.R. Steacie's group, the subject of which was gas phase reaction rates (1946–49). His first paper was published during this time concerning the mercury photosensitized reaction of dimethyl ether [5]. A second postdoctoral position followed at the University of North Carolina with O. K. Rice concerning reaction rate theory (1949–51).

The Career

Year	*Post*	*Institute*
1951–54	Assistant Professor	Polytechnic Institute of Brooklyn
1954–58	Associate Professor	Polytechnic Institute of Brooklyn
1958–64	Professor	Polytechnic Institute of Brooklyn
1964–78	Professor	University of Illinois
1978–	Arthur Amos Noyes Professor of Chemistry	California Institute of Technology

Marcus was also Acting Head of the Division of Physical Chemistry, Polytechnic Institute of Brooklyn (1961–62) and Head of the Division of Physical Chemistry, University of Illinois (1967–68). He was a Member of the Courant Institute of Mathematical Sciences, New York University (1960–61) and NSF Senior Postdoctoral Fellow (1960–61), an Alfred P. Sloan Fellow (1960–63) and, following a period of Visiting Professor at the University of Oxford, he was a Humboldt Awardee at the Technical University of Munich (1975–76 for both activities).

Marcus has had many research interests including:

- electron transfer processes in:
  - solutions
  - proteins
  - at interfaces

- reaction dynamics
- work concerning natural collision coordinates
- action-angle variables
- unimolecular reactions (eg ozone to atomic and molecular oxygen)
- intramolecular dynamics
- the effect of donor/acceptor electronic coupling on the reaction rates
- the use of artificial intelligence (AI) in problem solving
- electron coupling studies
- rates of electron transfer at semiconductor electrode/liquid interfaces
- electron transfers at the electrode/liquid surface including different electrodes
- studies of intramolecular vibrational energy

Marcus is or has been associated with the publication of many journals including: the Journal of Chemical Physics (1964–66); the Annual Review of Physical Chemistry (1964–69); the Journal of Physical Chemistry (1968–72 1980–84); Accounts of Chemical Research (1968–73); International Journal of Chemical Kinetics (1976–1980); Molecular Physics (1977–1980); Chemical Physics Letters (1980–1990); Laser Chemistry (1982 to date); Advances in Chemical Physics (1984 to date); Theoretica Chimica Acta (1985–1988); International Advisor in Chemistry, World Scientific Publishing (1987 to date); International Reviews in Physical Chemistry (1988 to date); Progress in Physics, Chemistry and Mechanics (China) (1989 to date); the Journal of the Chemical Society. Faraday Transactions (1990–1995); Journal of the Chemical Society. Perkin Transactions 2, (1992 to date); Chemical Physics Research (India) (1992 to date); Trends in Chemical Physics (India) (1992 to date). Marcus was created an Honorary Editor, International Journal of Quantum Chemistry (1996).

He has also published the following books: *Studies in the reactions of high energy atoms and molecules (WADC technical report)*, Pub: Wright Air Development Center, Air Research and Development Command, U.S. Air Force, (1959), ASIN: B0007EL7M4; *The Hill reaction as a model for chemical conversion of solar energy*, Pub: U.S. Dept. of Commerce, Office of Technical Services, (1959), ASIN: B0007HOS68; *Interview with Rudolph A. Marcus (California Institute of Technology Oral History Project)*, Pub: Caltech Archives, (1995), ASIN: B0006R48D2; *Pneumatic Conveying of Solids: A theoretical and practical approach (Particle Technology Series)*, with G.E. Klinzing, F. Rizk and L.S. Leung, Pub: Springer; 2nd edition, (1997), ISBN: 0412724405; *Electron Transfer in Chemistry, Principles, Theories, Methods, and Techniques* with Vincenzo Balzani, Piotr Piotrowiak, Michael A. J. Rodgers, Jochen Mattay, Didier Astruc, H. B. Gray, Jay Winkler, Shunichi Fukuzumi, Thomas E. Mallouk, Yehuda Haas, A. P. de Silva and Ian Gould, Pub: Wiley-VCH; Slipcase edition, (2001), ISBN: 3527299122; *Theories of Chemical Reactions Rates: Selected Papers of Rudolph A. Marcus (Series on 20th Century Chemistry)* (Editor), Pub: World Scientific Pub Co Inc, (2004), ISBN: 9810215053 and *Intramolecular Motion and Chemical Reaction (Royal Society Discussion Volumes)* with by Ian M. Mills and M. S. Child, Pub: Cambridge University Press, (1991), ISBN: 0521394996.

Marcus' Nobel Prize lecture is dated 8th December 1992 with the title 'Electron Transfer Reactions in experimental chemistry: Theory and Experiment'. Marcus explained his entry into the field of electron transfer theory thus [2]:

> *...My first contact with electron transfers came in 1955 as a result of' chancing upon a 1952 symposium issue on the subject in the Journal of Physical Chemistry. An article by Bill Libby caught my eye-a use of the Franck-Condon principle to explain some experimental results, namely,*

why some isotopic exchange reactions which involve electron transfer between pairs of small cations in aqueous solution, are relatively slow, whereas electron transfers involving larger ions, such as

$$Fe(CN)_6^{3-} \text{ to } Fe(CN)_6^{4-} \text{ and } Mn0_4^{-} \text{ to } Mn0_4^{2-}$$

are relatively fast. Libby explained this observation in terms of the Franck-Condon principle...

Further data concerning electron transfer reactions and separately the Franck-Condon principle, can be found, for example, in references [6,7] respectively.

Honours and Awards

Marcus has received Honorary D.Sc. degrees from: the University of Chicago (1983); the Polytechnic University (1986); McGill University, Montreal, Canada (1988); the University of New Brunswick, St. John, Canada (1993); Queen's University, Kingston, Canada (1993); the University of Oxford (1995); the University of North Carolina at Chapel Hill (1996); the University of Illinois at Urbana-Champaign (1997); the Technion-Israel Institute of Technology, Israel (1998); the Universidad Politecnica de Valencia (1999): Northwestern University (2000); the University of Waterloo (2002). He was also received the Honorary degree of Fil.dr. from the University of Göteborg (1987) and is an Honorary Doctor of Yokohama National University, Japan (1996).

Marcus has received many awards and prizes including: the Anne Molson Prize in Chemistry (1943); the Alexander von Humboldt Foundation Senior U.S. Scientist Award (1976); the Irving Langmuir Award in Chemical Physics (1978); the Robinson Medal, Royal Society of Chemistry (1982); the Chandler Medal (1983); the Wolf Prize in Chemistry (1985); the Peter Debye Award in Physical Chemistry (1988); the Willard Gibbs Medal, American Chemical Society (1988); the Centenary Medal, Royal Society of Chemistry (1988); the National Medal of Science (1989); the William Lloyd Evans Award (1990); the Theodore William Richards Medal, American Chemical Society (1990); the Edgar Fahs Smith Award, American Chemical Society (1991); the Linus Pauling Award, American Chemical Society (1991); the Remsen Award, American Chemical Society (1991); the Nobel Prize in Chemistry (1992); the American Academy of Achievement Golden Plate Award (1993); the Joseph O. Hirschfelder Prize in Theoretical Chemistry (1993); the Lavoisier Medal, Société Française de Chimie (1994); the Treasure of Los Angeles Award, Central City Association of Los Angeles (1995); the Auburn-G. M. Kosolapoff Award, American Chemical Society (1996); the Award in Theoretical Chemistry, American Chemical Society (1997); the Oesper Award, American Chemical Society (1997); the Top 75 Award, Chemical and Engineering News, American Chemical Society (1998).

He is a Member of the National Academy of Sciences (1970); a Fellow of the American Academy of Arts and Sciences (1973); a Member of the International Academy of Quantum Molecular Science (1987); a Foreign Member of the Royal Society (1987); a Member of the American Philosophical Society, 1990- (Member of Council, 1999); a Honorary Fellow of the Royal Society of Chemistry (1991); a Foreign Fellow of the Royal Society of Canada (1993); a Honorary Member of the International Society of Electrochemistry (1994); a Honorary Member of the Korean Chemical Society (1996); a Foreign Member of the Chinese Academy of Sciences (1998); a Honorary Member of the European Academy of Sciences (2004); First Honorary Scientific Fellow, Literary & Historical Society, University College, Dublin (2004). He is an Honorary Citizen of Winnipeg (1994) and holds the Key to the city of Taipei, Taiwan (1999). He is a Member of: Alpha Chi Sigma; the American

Association for the Advancement of Science; the American Chemical Society; American Physical Society.

Marcus is also an Honorary Professor of Fudan University (1994–); an Honorary Professor at the Institute of Chemistry, Chinese Academy of Sciences (1995–); an Honorary Fellow of University College, University of Oxford (1995–); Linnett Visiting Professor of Chemistry of the University of Cambridge (1996); an Honorable Visitor to the National Science Council, Republic of China (1999); a Visiting Professor of Theoretical Chemistry, IBM, University of Oxford, England (1975–76); a Honorary Professor of Ocean University of Qingdao, China (2002); an Honorary Professor of Tianjin University, China (2002); an Honorary Board Member of the International Society of Theoretical Chemical Physics (1995); an Honorary Board Member of The International Raoul Wallenberg Foundation (2003); an Honorary Board Member of The Angelo Roncalli International Committee (2003).

Marcus was a Chair or Member of many committees starting with Chairman of the Division of Physical Chemistry, American Chemical Society (1964–65) through Member of the National Research Council-National Academy of Sciences, Climatic Impact Committee, Panel on Atmospheric Chemistry (1975–78) to more recent work on the following: Member of the Mathematics Panel of the International Benchmarking of U.S. Research Fields, NAS-NRC Committee on Science, Engineering and Public Policy (1996–1998); Panel Member of the Accountability of Federally Funded Research, COSEPUP (2000-to date); Co-organizer of the First Gordon Research Conference Summer School, "Analytical Approaches to Rate Processes and Time-Resolved Spectroscopy in Condensed Phases" (2000).

Bibliography

1. Reproduced with permission.
2. From Nobel Lectures, Chemistry 1991–1995, Editor Bo G. Malmström, World Scientific Publishing Co., Singapore, 1997.
3. Who's who: 2005.
4. http://www.cce.caltech.edu/faculty/marcus/.
5. R. A. Marcus, B. de B. Darwent, and E. W. R. Steacie, J. Chem. Phys., 1948, 16, 987.
6. Electron Transfer Reactions: Inorganic, Organometallic, and Biological Applications (Advances in Chemistry Series), by Stephan S. Isied, An American Chemical Society Publication, 1997, ISBN: 0841234566.
7. Principles of Lasers by Orazio Svelto, Springer, 4th ed. 2004, ISBN: 0306457482.

1993

Kary B. Mullis (0.5)

"for contributions to the developments of methods within DNA-based chemistry particularly for his invention of the polymerase chain reaction (PCR) method" [2]

During his presentation speech, Carl-Ivar Brändén, of the Royal Swedish Academy of Sciences commented of Mullis and his work [2]:

...Our genetic material, which gives every living organism its unique characteristics, is built up from large and complex DNA molecules, each comprising hundreds of millions of atoms. For a long time it was believed that these molecules were outside the realm of the chemical laboratory and that their manipulation could only be achieved through the complex machinery of a living cell. This year's Nobel Laureates in Chemistry, Kary B. Mullis and Michael Smith, have drastically changed this notion by providing research tools that chemists can use outside a living cell to amplify and specifically modify any given gene...

...Isolation and amplification of a specific gene was one of the outstanding problems in DNA technology, including site-directed mutagenesis, until 1985 when Kary Mullis presented the Polymerase Chain Reaction, now commonly known as PCR. Using this method it is possible to amplify and isolate in a test tube a specific DNA segment within a background of a complex gene pool. In this repetitive process the number of copies of the specific DNA segment doubles during each cycle. In a few hours it is possible to achieve more than 20 cycles, which produces over a million copies.

The PCR method has already had a profound influence on basic research in biology. Cloning and sequencing of genes as well as site-directed mutagenesis have been facilitated and made more efficient. Genetic and evolutionary relationships are easily studied by the PCR method even from ancient fossils containing only fragments of DNA. Biotechnology applications of PCR are numerous. In addition to being an indispensable research tool in drug design, the PCR method is now used in diagnosis of viral and bacterial infections including HIV. The method is so sensitive that it is used in forensic medicine to analyze the DNA content of a drop of blood or a strand of hair...

Kary Banks Mullis was born on 28th Dec. 1944, in Lenoir, N.C., the son of Cecil Banks Mullis and Bernice Alberta Fredericks, née Barker. Mullis married Richards Train Haley (1963), with whom he had one daughter, Louise. The marriage was dissolved in 1975. He married Cynthia Gibson (1975), and had two sons Christopher and Jeremy, the marriage was dissolved. He married Nancy Lier Cosgrove in 1998. Mullis lives in Newport Beach, California, and in Anderson Valley, California [2–4].

The Early Years

Mullis attended high school in Columbia, followed by a B.S. degree in chemistry from the Georgia Institute of Technology (1966). His Ph.D. in biochemistry was under J.B. Neilands, at the University of California, Berkeley (1972), where he stayed as a lecturer in biochemistry until 1973. From there Mullis moved to the University of Kansas Medical School, where he undertook postdoctoral work in pediatric cardiology (1973–76), prior to postdoctoral work in pharmaceutical chemistry at the University of California, San Francisco (1977–79).

The Career

Year	*Post*	*Institute*
1979	Scientist	Cetus Corporation
1986	Director of Molecular Biology	Xytronyx Inc
1988 – to date	Consultant	Various companies

Mullis is a Distinguished Researcher at the Children's Hospital and Research Institute in Oakland, California. He was a Vice President and Director of Molecular Biology at Burstein Technologies in Irvine, California and is Founder and Chief Executive of Altermune LLC.

Mullis has spent the majority of his adult life as an author, consultant and inventor. The research work that he undertook during the 1979–86 included:

Cetus Corp

- invented the polymerase chain reaction (PCR)
- oligonucleotide synthesis

Writing in Scientific American, Mullis wrote of PCR:

...Beginning with a single molecule of the genetic material DNA, the PCR can generate 100 billion similar molecules in an afternoon. The reaction is easy to execute. It requires no more than a test tube, a few simple reagents and a source of heat. The DNA sample that one wishes to copy can be pure, or it can be a minute part of an extremely complex mixture of biological materials. The DNA

> *may come from a hospital tissue specimen, from a single human hair, from a drop of dried blood at the scene of a crime, from the tissues of a mummified brain or from a 40,000-year-old wooly mammoth frozen in a glacier...*

This invention has been hailed by some to be one of the 'monumental scientific techniques of the twentieth century', having applications in biotechnology, forensic science, genetic engineering and medicine. It is the basis of the new scientific discipline of paleobiology.

Xytronyx, Inc.

- DNA technology
- photochemistry

Mullis has authored several major patents, including:

- PCR technology
- A technique for the instant mobilizing of the immune system to neutralize invading pathogens and toxins
- UV-sensitive plastic that changes color in response to light

His many publications include:

- *The Cosmological Significance of Time Reversal (Nature)*
- *Primer-directed Enzymatic Amplification of DNA with a Thermostable DNA Polymerase* (Science)
- *Specific Synthesis of DNA in vitro via a Polymerase Catalyzed Chain Reaction* (Methods in Enzymology)
- *The Unusual Origin of the Polymerase Chain Reaction* (Scientific American).

And the books (the first being his autobiography):
Dancing Naked in the Mind Field (Vintage), Pub: Vintage, (2000), ISBN: 0679774009; *The Polymerase Chain Reaction*, Pub: Birkhauser, (1994), ISBN: 3764336072

Mullis' nobel prize lecture is dated 8th December 1993 with the title 'The Polymerase Chain Reaction'. Of the invention of PCR technology, Mullis commented on his literature search following his idea during a car journey [2]:

> *...Monday morning I was in the library. The moment of truth. By afternoon it was clear. For whatever reasons, there was nothing in the abstracted literature about succeeding or failing to amplify DNA by the repeated reciprocal extension of two primers hybridized to the separate strands of a particular DNA sequence. By the end of the week I had talked to enough molecular biologists to know that I wasn't missing anything really obvious. No one could recall such a process ever being tried.*
>
> *However, shocking to me, not one of my friends or colleagues would get excited over the potential for such a process. True. I was always having wild ideas, and this one maybe looked no different than last week's. But it WAS different. There was not a single unknown in the scheme. Every step involved had been done already. Everyone agreed that you could extend a primer on a DNA*

template, everyone knew you could melt double stranded DNA. Everyone agreed that what you could do once, you could do again. Most people didn't like to do things over and over, me in particular. If I had to do a calculation twice, I preferred to write a program instead. But no one thought it was impossible. It could be done, and there was always automation. The result on paper was so obviously fantastic, that even I had little irrational lapses of faith that it would really work in a tube, and most everyone who could take a moment to talk about it with me, felt compelled to come up with some reason why it wouldn't work. It was not easy in that post-cloning, pre-PCR year to accept the fact that you could have all the DNA you wanted. And that it would be easy.

I had a directory full of untested ideas in the computer. I opened a new file and named this one polymerase chain reaction...

Many articles and a few books have been written concerning the PCR reaction, see for example [5–7].

Honours and Awards

Mullis was presented the Honorary degree of Doctor of Science from the University of South Carolina (1994).

He has been presented with many awards and prizes including: the William Allan Memorial Award of the American Society of Human Genetics (1990); the Preis Biochemische Analytik of the German Society of Clinical Chemistry and Boehringer Mannheim (1990); the Gairdner Award, Toronto, Canada (1991); the National Biotechnology Award (1991); the R&D Scientist of the Year (1991); the California Scientist of the Year Award (1992); the Robert Koch Award (1992); the Japan Prize (1993); the Nobel Prize in Chemistry, jointly with M. Smith (1993); the Thomas A. Edison Award (1993).

Mullis was inducted into the National Inventors Hall of Fame in 1998.

He has consulted for a large number of companies including: Abbott Labs; Amersham Inc; Angenics; Cytometrics; Eastman Kodak; Milligen/Biosearch; Specialty Laboratories.

He serves on the board of scientific advisors of several companies, provides expert advice in legal matters involving DNA, and is a frequent lecturer at academic meetings, colleges, corporations and universities around the world.

Bibliography

1. Reproduced with permission.
2. From Nobel Lectures, Chemistry 1991–1995, Editor Bo G. Malmström, World Scientific Publishing Co., Singapore, 1997.
3. Who's who: 2005.
4. http://www.karymullis.com/.
5. PCR Applications: Protocols for Functional Genomics (Plastic Comb), by Michael A. Innis (Ed), David H. Gelfand, John J. Sninsky, Academic Press, 1999, ISBN: 0123721865.
6. The Polymerase Chain Reaction (PCR) for Human Viral Diagnosis, by Jonathan P. Clewley, CRC, 1994, ISBN: 0849348331.
7. Quantitation of mRNA by Polymerase Chain Reaction: Nonradioactive PCR Methods (Springer Lab Manuals) by Thomas Köhler, Dirk Laßner, Anne-Katrin Rost, Barbara Thamm, Barbara Pustowoit and Harald Remke, Springer, 1st ed. 1995, ISBN: 3540591923.

1993

Michael Smith (0.5)

"for contributions to the developments of methods within DNA-based chemistry particularly for his fundamental contributions to the establishment of oligonucleotide-based, site-directed mutagenesis and its development for protein studies" [2]

During his presentation speech, Carl-Ivar Brändén, of the Royal Swedish Academy of Sciences, commented of Smith and his work [2]:

...The foundations of the conceptual framework for this development were laid down by the discovery of the double-helical nature of the DNA structure for which the Nobel Prize in Physiology or Medicine was given to Francis Crick, James Watson and Maurice Wilkins in 1962. The four building blocks of the DNA molecule, the nucleotides, are arranged in a specific order along the molecule to form a genetic code for the sequence of amino acids in the corresponding protein molecules. This genetic information is present in a complementary fashion in the two strands of the DNA molecule. Hence, one strand can serve as a template for the synthesis of the second strand. Mutations occur when the sequence of nucleotides is changed. In living cells such mutations occur randomly and evolution is therefore a trial and error process. It can easily be calculated that for a given gene only a tiny fraction of all possible combinations of these nucleotides have been made and tested during the 3.5 billion years that life has been present on Earth. Consequently, there is scope for the design of novel and interesting protein molecules provided we have the knowledge and methodology to obtain and amplify existing genes and to make appropriate changes to them.

In the 1970s, Michael Smith developed a general method for producing mutations in a gene, not in a random fashion but at specific positions determined in advance from the sequence of the nucleotides in the gene. This method of site-directed mutagenesis has created completely new opportunities to study the properties of protein molecules: how they function as catalysts or as signal transmitters through membranes, which factors determine how they fold into specific three-dimensional structures and how they interact with other molecules in the cell. Such protein engineering is also of importance in modern biotechnology and drug design. Novel antibodies have been created that can kill certain cancer cells. Plants that produce proteins enriched in essential amino acids are being field tested and in the future this method might produce engineered wheat and corn flour that has the same nutritional value as meat...

Michael Smith was born on 26th April 1932 in South Shore, Blackpool, England, the elder son of Rowland Smith (a market gardener) and Mary Agnes Smith, who worked in her mother's boarding house. Smith had a younger brother, Robin. He married Helen Wood (married 1960, separated 1983) and they had three children, Tom, Ian and Wendy. Smith died on the evening of the 4th October 2000 of leukeamia. He was survived by his partner Elizabeth Raines [2–4].

The Early Years

Smith attended Marton Moss Church of England School (1937–43) prior to obtaining a scholarship to the local private school, Arnold School (1943–50). Smith attributes his interest in chemistry to Sidney Law, his chemistry teacher whilst at Arnold School. He became a boy scout whilst at school through the influence of Mr. Barnes, and enjoyed camping and the countryside of British Columbia for the rest of his life. He obtained a Blackpool Education Committee Scholarship, through which he attended the University of Manchester (1950–53), graduating with an upper second chemistry degree. Smith continued at the University of Manchester for his Ph.D. under H.B. Henbest, undertaking research concerning cyclohexane diols, graduating in 1956.

Smith's postdoctoral work was with Gobind Khorana (The Nobel Prize Laureate in Physiology or Medicine 1968) in Vancouver, initially on the synthesis of nucleoside-5′ triphosphates. Berg (Nobel Laureate for Chemistry, 1980) visited Khorana's group in the summer of 1956, during which time Smith started to work for Khorana. He later helped to develop the use of the methoxyl-trityl family of protecting groups, which is still in use in the automated synthesis of RNA and DNA fragments. Smith moved with the Khorana group when it relocated to the Institute for Enzyme Research at the University of Wisconsin.

The Career

Year	*Post*	*Institute*
1960–61	Research Associate	Institute for enzyme research, University of Wisconsin
1961–66	Head of Section, Vancouver Laboratory	Fisheries Research Board of Canada
1966–70	Associate Professor	University of British Colombia
1970–97	Professor	University of British Colombia
1997	Retired	University of British Colombia
1997	Director, Genome Sequencing Centre	Cancer Agency, Vancouver

Smith also held a number of posts concurrent with those listed in the above table. They were: Medical Research Associate, Medical Research Council of Canada (1966–); Co-founder of biotechnology company, Zymos (1981); founded Biotechnology Laboratory, University of British Columbia (1986), Director until 1996; Director, Genome Sequencing Centre, Cancer Agency, Vancouver (1997). Smith took sabbaticals at Rockefeller University, the Laboratory of Molecular Biology of the Medical Research Council in Cambridge and Yale University.

Smith's research interests included:

Fisheries Research Laboratory

- marine biology
- new synthetic method for nucleoside-3′,5′ cyclic phosphates
- nucleic acid chemistry

Zymos

- the development of a process for producing human insulin from yeast

University of British Columbia

- sequence determination
- gene isolation
- site-directed mutagenesis

Smith's Nobel Prize lecture is dated 8th December 1993 with the title 'Synthetic DNA and Biology'. Smith commented on the start of his work on synthetic DNA [2]:

> *...I suggested to Caroline Astell, who was studying for her Ph.D., a series of model studies using chemically synthesized oligonucleotides directed at defining the stability and specificity of oligonucleotide duplexes of defined sequence. The objective was to see if a synthetically accessible oligonucleotide could be a tool for identifying and isolating a specific messenger RNA based on Watson-Crick hydrogen-bonding. From the statistical point of view it should be possible, since the length required of an oligonucleotide in order for it to be unique in a given genome size ranges from 9 nucleotides for phage lambda with a genome of* 4.6×10^4 *base pairs to 18 nucleotides for the higher plants...*
>
> *...Firstly, it was clear that, under appropriate conditions of ionic strength, stable duplexes containing as few as six dA-dT base-pairs can be formed. Further, additional base-pairs increase duplex stability significantly, but in a non-linear manner with the incremental increase in stability diminishing with increased length. Also, a single base-pair mismatch of dT with dT or of dT with dG decreases duplex stability by the same amount (equivalent to removing about two base-pairs), but duplex formation is still possible at low temperature. In addition, it was obvious that oligonucleotide duplexes form rapidly at low temperatures, conditions under which denatured double strand DNA does not reanneal.*
>
> *These experiments convinced me that it would be possible to use a synthetic oligonucleotide to identify an RNA or a DNA containing the exact Watson-Crick complement of its sequence and to differentiate it from similar but not identical sequences...*

The study of mutagenesis is an ongoing topic, see for example [5–7].

Honours and Awards

Smith was presented with many honorary degrees including the DCL from Bishop's University (1995), the D.L. University of Windsor (1998) and a Doctorate from the Université de Montréal (1996). He was also presented with Honorary D.Sc. degrees from the: University of Guelph (1988); University of British Columbia (1994); University of Manchester (1994); York University (1994); McGill University (1995); Mount Allison University (1995); Laval University (1995); University of Ottowa (1995); Queen's University (1995); University of Victoria (1995); University of Western Ontario (1995); Carleton University (1996); University of Toronto (1996); Acadia University (1997); McMaster University (1998); University of New Brunswick (1997); University of Waterloo (1997). He was also presented with Honorary LL.D. degrees from the: Simon Fraser University (1994); St Francis Xavier University (1995); University of Alberta (1996). He also received an Honorary M.D. from the Université de Sherbrooke (1995).

Smith was presented with many awards and prizes including the: Jacob Biely Faculty Research Prize (1977); Boehringer Mannheim Prize of the Canadian Biochemical Society (1981); Gold Medal of the Science Council of British Columbia (1984); Award of Excellence, Genetics Society of Canada (1988); Gairdner Foundation International Award (1986); Killam Research Prize (1986–87); G. Malcolm Brown Award, Canadian Federation of Biological Sciences (1989); Flavelle Medal of the Royal Society of Canada (1992); Nobel Prize for Chemistry, jointly with Mullis (1993); American Academy of Achievement Golden Plate Award (1994); Henry Friesen Award (1994); Canadian Medical Association Medal of Honour (1994); Nathan Rosen Award (1994); Principal Award of the Earnest C. Manning Awards Foundation (1994); Canadian Medical Hall of Fame (1995); LeSueur Medal (1995); Reconnaissance de Mérite Scientifique of the University of Quebec (1995); R.M. Taylor Medal and Award (1995); Royal Bank Award (1999).

He was: a Fellow of the Royal Society of Canada (1981); a Fellow of the Royal Society (1986); an Honorary Member of the Canadian Society for Chemical Investigation (1994); a Foreign Associate of the National Academy of Sciences (1996); an Honorary Fellow of the Royal Society of Physicians (1997).

He received the Order of British Columbia (1994) and the Companion of the Order of Canada (1995).

Smith presented the Drummond Lecture at the University of Calgary (1989) and the Darwin Lecture also at the University of Calgary (1989).

Smith's citation for election (date: 20/03/1986) as Fellow of the Royal Society reads [8]:

He has made many contributions to the chemistry and molecular biology of nucleotides and polynucleotides. With nucleotides these include developing general procedures for synthesis of nucleoside 5'-mono- and polyphosphates, and 3'–5' cyclic phosphates. With polynucleotides he has developed new methods for chemical and enzymatic synthesis of oligodeoxyribonucleotides; devised more convenient strategies for sequence determination and determined the sequences of phage and yeast genes; pioneered application of synthetic oligodeoxyribonucleotides to problems in molecular biology including their use as probes in gene isolation, as primers in sequencing double-stranded DNA and mRNA, and most importantly, as specific mutagens in site-directed mutagenesis.

No proposers are listed in the Sackler database.

Bibliography

1. Reproduced with permission.
2. From Nobel Lectures, Chemistry 1991–1995, Editor Bo G. Malmström, World Scientific Publishing Co., Singapore, 1997.
3. Who's who: 2005.
4. Biographical Memoirs of Fellows of the Royal Society, 2001, vol. 47, pp. 429–441.
5. DNA Repair And Mutagenesis by Errol C. Friedberg, Graham C. Walker, Wolfram Siede, Richard D. Wood, Roger A. Schultz and Tom Ellenberger, ASM Press, 2nd ed. 2005, ISBN: 1555813194.
6. Exocyclic DNA Adducts in Mutagenesis and Carcinogenesis (Iarc Scientific Pubcations, No. 150), by Bea Singer and Helmut Bartsch, IARC Scientific Publications, 2005, ISBN: 9283221508.
7. Environmental Mutagenesis (A Volume in the Human Molecular Genetics Series) (Human Molecular Genetics) by David H. Phillips and Stanley Venitt, Academic Press, 1995, ISBN: 0122204328.
8. Sackler database, Royal Society.

1994

© Nobel Foundation [1]

George A. Olah

"for his contribution to carbocation chemistry"[2]

During his presentation speech, Salo Gronowitz, of the Royal Swedish Academy of Sciences, commented of Olah and his work [2]:

> *...created methods to prepare long-lived carbocations in high concentrations, which made it possible to study their structure, stability and reactions with spectroscopic methods. He achieved this by using special solvents, which did not react with the cations. He observed that in these solvents at low temperatures, carbocations could be prepared with the aid of superacids, acids eighteen powers of ten stronger than concentrated sulfuric acid. Through Olah's pioneering work he and the scientists who followed in his footsteps could obtain detailed knowledge about the structure and reactivity of carbocations...*
>
> *...Olah found that there are two groups of carbocations, the trivalent ones called carbenium ions, in which the positive carbon atom is surrounded by three atoms, and those in which the positive carbon is surrounded by five atoms, called carbonium ions. The disputed existence of these penta-coordinated carbocations was the reason for the scientific feud. By providing convincing proof that penta-coordinated carbocations exist, Olah demolished the dogma that carbon in organic compounds could at most be tetra-coordinated, or bind a maximum of four atoms. This had been one of the cornerstones of structural organic chemistry since the days of Kekulé in the 1860s.*
>
> *Olah found that the superacids were so strong that they could donate a proton to simple saturated hydrocarbons, and that these penta-coordinated carbonium ions could undergo further reactions. This fact has contributed to a better understanding of the most important reactions in petrochemistry. His discoveries have led to the development of methods for the isomerization of straight chain alkanes, which have low octane numbers when used in combustion engines, to produce branched alkanes with high octane numbers. Furthermore, these branched alkanes are important as starting materials in industrial syntheses. Olah has also shown that with the aid of superacids it is possible to prepare*

larger hydrocarbons with methane as the building block. With superacid catalysis it is also possible to crack heavy oils and liquefy coal under surprisingly mild conditions...

George A. Olah was born on 22nd May 1927 in Budapest, Hungary, the son of Julius Olah (a lawyer) and Magda, née Krasznai. He married Judith Lengyel (1949) and they have two sons George (1954) and Ronald (1959). Olah is the Distinguished Donald P. and Katherine B. Loker Professor of Chemistry at the University of Southern California [2–5].

The Early Years

Olah attended a school in Budapest run by the Roman Catholic Piarist Fathers, where he studied Latin (for eight years) along with German and French. His favorite subjects at the time were history and literature. He entered the Technical University of Budapest, where he studied chemistry for his first degree, and enrolled for a Ph.D. under G. Zemplen, a former student of Emil Fischer, where he worked on organo fluorine compounds. Fluorine containing starting materials were hard to come by in Hungary at that time. Olah was particularly grateful for compounds that Hans Meerwein sent, a cylinder of boron trifluoride for Olah to use in his work.

The Career

Year	*Post*	*Institute*
1949	Assistant Professor of Organic Chemistry	Technical University, Budapest
1954	Head of Department of Organic Chemistry and Associate Scientific Director	Central Research Institute of the Hungarian Academy of Sciences, Budapest
1957	Research Scientist	Dow Chemical Company
1965	Professor and Chairman	Western Reserve University, Cleveland, Ohio
1967	Chairman of combined Departments of Chemistry	Case Institute of Technology and case Western Reserve University
1967	C. F. Mabery Distinguished Professor of Research in Chemistry	Case Western Reserve University
1977	Professor of Chemistry and Scientific Director	Hydrocarbon Research Institute, University of Southern California
1980– to date	Distinguished Donald P. and Katherine B. Loker Professor of Chemistry	University of Southern California
1991– to date	Director, Loker Hydrocarbon Research Institute	University of Southern California

Olah fled from Hungary in early December of 1956, and moved to Montreal via London. He was a Guggenheim Fellow in 1972 and again in 1988. His research has included:

Dow Chemicals

- Initial work on stable carbocations

Academic work

- Synthetic and Mechanistic Organic Chemistry
- Carbocations and Onium Ions
- Ionic Organic Reactions and Their Intermediates
- Hydrocarbon Chemistry - Methane and Carbon Dioxide Conversion. Methanol economy
- Synthetic Reagents and Methods

Olah has published over 1200 published scientific papers and 115 patents.

In addition to his autobiography, *A Life Of Magic Chemistry: Autobiographical Reflections of a Nobel Prize Winner*, Publisher: Wiley-Interscience, (2000), ISBN: 0471157430, Olah has written the following books: *Carbonium Ions Volume Methods of Formation (Reactive Intermediates in Organic Chemistry S.)*, Pub: Wiley, John Sons, (1970), ISBN: 0471653330; *Carbonium Ions (Reactive Intermediates in Organic Chemistry*, Pub: John Wiley & Sons Inc, ISBN: 0471653349; *Friedel-Crafts Chemistry (Interscience monographs on organic chemistry)*, Pub: John Wiley & Sons Inc, (1973), ISBN: 0471653152; *Carbocations and Electrophilic Reactions*, Pub: John Wiley & Sons Inc, (1974), ISBN: 0471653144; *Chemistry of Energetic Materials* with David R. Squire (Ed), Pub: Academic Press, (1991), ISBN: 0125254407; *Synthetic Fluorine Chemistry* with Richard D. Chambers (Ed) and G. K. Surya Prakash (Ed), Pub: Wiley-Interscience, (1992), ISBN: 0471543705; *Across Conventional Lines: Selected Papers of George A Olah (in 2 Vols)* with G. K. Surya Prakash (Ed), Pub: World Scientific Publishing Company, (2003), ISBN: 9810227698; *Hydrocarbon Chemistry* with Árpád Molnár, Pub: Wiley-Interscience; 2 edition, (2003), ISBN: 0471417823; *Carbocation Chemistry* G. K. Surya Prakash (Ed), Pub: Wiley-Interscience, (2004), ISBN: 0471284904; *Beyond Oil and Gas: The Methanol Economy* with Alain Goeppert and G. K. Surya Prakash, Pub: John Wiley & Sons, (2006), ISBN: 3527312757.

Olah's Nobel Prize lecture is dated 8th December 1994 with the title 'My search for carbocations and their role in chemistry'. Olah commented [2]:

> *...my involvement with the study of carbocations dates back to the fifties and resulted in the first direct observation of alkyl cations and subsequently the whole spectrum of carbocations as long lived species in highly acidic (superacidic) solutions. The low nucleophilicity of the involved counter anions (SbF_6, Sb_2F_{11}, etc.) greatly contributed to the stability of the carbocations, which were in some instances even possible to isolate as crystalline salts. The developed superacidic, "stable ion" methods also gained wide application in the preparation of other ionic intermediates (nitronium, halonium, oxonium ions, etc.). At the same time the preparation and study of an ever increasing number of carbocations allowed the evolvement of a general concept of carbocations, which I suggested in a 1972 paper, as well as naming the cations of carbon compounds as "carbocations"...*

During his presentation, Olah discussed the preparation, some properties and some chemistry of carbocations using the following headings:

- *From acyl cations to alkyl cations*
- *Earlier unsuccessful attempts to observe alkyl cations in solution*

- *Long lived alkyl cations from alkyl fluorides in antimony pentafluoride and related conjugated superacids*
- *The general concept of carbocations*
    - *The role of carbocations in electrophilic reactions*
- *Protosolvolytic activation of carbocationic systems*
- *Activation by solid superacids and possible relevance to enzymatic systems*

There have been several recent texts devoted to carbocations as well as that written by Olah and mentioned above [6–8].

Honours and Awards

Olah has received Honorary D.Sc. degrees from: the University of Durham (1988); the Technical University of Budapest (1989); the University of Munich (1990); the University of Crete (1994); Case Western Reserve University, Cleveland (1995); the University of Jozsef Attila, Szeged, Hungary (1995); the University of Southern California, Los Angeles (1995); University of Veszprém, Veszprém (1995); the University of Montpellier, Montpellier (1996); State University of New York (1998); the University of Pecs (2001).

He has received the following awards and prizes: the American Chemical Society Award in Petroleum Chemistry (1964); the Leo Hendrick Baekeland Award (1967); the Morley Medal (1970); the Alexander von Humbolt-Stiftung Award for Senior U.S. Scientists (1979); the USC Associates Award For Creativity In Research And Scholarship (1985); the Michelson-Morley Award (1987); the Roger Adams Award in Organic Chemistry (1989); the Richard C Tolman Award (1992); the Chemical Pioneers Award (1993); the William Lloyd Evans Award (1993); the Nobel Prize for Chemistry (1994); the George Washington Award, American Hungarian Foundation (1995); the USC Presidential Medallion (1995); the USC Raubenheimer Outstanding Senior Faculty Award (1995); the Cotton Medal, American Chemical Society (1996); the Golden Plate Award (1996); the Inventor of the Year Award (1996); the Kapitsa Gold Medal, Russian Academy of Natural Sciences (1996); the Government of Hungary Award (1997); the Golden Medal of Charles University (1999); the Hanus Medal (1999); the Arthur C. Cope Award, American Chemical Society (2001); the Albert Einstein Medal, Russian Academy of Natural Sciences (2002); the Boron Americas (BUSA) Award (2002); the Bolyai Prize, Hungarian Academy of Sciences (2002); the Grand Cordon of the Order of the Rising Sun of Japan (2003); the Kunsho Medal (2003); the Grand Cordon of the Order of the Rising Sun from the Japanese Chamber of Commerce of Southern California (2004); the Priestley Medal, the American Chemical Society (2005).

Olah is a: Fellow of the Society for the Promotion of Science, Japan (1974); Member of the U.S. National Academy of Sciences (1976); Foreign Member of the Italian National Academy dei Lincei (1982); Honorary Member of the Italian Chemical Society (1988); Member of the European Academy of Arts, Sciences & Humanities (1989); Honorary Member of the Hungarian Academy of Sciences (1990); Foreign Member of the Russian Academy of Natural Sciences (1996); Member of the American Academy of Achievement (1996); Foreign Member of the Royal Society (1997); Foreign Member of the Royal Society of Canada (1997); Honorary Fellow of the Royal Society of Chemistry (1999); Honorary Member of the Chemical Society of Japan (2002); Fellow of the American Academy of Arts and Sciences (2002); Corresponding Member of the Royal Academy of Sciences and Arts of Barcelona (2002); Member of the American Philosphical Society (2002); Member of the Royal Spanish Academy, Arts & Sciences.

Olah became a naturalized U.S. citizen in 1971 and holds the Order of the Corvin Chain of Hungary (2001).

He has been a Visiting Professor at the following institutions: the University of Colorado; the University of Heidelberg; the University of London, King's College; the University of Montpellier; the University of Munich; Ohio State University; the University of Paris; the University of Strasbourg; the Swiss Federal Institute of Technology.

Bibliography

1. Reproduced with permission.
2. From Nobel Lectures, Chemistry 1991–1995, Editor Bo G. Malmström, World Scientific Publishing Co., Singapore, 1997.
3. Who's who: 2005.
4. http://chem.usc.edu/faculty/Olah.html.
5. http://www.usc.edu/schools/college/faculty/faculty1003584.html.
6. Stable Carbocation Chemistry ed by G. K. Surya Prakash and P. V. Schleyer, Wiley-Interscience, 1996, ISBN: 0471594628.
7. Dicoordinated Carbocations ed by Zvi Rappoport and Peter J. Stang, John Wiley & Sons, 1998, ISBN: 0471967513.
8. The Borane, Carborane, Carbocation Continuum ed by Joseph Casanova, Wiley-Interscience, 1st ed. 1998, ISBN: 0471180750.

1995

© Nobel Foundation [1]

Paul J. Crutzen (0.33)

"for their work in atmospheric chemistry, particularly concerning the formation and decomposition of ozone" [2]

During his presentation speech, Ingmar Grenthe, of the Royal Swedish Academy of Sciences, commented on climate change in general and Crutzen's work in particular [2]:

...The sun is the engine of life. Solar radiation is the source of energy for nearly all living organisms. But only some of the sun's rays are beneficial. It also emits ultraviolet radiation that harms living beings. Many of us have painful experience of excessive sunbathing. Life in the forms we are familiar with is the result of photosynthesis in green plants, which transforms the carbon dioxide in the air into biomass and oxygen. It has taken hundreds of millions of years for the biosphere to develop the atmospheric composition we have today. In the upper atmosphere, or stratosphere, solar radiation can transform oxygen into ozone. The highest ozone concentrations are found at an altitude of between 15 and 50 km. This ozone layer absorbs the sun's ultraviolet radiation very effectively, thereby reducing hazardous radiation on the earth's surface. This, in turn, makes efficient photosynthesis possible. Here we see an example of a feedback mechanism between the chemistry of the biosphere and the atmosphere. If it is disrupted, there may be serious consequences for life on our planet.

This year's laureates have made a series of major contributions to our knowledge of atmospheric chemistry. This has included studying how ozone is formed and decomposes and how these processes can be affected by chemical substances in the atmosphere, many of them the result of human activity. In 1970 Paul Crutzen demonstrated that nitrogen oxides, formed during combustion processes, could affect the rate of ozone depletion in the stratosphere. He suggested that dinitrogen monoxide, popularly known as "laughing gas" and formed through microbiological processes in the ground, could have the same effect. He has also studied the formation of ozone in the lower atmosphere. Ozone is one ingredient of "smog," which is formed by the influence of solar radiation on air pollutants, especially exhaust gases from motor vehicles and other combustion systems. Whereas stratospheric ozone is a prerequisite for life, tropospheric ozone is strongly toxic and harmful to most organisms, even in small quantities...

Paul J. Crutzen was born in Amsterdam on 3rd December 1933, the son of Anna Gurk and Jozef Crutzen. He married Terttu Soininen (1958) and they have two daughters Ilona (1958) and Sylvia (1964). Crutzen can be contacted through the Max-Planck-Institute for Chemistry, Department of Atmospheric Chemistry, P.O. Box 3060 [2,3,4].

The Early Years

Crutzen began his primary school in September 1940, just a few months after the Netherlands were overrun by the German Army, causing the school class to move between different buildings in Amsterdam, as their school building was confiscated. He entered Higher Citizen School or "Hogere Burgerschool" (HBS) in 1946, where he concentrated on natural sciences alongside English, French and German.

He attend the Middelbare Technische School (MTS) where he undertook study to be a civil engineer (1951–1954).

Year	*Post*	*Institute*
1954–1958	Engineer – interrupted for military service (Dutch army)	Bridge Construction Bureau of the City of Amsterdam
1958–1959	House Construction Bureau (HKB)	Gaevle, Sweden
1959–1974	Computer consulting teaching and research positions	Department of Meteorology of the University of Stockholm

During his time at the University of Stockholm, Crutzen obtained an M.Sc. (1963), a Ph.D. in Meteorology (1968) with the thesis "*Determination of parameters appearing in the 'dry' and the 'wet' photochemical theories for ozone in the stratosphere*". Crutzen undertook work at the European Space Research Organization (1969 – 1971), part of the Clarendon Laboratory of the University of Oxford. Crutzen was also a visiting Fellow of St. Cross College, Oxford during this time (1969–1971). He was awarded a D.Sc. from the University of Stockholm (1973) for work submitted under the title "*On the photochemistry of ozone in the stratosphere and troposphere and pollution of the stratosphere by high-flying aircraft*"

The Career

Year	*Post*	*Institute*
1974–1977	Research Scientist in the Upper Atmosphere Project and Consultant at the Aeronomy Laboratory	National Center for Atmospheric Research (NCAR), Boulder, Colorado and at the National Oceanic and Atmospheric Adminisitration (NOAA), Boulder, Colorado

(*continued*)

(*continued*)

Year	*Post*	*Institute*
1977–1980	Senior Scientist and Director of the Air Quality Division	National Center for Atmospheric Research (NCAR), Boulder, Colorado
1980–2000	Member of the Max-Planck-Society for the Advancement of Science and Director of the Atmospheric Chemistry Division	Max-Planck-Institute for Chemistry, Mainz
1987–1991	Professor (part-time)	Department of Geophysical Sciences, University of Chicago
1992– to date	Professor (part-time)	Scripps Institution of Oceanography, University of California, La Jolla
1997–2000	Professor (part-time)	Utrecht University, Institute for Marine and Atmospheric Sciences

Crutzen was also Adjunct Professor at the Atmospheric Sciences Department, Colorado State University, Fort Collins, Colorado (1976 – 1981) and Executive Director, Max-Planck-Institute for Chemistry, Mainz (1983 – 1985).

Crutzen's main research work has involved:

- Global modelling of atmospheric chemical processes
- Interactions of atmospheric chemistry with climate
- Tropospherical chemistry
- Stratospheric and tropospheric chemistry
- Studies of the potential role of halogen photochemistry with ozone in the marine boundary layer

Crutzen has written many books including: *A two-dimensional photochemical model of the atmosphere below 55 km. Estimates of natural and man-caused ozone perturbations due to NOx*, Pub: Environmental Research Laboratories, Aeronomy Laboratory; National Center for Atmospheric Research, Upper Atmosphere Project, (1975), ASIN: B0006WFCTQ; *Numerical investigations of the photochemical and transport processes which affect ozone and other trace constituents in the atmosphere*: [1979 annual summary of research], Pub: Dept. of Atmospheric Science, Colorado State University, (1979), ASIN: B00070TD1K; *Atmospheric Change: An Earth System Perspective* with T. E. Graedel, Pub: W H Freeman & Co, (1992), ISBN: 0716723344; *Fire in the Environment: The Ecological, Atmospheric, and Climatic Importance of Vegetation Fires: Report of the Dahlem Workshop Held in Berlin 15- (Dahlem Environmental Sciences Research Reports)* with P. J. Crutzen and J. G. Goldammer (Ed), Pub: John Wiley & Sons Inc, (1993), ISBN: 0471936049; *Clouds, Chemistry and Climates (Nato Asi Global Environmental Change; Series 1 Volume 35)*, by NATO Scientific Affairs Division (Corporate Author), Chemistry and Climate NATO Advanced Research Workshop Clouds, (Corporate Author) and Veerabhadran Ramanathan (Ed), Pub: Springer, (1995), ISBN: 3540604332; *Atmosphere, Climate, and Change (Scientific American Library)* with Thomas E. Graedel, Pub: Scientific American Library, (1995), ISBN: 071675049X and *Earth System Analysis for Sustainability (Dahlem Workshop Reports)* with Hans Joachim Schellnhuber (Ed), William C. Clark (Ed), Martin Claussen (Ed) and Hermann Held (Ed), Pub: The MIT Press, (2004), ISBN: 0262195135

Crutzen's Nobel Prize lecture is dated 8th December 1995 with the title 'My life with O_3, NO_x and other YZO_xs'.

The environmental impact of CFC's has led to many publications including [5–8].

Honours and Awards

Crutzen has been awarded Honorary D.Sc. degrees from many institutes including: York University, Canada (1986); the Université Catholique de Louvain (1992); the University of East Anglia (1994); the Aristotle University, Thessaloniki (1996); the Université de Bourgogne, Dijon (1997); the University of Chile, Santiago (1997); the University of Liège (1997); Oregon State University, Corvallis (1997); the University of San José, Costa Rica (1997); Tel Aviv University (1997); the University of Athens (1998); Democritus University of Thrace (2001); Nova Gorica Polytechnic, Slovenia (2002); University of Hull (2002).

He has been presented with many awards and prizes including: Outstanding Publication Award, Environmental Research Laboratories, National Oceanic and Atmospheric Administration (NOAA), Boulder, Colorado (1976); Special Achievement Award, Environmental Research Laboratories, NOAA, Boulder (1977); the Rolex-Discover Scientist of the Year (1984); the Leo Szilard Award for "Physics in the Publics Interest" of the American Physical Society (1985); the Tyler Prize for the Environment (1989); the Volvo Environmental Prize (1991); the Deutscher Umweltpreis of the Umweltstiftung (German Environmental Prize of the Federal Foundation for the Environment) (1994); the Max-Planck Research Prize jointly with Dr. M. Molina (1994); the Nobel Prize in Chemistry jointly with Dr. M. Molina and Dr. F. S. Rowland (1995); the Global Ozone Award for "Outstanding Contribution for the Protection of the Ozone Layer" by UNEP (United Nations Environment Programme) (1995); the Minnie Rosen Award for "High Achievement in the Service to Mankind" of Ross University, New York (1996); the Louis J. Battan Author's Award jointly with T.E. Graedel, the American Meteorological Society (1996); Election to the Global 500 Roll of Honour of the United Nations Environmental Programme (UNEP) (1996); the Bundesverdienstorden (Order of Merit of the Federal Republic of Germany) (1996) the H. Julian Allen Award 1998, for outstanding scientific paper (co-authored) of 1998 at NASA Ames Research Center (1999); honored by the Karamanlis Institute for Democracy Athens, Greece, for outstanding contributions to Science and Society (2001).

Crutzen has also received the Medail d'Or de la Ville de Grenoble, France (1996) and the "Commandeur in de Orde van de Nederlandse Leeuw" (appointed by the Queen of the Netherlands) (1996).

He is a Fellow of the American Geophysical Union (1986) and a Member of: the Royal Swedish Academy of Sciences (1991); the Royal Swedish Academy of Engineering (1991); the Leopoldina, Halle (1991); the Pontifical Academy of Sciences (Vatican) (1996); the Accademia Nazionale dei Lincei, Roma, Italy (1997) and of the Council of the Pontifical Academy of Sciences (2001). He is also Honorary Member of the International Ozone Commission (1996); the European Geophysical Society (EGS) (1996); the American Meteorological Society (1997); the European Geophysical Society (EGS) (1997); the Swedish Meteorological Society (2000); the World Innovation Foundation, WIF (2002). Crutzen is a Foreign Honorary Member of the American Academy of Arts and Sciences, Cambridge, U.S.A. (1986); a Titular Member of the European Academy of Arts, Sciences and Humanities, Paris

(1996); a Foreign Member of the Russian Academy of Sciences (1999); a Foreign Associate of the U.S. National Academy of Sciences (1994); a Founding Member of Academia Europaea (1988); a Corresponding Member of The Royal Netherlands Academy of Science (1990).

Crutzen was the: Raymond and Beverly Sackler Distinguished Lecturer in Geophysics and Planetary Sciences, Tel Aviv University, Israel (1994); Erasmus Lecturer and Medalist of the Academia Europeae, Gent, Belgium (1997); Ceremonial Lecturer at the Annual Assembly of the Deutsche Physikalische Gesellschaft (German Physical Society), Regensburg (1998).

He is Honorary Professor at the Johannes Gutenberg-University of Mainz (1993). He was Honorary Member of the Commission on Atmospheric Chemistry and Global Pollution (CACGP) (1998) and Honorary Chairman "Climate Conference 2001", 20.–24. August, Utrecht, The Netherlands (2001). Crutzen was the most cited author in the Geosciences (worldwide) with 2911 citations from 110 publications during the decade 1991–2001, ISI (Institute for Scientific Information, Philadelphia, USA), issue Nov./Dec. 2001 (2002). Crutzen is a Member of the Advisory Council of the Volvo Environment Prize (1993–); a Member of the Prix Lemaitre Committee, Belgium (1994–); a Member of the Scientific Advisory Group of the School of Environmental Sciences, University of East Anglia, Norwich, Britain (1995–); Vice-Chairman of the Scientific Committee for the International Geosphere-Biosphere Project (SC-IGBP) (1998–); a Member of the Steering Committee of the Center for Atmospheric Sciences, University of California, Berkeley (2000–); a Member of the Framework Programme Expert Advisory Group (EAG) on "Global change, climate and biodiversity", European Commission, Brussels (2001–); a Member of the advisory committee of the Institute: Urbanization, Emissions, and the Global Carbon Cycle, START, Washington DC (2002–); a Member of the Founders' Assembly, Foundation Lindau Nobelprizewinners Meetings at Lake Constance, Lindau (2003–); a Member of the International Polar Foundation, Brussels (2003–).

Bibliography

1. Reproduced with permission.
2. From Nobel Lectures, Chemistry 1991–1995, Editor Bo G. Malmström, World Scientific Publishing Co., Singapore, 1997.
3. Who's who: 2005.
4. http://www.mpch-mainz.mpg.de/˜air/crutzen/vita.html.
5. BETWEEN EARTH AND SKY: How CFCs Changed Our World and Endangered the Ozone Layer, by Seth Cagin, Pantheon, 1st ed. 1993, ISBN: 0679420525.
6. Stratospheric Ozone Depletion (Global Change Instruction Program), by Ann M. Middlebrook, Margaret A. Tolbert, University Science Books, 2000, ISBN: 1891389106.
7. Protecting the Ozone Layer: Science and Strategy (Environmental Science), Oxford University Press, 2003, ISBN: 0195155491.
8. Protecting The Ozone Layer: The United Nations History by Stephen O. Andersen, K. Madhava Sarma, Lani Sinclair, Earthscan Publications, 2005, ISBN: 1844071723.

1995

Mario J. Molina (0.33)

"for their work in atmospheric chemistry, particularly concerning the formation and decomposition of ozone" [2]

During his presentation speech, Ingmar Grenthe, of the Royal Swedish Academy of Sciences, commented of Molina's work [2]:

> *...In 1974 Mario Molina and Sherwood Rowland showed that chlorine compounds formed by the photochemical decomposition of chlorofluorocarbons (CFC or "Freon" gases) could decompose the stratospheric ozone. They presented detailed hypotheses on how these complicated processes occurred.*
>
> *The discoveries of the three researchers have an unusually close connection with the consequences of modern technology. Supersonic aircraft release nitrogen oxides in the stratosphere. Motor vehicles and stationary combustion plants release the same substances into the lower atmosphere. CFC gases from refrigerators and air conditioners, and in the form of aerosol spray propellants-combined with a "throwaway culture"-result in large-scale emissions of chlorine compounds into the atmosphere. The findings presented by this year's laureates in chemistry have had an enormous political and industrial impact. This was because they clearly identified unacceptable environmental hazards in a large, economically important sector. Their models were also subjected to very rigorous examination which eventually confirmed the main features of their original hypotheses. One obvious result is an international agreement known as the Montreal Protocol, which regulates the manufacture and use of CFCs.*
>
> *Perhaps the most spectacular observation of changes in the stratospheric ozone content was made in 1985 over Antarctica by Joseph Farman and his colleagues. They observed a rapid and dramatic depletion of ozone in the polar region when sunlight returned after the polar night. The ozone content then built up to more normal levels during the subsequent polar summer and winter, after which the process was repeated. This recurring "ozone hole" was completely unexpected.*

Eventually a scientific explanation was found, mainly through the research of Susan Solomon, with important contributions from this year's laureates in chemistry as well...

Mario José Molina was born in Mexico City on the 19th March 1943, the son of Roberto Molina Pasquel (a lawyer) and Leonor Henriquez de Molina. He married Luisa Tan (1973) and they have one son, Felipe (1977). Molina is currently a Professor in the Department of Earth, Atmospheric and Planetary Sciences at the Massachusetts Institute of Technology and can be contacted through them [2–4].

The Early Years

Molina attended schools in Mexico City and a boarding school in Switzerland (from 11 years old). He was already interested in science before he entered high school. He converted one of the bathrooms in his house into a laboratory, where he worked through some of the experiments in a chemistry set. His aunt, Esther Molina helped him to work through more challenging experiments, as she herself was a chemist.

His undergraduate education was taken at Universidad Nacional Autonoma de Mexico (1960), from which he graduated with a chemical engineering BS degree in 1965. At this point, Molina decided to study for a Ph.D. in physical chemistry, but realized that he was weak in mathematics and physics in his undergraduate studies. He decided to undertake some periods of study prior to enrolling for a Ph.D., part of the time for which was spent at the University of Freiburg, and several months in Paris. He briefly returned to Mexico as an Assistant Professor at the Universidad Nacional Autonoma de Mexico, and in 1968, enrolled at the University of California at Berkeley for his graduate studies in physical chemistry, under George C. Pimentel, where his work involved the distribution of internal energy in the products of chemical and photochemical reactions. Molina completed his Ph.D. degree in 1972 and stayed on for postdoctoral work until the autumn of 1973, at which time he joined F. Sherwood Rowland's group as a postdoctoral fellow [2–5].

Molina commented of his ozone depletion work whilst in Rowland's group during his Nobel Autobiography entry [2]:

...Three months after I arrived at Irvine, Sherry (Rowland – see next entry) and I developed the "CFC-ozone depletion theory." At first the research did not seem to be particularly interesting - I carried out a systematic search for processes that might destroy the CFCs in the lower atmosphere, but nothing appeared to affect them. We knew, however, that they would eventually drift to sufficiently high altitudes to be destroyed by solar radiation. The question was not only what destroys them, but more importantly, what are the consequences. We realized that the chlorine atoms produced by the decomposition of the CFCs would catalytically destroy ozone. We became fully aware of the seriousness of the problem when we compared the industrial amounts of CFCs to the amounts of nitrogen oxides which control ozone levels; the role of these catalysts of natural origin had been established a few years earlier by Paul Crutzen. We were alarmed at the possibility that the continued release of CFCs into the atmosphere would cause a significant depletion of the Earth's stratospheric ozone layer. Sherry and I decided to exchange information with the atmospheric sciences community: we went to Berkeley to confer with Professor Harold Johnston, whose work on the impact of the release of nitrogen oxides from the proposed supersonic transport (SST) aircraft on the stratospheric ozone layer was well known to us. Johnston informed us that months earlier Ralph Cicerone and Richard Stolarski had arrived at similar conclusions concerning

the catalytic properties of chlorine atoms in the stratosphere, in connection with the release of hydrogen chloride either from volcanic eruptions or from the ammonium perchlorate fuel planned for the space shuttle.

We published our findings in Nature, in a paper which appeared in the June 28, 1974 issue...

The Career

Year	*Post*	*Institute*
1975	Assistant Professor	University of California at Irvine
1979	Associate Professor	University of California at Irvine
1982	Senior Research Scientist, Molecular Physics and Chemistry Section	NASA Jet Propulsion Laboratory
1989– to date	Professor in the Department of Earth, Atmospheric and Planetary Sciences	Massachusetts Institute of Technology

Molina's work has involved:

Work at the Jet Propulsion Laboratory included:

- investigations concerning the chemistry which is promoted by polar stratospheric clouds
- experiments with chlorine peroxide

Massachusetts Institute of Technology

- the chemistry of atmospheric aerosols
- the chemistry of air pollution of the lower atmosphere
- gas phase chemical kinetics and photochemistry
- global air pollution
- global atmospheric chemistry issues
- the stratospheric ozone layer and its susceptibility to human-made perturbations
- the threat to the ozone layer from chlorofluorocarbon (CFC) gases
- tropospheric pollution issues

Molina has written many publications in scientific journals and also the book *Urban Air Pollution and Forests: Resources at Risk in the Mexico City Air Basin (Ecological Studies)*, with Mark E. Fenn (Ed), L. I. De Bauer (Ed), Tomas Hernandez-Tejeda (Ed), Ma. De Lourdes De LA I. De Bauer (Ed), Pub: Springer-Verlag Telos; (2002) ISBN: 038795337X

Molina's Noble Prize lecture is 8th December 1995 with the title 'Polar Ozone Depletion'. Molina commented [2]:

...In early 1974 F. Sherwood (Sherry) Rowland and I proposed that chlorofluorocarbons (CFCs) would decompose in the stratosphere, releasing chlorine atoms which would catalytically destroy

ozone. The CFCs are industrial compounds which have been used as refrigerants, solvents, propellants for spray cans, blowing agents for the manufacture of plastic foams, etc. The two important properties which make these compounds very useful are (1) they can be readily transformed from a liquid into a vapor under mild temperature and pressure conditions... and (2) they are chemically very inert, and hence are non-toxic, non-flammable, and do not decompose inside a spray can or a refrigerator...

...The rapid seasonal decline of stratospheric ozone over the South Pole-the so called "Antarctic ozone hole" - is a startling phenomenon: a large fraction of the total column ozone-more than a third-disappears in the spring months over an area coinciding largely with the Antarctic continent... In recent years it has become more severe: in October 1992 and 1993 more than 99% of the ozone disappeared at altitudes between about 14 and 19 km, where concentrations of this species are usually largest; and in 1995, in mid-September, the ozone hole was the largest ever recorded that early in the austral spring...

Molina concluded his presentation:

...Much remains to be learned about stratospheric chemistry - and, in more general terms, about the physics and chemistry of the global atmosphere. On the other hand, the cause-effect relationship between human-produced chemicals and ozone depletion is rather well established now: the signals connected with the Antarctic ozone hole are very large indeed. The stratospheric ozone issue has shown us that mankind is quite capable of significantly affecting the atmosphere on a global scale: the most striking effects of CFCs, which are emitted mostly in the north, are seen as far away as possible from the sources, namely over the South Pole. This global problem has also shown us that different sectors of society can work together-the scientific community, industry, environmental organizations, government representatives and policy makers-to reach international agreements: the Montreal Protocol on Substances that Deplete the Ozone Layer has established a very important precedent for the solution of global environmental problems...

The environmental impact of CFC's has led to many publications including [5–8].

Honours and Awards

Molina has been presented with several awards including the Tyler Ecology and Energy Prize (1983) and the Nobel Prize for Chemistry jointly with Crutzen and Rowland (1995).

He is a Member of the US National Academy of Science (1993), of the Institute of Medicine and of the Pontifical Academy of Sciences.

Molina has served on the National Research Council Board on Environmental Studies and Toxicology; the US President's Committee of Advisors in Science and Technology and the Secretary of Energy Advisory Board.

He has also been or is currently a member of various Boards of US-Mexico Foundation of Science and other non-profit environmental organizations.

Bibliography

1. Reproduced with permission.
2. From Nobel Lectures, Chemistry 1991–1995, Editor Bo G. Malmström, World Scientific Publishing Co., Singapore, 1997.

3. Who's who: 2005.
4. http://www-eaps.mit.edu/molina/.
5. BETWEEN EARTH AND SKY: How CFCs Changed Our World and Endangered the Ozone Layer, by Seth Cagin, Pantheon, 1st ed. 1993, ISBN: 0679420525.
6. Stratospheric Ozone Depletion (Global Change Instruction Program) by Ann M. Middlebrook, Margaret A. Tolbert, University Science Books, 2000, ISBN: 1891389106.
7. Protecting the Ozone Layer: Science and Strategy (Environmental Science), Oxford University Press, 2003, ISBN: 0195155491.
8. Protecting The Ozone Layer: The United Nations History by Stephen O. Andersen, K. Madhava Sarma, Lani Sinclair, Earthscan Publications, 2005, ISBN: 1844071723.

1995

© Nobel Foundation [1]

F. Sherwood Rowland (0.33)

"for their work in atmospheric chemistry, particularly concerning the formation and decomposition of ozone" [2]

During his presentation speech, Ingmar Grenthe, of the Royal Swedish Academy of Sciences, commented of Rowland's work [2]:

> *...In 1974 Mario Molina and Sherwood Rowland showed that chlorine compounds formed by the photochemical decomposition of chlorofluorocarbons (CFC or "Freon" gases) could decompose the stratospheric ozone. They presented detailed hypotheses on how these complicated processes occurred.*
>
> *The discoveries of the three researchers have an unusually close connection with the consequences of modern technology. Supersonic aircraft release nitrogen oxides in the stratosphere. Motor vehicles and stationary combustion plants release the same substances into the lower atmosphere. CFC gases from refrigerators and air conditioners, and in the form of aerosol spray propellants-combined with a "throwaway culture" - result in large-scale emissions of chlorine compounds into the atmosphere. The findings presented by this year's laureates in chemistry have had an enormous political and industrial impact. This was because they clearly identified unacceptable environmental hazards in a large, economically important sector. Their models were also subjected to very rigorous examination which eventually confirmed the main features of their original hypotheses. One obvious result is an international agreement known as the Montreal Protocol, which regulates the manufacture and use of CFCs.*
>
> *Perhaps the most spectacular observation of changes in the stratospheric ozone content was made in 1985 over Antarctica by Joseph Farman and his colleagues. They observed a rapid and dramatic depletion of ozone in the polar region when sunlight returned after the polar night. The ozone content then built up to more normal levels during the subsequent polar summer and winter, after which the process was repeated. This recurring "ozone hole" was completely unexpected.*

Eventually a scientific explanation was found, mainly through the research of Susan Solomon, with important contributions from this year's laureates in chemistry as well...

F. Sherwood Rowland was born in Delaware, the small central Ohio town, on 28th June 1927, the second of three sons. He married Joan Lundberg (1952) and they have two children, Ingrid Drake Rowland and Jeffrey Sherwood Rowland. Rowland is the Bren Research Professor of Chemistry in the School of Physical Sciences at the University of California [2–5].

The Early Years

Rowland attended public schools in Delaware. For several summer holidays in his early teens, Rowland operated the local volunteer weather station, an auxiliary part of the U.S. Weather Service. Too young for compulsory military service on completion of his high school studies (1943), Rowland enrolled at Ohio Wesleyan University. However, he enlisted in the Navy program to train radar operators in 1945, spending the next fourteen months as a non commissioned officer with a rating of Specialist (Athletics) 3rd class. Discharge from the Navy was followed by completion of his B.S. degree (1948) from Ohio Wesleyan University. Rowland was admitted to the Department of Chemistry at the University of Chicago (1948), graduating with an M.S. degree in 1951 and a Ph.D. in 1952, for a thesis concerning the chemical state of cyclotron-produced radioactive bromine atoms. His mentor at the University of Chicago was Willard F. Libby (1960 Nobel Chemistry Laureate), and his teachers included Fermi, Harold Urey (1934 Nobel Chemistry Laureate) and Henry Taube (1983 Nobel Chemistry Laureate). Rowland was awarded the D.Sc. from the University of Chicago in 1989.

The Career

Year	*Post*	*Institute*
1952–1956	Instructor in Chemistry	Princeton University
1956–1958	Assistant Professor of Chemistry	University of Kansas
1958–1963	Associate Professor of Chemistry	University of Kansas
1963–1964	Professor of Chemistry	University of Kansas
1964 – to date	Professor of Chemistry and Department Chairman	University of California at Irvine

Rowland was the first Department Chairman at the University of California at Irvine (1964–1970). He has been Daniel G. Aldrich Jr. Professor of Chemistry (1985–1989), Donald Bren Professor of Chemistry (1989–1994) and is currently Donald Bren Research Professor of Chemistry (1994 – to date).

Rowland has many research interests that have included studies of ozone layer depletion from the stratosphere by chlorofluorocarbon gases (in conjunction with Mario J. Molina). They have included (in approximately date order) [3–5]:

- a one-step synthesis of radioactive tritium-labeled glucose
- investigations into the chemical reactions of energetic tritium atoms
- "Hot atom" chemistry in general
- chlorine and fluorine chemistry using the radioactive isotopes ^{38}Cl and ^{18}F

- atmospheric chemistry and chemical kinetics
- studying perturbations of the upper troposphere/lower stratosphere caused by commercial aircraft
- studying the effects of biomass burning over tropical asia
- tropospheric ozone production
- investigations of photochemical transformations within snow
- studying the effect of snows on atmospheric composition

Rowland has published more than 330 scientific publications in the areas of atmospheric chemistry, radiochemistry and chemical kinetics. He has also authored the following books: *The CFC-ozone puzzle: Environmental science in the global arena*, Pub: National Council for Science and the Environment, (2001), ISBN: 0971043914; *The Changing Atmosphere: Report of the Dahlem Workshop on the Changing Atmosphere, Berlin, 1987,* November 1–6, (Physical, Chemical, and Earth Sciences Research Reports) with I. S. A. Isaksen (Ed), Guy Brasseur, Stifterverband, Fur Die Deutsche Wissenschaft (Corporate Author), Berlin (Corporate Author), Pub: John Wiley & Sons Ltd, (1988), ISBN: 0471920479 and *Earth's atmosphere in the twenty-first century: Chlorofluorocarbons and ozone (Carolina environmental essay series)*, Pub: Institute for Environmental Studies, University of North Carolina at Chapel Hill, (1987), ASIN: B0007190HQ.

Rowland's Nobel Prize lecture is titled 'Nobel Lecture in Chemistry' and is dated 8th December 1995. During the course of his lecture, Rowland commented [2]:

> *...Numerous theories were suggested for this new observation of ozone loss, and can generally be classified into three groups: (a) natural change in the dynamics of the Antarctic stratosphere; (b) change in the natural chemistry, e.g. NO_x, of the stratosphere; or (c) chemical changes induced by mankind, especially through the introduction of artificial chlorine-containing compounds such as the CFCs. A decrease in the springtime stratospheric temperatures was observed to accompany the observations of lessened quantities of ozone, and was briefly considered as a possible cause for its destruction. However, the lowered temperatures were soon established to be occurring after the ozone loss, and to be the consequence of reduced UV absorption by reduced amounts of ozone, not a cause of the ozone loss. The observed temperature decreases are in fact a slower warm-up from the wintertime temperature minimum.*
>
> *Three polar expeditions in 1986 and 1987 provided the scientific basis for the conclusion that the ozone losses over Antarctica were indeed the consequence of chemical reactions driven by the much higher stratospheric chlorine concentrations of the mid-1980's versus those of 1950–1970. Two of these were ground-based expeditions to McMurdo, Antarctica, led by Susan Solomon. Two maxima were detected in 1986 in the vertical profile of the important free radical, ClO, by ground-based millimeter wave spectroscopy...*

The environmental impact of CFC's has led to many publications including [6–9].

Honours and Awards

Rowland has been presented with Honorary D.Sc. degrees from the: Duke University (1989); Whittier College (1989); Princeton University (1990); Haverford College (1992); Clark University (1996);

University of East Anglia University (1996). He has also been presented with Honorary L.L.D. degrees from Ohio Wesleyan University (1989) and Simon Fraser University (1991).

Rowland has received many awards and prizes including: the Tyler World Prize in Environment and Energy (1983); the Charles Dana Award for Pioneering Achievements in Health (1987); the Japan Prize in Environmental Science and Technology (1989); the Peter Debye Award in Physical Chemistry from the American Chemical Society (1993); the Roger Revelle Medal of the American Geophysical Union (1994); the Nobel Laureate in Chemistry jointly with Crutzen and Molina (1995).

Rowland is a Member of many learned societies including: the Academy of Arts and Sciences (1991); the American Chemical Society (1953); the American Physical Society (1960); the American Philosophical Society; the National Academy of Sciences.

He has also served on many committees and government bodies including: Member of the California Council on Science and Technology (1989–1997); Member of the Advisory Group of the International Institute for Applied Systems Analysis (IIASA); American Member of the Executive Committee of the National Institute for Global Environmental Change and Co-Director, Western Region, (1990–1993); Member of the Board of Directors of the American Association for the Advancement of Science (1991–94); Member of the Executive Committee for the Tyler Prize (1992 – to date); Chairman of the Committee on International Organizations and Programs of the National Academy of Sciences (1993–1994); Member of the General Committee of the International Council of Scientific Unions (1993–1998); Member of the German-American Academic Council (1994 – to date); Foreign Secretary of the United States National Academy of Sciences (1994 – to date); Former President of the American Association for the Advancement of Science; Chairman of the Committee on International Affairs, National Academy of Sciences (1994 – date); Co-Chair of the Interacademy Panel on International Issues (1995–present); Member of the Ozone Commission of the International Association of Meteorology and Atmospheric Physics (IAMAP) (1980–1988, Honorary Member, 1996 – to date); Member of the Nominating Committee of the American Association for the Advancement of Science (1998–1999).

Rowland has also given many lectures around the world [10].

Bibliography

1. Reproduced with permission.
2. From Nobel Lectures, Chemistry 1991–1995, Editor Bo G. Malmström, World Scientific Publishing Co., Singapore, 1997.
3. Who's who: 2005.
4. http://www.chem.uci.edu/people/faculty/rowland/.
5. http://www.physsci.uci.edu/~rowlandblake/people/drowland.html.
6. BETWEEN EARTH AND SKY: How CFCs Changed Our World and Endangered the Ozone Layer, by Seth Cagin, Pantheon, 1st ed. 1993, ISBN: 0679420525.
7. Stratospheric Ozone Depletion (Global Change Instruction Program) by Ann M. Middlebrook, Margaret A. Tolbert, University Science Books, 2000, ISBN: 1891389106.
8. Protecting the Ozone Layer: Science and Strategy (Environmental Science), Oxford University Press, 2003, ISBN: 0195155491.
9. Protecting The Ozone Layer: The United Nations History by Stephen O. Andersen, K. Madhava Sarma, Lani Sinclair, Earthscan Publications, 2005, ISBN: 1844071723.
10. http://www.vigyanprasar.com/comcom/rowlandcv.htm.

1996

Robert F. Curl Jr. (0.33)

"for their discovery of fullerenes" [2]

During his presentation speech, Lennart Eberson of the Royal Swedish Academy of Sciences, commented of Curl and his work [2]:

> *...The discovery of buckminsterfullerene-or C_{60}-was made by using an advanced instrument, in which a laser vaporized a very small quantity of carbon in one five billionth of a second. When the hot carbon gas condensed, it formed clusters containing different numbers of carbon atoms. The one with 60 carbon atoms was the most common. Many of these various carbon molecules were shown to have the same stability as C_{60} and were therefore also assumed to be closed; the collective name for such clusters was fullerenes It was also possible to produce fullerenes that enclosed a metal atom inside, for example potassium or cesium.*
>
> *The problem with these experiments was that fullerenes were available in such small quantities that their postulated structure could not be rigorously verified. The years from 1985 to 1990 were filled with scientific disputes, in which the stubbornness, ingenuity and enthusiasm of the fullerenes' discoverers kept their hypothesis alive despite rather severe criticism. Only in 1990 were physicists Donald Huffman and Wolfgang Krätschmer able to produce gram-sized quantities of C_{60} using a method that could be quickly and inexpensively duplicated in any laboratory. This made it possible to apply the whole battery of structural determination methods and show that C_{60} really had the structure its discoverers had hypothesized. Chemists now quickly went to work studying the chemistry of fullerenes. They were able to try out various applications of the chemistry and physics of fullerenes.*
>
> *So why are the fullerenes so interesting? To understand this, we must look at the structure of other forms of carbon. Graphite consists of carbon atoms attached together in very large flat*

networks that are piled on top of each other, whereas diamonds consist of carbon atoms bound into endless three-dimensional networks. Both are examples of what we usually call polymers. The chemistry that can be done using these forms of carbon is rather limited-and not entirely inexpensive, in the case of diamonds! A fullerene, on the other hand, has a closed, low-molecular structure that can be chemically processed and modified in an almost infinite number of ways...

Robert Floyd Curl, Jr. was born in Alice, Texas on 23rd August 1933, the only son of two children of Robert Floyd Curl (a Methodist minister) and Lessie Waldeen Curl (a housewife). Curl married Jonel Whipple (1955) and they have two sons. Curl is the Kenneth S. Pitzer-Schlumberger Professor of Natural Sciences and University Professor at Rice University [2–4].

The Early Years

The nature of Curl senior's employment resulted in all of the Curl family living in many small south Texan towns including Alice, Austin, Brady, Brownsville, Del Rio, Kingsville, McAllen and San Antonio. Curl was given a chemistry set from his parents when he was nine. Within a week he had decided to become a chemist. His high school chemistry teacher, Mrs. Lorena Davis did all that she could to foster this enthusiasm. He attended Rice Institute (later named Rice University), from which he graduated with a B.A. in chemistry (1954). He enrolled at the University of California at Berkeley for a Ph.D. under Pitzer, where he investigated the matrix isolation infrared spectrum of disiloxane in order to establish whether the SiO–Si bond was linear or bent. Graduating in 1957, Curl went to work with E. Bright Wilson at Harvard (1957–58) on internal rotation barriers, during which time he met Dudley Herschbach (Nobel Chemistry Laureate 1986) – a fellow student.

The Career

Year	*Post*	*Institute*
1958	Assistant Professor	Rice University
1963	Associate Professor	Rice University
1967	Professor of Chemistry	Rice University
1996–2003	Harry C and Olga K Wiess Professor of Natural Science	Rice University
2003– to date	Kenneth S. Pitzer-Schlumberger Professor of Natural Sciences and University Professor	Rice University

Curl was Master of Lovett College (1968–72) and Chairman of the Chemistry Department (1992–96).

Whilst Curl is best known for his involvement in the discovery of fullerenes, he has been involved in many other aspects of physical chemistry, in particular the development of spectroscopic techniques, for example:

- infrared laser spectroscopy
- kinetic spectroscopy
- the development of infrared proble lasers
- the study of the spectra of larger free radicals (using slit jet supersonic expansion cooling)
- excimer laser flash photolysis

These techniques have been used to study the kinetics of free radicals amongst other phenomenon.

Curl's Nobel Prize lecture is dated 7th December 1996 with the title 'Dawn of the Fullerenes: Experiment and Conjecture'. Curl commented [2]:

> *...I won't review in detail here the experiments we carried out in August and September of 1985 which resulted in our proposal that truncated icosahedron C_{60} (buckminsterfullerene) is formed spontaneously in condensing carbon vapor. Accounts of this have been given by two of the five people involved and in two books on the discovery of the fullerenes. The authors of these books consulted all five of us and, in my opinion, made the best effort possible to use the recollections of those involved to recreate these events.*
>
> *However, in order to understand what we did it is necessary to learn something about the apparatus invented by Richard Smalley to investigate compounds and clusters formed from refractory elements. He and his students used it to investigate the high resolution electronic spectra of a number of metal dimers, copper trimer, and SiC_2...*
>
> *...The heart of the experiment is the laser vaporization supersonic molecular beam source. The source went through several design variations primarily to accommodate the physical form of the available sample material to be vaporized. For these experiments, it was a disk vaporization source because the semiconductor samples that we had been investigating were more readily available in sheet form. The figure below shows a cross-section of this pulsed molecular bean source. In operation, the solenoid-actuated pulsed valve was fired to release through the 1 mm orifice a pulse of He gas over the sample lasting somewhat less than a millisecond. The backing pressure could be as high as 10 atm. At some point during the gas pulse usually near its middle, the Q-switched frequency-doubled vaporization laser was fired generating a 5 nsec long pulse of green light (532 nm) with an energy of roughly 30–40 mJ. This laser was focused onto the rotating-translating graphite disk (to avoid digging pits into the sample) vaporizing a plume of carbon vapor into the gas stream. Multiphoton ionization and the subsequent heating of the resulting plasma limited the amount of material vaporized in a single shot and insured that the species initially contained in it were atoms or very small molecules such as C2 and C3.*
>
> *The material vaporized was caught up in the helium gas flow, was mixed with it and cooled by it. The cooling vapor then began to condense into clusters. The extent of clustering could be varied by changing the backing pressure, the timing of the firing of the vaporization laser with respect to the center of the gas pulse, and by varying the length and geometry of the channel downstream from the vaporization point. In the configuration shown, an integration cup has been added to the end of the gas channel to provide more time for clustering and reaction before supersonic expansion...*

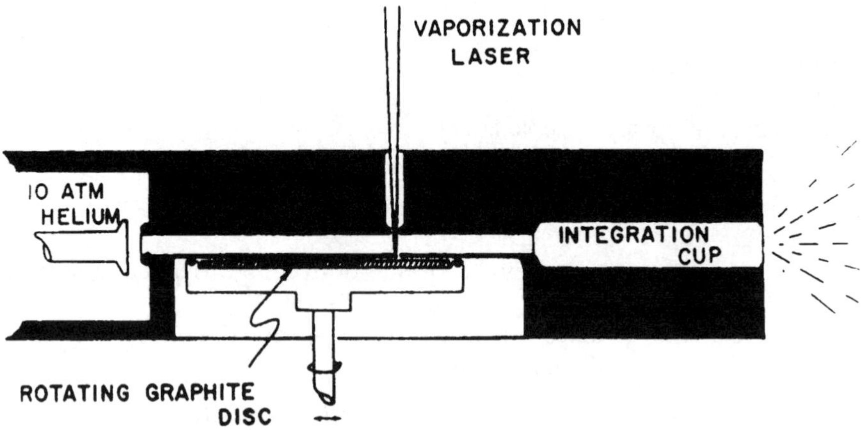

Fig. 2. Laser vaporization source for producing molecular beams of clusters of refractory materials. The integrating cup can be removed. For the carbon experiments, the target is a graphite disk which is rotated slowly to provide fresh vaporization surface. The vaporization laser, a 5 nsec pulse at 532 nm of about 30–40 mJ, is focused onto the surface of the graphite. The pulsed nozzle passes helium over this vaporization zone.

Reproduced from Curl's Nobel lecture

Fullerenes per se are now only a small part of many investigations into the more general field of nanotechnology, of which they were the founding technology, for example see [5–9].

Honours and Awards

Curl was presented with Honorary degrees from the University of Buenos Aires (1997) and the Universite du Littoral Cote d'Opale (2002).

He has been awarded many prizes including: the Clayton Prize (1958); the APS Prize for new materials (1992); the Nobel Prize for Chemistry, jointly with Kroto and Smalley (1996); the Centenary Medal of the Royal Society of Chemistry (1999); Texas Distinguished Scientist (1997); the Achievement in carbon science (1997); the Order of the Gold Plate (1997); the University of Bochum Research Prize, University of Bochum (2004).

He was elected Fellow of the Royal Society of New Zealand (2001) and is an Honorary Professor of the University of Science & Technology of China (2002).

Curl has travelled the world presenting lectures concerning fullerenes.

Bibliography

1. Reproduced with permission.
2. From Les Prix Nobel. The Nobel Prizes 1996, Editor Tore Frängsmyr, Nobel Foundation, Stockholm, 1997.
3. Who's who: 2005.
4. http://www.ruf.rice.edu/~rfcurl/.
5. Putting the Nano into nanotechnology, Chemistry World, Dec 2005.

6. An Introduction to Nanotechnology by Charles P. Poole Jr and Frank J. Owens, Wiley-Interscience, 2003, ISBN: 0-471-07935-9.
7. Clusters and Nanomaterials eds Y. Kawazoe, T. Kondow and K. Ohno, Springer, 2002, ISBN: 3-540-41805-9.
8. Fullerenes: Chemistry, Physics, and Technology ed by Karl M. Kadish, Rodney S. Ruoff, Karl M. Kadish and Rodney S. Ruoff, Wiley-Interscience, 1st ed. 2000, ISBN: 0471290890.
9. Fullerenes: Chemistry and Reactions by Andreas Hirsch, Michael Brettreich and Fred Wudl, John Wiley & Sons, 2005, ISBN: 3527308202.

1996

© Nobel Foundation [1]

Sir Harold W. Kroto (0.33)

"for their discovery of fullerenes" [2]

During his presentation speech, Lennart Eberson, of the Royal Swedish Academy of Sciences, commented of Kroto and his work [2]:

> *...The discovery of buckminsterfullerene - or C_{60} - was made by using an advanced instrument, in which a laser vaporized a very small quantity of carbon in one five billionth of a second. When the hot carbon gas condensed, it formed clusters containing different numbers of carbon atoms. The one with 60 carbon atoms was the most common. Many of these various carbon molecules were shown to have the same stability as C_{60} and were therefore also assumed to be closed; the collective name for such clusters was fullerenes It was also possible to produce fullerenes that enclosed a metal atom inside, for example potassium or cesium.*
>
> *The problem with these experiments was that fullerenes were available in such small quantities that their postulated structure could not be rigorously verified. The years from 1985 to 1990 were filled with scientific disputes, in which the stubbornness, ingenuity and enthusiasm of the fullerenes' discoverers kept their hypothesis alive despite rather severe criticism. Only in 1990 were physicists Donald Huffman and Wolfgang Krätschmer able to produce gram-sized quantities of C_{60} using a method that could be quickly and inexpensively duplicated in any laboratory. This made it possible to apply the whole battery of structural determination methods and show that C_{60} really had the structure its discoverers had hypothesized. Chemists now quickly went to work studying the chemistry of fullerenes. They were able to try out various applications of the chemistry and physics of fullerenes.*
>
> *So why are the fullerenes so interesting? To understand this, we must look at the structure of other forms of carbon. Graphite consists of carbon atoms attached together in very large flat networks that are piled on top of each other, whereas diamonds consist of carbon atoms bound into endless*

three-dimensional networks. Both are examples of what we usually call polymers. The chemistry that can be done using these forms of carbon is rather limited - and not entirely inexpensive, in the case of diamonds! A fullerene, on the other hand, has a closed, low-molecular structure that can be chemically processed and modified in an almost infinite number of ways...

Harold W. Kroto born on 7th October 1939 in Wisbech, Cambridgeshire, the son of Heinz and Edith Kroto. He married Magaret Henrietta Hunter (1963) and they have two sons. Kroto is currently the Epps Professor of Chemistry and Biochemistry at Florida State University in Tallahassee, Florida [2–5].

The Early Years

Kroto's parents originated from Silesia but moved to Berlin, and set up a business printing faces on balloons. Kroto senior fled from Berlin (1937) and eventually settled in London, only to be interned on the Isle of Man during the War. His mother moved Kroto from London to Bolton (1940), where his father joined them after the War. Kroto was educated in Bolton School (1947 – 1958) where he concentrated on science in the sixth form (chemistry, physics and maths). Ian McKellen, the actor, was in the same year at school. He entered Sheffield University (1958), graduating B.Sc. degree (first class honours) in 1961. He stayed at Sheffield University for a Ph.D. under R. Dixon working on the '*Spectroscopy of Free Radicals produced by Flash Photolysis*'. G. Porter (Chemistry Nobel Laureate 1967) was Professor of Physical Chemistry at the time. Kroto graduated with his Ph.D. in 1964 and went to Ottawa for postdoctoral work with D. Ramsay at the National Research Council (1964 – 1966), working on flash photolysis/spectroscopy. Gerhard Herzberg (Chemistry Nobel Laureate 1971) was in the department at the time. A further postdoctoral position followed at the Bell Telephone Laboratories NJ (1966 – 1967) with Yoh Han Pao where he worked on studies of liquid phase interactions by laser Raman spectroscopy.

The Career

Year	*Post*	*Institute*
1967	Tutorial Fellow	University of Sussex
1968	Lecturer	University of Sussex
1977	Reader	University of Sussex
1985	Professor	University of Sussex
1991	Royal Society Research Professor	University of Sussex
2004–date	Epps Professor of Chemistry and Biochemistry	Florida State University in Tallahassee

Although world renowned for his work on fullerenes, Kroto has been involved with several other areas of spectroscopy and organic chemistry including:

- gas phase spectroscopy of unstable intermediates and free radicals using electronic spectroscopy
- liquid phase intermolecular interactions using Raman spectroscopy
- theoretical studies of free radicals and small molecules
- using third row elements (P, S and Se) to form new compounds with carbon which contained multiple bonds (eg C=P)

- the study of the cyanopolyynes (long linear carbon chain molecules) and their discovery in space and in stars
- Buckminsterfullerene
- nano structures:
  - nanotubes
  - nanowires

Kroto wrote (in his Nobel autobiography [2]):

> *...I have been asked many questions about our Nobel Prize and have many conflicting thoughts about it. I have particular regrets about the fact that the contributions of our student co-workers Jim Heath, and Sean O'Brien as well as Yuan Liu receive such disparate recognition relative to that accorded to ours (e.g. Bob, Rick and me). I also have regrets with regard to the general recognition accorded to the amazing breakthrough that Wolfgang Krätschmer and Don Huffman made with their students Kostas Fostiropoulos and Lowell Lamb in extracting C_{60} using the carbon arc technique and which did so much to ignite the explosive growth of Fullerene Science...*

Kroto, in conjunction with Patrick Reams, a BBC producer, set up the Vega Science Trust (1995), the remit for which was the creation and broadcast of high quality scientific films. Some of these films have been broadcast on BBC2 and BBC Prime. The Trust also records interviews with leading scientists so that our scientific heritage may be recorded. Some of the scientists who have been filmed include: Bell-Burnell; Blakemore; Dresselhouse; Feynman; Greenfield; Iijima; Klemperer; Lucas; McGlade; Perutz; Rotblat; Sanger; Sulston; Tonomura; Walker. The trust has also made films of lectures in the Royal Institution which have been coordinated by Gill Watson [5].

Kroto has also published the book *The Fullerenes*, with John E. Fischer and David E. Cox, Pub: Pergamon; 1st ed edition (1993) ISBN: 0080421520.

Kroto's Nobel Prize lecture is dated 7th December 1996 with the title 'Symmetry, Space, Stars and C_{60}'. Kroto described the early discussions of the structure of a C_{60} peak that appeared in the time of flight mass spectrum thus [2]:

> *...That evening/night Rick (Smalley) experimented with sheets of hexagons, Jim together with Carmen experimented with toothpicks and jelly beans and Bob (Curl) and I discussed the stardome solution again. Rick could not make any progress until he remembered our discussion of the pentagons in the stardome. Rick discovered that the structure started to curve into a saucer shape as soon as pentagons were included among the hexagons and finally closed after the twelfth pentagon. The next morning when he revealed his paper model, I remember being ecstatic. It was beautiful and looked just like the stardome as I remembered it. It is of course a truncated icosahedron and the fact that it turned out to be a football too was most appropriate. After all, the whole discovery story is an archetypal example of team effort. I remember thinking that the molecule was so beautiful that it just has to be right – and anyway even if it were not, everybody would surely love it, which they did – eventually! My suggestion that we call the molecule Buckminsterfullerene (the –ene ending fitted perfectly) was after some discussion accepted and we sent the paper to Nature – the date of receipt was 13th September (NB – the experiments started on 1st September)...*

The structure of C_{60} was confirmed as being:

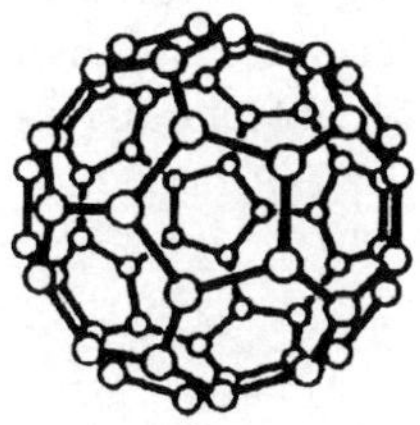

C_{60}

reproduced from Kroto's Nobel presentation [2]

Fullerenes per se are now only a small part of many investigations into the more general field of nanotechnology, of which they were the founding technology, for example see [6–9].

Honours and Awards

Kroto has been presented with honorary degrees from the universities of: Aberdeen; Aveiro (Portugal); Bielefeld Germany); Exeter; Gustavus Adolphus College (Minnesota, USA); Halifax (NovaScotia, Canada); Helsinki (Finland); Hong Kong City (China); Hull; Kingston; Université Libre (Bruxelles); Limburg (Belgium); Leicester; Manchester Metropolitan; Nottingham; Patras (Greece); Sheffield; Sheffield-Hallam; Stockholm (Sweden); Strathclyde; Sussex; University College London; Yokohama City (Japan).

He has been presented with many awards and prizes including: the International Prize for New Materials by the American Physical Society, jointly with R. Curl and R. Smalley (1992); the Italgas Prize for Innovation in Chemistry (1992); the Royal Society of Chemistry Longstaff Medal (1993); the Hewlett Packard Europhysics Prize jointly with W. Kraetschmer, D. Huffman and R. Smalley (1994); the Nobel Prize for Chemistry jointly with R. Curl and R. Smalley (1996); the American Carbon Society Medal for Achievement in Carbon Science jointly with R. Curl and R. Smalley (1997); the Dalton Medal 1998; the Ioannes Marcus Marci Medal 2000 (Prague); the Erasmus Medal of Academia Europaea (2002); the Copley medal (2004).

Kroto is a Fellow of the Royal Society of Chemistry and of the Royal Society (1990) and the Mexican Academy of Science. He is a Member of the Academia Europaea (1993), an Honorary Foreign Member of the Korean Academy of Science and Technology (KAST) (1997), an Honorary Fellow of the Royal Microscopical Society (1998), an Honorary Fellow of the Royal Society of Edinburgh (1998) and an Honorary Fellow of the Royal Society of Chemistry (2000). He is also Honorary Fellow of the Bolton Institute.

He was knighted for his contributions to chemistry in 1996.

Kroto was: President of the Royal Society of Chemistry (2002–2004); Tilden Lecturer of the Royal Society of Chemistry (1981); Blackett Lectureship of the Royal Society (1999); Faraday Lecturer with award (2001). He was a Visiting Professor of UCLA (1990 – 1995) and is a Visiting Professor UCSB (1996 – to date). Kroto is also a Member of National Advisory Committee on Cultural and Creative Education.

Kroto's work through the Vega Trust has led to international recognition with the Chemical Industries Association President's prize short list for the 1998 and 1999, and the award of Prix Leonardo Bronze Medal (2001).

Kroto has also had an interest in graphic design. This interest has resulted in the design work of many book and magazine covers leading to awards for the Sunday Times Book Jacket Design competition (1964). A citation in the international design annual "Modern Publicity" (1979) for the cover of "Chemistry at Sussex" and the Moet Hennessy/Louis Vuitton Science pour l'Art Prize (1994).

Bibliography

1. Reproduced with permission.
2. From Les Prix Nobel. The Nobel Prizes 1996, Editor Tore Frängsmyr, Nobel Foundation, Stockholm, 1997.
3. Who's who: 2005.
4. http://www.kroto.info/.
5. http://www.vega.org.uk/.
6. An Introduction to Nanotechnology by Charles P. Poole Jr and Frank J. Owens, Wiley-Interscience, 2003, ISBN: 0-471-07935-9.
7. Clusters and Nanomaterials eds. Y. Kawazoe, T. Kondow and K. Ohno, Springer, 2002, ISBN: 3-540-41805-9.
8. Fullerenes: Chemistry, Physics, and Technology ed. by Karl M. Kadish, Rodney S. Ruoff, Karl M. Kadish and Rodney S. Ruoff, Wiley-Interscience, 1st ed. 2000, ISBN: 0471290890.
9. Fullerenes: Chemistry and Reactions by Andreas Hirsch, Michael Brettreich and Fred Wudl, John Wiley & Sons, 2005, ISBN: 3527308202.

1996

Richard E. Smalley (0.33)

"for their discovery of fullerenes" [2]

During his presentation speech, Lennart Eberson of the Royal Swedish Academy of Sciences, commented of Smalley and his work [2]:

...The discovery of buckminsterfullerene - or C_{60} - was made by using an advanced instrument, in which a laser vaporized a very small quantity of carbon in one five billionth of a second. When the hot carbon gas condensed, it formed clusters containing different numbers of carbon atoms. The one with 60 carbon atoms was the most common. Many of these various carbon molecules were shown to have the same stability as C_{60} and were therefore also assumed to be closed; the collective name for such clusters was fullerenes It was also possible to produce fullerenes that enclosed a metal atom inside, for example potassium or cesium.

The problem with these experiments was that fullerenes were available in such small quantities that their postulated structure could not be rigorously verified. The years from 1985 to 1990 were filled with scientific disputes, in which the stubbornness, ingenuity and enthusiasm of the fullerenes' discoverers kept their hypothesis alive despite rather severe criticism. Only in 1990 were physicists Donald Huffman and Wolfgang Krätschmer able to produce gram-sized quantities of C_{60} using a method that could be quickly and inexpensively duplicated in any laboratory. This made it possible to apply the whole battery of structural determination methods and show that C_{60} really had the structure its discoverers had hypothesized. Chemists now quickly went to work studying the chemistry of fullerenes. They were able to try out various applications of the chemistry and physics of fullerenes.

So why are the fullerenes so interesting? To understand this, we must look at the structure of other forms of carbon. Graphite consists of carbon atoms attached together in very large flat networks that are piled on top of each other, whereas diamonds consist of carbon atoms bound into endless three-dimensional networks. Both are examples of what we usually call polymers. The chemistry that can be done using these forms of carbon is rather limited - and not entirely inexpensive, in the case of diamonds! A fullerene, on the other hand, has a closed, low-molecular structure that can be chemically processed and modified in an almost infinite number of ways...

Richard Errett Smalley was born in Akron, Ohio, on 6th June 1943, the youngest of four children of Frank Dudley Smalley and Virginia Smalley, née Rhoads. He married Judith Grace Sampieri (1968) and they had one son Chad R. Smalley (1969). The marriage was dissolved in 1979. He married Marry Lynn Chapieski (1980), marriage dissolved in 1994. He married JoNell Marie Chauvin (1997) by whom he has one son, Preston C. Smalley (1997). He later married Deborah. Smalley died on the afternoon of the 28th October 2005 at M.D. Anderson Cancer Center [2–4].

The Early Years

Smalley moved to Kansas City when aged three and lived there until he left home to study at eighteen. He developed an interest in science from the Sputnik launch (1957) and thereafter concentrated his efforts studying physics and chemistry. He enrolled at Hope College, Holland, Michigan in 1961, but left in 1963 after the death of his favorite professor, J. Harvey Kleinheksel (of a heart attack), and the retirement of Gerrit Van Zyl, for whom Smalley wanted to work. Smalley completed his chemistry B.S. from the University of Michigan, Ann Arbor, in 1965 and started work for Woodbury, a polypropylene manufacturing plant (owned by the Shell Chemical Company) as a Research Chemist. After two years he moved to the Plastics Technical Center, where he developed various analytical methods for polyolefins. Smalley left the Shell organization in the Autumn of 1968 and enrolled at Princeton University. He was assessed for military service (late 1968), was reclassified 1A for the draft, and reported to the processing center in Newark for his physical, which he passed. He was later reclassified and was not drafted.

He began his Ph.D. at Princeton under Elliot R. Bernstein (1969), the subject for which was the study of 1,3,5-triazine, particulary studying the the Jahn Teller effect. During the course of his work, he received an M.A. from Princeton University (1971), graduating with his Ph.D. in 1973. Smalley spent the next three years (1973–76) at the James Franck Institute, University of Chicago as a postdoctoral research associate with D. H. Levy. After some thought, he began his work on the jet cooled spectrum of NO_2.

The Career

Year	*Post*	*Institute*
1976 – 1980	Assistant Professor, Department of Chemistry	Rice University
1980 – 1981	Associate Professor, Department of Chemistry	Rice University
1981 – 1982	Professor, Department of Chemistry	Rice University
1982 – 2005	Gene and Norman Hackerman Professor of Chemistry	Rice University
1990 – 2005	Professor of Physics	Rice University
2002 – 2005	University Professor	Rice University

Smalley was a Harold W. Dodds Fellow of Princeton University (1973) and an Alfred P. Sloan Fellow (1978 – 1980).

Smalley has had many research interests including:

- the construction and use of a supersonic beam apparatus adapted to use pulsed dye lasers in the ultraviolet
- the study of a large uranium containing molecule (a hexafluoroacetylacetonate-, tetahydrofur-ancomplexed form of UO_2).
- the study of the properties of nanometer-scale particles consisting of a precise number of atoms
- work on metal and semiconductor clusters
- single wall carbon nanotube single-crystal growth
- the synthesis of carbon single wall nanotube fibers
- fullerene nanotubes
- fullerene nanostructures

Smalley's Nobel Prize lecture is dated 7th December 1996 with the title 'Discovering the Fullerenes'. Smalley commented [2]:

> *...The laser vaporization supersonic beam source and the associated probe techniques that we developed to study the 2-200 atom metal and semiconductor clusters was effectively a new sort of microscope. It allowed one to "see" something of the nature of nanoscopic aggregates of atoms in a way that was entirely new, and very poignant. Whatever we measured for the cluster in the supersonic beam, we knew that it was the true property of that cluster traveling free in space. We developed the technique in order to bring a sort of intellectual tension to surface science: to make measurements so fundamental that theorists stayed awake at night trying to understand them. Although we had no notion of it at the time we were engaged in this enterprise in the early 1980's, we were building the instrument and the line of research that would discover the fullerenes. All one had to do is put carbon in this new "microscope", adjust the focus a little, and "see" the fullerenes revealed plainly for the first time.*
>
> *As it happened, we at Rice were not the first to put carbon into the new microscope. In 1984 a group headed by Andrew Kaldor at Exxon used such an apparatus (actually one that had been designed and built at Rice) in a study of carbon clusters that was motivated by the desire to study coke buildup on reforming catalyst. Now ...carbon clusters were evident extending out beyond 100 atoms, and it was immediately clear that a whole new world of interesting carbon clusters could exist that had never been seen before.*
>
> *Three distinct regions characterized the mass spectrum: first the small clusters, containing fewer than 25 atoms, consisting of the chains and monocyclic rings so well known from the earlier studies; second a new region between about 25 and 35 atoms in which few species of any sort were observeda region that the Rice group came to call the "forbidden zone"; and, third, an even-numbered clusters distribution extending from the high 30's to well over 150 atoms. By early 1984 the laservaporization supersonic cluster beam technique had been used to study clusters broadly throughout the periodic table, and many "magic number" cluster distributions had been found and studied, but nothing remotely like this even-numbered distribution of carbon had ever been seen for any other element. Today this is still true: carbon is unique...*

Fullerenes per se are now only a small part of many investigations into the more general field of nanotechnology, of which they were the founding technology, for example see [5–8].

Honours and Awards

Smalley has been presented with the degree of Honorary Doctor from the University of Liege (1991) and Honorary D.Sc. degrees from the: University of Chicago (1995); University of Michigan (1997); the University of Pennsylvania (2002); Rensselaer Polytechnique Institute (2003); University of Richmond (2004); Technion, Israel Institute of Technology (2004); Tuskegee University (2004).

He received the following awards and prizes: the Irving Langmuir Prize in Chemical Physics (1991); the Popular Science Magazine Grand Award in Science & Technology (1991); the APS International Prize for New Materials, jointly with R. F. Curl & H. W. Kroto (1992); the Jack S. Kilby Award (1992); the Ernest O. Lawrence Memorial Award (1992); the Welch Award in Chemistry (1992); the Auburn-G.M. Kosolapoff Award (1992); the Southwest Regional Award (1992); the William H. Nichols Medal (1993); the John Scott Award (1993); the Hewlett-Packard Europhysics Prize (1994); the Harrison Howe Award (1994); the Madison Marshall Award (1995); the Franklin Medal (1996); the Nobel Prize in Chemistry, jointly with R. F. Curl & H. W. Kroto (1996); the American Carbon Society Medal (1997); Distinguished Public Service Medal from the US Department of the Navy (1997); the Highly Cited Researcher Certificate, ISI Thompson Scientific (2002); the Lifetime Achievement Award, Small Times Magazine (2003); the Glenn T. Seaborg Medal, University of California at Los Angeles (2002); the Distinguished Alumni Award, Hope College (2005); the 50th Anniversary Visionary Award, SPIE, Int'l Society for Optical Engineering (2005); the Mariluisa Ferrari Lectureship for Life Award, Ohio State University (2005). He was one of the Top 75 Distinguished Contributors, Chemical & Engineering News (1998) and was Rice University Homecoming Queen (1996) which was presented by Rice University Undergraduates.

He was a Fellow or Member of: the AAAS (2003); the American Academy of Arts and Sciences (1991); the American Chemical Society; the American Physical Society (1987); the American Institute of Physics; the American Association for the Advancement of Science; the National Academy of Sciences; the Materials Research Society; the National Academy of Sciences (1990); of the Sigma Xi.

Smalley was Chairman, Rice Quantum Institute (1986 – 1996, Founder (1979)) and Founding Director of the Rice Center for Nanoscale Science & Technology (CNST) (1996 – 2002). He was Chairman of Carbon Nanotechnologies, Inc. (2000 – 2005) and a Director of the Carbon Nanotechnology Laboratory, Rice University (2003 – 2005).

Bibliography

1. Reproduced with permission.
2. From Les Prix Nobel. The Nobel Prizes 1996, Editor Tore Frängsmyr, Nobel Foundation, Stockholm, 1997.
3. Who's who: 2005.
4. http://smalley.rice.edu/smalley.cfm?doc_id=5111.
5. An Introduction to Nanotechnology by Charles P. Poole Jr and Frank J. Owens, Wiley-Interscience, 2003, ISBN: 0-471-07935-9.

6. Clusters and Nanomaterials eds Y. Kawazoe, T. Kondow and K. Ohno, Springer, 2002, ISBN: 3-540-41805-9.
7. Fullerenes: Chemistry, Physics, and Technology ed by Karl M. Kadish, Rodney S. Ruoff, Karl M. Kadish and Rodney S. Ruoff, Wiley-Interscience, 1st ed. 2000, ISBN: 0471290890.
8. Fullerenes: Chemistry and Reactions by Andreas Hirsch, Michael Brettreich and Fred Wudl, John Wiley & Sons, 2005, ISBN: 3527308202.

1997

© Nobel Foundation [1]

Paul D. Boyer (0.25)

"for their elucidation of the enzymatic mechanism underlying the synthesis of adenosine triphosphate (ATP)" [2]

During his presentation speech on 10th December 1997, Bertil Andersson, of the Royal Swedish Academy of Sciences, commented of Boyer's work [1,2]:

...All energy on earth originates from the sun. Green plants can absorb sunlight and convert it into chemical energy through the process of photosynthesis, in which carbon dioxide and water form sugar, starch and other complex carbon compounds. Other organisms, such as humans and animals, are in turn dependent on these carbon compounds as sources of energy, and they burn them with the help of oxygen. That is why we breathe. Nature can thus be said to have chosen a combination of solar and coal-fired power plants for its energy supply. Although these two energy conversion systems may seem different in purely technical terms, in many respects they operate in the same way in living cells. The most important similarity is that the energy released is utilized with the help of the ATP molecule.

According to Peter Mitchell, the 1978 Nobel Laureate in Chemistry, the energy released in photosynthesis and cell respiration initiates a stream of positively charged hydrogen ions. These hydrogen ions, in turn, drive the production of ATP with the aid of a membrane-bound enzyme called ATP synthase. Two of this year's laureates, Paul Boyer and John Walker, have studied this important enzyme and have shown that it functions in a unique way. Among other things, they have demonstrated that ATP synthase can be compared to a molecular machine, whose rotating bent axle is driven in a step-wise process by "biological electricity" - that is, the flow of hydrogen ions. Because of the asymmetry of the rotating axle, three subunits of the enzyme assume different forms and functions: a first form that binds adenosine diphosphate (or ADP) and phosphate building blocks, a second form where these two molecules are chemically combined into a new ATP molecule, and a third form where the ATP that has been formed is released. In the next twist of the axle, the three subunits switch form and thus also function with each other, and another ATP molecule can be formed, and so on. This "binding change mechanism" was put forward by Boyer in the late 1970s, but only in 1994 did his ingenious model gain general acceptance among researchers. In August of that year, Walker and his colleagues published three-dimensional images of ATP synthase that had been obtained by X-ray analysis of enzyme crystals. These X-ray images, magnified several million times, showed how an asymmetrically elongated protein molecule interacted with three other protein units that all showed mutually different forms. Walker had finally revealed the detailed blueprint of the molecular machine and shown that Boyer's theory of ATP formation was correct...

Paul Delos Boyer was born in Provo, Utah, on 31st July 1918, one of two sons of Dell Delos Boyer and Grace Guymon. He married Lyda Whicker (1939) and they have three children Gail, Hali and Douglas. Boyer retired in 1989 but can be contacted through the University of Los Angeles Molecular Biology Institute [2–4].

The Early Years

Boyer grew up in Provo, Utah where he attended Parker Elementary School, the Farrer Junior High School and the local Provo High School, where he was taught chemistry by Rees Bench. He enrolled in the nearby Brigham Young University, from which he graduated with a B.S. degree in chemistry (1939), and obtained a Wisconsin Alumni Research Foundation Scholarship for graduate studies. He used the scholarship to study in the Biochemistry Department at the University of Wisconsin in Madison, under Professor Paul Phillips. He attended lectures by Otto Meyerhof (Nobel Laureate for Medicine 1922), Carl Cori (Nobel Laureate for Medicine 1947) and Fritz Lipmann (Nobel Laureate for Medicine 1953). Boyer graduated with his Ph.D. in 1943 and moved to Stanford University (1943–45) where he undertook war-related research looking at the stabilization of serum albumin for transfusions. He was a member of the US National Guard (Navy) during his military service, which was undertaken in the Navy Medical Research Institute in Bethesda, Maryland (1945–46).

The Career

Year	*Post*	*Institute*
1946	Associate Professor	University of Minnesota
1953	Professor	University of Minnesota
1956–63	Hill Professor of Biochemistry	University of Minnesota
1963–89	Professor in the Department of Chemistry and Biochemistry	UCLA
1965–83	Director, Molecular Biology Institute	UCLA
1985–89	Biotechnology program	UCLA
1989	Retired	

Boyer was a Guggenheim Fellow (1955), during which he worked with both Professor Hugo Theorell (Nobel Laureate for Medicine 1955) in Sweden on the mechanism of alcohol dehydrogenase, and at the Wenner-Gren Institute of the University of Stockholm with Olov Lindberg.

Boyer's work has centered on the three postulates for the binding mechanism for ATP synthesis, namely [2]:

- that energy input was not used primarily to form ATP but to promote the binding of phosphate and mostly the release of tightly bound ATP;
- that three identical catalytic sites went through compulsory, sequential binding changes;
- that the binding changes of the catalytic subunits, circularly arranged on the periphery of the enzyme, were driven by the rotation of a smaller internal subunit.

During the course of the investigations on the three postulates above, he also undertook some key studies of:

- studies to detect a possible phosphorylated intermediate in ATP synthesis using ^{32}P
- the discovery of a new type of phosphorylated protein
- the discovery of an intermediate in the substrate level phosphorylation of the citric acid cycle
- the binding change mechanism for ATP synthesis
- aspects of the complex ATP synthase were explored

Boyer was editor or associate editor of Annual Biochemistry Reviews (1964–89) and of the series *The Enzymes*. He has also written several books including:

The Enzymes (Ed), Pub: Academic Pr; 3rd edition, (1970), ISBN: 0121227022; *Enzymes: Peptide Bond Hydrolysis* (Ed), Publisher: Academic Pr; 3rd edition, (1971), ISBN: 0121227030; *The Enzymes: Oxidation-Reduction (Part B)* (Ed), Pub: Academic Pr; 3rd edition, (1975), ISBN: 012122712X; *Enzymes: Nucleic Acids/Part B (Enzymes)*, Pub: Academic Press; 3rd edition, (1982), ISBN: 0121227154; *Enzymes: Lipid Enzymology (Enzymes)* (Ed), Pub: Academic Press; 3rd edition, (1983), ISBN: 0121227162 and *The Enzymes: Control by Phosphorylation, Part A: General Features, Specific Enzymes (I)* with Edwin G. Krebs (Ed), Pub: Academic Press, 3rd edition, (1986), ISBN: 0121227170.

Boyer's Nobel Prize lecture is dated 8th December 1997 with the title 'Energy, Life and ATP'. Boyer commented [2]:

> *...Of more relevance to ATP synthase were experiments with $^{18}0$ and ^{32}P, initiated because of the demonstration by Mildred Cohn that mitochondria would catalyze a rapid exchange of phosphate oxygens with those of water. We found from the ^{32}P experiments that the overall reaction of oxidative phosphorylation was dynamically reversible. The $^{18}0$ experiments revealed the striking finding that the exchange of inorganic phosphate oxygens with water was occurring even more rapidly. We attributed this to the formation of a covalent intermediate, which was then cleaved by inorganic phosphate. We tried unsuccessfully to separate out fractions from mitochondria that would catalyze the first step leading to the formation of an intermediate in oxidative phosphorylation. It was some sixteen years later that we found the simple explanation that no intermediate was formed, and that the rapid $^{18}0$ exchange resulted from the rapid and reversible formation of a tightly bound ATP...*

> *...In the 1960s we embarked on another, only partially successful, series of experiments. By using ^{32}P as a sensitive tracer, we found in mitochondria a labeled protein that was an intermediate between inorganic phosphate and ATP. We identified this as a previously unrecognized*

phosphorylated protein, with a phosphoryl group attached to a histidine residue. We mistakenly thought we had identified an intermediate in oxidative phosphorylation, but subsequently found it to be an intermediate in GTP or ATP formation by the succincyl CoA synthetase of the citric acid cycle. We were reaching for a gold but got a bronze instead...

CATALYTIC COOPERATIVITY

...We were now launched on an exciting period of research. As we probed mitochondrial oxidative phosphorylation further by ^{32}P *and* $^{18}0$ *isotope ex changes, some puzzling aspects emerged. For example, when submitochond rial particles capable of oxidative phosphorylation were hydrolyzing ATP, a lively medium*

$$\text{ATP} \rightleftharpoons \text{HOH}$$

exchange occurred. Removal of product ADP stopped this exchange, although the reversal of ATP hydrolysis was still occurring on the enzyme. Somehow the lack of medium ADP to bind to the enzyme was stopping the release of ATP. It was not apparent how this could occur if the simple scheme was used to explain the oxygen exchanges. Similarly, during net synthesis of ATP, removal of the medium ATP stopped the medium

$$\text{P}_\text{i} \rightleftharpoons \text{HOH}$$

exchange. An explanation for why these oxygen exchanges were blocked, and for other related observations, was suggested by one of my graduate students, Celik Kayalar from Turkey. Celik said he could account for these results if the catalytic sites had to work cooperatively, so that ATP could not be released from one site unless ADP and P_i*, were available to bind at another site, or that* P_i *could not be released from one catalytic site unless ATP were available to bind at another catalytic site. Celik, together with Jan Rosing from Holland, demonstrated and characterized sequential and cooperative participation of catalytic sites with the synthase in submitochondrial particles capable of or during oxidative phosphorylation. In addition, their results gave evidence that the binding changes accompanying proton translocation also promoted the tight binding of* P_i*...*

A recent text describes the influence of the phosphate tail on energy transport [5].

Honours and Awards

Boyer has received honorary doctorates from the Universities of Stockholm (1974), Minnesota (1996), and Wisconsin (1998).

He has received many awards and prizes including: the Award in Enzyme Chemistry of the American Chemical Society (1955); the Rose Award of the American Society of Chemistry and Molecular Biology (1989); the Nobel Prize in Chemistry, jointly with Skou and Walker (1997).

He is a Fellow of the American Chemical Society.

Boyer has served as Chairman of the Biochemistry Section of the American Chemical Society (1959–60) and as President of the American Society of Biological Chemists (1969–70).

Bibliography

1. Reproduced with permission.
2. From Les Prix Nobel. The Nobel Prizes 1997, Editor Tore Frängsmyr, [Nobel Foundation], Stockholm, 1998.
3. Who's who: 2005.
4. http://www.chem.ucla.edu/dept/alumni/Boyer_bio.html.
5. Nucleoside Triphosphates and Their Analogs: Chemistry, Biotechnology and Biological Applications by Morteza Vaghefi, CRC, 2005, ISBN: 1574444980.

1997

© Nobel Foundation [1]

John E. Walker (0.25)

"for their elucidation of the enzymatic mechanism underlying the synthesis of adenosine triphosphate (ATP)" [2]

During his presentation speech on 10th December 1997, Bertil Andersson, of the Royal Swedish Academy of Sciences, commented of Walker's work [2]:

...All energy on earth originates from the sun. Green plants can absorb sunlight and convert it into chemical energy through the process of photosynthesis, in which carbon dioxide and water form sugar, starch and other complex carbon compounds. Other organisms, such as humans and animals, are in turn dependent on these carbon compounds as sources of energy, and they burn them with the help of oxygen. That is why we breathe. Nature can thus be said to have chosen a combination of solar and coal-fired power plants for its energy supply. Although these two energy conversion systems may seem different in purely technical terms, in many respects they operate in the same way in living cells. The most important similarity is that the energy released is utilized with the help of the ATP molecule.

According to Peter Mitchell, the 1978 Nobel Laureate in Chemistry, the energy released in photosynthesis and cell respiration initiates a stream of positively charged hydrogen ions. These hydrogen ions, in turn, drive the production of ATP with the aid of a membrane-bound enzyme called ATP synthase. Two of this year's laureates, Paul Boyer and John Walker, have studied this important enzyme and have shown that it functions in a unique way. Among other things, they have demonstrated that ATP synthase can be compared to a molecular machine, whose rotating bent axle is driven in a step-wise process by "biological electricity"-that is, the flow of hydrogen ions. Because of the asymmetry of the rotating axle, three subunits of the enzyme assume different forms and functions: a first form that hinds adenosine diphosphate (or ADP) and phosphate building blocks, a second form where these two molecules are chemically combined into a new ATP molecule, and a third form where the ATP that has been formed is released. In the next twist of the axle, the three subunits switch form and thus also function with each other, and another ATP molecule can be formed, and so on. This "binding change mechanism" was put forward by Boyer in the late 1970s, but only in 1994 did his ingenious model gain general acceptance among researchers. In August of that year, Walker and his colleagues published three-dimensional images of ATP synthase that had been obtained by X-ray analysis of enzyme crystals. These X-ray images, magnified several million times, showed how an asymmetrically elongated protein molecule interacted with three other protein units that all showed mutually different forms. Walker had finally revealed the detailed blueprint of the molecular machine and shown that Boyer's theory of ATP formation was correct...

John Ernest Walker was born in Halifax, Yorkshire, on 7th January 1941, the only son of three children (two younger sisters Jennifer and Judith) of Thomas Ernest Walker (a stone mason) and Elsie Walker, (née Lawton). He married Christina Jane Westcott (1963) and they have two daughters Esther (1976) and Miriam (1978). Sir John is currently the Director of the Dunn Human Nutrition Unit and Professor of Molecular Bioenergetics at the Dunn Human Nutrition Unit Cambridge [2–4].

The Early Years

Walker lived in Elland and attended the local Rastrick Grammar School, where he concentrated on mathematics and the physical sciences during the last few years, whilst also being captain of the school cricket and football teams. He went to St. Catherine's College, Oxford (1960), obtaining the B.A. degree in Chemistry in 1964. His doctorate work was under E. P. Abraham in the Sir Willian Dunn School of Pathology, Oxford (from 1965). He was influenced by the "The Thread of Life" series of programmes on the BBC, which were given by John Kendrew (Nobel Laureate in Chemistry, 1962) and by the books "Bacterial Genetics" (William Hayes) and "Molecular Biology of the Gene" (J. D. Watson, Nobel Laureate for Medicine, 1962). Graduation with a D. Phil. degree (1969) was followed by five years working abroad, as a Visiting Research Fellow at the University of Wisconsin (1969–71), a NATO Research Fellow CNRS Gif-sur-Yvette (1971–72) and as a EMBO Research Fellow at the Pasteur Institute (1972–74).

Walker attended a research workshop in Cambridge entitled "Sequence Analysis of Proteins" (Easter 1974), during the banquet for which he was seated next to F. Sanger (Nobel Laureate in Chemistry, 1958 and 1980). This conversation ultimately led to Walker's appointment at the MRC in Sanger's department (1994) [2–4].

The Career

Year	*Post*	*Institute*
1974–98	Staff Scientist, Laboratory of Molecular Biology	MRC, Cambridge
1998 – to date	Director,	Dunn Human Nutrition Unit, MRC, Cambridge
2002–to date	Professor of Molecular Bioenergetics	Dunn Human Nutrition Unit, MRC, Cambridge

During his time at the MRC in Cambridge, Walker has worked for or alongside many Nobel Laureates including: Max Perutz (Chemistry Nobel Laureate, 1962); Frederick Sanger (Chemistry Nobel Laureate, 1958 and 1980); Aaron Klug (Chemistry Nobel Laureate, 1982); Francis Crick (Medicine Nobel Laureate, 1962); César Milstein (Medicine Nobel Laureate, 1984).

Walker has worked on many aspects of biochemistry including:

- mass spectrometric methods for analysing proteins
- the modified genetic code in mitochondria
- a structural study of the ATP synthase from bovine heart mitochondria and from eubacteria.
- the structure of the F catalytic domain of the enzyme from bovine mitochondria,
- validating Sanger's methods for sequencing DNA
- the use of protein information to interpret the DNA sequences
- demonstrating the existence of triple overlapping genes (in bacteriophage G4)
- DNA sequencing of bovine and human mitochondria genomes
- studying transport proteins in the inner membranes of mitochondria.
- nutritional processes in man
- ATP synthase.

Walker's Nobel Prize lecture is dated 8th December 1997 with the title 'ATP Synthesis by Rotary Catalysis'. Walker commented [2]:

> *...Since 1978, my colleagues and I have concentrated on analyzing the structure of the ATP synthase, another multisubunit complex from mitochondria, where it is found in the inner membrane alongside the three proton pumping enzymes. Similar complexes are found in chloroplast and eubacterial membranes. Throughout our endeavours, we have been motivated by the expectation that detailed knowledge of its structure would lead to a deeper understanding of how ATP is made. A substantial part of our efforts has been directed at establishing the subunit compositions of the ATP synthesizing enzymes from various sources, and with determining the primary sequences of the subunits. These rather extensive analyses helped to show that the overall structure of the ATP synthase, and hence the general principles governing its operation, are very similar in mitochondria, chloroplasts and eubacteria, although the enzymes from the various sources differ in the details of both their sequences and subunit compositions. It is also known that the ATPases from various sources differ in the mechanisms that regulate their catalytic activities.*
>
> *In 1962, the inside surface of the inner membranes of bovine heart mitochondria was found by electron microscopic examination to be lined with mushroom shaped knobs about 100 Å in diameter. Later on, similar structures were found to be associated with the thylakoid membranes of chloroplasts, and with the inner membranes of eubacteria. At the time of their discovery, the function of these membrane bound knobs was not known, but they were thought to be probably important in biological energy conversion, and hence they were named "the fundamental particles of biology". In a brilliant series of biochemical reconstitution experiments conducted in the 1960s, Efraim Racker established that they were the ATP synthesizing enzyme complex, and hence that these early micrographs were the first glimpses of its structure.*
>
> *We now know that the head of the mushroom is a globular protein complex (known as the F_1 domain), where the catalytic sites of the ATP synthase lie. The F_1 part is attached to the membrane sector by a slender stalk about 45 Å long. The hydrophobic membrane domain (known as F_0)*

transports protons back through the energized membrane into the matrix, somehow releasing energy in this process and making it available to drive ATP synthesis in the catalytic F_1 *domain. For some years, it has been accepted that three protons are transported back through the* F_0 *membrane sector for each ATP molecule that is formed in* F_1*. However, recent experiments suggest that the chloroplast enzyme transports four protons through the membrane for each ATP that is made. Therefore, it appears that either one or, less likely, both of these values of the* H^+*:ATP ratio are incorrect, or that the mechanisms and structures of the mitochondrial and chloroplast enzymes differ significantly in their* F_0 *domains. This point will be elaborated later in a consideration of the possible structure and mechanism of* F_0*...*

A recent text describes the influence of the phosphate tail on energy transport [5].

Honours and Awards

Walker has received honorary D.Sc. degrees from the Universities of: Oxon (1999); London (2002); Sussex (2003); Liverpool (2004).

He has been presented with many awards and prizes including: the A. T. Clay Gold Medal (1959); the Johnson Foundation Prize by the University of Pennsylvania (1994); the Ciba Medal and Prize (1995); the Peter Mitchell Medal of the European Bioenergetics Congress (1996); the Gaetano Quagliariello Prize for Research in Mitochondria by the University of Bari, Italy (1997); the Nobel Prize for Chemistry joint with Paul D. Boyer and Jens C. Skou (1997); the Messel Medal (2000); the RSC Award for Biomembrane Chemistry (2003). Walker was elected a: Member EMBO (1983); Fellow of the Royal Society (1995); Honorary Fellow of the Institute of Biology (2002); Foreign Member of l'Accademia Nazionale dei Lincei, Rome (2003); Foreign Associate of NAS (2004). He was made a Fellow of Sidney Sussex College, Cambridge (1997); was a Founder FMedSci (1998); became a Honorary Fellow of St. Catherine's College, Oxford (1998).

Walker was knighted in 1999.

He is an Honorary Professor of Peking Union Medical College, Beijing (2001 – date).

Bibliography

1. Reproduced with permission.
2. From Les Prix Nobel. The Nobel Prizes 1997, Editor Tore Frängsmyr, [Nobel Foundation], Stockholm, 1998.
3. Who's who: 2005.
4. http://www.mrc-dunn.cam.ac.uk/research/walker.html.
5. Nucleoside Triphosphates and Their Analogs: Chemistry, Biotechnology and Biological Applications by Morteza Vaghefi, CRC, 2005, ISBN: 1574444980.

1997

© Nobel Foundation [1]

Jens C. Skou (0.5)

"for the first discovery of an ion-transporting enzyme, Na+, K+ -ATPase" [2]

During his presentation speech on 10th December 1997 Bertil Andersson, of the Royal Swedish Academy of Sciences, commented of Skou's work [2]:

> *...Let me now leave ATP production and instead turn to the use of ATP. In 1957, Jens Christian Skou discovered an enzyme called sodium, potassium ATPase (or Na+,K+-ATPase), which maintains the right ion balance in living cells. This enzyme, too, can be described with a technical analogy: It functions as a biological "pump" that transports potassium ions into a cell, while transporting sodium ions in the opposite direction out of the cell. This is a process that requires a lot of energy, and up to one third of the ATP formed in the body may be used to drive the Na+,K+ pump. Today, we are also aware of a number of other ion pumps, which have been discovered as a consequence of Skou's pioneering work. All these ion pumps are prerequisites for various important life functions such as transmission of nerve impulses, muscle contraction and digestion. The effects of many pharmaceuticals, such as heart and ulcer medicines, are related to the action of cellular ion pumps.*
>
> *Skou's discovery clearly illustrates the unpredictability of basic research - in 1957, no one could have imagined that his somewhat odd experiments, which consisted of studying the effects of various salts on tissue from a shore crab, would be important 40 years later to industrial operations and the production of new pharmaceuticals. Research breakthroughs and their applications cannot be custom-ordered. Instead they emerge from a combination of curiosity, scientific excellence and far-sightedness...*

Jens Christian Skou was born in Lemvig, Denmark, on the 8th October 1918, the oldest son of Magnus Martinus Skou (a timber and coal merchant) and An Margerethe, née Knak. There were four children, three sons and one girl – Skou was the eldest. He married Ellen Margrethe Nielsen (1948).

They had a daughter who died at the age of 18 months and two other daughters Hanne (1952) and Karen (1954). Skou retired in 1988, but can be contacted through the Institute of Biophysics at Aarhus University [2,3,4].

The Early Years

Skou's childhood was spent in a large house in Lemvig, with a summer house on the North Sea coast. His father died from pneumonia when Skou was 12 years old, leaving an uncle to run the family business, with his mother as a silent partner. He attended a high school in Haslev, Zealand, where he was a boarder (from aged 15), during which time he became involved in scouting and various sports activities, whilst enjoying mathematics more than the other science subjects. He enrolled at the University of Copenhagen (1937) where he studied medicine. The seven year course involved three years studying anatomy, biochemistry, chemistry, physics and physiology and a further four years on forensic medicine, pathology and public health. The German occupation of Denmark started in April 1940, which was when Skou was half way through his training. He was able to sit his exams on time in May and June of 1944 and graduated with an MD from the University of Copenhagen (1944).

Skou spent the following two years in hospitals, initially as an intern and then as a resident.

Year	*Post*	*Institute*
1944	Intern	Hjørring hospital
1945	Resident	Hjørring hospital
1946	Resident	Orthopaedic hospital, Aarhus
1947	Assistant Professor	Institute for Medical Physiology at Aarhus University

He took the post at university so that he could study and write up his doctorate, graduating with a DrMedSci from the University of Aarhus in 1954. He took an extra job in 1949, becoming a doctor on call one night a week, which allowed him to have a telephone and a car – which were difficult to obtain by other ways under the war restrictions.

The Career

Year	*Post*	*Institute*
1954	Associate Professor	Institute of Physiology, Aarhus University
1963	Professor and Chairman	Institute of Physiology, Aarhus University
1978	Professor of Biophysics	Institute of Biophysics, Aarhus University
1988	Retired	

Skou was nominated to become the Professor of Biophysics at Aarhus University without the post being advertised.

He has many biochemical interests including:

- local anaesthetics
- the sodium-potassium pump
- active transport of cations
- membrane physiology
- the structure and function of the active transport system, the Na^+, K^+-ATPase

He published the discovery of the Na+,K+-ATPase in 1957 [5].

Skou's Nobel Prize lecture is dated 8th December 1997 with the title 'The identification of the sodium-potassium pump'. During the course of his presentation, Skou reflected [2]:

...In 1959, after I had given my first paper on the crab nerve enzyme at the 21st International Congress for Physiology in Buenos Aires, Professor Hodgkin, who was the great name in neurophysiology and came from the famous Cambridge University, invited me for lunch to hear more about the enzyme. His interest suggested to me that the observation was of a certain importance.

Looking back, it was a very simple experiment to identify the pump. Just break the membrane and by this gain access to the Na^+ site on the inside and the K^+ site on the outside, add some ATP and test for the combined effect of Na^+ and K^+. It ought to have been done by someone who worked in the transport field and knew about active transport. I felt like an intruder in a field that was not mine.

There was however, much luck involved. First, that Nachmansohn had invited me to Woods Hole, where I learned about the giant axons and read about the observation by Libet. Next, that from the monolayer experiments I became interested in membrane proteins, especially lipoproteins, and therefore took notice of Libet's observation. Finally, that I chose crab nerves as a test object. I learned later, that after homogenization of most other tissues in order to break the membrane, the membrane pieces form vesicles, which must be opened by treatment with detergent, in order to get access to both sides of the membrane, and thereby to see the combined effect of Na^+ and K^+. This is not the case with the crab nerve membranes.

From then on the Na^+, K^+-ATPase took me away from the monolayer experiments. I never did the planned experiments on the effect of the local anaesthetics on a monolayer of a mixture of lipids and proteins; and I never used the Na^+, K^+-ATPase for monolayer experiments, first of all because it was not until 1980 that it became possible to extract the enzyme from a memin a pure stable water-soluble form, and secondly that was only possible with the use of detergent. A problem would have been the detergent, but I had lost interest in monolayers...

Skou concluded his lecture:

...It may seem disappointing that 30 years of work, since the conclusion that the membrane bound Na^+, K^+-ATPase is identical with the Na^+, K^+-pump, has not given us an understanding of the basic molecular events behind the transport. However, considering that the problem is to reveal how 1320 amino acids inside a volume of 60 X 60 X 100 $Å^3$ can be assembled to a very efficient machine, which can convert the chemical energy from the hydrolysis of ATP into work, namely the transport

of cations against their electrochemical gradient, and which can distinguish between so closely related cations as Na^+ and K^+, it can be of no surprise that progress is slow.

Thirty years ago it seemed impossible to find a way to purify the enzyme not to speak of getting it into solution and reconstitute it into liposomes. Nobody dared to dream about knowing the sequence, and yet this and much much more have been accomplished...

A recent text provides collective information on the chemical, physiochemical, and biological properties of both natural and modified ATP [6].

Honours and Awards

Skou was presented with an honorary DrMedSci from the University of Copanhagen in 1986.

He has received many awards and prizes including: the Leo Prize (1954); the Novo Prize (1965); the Consul Carlsen Prize (1973); the A. Retzius Gold Medal (1977); the Fernström Foundation Prize (1985); the Prakash Datta Medal (1985) and the Nobel Prize in Chemistry jointly with Paul D. Boyer; John E. Walker (1997).

Skou is a: Member of the Danish Royal Academy of Science (1965); Member of the Deutsche Academic der Naturforscher Leopoldina (1977); Member of the EMBO (1978); Foreign Associate of the US academy of Science (1988); Honorary Member of the Japanese Biochemical Society (1988); Foreign Honorary Member of the American Academy of Arts and Science (1999); Member of the American Physiology Society (1990); Member of the Academia Europaea (1993); Member of the International Academy of Humanism (2002).

Bibliography

1. Reproduced with permission.
2. From Les Prix Nobel. The Nobel Prizes 1997, Editor Tore Frängsmyr, [Nobel Foundation], Stockholm, 1998.
3. Who's who: 2005.
4. http://www.fi.au.dk/nobel.html.
5. Biochimica et Biophysica Acta 1957, 23, 394–401.
6. Nucleoside Triphosphates and Their Analogs: Chemistry, Biotechnology and Biological Applications, by Morteza Vaghefi, CRC, 2005, ISBN: 1574444980.

1998

© Nobel Foundation [1]

Walter Kohn (0.5)

"for his development of the density-functional theory" [2]

During his presentation speech on the 10th December 1998, Björn Roos, of the Swedish Academy of Sciences, commented of Kohn [1, 2]:

> *...Due to extraordinary circumstances, Professor Walter Kohn is not with us today. We hope to see him at the Prize Ceremony next year, instead...*

The following year on the 10th December 1999, Bengt Nordén, of the Royal Swedish Academy of Sciences, commented of Kohn [3]:

> *...Professor Kohn. You were awarded last year's Nobel Prize together with Professor John Pople for your contributions to Computational Quantum Chemistry. They have been of fundamental importance also in the context of this year's Nobel Prize, as calculations of energies and structures of the molecules in their various states are crucial for the interpretation of the experiments.*
>
> *On behalf of the Royal Swedish Academy of Sciences, may I warmly congratulate you and ask you to come forward to receive the 1998 Nobel Prize in Chemistry from the hands of His Majesty the King...*

Walter Kohn was born in Vienna, Austria on the 9th March 1923, the son of Salomon and Gittel Kohn. He married his first wife Lois Mary Adams in 1948 and they had three daughters. He married his second wife Mara Schiff in 1978. Kohn is currently Emeritus and Research Professor in the Department of Physics at the University of California at Santa Barbara. He was naturalized as an American citizen in 1957, having already been a naturalized Canadian citizen [2, 4, 5].

The Early Years

Kohn was born into a middle class Jewish family, which moved to Vienna a few years after the end of World War I. The family owned a summer property in Heringsdorf on the Baltic Sea, not far from Berlin, and spent summer holidays there until Hitler came to power in 1933. His mother enrolled Kohn in the Akademische Gymnasium (a public high school in Vienna's inner city) following public elementary school. He had little interest in mathematics at high school, so much so that he received the only 'C' in high school. The family business was confiscated without compensation, and so Kohn was expelled from school, as his family could not pay the fees. He was then enrolled in the Jewish school, the Chajes Gymnasium, during which time he was taught physics by Dr. Emil Nohel and mathematics by Dr. Victor Sabbath.

Both of Kohn's parents and several of his other relatives and teachers were murdered during the Holocaust. Kohn found himself, and his sister Minna, taken to the home of Charles and Eva Hauff in Sussex, England, where he studied at the East Grinstead County School. Kohn's original plan was to study to become a farmer. He contracted meningitis in 1939, and so concentrated his school work on chemistry, mathematics and physics.

Holders of enemy passports were ordered to be interned by Churchill, so at the age of seventeen (May 1940), Kohn spent two months in various British camps, before being shipped to Canada (July 1940) to a camp in Trois Rivieres. He was moved around among various other camps in New Brunswick and Quebec, and with the help of a fellow internee, A. Heckscher, he passed the junior matriculation exam and exams in chemistry, mathematics and physics at McGill University.

Kohn was cleared by Scotland Yard of being a potential spy in January 1942, and was welcomed into the home of Bruno and Hertha Mendel of Toronto. A friend of the Mendels, Leopold Infeld, suggested to Kohn that he enrol for a physics degree at the University of Toronto, which he did, although as a German passport holder he was not allowed into the chemistry building. One of his fellow students at the time was Arthur Schawlow (Nobel Laureate for Physics, 1981).

Kohn took up military service in the Canadian Infantry Corps (1944–1945) and graduated from the University of Toronto with a BA in Mathematics and Physics (1945), followed by an MA in Applied Mathematics, also from the University of Toronto (1946). During his spare time at university, Kohn was an Industrial Physicist (part time) with Sutton Horsley Co (1941–43), and later a Geophysicist (part-time) with Koulomzine, Quebec (1944–46). Kohn obtained a Lehman Fellowship, which he used to work on his Ph.D. under Julian Schwinger (Nobel Laureate for Physics, 1965) at Harvard University, working on what was later to be known as Kohn's variational principle for scattering, and useful for nuclear, atomic and molecular problems. Kohn graduated with a Ph.D. from Harvard University in 1948. Several future Nobel Laureates were working in Schwinger's group at that time: P.W. Anderson (1997); N. Bloembergen (1981); B. Mottelson (1975). Kohn stayed on in Schwinger's group for a postdoctoral degree, which was a mixture of teaching and work on quantum electrodynamics and the field theory of strong interactions between mesons and nucleons.

The Career

Year	*Post*	*Institute*
1948–50	Instructor, Department of Physics	Harvard University
1950–53	Assistant Professor	Carnegie Mellon University

(*continued*)

(*continued*)

Year	*Post*	*Institute*
1953–56	Associate Professor	Carnegie Mellon University
1956–60	Professor	Carnegie Mellon University
1960–79	Professor, Department of Physics	University of California at San Diego
1979–84	Founding Director, Institute of Theoretical Physics	University of California at Santa Barbara
1984–91	Professor of Physics	University of California at Santa Barbara
1991 – date	Emeritus and Research Professor	University of California at Santa Barbara

Kohn held the additional posts of: Chairman of the Department of Physics UCSD, San Diego, California (1961–1963); Chairman of the San Diego Division of the Academic Senate (1968–1969); Member of the Advisory Board of the Statewide Institute of Global Conflict and Cooperation (1982–1992); Member of the Senate Review Committee of UC-Managed National Laboratories (1986–1989).

He was a: National Research Council Fellow in Copenhagen (1950–1951); Oersted Fellow in Copenhagen (1951–1952); Senior NSF Fellow at Imperial College (1958); Guggenheim Fellow in Paris (1963); NSF, Senior Postdoctoral Fellow in Paris (1967).

Kohn has been involved in many problems, including (in approximately chronological order):

- calculations of the Knight shift (of nuclear magnetic resonance)
- the development of a new variational approach to the study of wavefunctions in periodic crystals
- a criticism of Froehlich's pre-BCS theory of superconductivity
- work on scattering theory
- the inverse scattering problem
- obtaining a complete solution for S-wave scattering by a spherical potential
- the development of a theory for the energy band structure of electrons for periodic potentials
- effective mass theory
- the development of an effective Hamiltonian in the presence of magnetic fields, for complex holes
- obtaining the first non-heuristic derivation of the Boltzman transport equation for quantum mechanical particles
- the development of density functional theory (DFT)
- the Hohenberg Kohn (HK) variational principle
- the theory of surfaces, mostly electronic structure

He has written over 200 research papers and several books including: *Local theory of disordered systems*, Pub: Dept. of Physics, University of California, San Diego, (1969), ASIN: B0007EBE04; *The university and the nuclear predicament (IGCC policy papers)*, Pub: IGCC, (1988), ISBN: 0934637075 and *Part I: Physical Chemistry. Part II: Solid State Physics* with N. Balazs and Arthur S. Wightman (Ed), Pub: Springer; 1 edition, (January 15, 1997), ISBN: 3540569855.

Kohn's Nobel Prize lecture is dated 28th January 1999 with the title 'Electronic structure of matter – wave functions and density functionals'. Kohn commented [2]:

> *...In my view DFT (density functional theory) makes two kinds of contribution to the science of multi-particle quantum systems, including problems of electronic structure of molecules and of condensed matter:*
>
> *The first is in the area of fundamental* ***understanding****. Theoretical chemists and physicists, following the path of the Schroedinger equation, have become accustomed to think in a truncated* ***Hubert space of single particle orbitals****. The spectacular advances achieved in this way attest to the fruitfulness of this perspective. However, when high accuracy is required...*
>
> *...The second contribution is practical. Traditional multiparticle wavefunction methods when applied to systems of* ***many particles*** *encounter what I call an exponential wall when the number of atoms, N, exceeds a critical value which, for "chemical accuracy", currently is in the neighbourhood of* $N_0 \sim 10$ *(to within a factor of about 2) for a system without symmetries. A major improvement along present lines in the analytical and/or computational aspects of these methods will lead to only modest increases in* N_0*. Consequently, problems requiring the simultaneous consideration of very many interacting atoms,* $N/N_0 \gg 1$*, such as large organic molecules, molecules in solution, drugs, DNA, etc. overtax these methods. On the other hand, in DVF, computing time T rises much more moderately with the number of atoms...*

For further texts concerning computational chemistry, see for example [5–7].

Honours and Awards

Kohn has been presented with many honorary degrees including: D. Natural Sciences from the Federal Institute of Technology (ETH), Zurich, Switzerland (1994); L.L.D. from the University of Toronto, Toronto, Canada (1967). He holds honorary PhD degrees from: the Hebrew University of Jerusalem (1981); Brandeis University (1981); the Weizmann Institute, Israel (1997); Tel Aviv University (1999). He has also been awarded Honorary D. Rerum Naturalium degrees from the Technical University of Dresden (2002) and the University of Wuerzburg (1995). Kohn was presented with honorary D.Sc. degrees from the: Universite de Paris-SUD, Paris (1980); Queen's University, Kingston, Canada (1986); Technical University of Vienna (1996); Oxford University (2001); Rutgers Unversity (2001); University of Sherbrooke, Canada (2002); Free University of Berlin (2003). Carnegie Mellon University awarded Kohn a honorary D. Science and Technology degree in 1999.

He has been presented with many awards and prizes including: the Oliver Buckley Prize (Solid State Physics) (1960); the Davisson-Germer Prize (Surface Physics) (1977); the National Medal of Science (1988); the Feenberg Medal (Many Body Physics) (1991); the Niels Bohr/UNESCO Gold Medal (1998); the Nobel Prize for Chemistry jointly with John A. Pople (1998).

Kohn is: a Member of the American Academy of Arts & Sciences (1963); a Member of the National Academy of Sciences (1969); a Member of the International Academy of Quantum Molecular Sciences (1991); a Member of the American Philosophical Society (1994); a Foreign Member of the Royal Society (1998); a Corresponding Member of the Bavarian Academy of Sciences (2003). He was a Member of the National Institute of Science and Technology (1946–98).

He has been a Visiting Scholar at many universities (1958–85): Pennsylvania; Michigan; Washington (Seattle); Paris; Copenhagen; Jerusalem; Imperial College (London); ETH (Zurich).

He has acted as a Consultant for: Bell Telephone Labs (1953–1966); Westinghouse Research Lab. (1953–1957); General Atomic Laboratory (1960–1972); IBM (1978).

He is: a Member Brookhaven National Laboratories, a Member of the Board of Governors of Tel Aviv University; a Member of the Board of Governors of the Weizmann Institute (1996 – date); a Member of the Board of the Population Institute.

Bibliography

1. Reproduced with permission.
2. From Les Prix Nobel. The Nobel Prizes 1998, Editor Tore Frängsmyr, [Nobel Foundation], Stockholm, 1999.
3. From Les Prix Nobel. The Nobel Prizes 1999, Editor Tore Frängsmyr, [Nobel Foundation], Stockholm, 2000.
4. Who's who: 2005.
5. http://www.physics.ucsb.edu/~kohn/.
6. Essentials of Computational Chemistry – theories and models by Christopher J. Cramer, J. Wiley and Sons Ltd, 2002, ISBN: 0-471-48552-7.
7. Mathematics and Computational Concepts in Chemistry ed N. Trinajstic, Ellis Horwood Ltd, ISBN: 0-85312-934-7.
8. Chemometrics: a textbook by D. L. Massart, B. G. M. Vandeginste, S. N. Deming, Y. Michotte and L. Kaufman, Elsevier, 1988, ISBN: 0-444-42660-4.

1998

© Nobel Foundation [1]

John A. Pople (0.5)

"for his development of computational methods in quantum chemistry" [2]

During his presentation speech on the 10th December 1998, Björn Roos, of the Swedish Academy of Sciences, commented of Pople and his work [2]:

> *...Today, we celebrate the fact that mathematics has invaded chemistry, that by means of theoretical calculations we can predict a large variety of chemical phenomena. Professors Walter Kohn and John Pople have individually made fundamental contributions to this development.*
>
> *An atom consists of a nucleus and electrons. The motion of the electrons is described by the laws of quantum mechanics. When these laws were formulated more than 70 years ago, researchers realized immediately that in them was contained the explanation of the chemical bond. It was realized that if the quantum mechanical equations could be solved, one would be able to explain how atoms are bound together to form molecules. It would be possible to explain why molecules look as they do, what their properties are, and how they react with each other to form new molecules. A theoretical description of all of chemistry was within reach...*
>
> *...He (Pople) realized that if quantum chemistry was going to become important in chemistry, one had to develop methods that were effective and which could be used to compute interesting properties like the structure of molecules and binding energies for the chemical bonds. They also had to be easy to use for the general chemist, who could not be expected to be knowledgeable about all the subtle details of quantum chemistry. Pople was able to fulfill these conditions through a series of crucial innovations and improvements. He designed a new tool, which could be used to study molecules and compute their properties. This tool is a computer program called GAUSSIAN. It contains a theoretical model chemistry where quantum mechanical equations are solved through a series of more and more refined approximations. Pople's methods are used today by thousands of scientists at universities and companies around the world, to study a large variety of problems in chemistry and biochemistry...*

John Anthony Pople was born in Burnham-on-Sea, Somerset, on the 31st October 1925, the son of Herbert Keith Pople (a clothing store proprietor) and Mary Frances, née Jones. He married Joy Cynthia Bowers (1952), and they had three sons and one daughter. Pople died in Chicago, Illinois, USA on the 15th March 2004 [2–4].

The Early Years

Pople lived in Burnham-on-Sea until shortly after the end of the Second World War (1946). He attended Bristol Grammar School, some thirty miles away as a boarding pupil, returning to his parents only for weekends. After a short while, he persuaded his parents to allow him to make the journey on a daily basis, which involved cycling, train and a mile walk each way each day. His interest in mathematics developed when he was twelve years old and was self taught in calculus from an old discarded text book. He studied for a scholarship and won a place at Trinity College, Cambridge University from October 1943, completing his mathematics degree in May 1945. Places at universities were hard to find just after the war as there were many ex-servicemen taking up their interrupted studies and so Pople joined the Bristol Aeroplane Company until he returned to Cambridge University in October 1947.

On arrival, Pople studied cosmology, fluid dynamics, quantum mechanics which was taught in part by Dirac (Nobel Laureate in Physics, 1933, who was Lucasian Professor of Mathematics at the time) and statistical mechanics. These courses were followed by work on a Ph.D. under J. Lennard-Jones working on liquid theory, specifically the bonding structures of water (1947–1951). During the course of this work, he won the Mayhew prize (1948) and the Smith Prize (1950) from the University. He followed his Ph.D. with a Research Fellowship at Trinity College.

The Career

Year	*Post*	*Institute*
1954–1958	Lecturer in Mathematics	Cambridge University
1958–1964	Head of Basic Physics Division	National Physical Laboratory
1964–1993	Carnegie Professor of Chemical Physics, later John Christian Warner Professor of Natural Sciences	Carnegie Insitute of Technology known as Carnegie Mellon University from 1967
1993	Trustees' Professor of Chemisty	Northwestern University

Pople held the additional post of Adjunct Professor of Chemistry of Northwestern University (1986–93).

During the course of his career, Pople became involved in many aspects of mathematical chemistry including:

- the development of many algorithms
- the development of one of the first Gaussian programmes for computational chemistry (PPP)
- the theory of approximate molecular orbital (MO) calculations

- the development of the methods of Complete Neglect of Differential Overlap (CNDO) (1965)
- the development of Intermediate Neglect of Differential Overlap (INDO)
- pioneering the use of *ab initio* calculations
- the theoretical background to nuclear magnetic resonance spectroscopy
- general fundamental problems of electronic structure

In his article concerning the Origin of PPP Theory [5], Pople commented:

> *...My principal contribution to PPP theory took place in the autumn of 1952. At that time, quantum chemistry was still at a primitive stage of development. Full computational treatments were almost entirely limited to two-electron or monatomic systems and were carried out tediously on manual (or sometimes electric-powered) calculators.*
>
> *Some important formal developments had taken place. The Hartree-Fock differential equations for molecular orbitals had been prepared and analyzed, but practical methods for their solution for polyatomic molecules were not available. Roothaan and also Hall had shown a year earlier how the Hartree-Fock problem could be cast in algebraic form with a finite basis set, but again no implementation had been achieved.*
>
> *Boys had just proposed the use of gaussian basis sets, but the idea was initially treated with skepticism as atomic orbitals were clearly "ungaussian" in character. The electronic computer had appeared in primitive form, but its application to electronic structure theory was still treated as a distant event.*
>
> *The most active branch of quantum chemistry in the early 1950s was the semi- empirical treatment of the π-electrons of aromatic systems, particularly hydrocarbons. This was primarily due to the initial work of Hückel, who developed the independent electron model with one π-type atomic orbital per carbon atom...*

Despite helping to develop the PPP programme and making it famous, the current owners of the programme banned him, and other researchers, from using it in his last decade.

Pople has published several books including: *High Resolution Nuclear Magnetic Resonance* with W. G. Schneider and H. J. Bernstein, Pub: McGraw Hill Text, (1959), ISBN: 0070505160; *Approximate Molecular Orbital Theory*, Pub: Mcgraw-Hill (Tx), (1970), ISBN: 0070505128; *Carnegie Mellon Quantum Chemistry Archive* (editor) with Robert A. Whiteside and Michael Fisch, Pub: Gaussian; 3rd edition, (1983), ISBN: 9990209979; *Strain Measurement Reference Book*, Pub: British Society for Strain Measurement, (1985), ISBN: 0950635103; *AB INITIO Molecular Orbital Theory* with Warren J. Hehre, Leo Radom and P. V. Schleyer, Pub: Wiley-Interscience; 1 edition, (1986), ISBN: 0471812412.

His Nobel Prize lecture is dated 8th December 1998 with the title 'Quantum Chemical Models'. During the course of his presentation, Pople commented on the necessary features of theoretical models [2]:

Target	*a target accuracy must be selected*
Formulation	*the approximate mathematical procedure must be precisely formulated*
Implementation	*the formulation method has to be implemented in a form which permits its application in reasonable times at reasonable cost*

(*continued*)

(*continued*)

Verification	*test the model against known chemical facts to determine whether the target has been achieved*
Prediction	*if the model has been properly validated according to some such criterion, it may be applied to chemical problems to which the answer is unknown or in dispute*

For further texts concerning computational chemistry, see for example [6–8].

Honours and Awards

Pople was presented with many awards and prizes including: the Marlow Medal of the Faraday Society (1958); the Irving Langmuir Award (1970); the Harrrison Howe Award (1971); the ACS Pittsburgh (1975); the Gilbert Newton Lewis Award (1973); the ACS Pauling award (1977); the Alexander von Humboldt Foundation Senior U.S. Scientist Award (1981); the G. Willard Wheland Award (1981); the Evans Award (1982); the Oespar Award (1984); the Davy Medal (1988); Computers in chemistry award (1991); the Wolf Prize in Chemistry (1992); the J.O. Hirschfelder Award in Theoretical Chemistry (1993); the Nobel Prize for Chemistry jointly with W. Kohn (1998); the Copley Medal 2002.

He was a Fellow: of the Royal Society of London (1961); of the American Physical Society; (1970) of the American Academy of Arts and Sciences (1971). He was a Foreign Associate of the National Academy of Science (1977) and a Corresponding Member of Australian Academy of Sciences (1993).

Pople was knighted in 2003.

He was Ford Visiting Professor at the Carnegie Institute of Technology (1961–62) and Visiting Professor at the Technion, Haifa.

Pople's citation for election (date: 16/03/1961) as Fellow of the Royal Society reads [9]:

'Distinguished for his work on the application of quantum and statistical mechanics to the explanation of the chemical and spectroscopic properties of individual molecules and the physical properties of assemblies of molecules in the gaseous and liquid states. His early work on the molecular orbital theory of chemical valency did much to clarify the fundamentals of that theory which he then applied with great success to the water molecule, showing for the first time the effect of the lone pair of electrons on its structure and associative properties. He was equally successful with many of the properties of aromatic molecules and radicals. He next tackled the properties of compressed gases, making notable contributions to our understanding of virial coefficients and of the dielectric and magnetic properties of gases. During the past three years he has become an acknowledge leader on the theory of nuclear magnetic resonance, especially the explanation of chemical shifts.'

No proposers are listed in the Sackler database.

Bibliography

1. Reproduced with permission.
2. From Les Prix Nobel. The Nobel Prizes 1998, Editor Tore Frängsmyr, [Nobel Foundation], Stockholm, 1999.

3. http://www.quantum-chemistry-history.com/Popl_Dat/Pople2.htm.
4. Who's who: 2004.
5. Int. J. Quant. Chem. 1990, 38, 349–371.
6. Essentials of Computational Chemistry – theories and models by Christopher J. Cramer, J. Wiley and Sons Ltd, 2002, ISBN: 0-471-48552-7.
7. Mathematics and Computational Concepts in Chemistry ed. N. Trinajstic, Ellis Horwood Ltd, ISBN: 0-85312-934-7.
8. Chemometrics: a textbook by D. L. Massart, B. G. M. Vandeginste, S. N. Deming, Y. Michotte and L. Kaufman, Elsevier, 1988, ISBN: 0-444-42660-4.
9. Sackler database, Royal Society.

1999

© Nobel Foundation [1]

Ahmed H. Zewail

"for his studies of the transition states of chemical reactions using femtosecond spectroscopy" [2]

During his presentation speech on the 10th December 1999, Bengt Nordén, of the Royal Swedish Academy of Sciences, commented of Zewail's work [2]:

> *...Science has always strived to see smaller and smaller things and faster and faster events. Since the time of Arrhenius a number of methods have been developed to measure increasingly faster reaction rates, many of them rewarded with Nobel Prizes. However, no one had, until recently, been able to observe what actually happens to the reacting molecule as it passes through its so-called transition state, a metaphor for a kind of intermediate state of the reaction in which bonds are broken and formed. This remained a misty no-man's land.*
>
> *The molecule passes the transition state as fast as the atoms in the molecule move. They move at a speed of the order of 1000 m/second – about as fast as a rifle bullet – and the time required for the atoms to move slightly within the molecule is typically tens of femtoseconds (1 fs = 10–15 seconds). Only few believed that such fast events would ever be possible to see.*
>
> *This, however, is exactly what Ahmed Zewail has managed to do. Twelve years ago he published results that gave birth to the scientific field called femtochemistry. This can be described as using the fastest camera in the world to film the molecules during the reaction and to get a sharp picture of the transition state. His "camera" is a laser technique with light flashes of only a few tens of femtoseconds in duration. The reaction is initiated by a strong laser flash and is then studied by a series of subsequent flashes to follow the events. The key to his success was that the first femtosecond flash or starting shot, excited all molecules in the sample at once, causing their atoms to swing in rhythm. The first experiments demonstrated in slow motion how bonds were stretched and broken in rather simple reactions, but soon studies of more complex reactions followed. The results were often surprising, and the dance of the atoms during the reaction was found to differ*

from what was expected. Zewail's use of the fast laser technique can be likened to Galilei's use of his telescope, which he directed towards everything that lit up the vault of heaven. Zewail tried his femtosecond laser on literally everything that moved in the world of molecules. He turned his telescope towards the frontiers of science...

Ahmed H. Zewail was born in Damanhur on the 26th February 1946, the son of Hassan A. Zewail and Rawhia, née Dar. He married Dema in 1989 and they have two daughters Maha and Amani, and two sons Nabeel and Hani. Zewail is the Linus Pauling chair Professor of Chemistry and Professor of Physics and Director of the NSF Laboratory for Molecular Sciences at the California Institute of Technology [2–4].

The Early Years

Zewail lived in Disuq in Egypt and attended Disuq's preparatory school and high school. He enrolled at Alexandria University from which he obtained a first class honours degree in chemistry (1967). He was appointed to the university post of Demonstrator and worked for an MS degree in the field of spectroscopy, particularly why the spectra of some molecules change with solvents (graduating in 1969). He transferred to the University of Pennsylvania under R. Hochstrasser for his Ph.D. the topic for which was the spectroscopy of dimers, from which he graduated in 1974. He went to the University of California at Berkeley as an IBM postdoctoral Fellow where he worked with C. Harris on the concept of coherence.

The Career

Year	*Post*	*Institute*
1976	Assistant Professor of Chemical Physics	Caltech
1978	Associate Professor of Chemical Physics	Caltech
1982	Professor of Chemical Physics	Caltech
1990	Linus Pauling Professor of Chemical Physics	Caltech
1995–to date	Linus Pauling Professor of Chemistry and Professor of Physics	Caltech
1996–to date	Director, Laboratory for Molecular Sciences	Caltech

Zewail was an Alfred P. Sloan Foundation Fellow (1978–1982); a Camille and Henry Dreyfus Teacher-Scholar Award (1979–1985) and a John Simon Guggenheim Memorial Foundation Fellow (1987).

Zewail has been or is involved with many aspects of chemistry including aspects of:

- nanostructures
- dense fluids
- transient structures
- polymers
- electron and proton transfer

He was or is a member of Advisory and Editorial Boards and is a current editor of Chemical Physics Letters. He has written his autobiography [5] and written or edited several text books including: *Laser Chemistry Applications*, Pub: Harwood Academic, (1984), ISBN: 3718601737; *Femtochemistry: Ultrafast Dynamics of the Chemical Bond, Vol. 1*, (World Scientific Series in 20th Century Chemistry, Vol. 3), Pub: World Scientific Publishing Company, (1994), ISBN: 9810217366; *Femtochemistry: Ultrafast Dynamics of the Chemical Bond, Vol. 2*, (World Scientific Series in 20th Century Chemistry, Vol. 3), Pub: World Scientific Publishing Company, (1994), ISBN: 9810217382; *Femtochemistry: Ultrafast Dynamics of the Chemical Bond, 2 Volume Set* (World Scientific Series in 20th Century Chemistry, Vol. 3), Pub: World Scientific Pub Co Inc, (1994), ISBN: 9810219407; *Rotational Spectroscopy of Diatomic Molecules (Cambridge Molecular Science)* with John M. Brown, Alan Carrington, Richard Saykally (Series Editor) and David King (Series Editor), Publisher: Cambridge University Press, (2003), ISBN: 0521530784, *Energy Landscapes: Applications to Clusters, Biomolecules and Glasses (Cambridge Molecular Science)* with David Wales, Richard Saykally (Series Editor) and David King (Series Editor), Pub: Cambridge University Press, (2004), ISBN: 0521814154; *Reise Durch Die Zeit*, Pub: Wiley-VCH, (2006), in German, ISBN: 3906390322.

Zewail's Nobel Prize lecture is dated 8th December 1999 with the title 'Femtochemistry Atomic-scale dynamics of the chemical bond using ultrafast lasers'. In his concluding remarks, Zewail outlined the concepts thus [2]:

> *...In studies of motion, we have exploited the concept of coherence among molecular wave functions to achieve atomic-scale resolution of dynamics - the change of molecular structures with time. Molecular wave functions are spatially diffuse and exhibit no motion. Superposition of a number of separate wave functions of appropriately chosen phases can produce a spatially localized and moving coherent superposition state, referred to as a wave packet; constructive and destructive interference (as in the interference of light waves) is the origin of such spatial localization. The packet has a well-defined (group) velocity and position which now makes it analogous to a moving classical marble, but at atomic resolution. The femtosecond light induces the coherence and makes it possible to reach atomic-scale spatial and temporal resolution, without violation of the uncertainty principle...*

Further details concerning femtosecond laser technology, principles and research can be found in references [6–8].

Honours and Awards

Zewail has received many honorary degrees including Doctor from: the University of Rome "La Sapienza", Italy (2000); the Université dc Liege, Belgium (2000); the Autonomous University of the State of Mexico, Toluca, Mexico (2004); Honorary D.Ph. from Lund University, Sweden (2003). He also has Honorary D.Sc. from the: American University in Cairo, Egypt (1993); Katholieke Universiteit, Leuven, Belgium (1997); Université de Lausanne, Switzerland (1997); University of Pennsylvania, U.S.A. (1997); University of New Brunswick, Canada (2000); Jadavpur University, India (2001); Heriot-Watt University, Scotland (2002); Boǵaziçi University, Istanbul, Turkey (2003); Ecole Normale Supérieure, Paris, France (2003); University of Dublin, Trinity College, Ireland (2004);

Oxford University, United Kingdom (2004). He also holds: The Honorary D.U. from Swinburne University, Australia (1999); the H.D.A.Sc from the Arab Academy for Science & Technology, Egypt (1999). A Honorary H.D.D. from Peking University, People's Republic of China (2004); a Honorary H.D.Sc. from Alexandria University, Egypt (1999); a Honorary LLD from Concordia University, Montréal, Canada (2002); a M.A from Oxford University. United Kingdom (1991); M.D. degrees from Queen of Angels-Hollywood Presbyterian Medical Center Honorary Member of the Medical Staff, U.S.A. (2000); Pusan National University, Korea (2003).

He has been presented with many awards and prizes including: the Alexander von Humboldt Award for Senior United States Scientists (1983); the National Science Foundation Award for especially creative research (1984; 1988; 1993); the Buck-Whitney Medal (1985); the Harrison Howe Award (1989); the Carl Zeiss International Award (1992); the Earle K. Plyler Prize (1993); the Medal of the Royal Netherlands Academy of Arts and Sciences (1993); the Wolf Prize in Chemistry (1993); the Bonner Chemiepreis (1994); the Herbert P. Broida Prize (1995); the Leonardo da Vinci Award of Excellence (1995); the College de France Medal (1995); the Peter Debye Award (1996); the National Academy of Sciences Award (1996); the J.G. Kirkwood Medal (1996); the Peking University Medal (1996); the Pittsburgh Spectroscopy Award (1997); the E. B. Wilson Award (1997); the Linus Pauling Medal Award (1997); the Robert A. Welch Award in Chemistry (1997); the Benjamin Franklin Medal (1998); the Paul Karrer Gold Medal (1998); the William H. Nichols Medal Award (1998); the Richard C. Tolman Medal Award (1998); the E.O. Lawrence Award (1998); the Merski Award (1999); the Nobel Prize in Chemistry (1999); the Röntgen Prize (1999); the Faye Robiner Award (2000); the Golden Plate Award (2000); the Medal of "La Sapienza" (2000); the Médaille de l'lnstitut du Monde Arabe (2000); the Honorary Medal, Universite Du Centre, Monastir (2000); the Distinguished Alumni Award (2002); the G.M.Kosolapoff Award (2002); the Distinguished American Service Award (2002); the Sir C. V. Raman Award (2002); the Arab American Award (2004); the Gold Medal, Burgos University (2004).

Zewail is a Fellow of the American Physical Society (1982), and of the American Association for the Advancement of Science, AAAS (2000) and of St. Catherine's College, Oxford, U.K. (1991). He is: Honorary Fellow of the Chemical Society of India (2001); Sydney Sussex College, Cambridge, (2002); the African Academy of Sciences, Nairobi, Kenya (2002); Royal Society of Chemistry, (2003); St. Catherine's College, Oxford, (2004); The Literary & Historical Society, University College, Dublin (2004). He is a Foreign Fellow of the Indian National Science Academy, New Delhi, India (2002). Zewail is a Member of: the National Academy of Sciences, U.S.A. (1989); the Third World Academy of Sciences, Italy (1989); Sigma Xi Society (1992); the American Academy of Arts and Sciences (1993); the Academic Européenne des Sciences, des Arts et des Lettres, France (1994); the American Philosophical Society (1998); the Pontifical Academy of Sciences (1999); the American Academy of Achievement (1999); the Royal Danish Academy of Sciences & Letters (2000); the Indian Academy of Sciences, Banglore, India (2001). He is a Foreign Member of: the Royal Society (2001); the Russian Academy of Sciences (2003); the Royal Swedish Academy of Sciences, Stockholm, (2003) and of the Royal Academy of Belgium, Brussels, (2003). Zewail is an Honorary Member of the European Academy of Sciences, Brussels (2004) and of the Board of Advisors of the National Society of High School Scholars, Atlanta, Georgia, U.S.A. (elected 2004). He is also an Honorary Foreign Member of the Korean Academy of Science and Technology (2002).

He has also been presented with the King Faisal International Prize in Science (1989), had stamp in his honour, Egypt Postage Stamps, with Portrait (1998), "The Fourth Pyramid" (1999), received the order of the Grand Collar of the Nile, Highest Honor of Egypt, conferred by President Mubarak

(1999), been presented with the City of Pisa Medal, City Mayor, Pisa, Italy (2000) and the Honorary Medal, City of Monastir, from the Mayor, Tunisia (2000). Ghana issued stamps in his honour in 2002.

Zewail has been or is a Visiting Professor at many institutes which started in 1979 when he was the John van Geuns Stichting Visiting Professor at the University of Amsterdam. Eighteen further Visiting Chairs have been conferred on Zewail since that time, some of which are still active.

He is a Member of: the World Scientific Advisory Board (1994 –); the Board of Advisors of the Max Planck Institute (1994 –); the Board of Trustees of the American University in Cairo (1999 –); the Board of Trustees of the Alexandria Library (2001 –); the Scientific Advisory Board of The Welch Foundation, (2002 –); the Board of Directors of the Qatar Foundation (2003 –); the Scientific Advisory Board of Chalmers University (2003 –); the Board of Trustees of TIAA-CREF (2004 –). He is also Patron of the Multilateral Initiative on Malaria (2003 –).

Zewail has given some three hundred named, plenary, and keynote lectures.

Bibliography

1. Reproduced with permission.
2. From Les Prix Nobel. The Nobel Prizes 1999, Editor Tore Frängsmyr, [Nobel Foundation], Stockholm, 2000.
3. Who's who.
4. http://www.its.caltech.edu/~femto/index.html.
5. *Voyage through time autobiography* of Ahmed Zewail, pub The American University in Cairo Press, 2002, ISBN: 977 424 677 2.
6. Femtosecond Laser Pulses: Principles and Experiments (Advanced Texts in Physics), ed by Claude Rulliere, Springer, 2nd ed. 2004, ISBN: 0387017690.
7. Ultrashort Laser Pulse Phenomena: Fundamentals, Techniques, and Applications on a Femtosecond Time Scale (Optics and Photonics Series) by Jean-Claude Diels, Wolfgang Rudolph, Paul F. Liao, Paul Kelley, Academic Press, 1996, ISBN: 0122154924.
8. Femtosecond Optical Frequency Comb: Principle, Operation and Applications ed by Jun Ye and Steven T. Cundiff, Springer, 1st ed. 2004, ISBN: 0387237909.

2000

© Nobel Foundation [1]

Alan J. Heeger (0.33)

"for the discovery and development of conductive polymers" [2]

During his presentation speech on the 10th December 2000 Bengt Nordén, of the Royal Swedish Academy of Sciences, commented on conductive polymers [2]:

...Polyacetylene can be seen as beads on a string made up of carbon atoms linked by chemical bonds, alternatingly single and double bonds. It is the electrons of the double bonds that give rise to the electrical conductivity. But this only happens after oxidizing the polymer chain a little here and there, for example using iodine. And why is that? The iodine removes one electron from a carbon atom, thus creating a hole in the electronic structure into which an electron from a neighboring atom can jump, whereupon a new hole is formed and so on. A hole, i.e. lack of electron, corresponds to a positive charge, and the movement of the hole along the chain gives rise to a current.

The exciting idea of being able to combine the flexibility and low weight of plastics with the electric properties of metals has stimulated scientists all over the world, resulting in a novel research field bordering physics and chemistry. Various theoretical models and new conductive, but also semi-conductive, polymers followed during the 1980s in the wake of the first discoveries. Today we can see several possible applications. How about electrically luminous plastic that may be used for manufacturing mobile phone displays or the flat television screens of the future? Or the opposite - instead using light to generate electric current: solar-cell plastics that can be unfolded over large areas to produce environmentally friendly electricity. Finally, lightweight rechargeable batteries may be necessary if we are to replace the combustion engines in today's cars with environmentally friendly electric motors-another application where electrical polymers might find use...

Alan Jay Heeger was born in Sioux City, Iowa, on 22nd January 1936, the son of Peter Jacob (a store owner, d. 1945) and Alice Minkin Heeger. He married Ruth Chudacroft and they have two sons Peter and David. Heeger is currently Professor of Materials (Engineering), Professor of Physics and Director of the Institute for Polymers and Organic Solids at the University of California at Santa Barbara [2–5].

The Early Years

Hceger spent his early years in Akron, Iowa, where he attended elementary school. He met his wife at high school, which he completed one year early. He enrolled at the University of Nebraska where he studied mathematics and physics, Theodore Jorgensen taught Heeger quantum physics. Heeger was awarded a high distinction for his B.S. degree (1957) and moved to the University of California at Berkeley, where his studied solid state theory in Alan Portis' group from which he graduated with a Ph.D. in 1961.

The Career

Year	*Post*	*Institute*
1962	Assistant Professor	University of Pennsylvania
1964	Associate Professor	University of Pennsylvania
1967	Professor	University of Pennsylvania
1974–1981	Director, Laboratory for Research on the Structure of Matter	University of Pennsylvania
1982 to date	Professor of Physics	University of California at Santa Barbara
1982–1999	Director, Institute for Polymers and Organic Solids	University of California at Santa Barbara
1987 to date	Professor of Materials (Engineering)	University of California at Santa Barbara

Heeger was Acting Vice-Provost for Research at the University of Pennsylvania (1981–82) and is Adjunct Professor of Physics at the University of Utah (1988 to date). He held the Presidential Chair from the University of California at Santa Barbara (2001). Heeger was a Alfred P. Sloan Foundation Fellow (1963–65) and was a John Simon Guggenheim Foundation Fellow (1968–1969). He was a National Academy Exchange Scholar-USSR (1977) and a Yamada Science Foundation Exchange Scholar - Japan (1978).

Heeger has many research interests including:

- magnetic impurities in metals
- the Kondo Effect
- the metal-physics of tetracyanoquinodimethane (TCNQ)
- semiconducting polymers
- polymer photodetectors
- photovoltaic cells
- metallic polymers
- transport in semiconducting polymers

- light emission from semiconducting polymers (LEDs)
- light-emitting electrochemical cells (LECs)

Heeger published the book *Nonlinear Optical Properties of Polymers (Materials Research Society, Vol 109)* with Joseph Orenstein and Donald R. Ulrich (Ed), Pub: Materials Research Society, (1988), ISBN: 0931837774.

Heeger's Nobel Prize lecture is dated 8th December 2000 with the title 'Semiconducting and metallic polymers: the forth generation of polymeric materials'. During his introduction, Heeger suggested some of the advantages to conductive polymers compared with conventional materials [2]:

> *...In 1976, Alan MacDiarmid, Hideki Shirakawa and I, together with a talented group of graduate students and post-doctoral researchers discovered conducting polymers and the ability to dope these polymers over the full range from insulator to metal. This was particularly exciting because it created a new field of research on the boundary between chemistry and condensed matter physics, and because it created a number of opportunities:*
>
> - *Conducting polymers opened the way to progress in understanding the funchemistry and physics of it-bonded macromolecules;*
> - *Conducting polymers provided an opportunity to address questions which had been of fundamental interest to quantum chemistry for decades:*
>   - *Is there bond alternation in long chain polyenes?*
>   - *What is the relative importance of the electron-elecron and the electron-lattice interactions in π-bonded macromolecules?*
> - *Conducting polymers provided an opportunity to address fundamental issues of importance to condensed matter physics as well, including, for example, the metal-insulator transition as envisioned by Neville Mott and Philip Anderson and the instability of one-dimensional metals discovered by RuPeierls (the "Peierls Instability").*
> - *Finally - and perhaps most important-conducting polymers offered the promise of achieving a new generation of polymers: Materials which exhibit the electrical and optical properties of metals or semiconductors and which retain the attractive mechanical properties and processing advantages of polymers.*
>
> *Thus, when asked to explain the importance of the discovery of conducting polymers, I offer two basic answers:*
>
> (i) *They did not (could not?) exist.*
> (ii) *They offer a unique combination of properties not available from any other known materials...*

Further material concerning electrically conducting polymers can be found in [6].

Honours and Awards

Heeger has been presented with many honorary degrees including Doctor of Humane Letters from the University of Massachusetts at Lowell(1999). He also holds Honorary D.Sc. degrees from the University of Mons (1992); the Abo Akademie University, Turku (Finland) (1998); the University of Nebraska (1999); the Japan Advanced Institute for Science and Technology (2001); the South

China Institute of Science and Technology (2001); Trinity College, Dublin (2005). He was presented with the Honorary Doctor of Philosophy from Bar-Ilan University, Israel (2001) and the Doctor of Technology from Abo Akademi University, Finland (1996) and from the University of Linköping, Sweden (1996).

He has been presented with many awards and prizes including: the Oliver E. Buckley Prize for Condensed Matter Physics (1983); the John Scott Award and Medal (1989); the Balzan Prize, "Science of New Materials", Bern, Switzerland (1995); the Nobel Prize in Chemistry joint, with MacDiarmid and Shirakawa (2000); the Chancellor's Medal from the University of California at Santa Barbara (2001); the President's Medal for Distinguished Achievement, University of Pennsylvania (2001).

He is a Fellow of the: American Physical Society (1968); Institute of Physics (2000); National Academy of Sciences (USA) (2001); National Academy of Engineering (USA) (2002). He is a Foreign Member of the Korean Academy of Science (2001).

Heeger was: Visiting Professor of Physics at the University of Geneva (1968–1969); the Morris Loeb Visiting Lecturer in Physics at Harvard University (1973) and the Albert Einstein Honorary Professor at the Chinese Academy of Sciences (2005). He was the Charles A. Stiefvater Lecturer at the University of Nebraska (1994) and Co-Chair of the Selection Committee for Distinguished "Part-Time" Professorships (2001–2002).

Heeger was Founder and President of the UNIAX Corporation (1990–1994) and is currently the Chief Scientist and Chairman of the Board, UNIAX Corporation (1994 to date). He is a Member of the Board of Directors Konarka Technologies, Lowell, MA (2002); Chairman and Board Member of Diode Solutions, Santa Barbara, CA (2003); RitDisplay, Taiwan (2003); the Ensemble Theatre Group, Santa Barbara, CA (2000–2005); the Heeger Center for Advanced Materials, Gwangju Institute of Science and Technology, Korea (2005). He is a Venture Partner of NGen Partners (2003); a Member of the Executive Committee-Materials Research Laboratory, Santa Barbara, CA (2005) and of the Institute for Collaborative Bio-Technology, Santa Barbara, CA (2005) and a Member of the Steering Committee for MC-CAM Research Programme, Santa Barbara, CA (2005).

Bibliography

1. Reproduced with permission.
2. From Les Prix Nobel. The Nobel Prizes 2000, Editor Tore Frängsmyr, [Nobel Foundation], Stockholm, 2001.
3. Who's Who.
4. http://www.ipos.ucsb.edu/ajh.html.
5. http://www.mrl.ucsb.edu/mrl/faculty/heeger.html.
6. Conductive Electroactive Polymers: Intelligent Materials Systems, Second Edition by Gordon G. Wallace, CRC, 2nd ed. 2002, ISBN: 1587161273.

2000

© Nobel Foundation [1]

Alan G. MacDiarmid (0.33)

"for the discovery and development of conductive polymers" [2]

During his presentation speech on 10th December 2000, Bengt Nordén, of the Royal Swedish Academy of Sciences, commented on conductive polymers [2]:

> *...Polyacetylene can be seen as beads on a string made up of carbon atoms linked by chemical bonds, alternatingly single and double bonds. It is the electrons of the double bonds that give rise to the electrical conductivity. But this only happens after oxidizing the polymer chain a little here and there, for example using iodine. And why is that? The iodine removes one electron from a carbon atom, thus creating a hole in the electronic structure into which an electron from a neighboring atom can jump, whereupon a new hole is formed and so on. A hole, i.e. lack of electron, corresponds to a positive charge, and the movement of the hole along the chain gives rise to a current.*
>
> *The exciting idea of being able to combine the flexibility and low weight of plastics with the electric properties of metals has stimulated scientists all over the world, resulting in a novel research field bordering physics and chemistry. Various theoretical models and new conductive, but also semi-conductive, polymers followed during the 1980s in the wake of the first discoveries. Today we can see several possible applications. How about electrically luminous plastic that may be used for manufacturing mobile phone displays or the flat television screens of the future? Or the opposite - instead using light to generate electric current: solar-cell plastics that can be unfolded over large areas to produce environmentally friendly electricity. Finally, lightweight rechargeable batteries may be necessary if we are to replace the combustion engines in today's cars with environmentally friendly electric motors-another application where electrical polymers might find use...*

Alan G. MacDiarmid was born in Masterton, New Zealand, on the 14th April 1927, one of five children (his siblings being Colin, Roderick, Sheila and Alice) of Archibald and Ruby MacDiarmid. He married Marian Mathieu (1954, deceased 1990). They had four children Heather, Dawn, Duncan and Gail. Although still a New Zealander by law, MacDiarmid became a naturalized United States citizen many years ago. He is currently the Blanchard Professor of Chemistry in the Department of Chemistry at the University of Pennsylvania [2–5].

The Early Years

MacDiarmid's interest in chemistry started when he found a chemistry text book of his father's. Also, the public library in Lower Hutt contained a book called *The Boy Chemist* which he kept out on almost permanent loan for a year during which he carried out most of the experiments. He attended a two-room school in Keri Keri, where the town population was 600 and Hutt Valley High School (1940–1943). He left due to the strained financial circumstances of his family after which he took a part-time job as "lab boy"/janitor in the chemistry department at Victoria University College which had a faculty of two at the time. He studied there part-time concentrating on mathematics and chemistry graduating with a B.Sc. and an M.Sc. during which he studied the chemistry of S_4N_4.

MacDiarmid obtained a Fullbright fellowship from the US State Department in 1950 with which he studied for his Ph.D. under Norris F. Hall at the University of Wisconsin, concentrating on the rate of exchange in ^{14}C-tagged complex metal cyanides. On completion of his Ph.D. (1953) he moved to the University of Cambridge to work under H.J. Emeléus for a second Ph.D., funded by a New Zealand Shell graduate scholarship. He graduated from these studies concerning silicon hydrides, in 1955.

The Career

Year	*Post*	*Institute*
1955	Junior faculty member, Queens College	University of St. Andrews
1955	Various positions in the Department of Chemistry	University of Pennsylvania
1988	Blanchard Professor of Chemistry	University of Pennsylvania

MacDiarmid has been involved in many aspects of 'metallic polymers', sometimes called 'synthetic metals' and various studies of their properties since the first discovery including:

- the preparation of (SN)x (as golden crystals)
- the preparation of Polyacetylene (as a silvery film)
- studies of (CH)x
- the electrochemical doping of polyacetylene (1977)
- the discovery of metallic conductivity in an organic polymer
- isomeric forms of polyaniline and its oligomers
- light-emitting organic polymers
- the study of magnetic and optical properties of 'metal polymers'

The polymers require alternating oxidized and reduced groups leading to paramagnetic properties, eg.

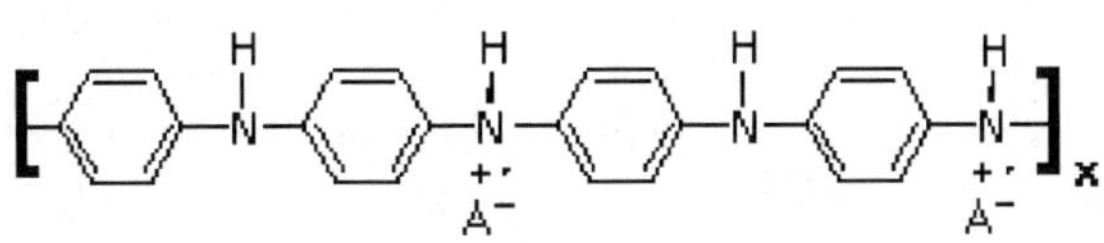

where A is a metal dopant, sometimes copper.

MacDiarmid has published several books including: *Bond to carbon*, Pub: Marcel Dekker, (1968), ASIN: B0007HR0C2; *Bond to Halogens and Halogenoids*, Pub: Marcel Dekker, (1972), ASIN: B00072TFPC; *Inorganic Syntheses, Vol. 17*, Pub: McGraw Hill Text, (June, 1977), ISBN: 0070443270; *The Polyanilines: A Novel Class of Conducting Polymers*, Pub: Univ. of Pennsylvania, (1992), ASIN: B0006R4ZV2.

His Nobel Prize lecture is dated 8th December 2000 with the title 'Synthetic metals: a novel role for organic polymers'. During his presentation, MacDiarmid commented [2]:

> *...Since the initial discovery in 1977, that polyacetylene $(CH)_x$, now commonly known as the prototype conducting polymer, could be p- or n-doped, either chemically or electrochemically to the metallic state, the development of the field of conducting polymers has continued to accelerate at an unexpectedly rapid rate and a variety of other conducting polymers and their derivatives have been discovered. This rapid growth rate has been stimulated by the field's fundamental synthetic novelty and importance to a cross-disciplinary section of investigators-chemists, electrochemists, biochemists, experimental and theoretical physicists, and electronic and electrical engineers and to important technological emerging applications of these materials.*
>
> *In the "doped" state, the backbone of a conducting polymer consists of a delocalized pi system. In the undoped state, the polymer may have a conjugated backbone such as in trans-$(CH)_x$ which is retained in a modified form after doping, or it may have a non-conjugated backbone, as in polyaniline (Icucoemeraldine base form), which becomes truly conjugated only after p- doping, or a non-conjugated structure as in the emeraldine base form of polyaniline which becomes conjugated only after protonic acid doping...*
>
> *...The basic purpose of this research is to blend the now well-established field of electronic/conducting polymers with the new, emerging field of nanoscience, by electrostatic fabrication ("electrospinning") to produce "nanoelectronics" - electronic junctions and devices significantly smaller than the diameter of a human hair ($\sim$50,000 nm). It is commonly accepted that a nanomaterial is defined as one consisting of a substance or structure which exhibits at least one dimension of less than 100 nm (0.1 μm).*

Our objectives were:

(i) *to develop a method by which nanofibers (diameter < 100 nm) of organic polymers could be controllably and reproducibly fabricated such that in one given preparation, all fibers would have a diameter < 100 nm and*

(ii) *to reproducibly and controllably fabricate, for the first lime, nanofibers of electronic polymers (in their semiconducting and metallic regimes) and/or their blends in conventional organic polymers for the purpose of ascertaining their applicability in the fabrication of nanoelectronic devices.*

We have made substantial progress in achieving these objectives by using a relatively little known, simple, convenient and inexpensive "electrospinning" method...

Further material concerning electrically conducting polymenrs can be found in [6].

Honours and Awards

MacDiarmid has received several honorary degrees including that of Ph.D. from Linkoping University (Sweden) (1990).

He has been presented with many awards and prizes including: the Philadelphia Section Award of the American Chemical Society (1967); the Frederic Stanley Kipping Award of the American Chemical Society (1970); the Marshall Award of the American Chemical Society (1982); the Doolittle Award of the American Chemical Society (1982); the Chemical Pioneer Award from the American Institute of Chemists (1984); the "Top 100" Innovation Award from Science Digest (1985); the John Scott Award, City of Philadelphia (1989); the Francis J. Clamer Award from the Franklin Institute (1993); the Chemistry of Materials Award from the American Chemical Society (1999); the Nobel Prize for Chemistry joint with Heeger, Shirakawa (2000); the Rutherford Medal of the Royal Society of New Zealand (2001).

He is a Member or Fellow of many societies including the Royal Society (2003).

MacDiarmid was a Visiting Professor at Kyoto University in Japan.

Among the named lectures that he has written and presented he was the recipient of the Royal Society of Chemistry Centenary Medal and Lectureship (1983).

Bibliography

1. Reproduced with permission.
2. From Les Prix Nobel. The Nobel Prizes 2000, Editor Tore Frängsmyr, [Nobel Foundation], Stockholm, 2001.
3. Who's who.
4. http://www.sas.upenn.edu/~macdiarm/.
5. http://www.sas.upenn.edu/chem/faculty/macdiarmid/macdiarmid.html.
6. Conductive Electroactive Polymers: Intelligent Materials Systems, 2nd Edition by Gordon G. Wallace, CRC, 2nd ed. 2002, ISBN: 1587161273.

2000

© Nobel Foundation [1]

Hideki Shirakawa (0.33)

"for the discovery and development of conductive polymers" [2]

During his presentation speech on the 10th December 2000, Bengt Nordén of the Royal Swedish Academy of Sciences, commented on conductive polymers [2]:

> *...Polyacetylene can be seen as beads on a string made up of carbon atoms linked by chemical bonds, alternatingly single and double bonds. It is the electrons of the double bonds that give rise to the electrical conductivity. But this only happens after oxidizing the polymer chain a little here and there, for example using iodine. And why is that? The iodine removes one electron from a carbon atom, thus creating a hole in the electronic structure into which an electron from a neighboring atom can jump, whereupon a new hole is formed and so on. A hole, i.e. lack of electron, corresponds to a positive charge, and the movement of the hole along the chain gives rise to a current.*
>
> *The exciting idea of being able to combine the flexibility and low weight of plastics with the electric properties of metals has stimulated scientists all over the world, resulting in a novel research field bordering physics and chemistry. Various theoretical models and new conductive, but also semi-conductive, polymers followed during the 1980s in the wake of the first discoveries. Today we can see several possible applications. How about electrically luminous plastic that may be used for manufacturing mobile phone displays or the flat television screens of the future? Or the opposite-instead using light to generate electric current: solar-cell plastics that can be unfolded over large areas to produce environmentally friendly electricity. Finally, lightweight rechargeable batteries may be necessary if we are to replace the combustion engines in today's cars with environmentally friendly electric motors-another application where electrical polymers might find use...*

Hideki Shirakawa was born in Tokyo, Japan, on 20th August 1936, the third child of five children of Hatsutarou (a physician) and Fuyuno Shirakawa (the daughter of a chief priest in a buddist temple). He married Chiyoko, née Shibuya, (1966) and they have two sons Chihiro and Yasuki. Shirakawa is currently a Emeritus Professor at the Institute of Materials Science, University of Tsukuba, Japan [2,3].

The Early Years

For the last ten years of his schooling (from the third year of elementary school until the end of high school), Shirakawa lived in the small city of Takayama (population less than 60,000) in the middle of Honshu, Japan. During this time, he collected plants and insects and made radios. Shirakawa wrote a short composition in the last year of his junior high school of his desire to become a scientist, particularly in the field of plastics. He entered the Tokyo Institute of Technology in April 1957 and studied applied chemistry. Graduation was followed with a Ph.D. in polymer physics from which he graduated with the degree of Doctor of Engineering (1966).

The Career

Year	*Post*	*Institute*
1966	Research Associate	Tokyo Institute of Technology
1979	Associate Professor, Institute of Materials Science	University of Tsukuba
1982	Professor, Institute of Materials Science	University of Tsukuba
2000	Retired	

Shirakawa was Chair of the Master's School in Sciences and Engineering, Graduate School, University of Tsukuba (1991–1993) and Provost of the 3rd Cluster of Colleges, University of Tsukuba, 1994–1997. He spent a sabbatical year at the University of Pennsylvania (1976–77) working with A.G. MacDiarmid and A. J. Heeger.

The majority of Shirakawa's work was in the field of 'synthetic metals' including:

- determining the polymerization mechanism of polyacetylene
  - a study of the doping mechanism
  - investigating the spectra of doped polyacetylene films
  - studying the electrical conductivity
- the synthesis of other conducting polymers
- the preparation of oriented films
- the synthesis of helical polyacetylene
- the synthesis of liquid crystalline polyacetylenes

Shirakawa wrote a book in 1990, published in Japanese, titled *Dodensei kobunshi kara nani ga mieru ka (Sutearingu shirizu kagaku gijutsu o sendosuru 30-nin)*, Pub: Mita Shuppankai, (1990), ISBN: 4895830705.

His Nobel Prize lecture is dated 8th December 2000 with the title 'The discovery of polyacetylene film: the dawning of an era of conducting polymers'. Shirakawa commented [2]:

> *...This lecture is not directly related to our discovery and development of conducting polymers to which the Nobel Prize in Chemistry 2000 was awarded. However, I would like to present my previous work that I carried out just before we reached the discovery of chemical doping...*
>
> *...The conventional method of polymerization in the laboratory is such that the catalyst solution is stirred thoroughly by any means to carry out the reaction under a homogeneous condition. The acetylene polymerization has not been an exception. It was customary for polymer chemists who synthesized polyacetylene to bubble acetylene gas into a catalyst solution with stirring. Unfortunately, the product was obtained as an intractable black powder from which it is very difficult to make into samples of a shape suitable for measurement of spectra arid physical properties because of its insolubility and infusibility.*
>
> *Soon after I joined Ikeda's group, we succeeded in synthesizing polyacetylene directly in the form of a thin film by a fortuitous error in 1967. After a series of experiments to reproduce the error, we noticed that we had used a concentration of the Ziegler-Natta catalyst nearly a thousand times greater than that usually used. It is worth noting that the insolubility of polyacetylene contributes to the formation of film. In addition, it was found that the film is composed of entangled micro-fibers called fibrils by a transmission electron microscope observation of an extremely thin film and by scanning electron microscope observation on a surface of a thick film. The fibril diameter is in the range of 20–100 nm depending upon the polymerization conditions. These inherent properties of polyacetylene are absolutely necessary for the film formation even under the higher concentration of the catalyst. One more important factor that should be added is that the catalyst used at the time, $Ti(O\text{-}n\text{-}C_4H_9)_4\text{-}(C_2H_5)_3Al$, was quite unique from the viewpoint of its good solubility in organic solvents such as hexane or toluene to give a homogeneous solution and its high activity to give exclusively high molecular weight and crystalline polymers...*

ELECTRICAL PROPERTIES OF AS-PREPARED POLYACETYLENE FILMS

> *The electrical resistivity of as-prepared films with various cis/trans contents was measured by the conventional two-probe method under vacuum in a temperature range of −120 to 20°C. The resistivity and energy gap of trans-rich polyacetylene were 1.0×10^4 Ωcm and 0.56 eV, respectively, whereas the values of a cis-rich (80 %) one were 2.4×10^8 Ωcm and 0.93 eV, respectively. Hatano and co-workers reported that the resistivity and energy gap measured on compressed pellets of powder polyacetylene synthesized by the same catalyst system are in the range of 1.4×10^4 and 4.2×10^5 Ωcm, and 0.46 eV, respectively, in good agreement with those for trans-rich polyacetylene film. In conclusion, it became apparent that the intrinsic electrical properties do not change very much between powder and film.*
>
> *The structures of pure cis and pure trans polyacetylene are given below.*

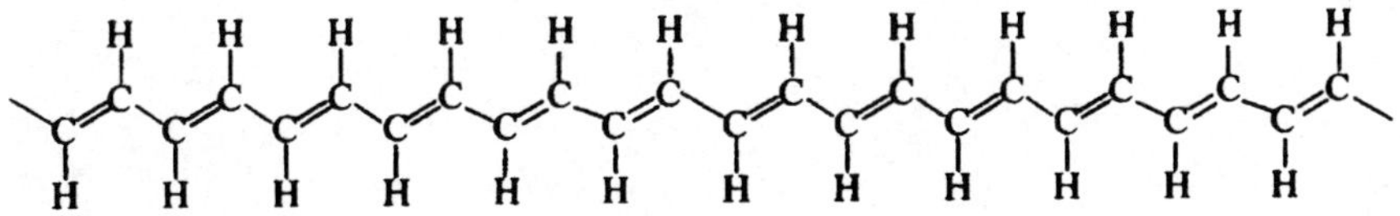

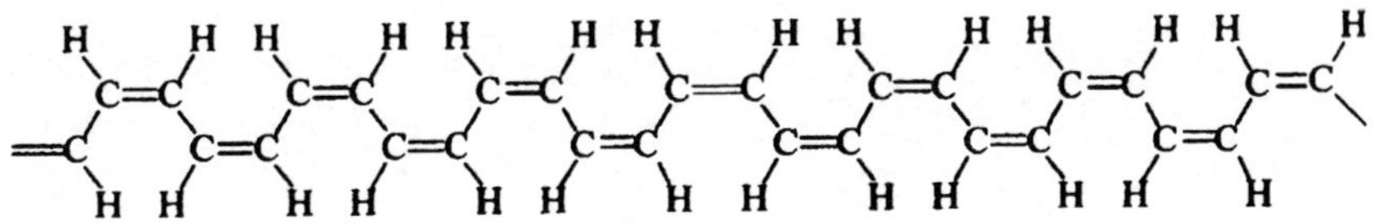

Further material concerning electrically conducting polymenrs can be found in [4].

Honours and Awards

Shirakawa was presented with many awards and prizes including:

the Award of the Society of Polymer Science, Japan (1983);

the Award for Distinguished Service in Advancement of Polymer Science, the Society of Polymer Science, Japan (2000);

the Person of Cultural Merits (2000);

the Order of Culture (2000);

the Nobel Prize for Chemistry joint with A. J. Heeger and A. G. MacDiarmid (2000).

Bibliography

1. Reproduced with permission.
2. From Les Prix Nobel. The Nobel Prizes 2000, Editor Tore Frängsmyr, [Nobel Foundation], Stockholm, 2001.
3. Who's who: 2005.
4. Conductive Electroactive Polymers: Intelligent Materials Systems, 2nd Edition by Gordon G. Wallace, CRC, 2nd ed. 2002, ISBN: 1587161273.

Section 3

Collaborations and Influences

Many scientists have been nominated for, but failed to be awarded, the Chemistry Nobel Prize. Records are released 50 years after the year in question and that data compiled into a book [1]. Listed below is some data that appears in the reference [1] where those scientists that are listed have been nominated for the Chemistry Nobel Prize for at least five years, but failed to be awarded it?

Name	*Year of birth*	*Years nominated*
Angelo Angeli	1864	1911, 1913, 1917, 1922, 1924, 1926, 1928, 1931
Ossian Aschan	1860	1920, 1922, 1923, 1924, 1925, 1931, 1936
Carl Auer von Welsbach	1858	1922, 1923, 1924, 1925, 1929
Gabriel Bertrand	1867	1909, 1928, 1929, 1931, 1932, 1933, 1934, 1935, 1937, 1942, 1944, 1946, 1947, 1948, 1949, 1950
Max Bodenstein	1871	1929, 1930, 1931, 1932, 1933, 1934, 1935, 1936, 1937
Giacomo Ciamician	1857	1905, 1907, 1908, 1911, 1912, 1914, 1916, 1919, 1921
Theodor Curtius	1857	1907, 1908, 1909, 1910, 1913, 1914, 1916, 1917, 1919, 1921, 1922, 1923, 1924
Victor Moritz Goldschmidt	1888	1929, 1930, 1931, 1932, 1933, 1934, 1936, 1940
Moses Gomberg	1866	1915, 1921, 1922, 1924, 1927, 1928, 1929, 1938, 1940
Phillipe-Auguste Guye	1862	1917, 1918, 1919, 1920, 1921, 1922
Albin Haller	1849	1907, 1912, 1916, 1919, 1920, 1921, 1922, 1923, 1924, 1925
Fritz Kögl	1897	1934, 1935, 1936, 1937, 1940, 1941, 1943, 1944, 1945, 1948, 1949, 1950
Henry Le Châtelier	1850	1905, 1907, 1908, 1909, 1914, 1915, 1920, 1921, 1922, 1926, 1927, 1928, 1929, 1930, 1931, 1934
Gilbert Newton Lewis	1875	1922, 1924, 1925, 1926, 1929, 1930, 1931, 1932, 1933, 1934, 1935, 1940, 1941, 1942, 1943, 1944, 1946
Lise Meitner	1878	1924, 1925, 1929, 1930, 1933, 1934, 1936, 1937, 1939, 1941, 1942, 1946, 1947, 1948
Carl Neuberg	1877	1929, 1930, 1931, 1932, 1933, 1934, 1935
Walter Noddack	1893	1932, 1933, 1934, 1935, 1937
Friederich Paneth	1887	1927, 1932, 1935, 1939, 1948
W. H. Perkin Jr.	1860	1911, 1912, 1913, 1914, 1918, 1919, 1920, 1925, 1929
Jean Perrin[a]	1870	1915, 1919, 1920, 1922, 1923, 1924, 1925, 1926
Amé Pictet	1857	1917, 1920, 1924, 1926, 1929, 1930, 1931, 1932, 1933
Albert Recoura	1862	1913, 1915, 1921, 1922, 1924
Thadeus Reichstein	1897	1943, 1945, 1947, 1948, 1949, 1950

(*continued*)

(*continued*)

Name	*Year of birth*	*Years nominated*
Otto Ruff	1871	1929, 1930, 1931, 1932, 1933, 1934, 1935, 1936, 1937
Wilhelm Schlenk	1879	1918, 1920, 1924, 1925, 1929
Sören Peter Lauritz Sörensen	1868	1921, 1922, 1923, 1924, 1932, 1933, 1934, 1935
Gustav Tammann	1861	1911, 1915, 1921, 1922, 1923, 1924, 1926, 1927, 1928, 1929, 1930, 1931, 1933, 1935, 1936, 1937
George Urbain	1872	1912, 1913, 1915, 1916, 1917, 1920, 1921, 1922, 1923, 1924, 1925, 1926, 1927, 1928, 1929, 1930, 1931, 1932, 1933, 1934, 1936, 1939
Paul Walden	1863	1914, 1920, 1927, 1934, 1947, 1948

[a] Laureate for Physics (1926)

Of the 135 scientists that have been awarded the Prize during the 20th Century, 7 were under 40 years of age when awarded the Prize, namely: Jean Frédéric Joliot (35); Adolf F. J. Butenandt (36); Ernest Rutherford (37); Iréne Joliot-Curie (38); Richard L. M. Synge (38); Richard Kuhn (38); Glenn T. Seaborg (39). Two scientists were over 80 years of age namely George Wittig (82) and Charles J. Pedersen (83). A scientist will be most likely to receive the award in the region of 50–55 years of age, see graph.

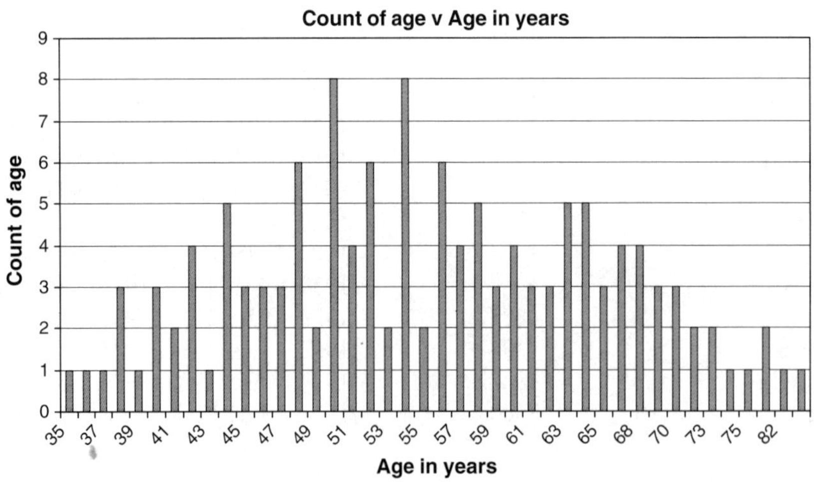

During the earlier half of the centuary the average age was closer to 50 whereas more recently the data would suggest that the average age at the time of the award may be increasing somewhat.

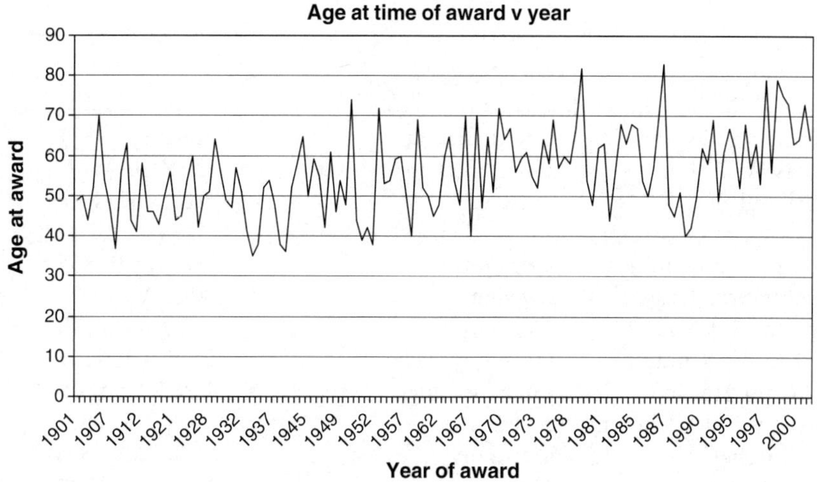

The table below shows the data for all of the scientists.

Year	*Name*	*Nationality*	*Birthdate*	*Age at Award*
1901	Jacobus H. van't Hoff	Netherlands	30th Aug 1852	49
1902	Hermann Emil Fischer	Germany	9th Oct 1852	50
1903	Svante A. Arrhenius	Sweden	19th Feb 1859	44
1904	Sir William Ramsay	Great Britain	2nd Oct 1852	52
1905	Adolf von Baeyer	Germany	31st Oct 1835	70
1906	Henri Moissan	France	28th Sept 1852	54
1907	Eduard Buchner	Germany	20th Mar 1860	47
1908	Ernest Rutherford	Great Britain	30th Aug 1871	37
1909	Wilhelm Ostwald	Germany	2nd Sept 1853	56
1910	Otto Wallach	Germany	27th Mar 1847	63
1911	Marie Curie	Poland - France	7th Nov 1867	44
1912	Victor Grignard	France	6th May 1871	41
	Paul Sabatier	France	5th Nov 1854	58
1913	Alfred Werner	Switzerland	12th Dec 1866	46
1914	Theodore W. Richards	United States	31st Jan 1868	46
1915	Richard M. Willstätter	Germany	13th Aug 1872	43
1916				
1917				
1918	Fritz Haber	Germany	9th Dec 1868	50
1919				
1920	Walther H. Nernst	Germany	25th June 1864	56

(*continued*)

(*continued*)

Year	Name	Nationality	Birthdate	Age at Award
1921	Frederick Soddy	Great Britain	2nd Sept 1877	44
1922	Francis W. Aston	Great Britain	1st Sept 1877	45
1923	Fritz Pregl	Austria	3rd Sept 1869	54
1924				
1925	Richard A. Zsigmondy	Germany	1st April 1865	60
1926	Theodor Svedberg	Sweden	30th Aug 1884	42
1927	Heinrich O. Wieland	Germany	4th June 1877	50
1928	Adolf O. R. Windaus	Germany	25th Dec 1876	51
1929	Sir Arthur Harden	Great Britain	12th Oct 1865	64
	Hans von Euler-Chelpin	Sweden	15th Feb 1873	56
1930	Hans Fischer	Germany	27th July 1881	49
1931	Friedrich Bergius	Germany	11th Oct 1884	47
	Carl Bosch	Germany	27th Aug 1874	57
1932	Irving Langmuir	United States	31st Jan 1881	51
1933				
1934	Harold C. Urey	United States	29th April 1893	41
1935	Jean Frédéric Joliot	France	19th March 1900	35
	Iréne Joliot-Curie	France	12th Sept 1897	38
1936	Peter J. W. Debye	Netherlands	24th March 1884	52
1937	Walter N. Haworth	Great Britain	19th March 1883	54
	Paul Karrer	Switzerland	21st April 1889	48
1938	Richard Kuhn	Germany	3rd Dec 1900	38
1939	Adolf F. J. Butenandt	Germany	24th March 1903	36
	Leopold Ruzicka	Switzerland	13th Sept 1887	52
1940				
1941				
1942				
1943	Georg de Hevesy	Hungary	1st Aug 1885	58
1944	Otto Hahn	Germany	8th March 1879	65
1945	Artturi I. Virtanen	Finland	15th Jan 1895	50
1946	James B. Sumner	United States	19th Nov 1887	59
	John H. Northrop	United States	5th July 1891	55
	Wendell M. Stanley	United States	16th Aug 1904	42
1947	Sir Robert Robinson	Great Britain	13th Sept 1886	61
1948	Arne W. K. Tiselius	Sweden	10th Aug 1902	46
1949	William F. Giauque	United States	12th May 1895	54
1950	Kurt Alder	Germany	10th July 1902	48
	Otto P. H. Diels	Germany	23rd Jan 1876	74
1951	Edwin M. McMillan	United States	18th Sept 1907	44
	Glenn Th. Seaborg	United States	19th April 1912	39

(*continued*)

Year	Name	Nationality	Birthdate	Age at Award
1952	Archer J. P. Martin	Great Britain	1st March 1910	42
	Richard L. M. Synge	Great Britain	28th Oct 1914	38
1953	Hermann Staudinger	Germany	23rd March 1881	72
1954	Linus C. Pauling	United States	28th Feb 1901	53
1955	Vincent du Vigneaud	United States	18th May 1901	54
1956	Sir Cyril N. Hinshelwood	Great Britain	19th June 1897	59
	Nikolai N. Semjonow	Soviet Union	3rd April 1896	60
1957	Sir Alexander R. Todd	Great Britain	2nd Oct 1907	50
1958	Frederick Sanger	Great Britain	13th Aug 1918	40
1959	Jaroslav Heyrovský	Czech Republic	20th Dec 1890	69
1960	Willard F. Libby	United States	17th Dec 1908	52
1961	Melvin Calvin	United States	8th April 1911	50
1962	John C. Kendrew	Great Britain	24th March 1917	45
	Max F. Perutz	Great Britain	19th May 1914	48
1963	Giulio Natta	Italy	26th Feb 1903	60
	Karl Ziegler	Germany	26th Nov 1898	65
1964	Dorothy C. Hodgkin	Great Britain	12th May 1910	54
1965	Robert B. Woodward	United States	10th April 1917	48
1966	Robert S. Mulliken	United States	7th June 1896	70
1967	Manfred Eigen	Germany	9th May 1927	40
	Ronald G. W. Norrish	Great Britain	9th Nov 1897	70
	George Porter	Great Britain	6th Dec 1920	47
1968	Lars Onsager	United States	27th Nov 1903	65
1969	Derek H. R. Barton	Great Britain	8th Sept 1918	51
	Odd Hassel	Norway	17th May 1897	72
1970	Luis F. Leloir	Argentina	6th Sept 1906	64
1971	Gerhard Herzberg	Canada	25th Dec 1904	67
1972	Christian B. Anfinsen	United States	26th March 1916	56
	Stanford Moore	United States	4th Sept 1913	59
	William H. Stein	United States	25th June 1911	61
1973	Ernst Otto Fischer	Germany	10th Nov 1918	55
	Geoffrey Wilkinson	Great Britain	14th July 1921	52
1974	Paul J. Flory	United States	19th June 1910	64
1975	Sir John Cornforth	Australia - Great Britain	7th Sept 1917	58
	Vladimir Prelog	Yugoslavia - Switzerland	23rd July 1906	69
1976	William N. Lipscomb	United States	9th Dec 1919	57
1977	Ilya Prigogine	Belgium	25th Jan 1917	60
1978	Peter Mitchell	Great Britain	29th Sept 1920	58

(*continued*)

(*continued*)

Year	Name	Nationality	Birthdate	Age at Award
1979	Herbert C. Brown	United States	22nd May 1912	67
	Georg Wittig	Germany	16th June 1897	82
1980	Paul Berg	United States	30th June 1926	54
	Walter Gilbert	United States	21st March 1932	48
	Frederick Sanger	Great Britain	13th Aug 1918	62
1981	Kenichi Fukui	Japan	4th Oct 1918	63
	Roald Hoffmann	United States	18th July 1937	44
1982	Aaron Klug	South Africa	11th August 1926	56
1983	Henry Taube	Canada	30th Nov 1915	68
1984	Bruce Merrifield	United States	15th July 1921	63
1985	Herbert A. Hauptman	United States	14th Feb 1917	68
	Jerome Karle	United States	18th June 1918	67
1986	Dudley Herschbach	United States	18th June 1932	54
	Yuan T. Lee	United States	19th Nov 1936	50
	John C. Polanyi	Canada	23rd Jan 1929	57
1987	Donald J. Cram	United States	22nd April 1919	68
	Charles J. Pedersen	United States	3rd Oct 1904	83
	Jean-Marie Lehn	France	30th Sept 1939	48
1988	Johann Deisenhofer	Germany	30th Sept 1943	45
	Robert Huber	Germany	20th Feb 1937	51
	Hartmut Michel	Germany	18th July 1948	40
1989	Thomas R. Cech	United States	8th Dec 1947	42
	Sidney Altman	United States	7th May 1939	50
1990	Elias James Corey	United States	12th July 1928	62
1991	Richard R. Ernst	Switzerland	14th August 1933	58
1992	Rudolph A. Marcus	Canada - United States	21st July 1923	69
1993	Kary B. Mullis	United States	28th Dec 1944	49
	Michael Smith	Great Britain - Canada	26th April 1932	61
1994	George A. Olah	United States	22nd May 1927	67
1995	Paul Crutzen	Netherlands	3rd Dec 1933	62
	Mario Molina	Mexico - United States	19th March 1943	52
	F. Sherwood Rowland	United States	28th June 1927	68
1996	Harold W. Kroto	Great Britain	7th Oct 1939	57
	Robert F. Curl, Jr.	United States	23rd August 1933	63
	Richard E. Smalley	United States	6th June 1943	53
1997	Paul D. Boyer	United States	31st July 1918	79
	John E. Walker	Great Britain	7th Jan 1941	56
	Jens C. Skou	Denmark	8th Oct 1918	79
1998	Walter Kohn	United States	9th March 1923	75
	John A. Pople	Great Britain	31st Oct 1925	73

(*continued*)

Year	Name	Nationality	Birthdate	Age at Award
1999	Ahmed H. Zewail	Egypt - United States	26th Feb 1946	63
2000	Alan J. Heeger	United States	22nd Jan 1936	64
	Alan G. MacDiarmid	United States - New Zealand	14th April 1927	73
	Hideki Shirakawa	Japanese	20th August 1936	64

Collaborations and Influences

There are many instances where the potential Chemistry Nobel Laureates visited or worked for other Chemistry Laureates. There are also instances where Nobel Prizes have been awarded to several members of the same family. For example:

Marie and Pierre Curie were jointly awarded half of the Physics Nobel Prize in 1903, Henri Becquerel received the other half. Marie went on to be awarded the Chemistry Nobel Prize in 1911 outright. Her daughter and son-in-law, Iréne Joliot-Curie and Frédéric Joliot, were jointly awarded the 1935 Chemistry Nobel Prize. Marie's other daughter Eve married H.R. Labouisse, who received the 1965 Nobel Peace Prize, on behalf of the United Nations Children's Fund.

Emil Fischer was a cousin of Otto Fischer.

Hans von Euler-Chelpin (Chemistry Nobel Prize 1929) attended the award of his son, Ulf von Euler (Physiology or Medicine 1970).

Lord Todd (Chemistry Nobel Prize 1957) was son-in-law to Sir Henry Hallett Dale (Physiology or Medicine 1936). Both Lord Todd and Sir Henry were Members of the Order of Merit and both were, in their time, President of the Royal Society.

The MRC laboratory in Cambridge has many notable laureates either as short term residents or career scientists including:

M. Perutz (Chemistry 1962)
J. Kendrew (Chemistry 1962)
F. Sanger (Chemistry 1958 and 1980)
Klug (1982)
J. Walker (Chemisty 1997)
F. Crick (Physiology or Medicine 1962)
Milstein (Physiology or Medicine 1984)

Many of the early Chemistry Laureates were either residents (for a period of time) or student of R. Bunsen or A. Kekulé. The following diagram illustrates the point. A single headed arrow has been used to denote a supervisor/student link and a double headed arrow to denote colleagues. The time frame for the collaborations ranges for a few months to many years. In addition, there are instances where a Chair has been occupied by a succession of eminent chemists. For example, the Chair in Munich had a succession of three Laureates flowing Liebig's tenure, namely A. von Baeyer followed by R. Willstätter followed by H. Wieland. The following diagram illustrates the early collaborations and influences.

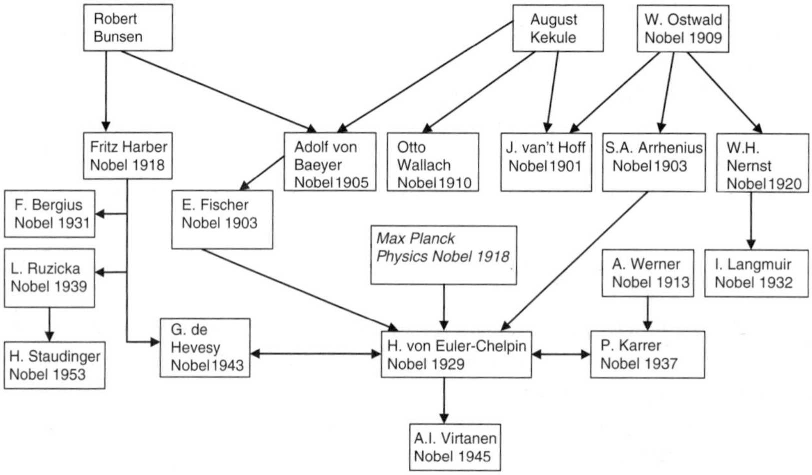

Of the chemists mentioned above, one Laureate Adolf von Baeyer, attracted many future laureates or scientists who were themselves nominated, but were not awarded the Nobel Prize. Those non-laureates who supervised future laureates are also included so that a more complete picture of the influence of the early laureates might be obtained.

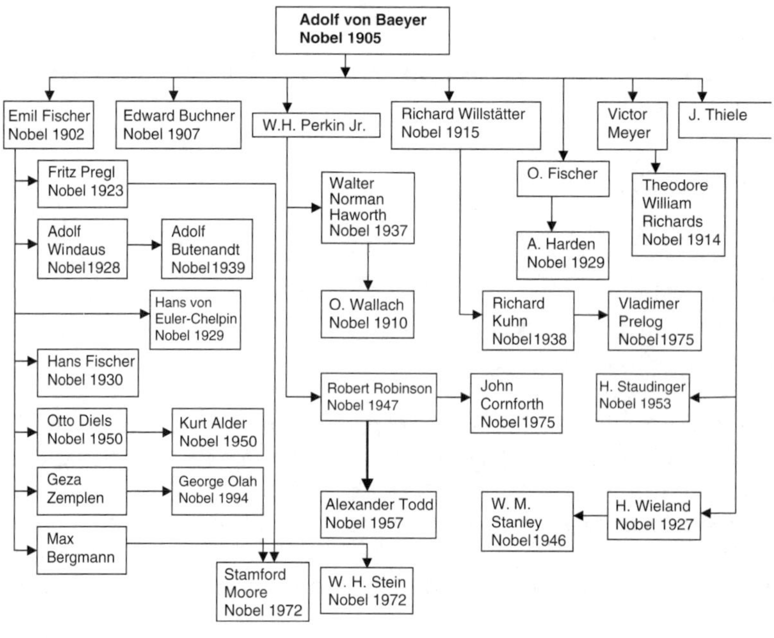

There are many more influence diagrams that could be drawn, and there are some simple relationships such as (where the date in brackets is the year that the laureate was awarded the Chemistry Nobel Prize):

Student	*Supervisor*
J. Sumner (1946)	The Svedberg (1926)
K. Alder (1950)	O. Diels (1950)
L. Onsanger (1968)	P. Debye (1936)
P. Flory (1974)	P. Debye (1936)
Prelog (1975)	Ruzicka (1939)
Deisenhofer (1988)	Huber (1988)
Altman (1989)	*Crick (Physiology or Medicine 1962)*
Smith (1993)	*Khorana (Physiology or Medicine 1968)*
Pople (1998)	*Dirac (Physics 1933)*

There are also some laureates whose supervisors worked for a laureate, e.g.:

Student	*Supervisor*
R. Synge (1952) worked for A. Tiselius (1948)	Tiselius (1948) worked for The Svedberg (1926)
G. Wilkinson (1973) worked for Briscoe	Briscoe worked for Richards (1914)
G. Wilkinson (1973) also worked for G. Seaborg (1951)	
R. Marcus (1992) worked for Winkler	Winkler worked for Hinshelwood (1956)

Where a particular university attracted many laureates there is the opportunity for one of the students to become a laureate thereby being influenced by several previous laureates. One such example is F. Sherwood Rowland (1995) who, whilst at the University of Chicago, was mentored by W. Libby (1960) whilst attending lectures by E. Fermi (Physics laureate 1938), H. Urey (1934) and H. Taube (1983). M. Molina (1995) was a postdoctoral fellow of Rowland's.

Of all the Nobel laureates who have worked on the interface between chemistry and physics, perhaps W. Kohn (1998) is more closely associated with other physicists than chemists having worked for J. Schwinger (Physics laureate 1965) for his Ph.D. at a time when B. Mottelson (physics 1975), N. Bloembergen (Physics 1981) and P. Anderson (Physics 1997) were in Schwinger's group. Kohn also worked for W. Shockley (Physics 1956) and was a colleague of A. Schawlow (1981).

During his work on atomic chemistry, E. Rutherford (1908) had three students who went on to be laureates namely G. de Hevesy (1943), O. Hahn (1944) and F. Soddy (1921) who incidentally also worked for W. Ramsay (1904) whilst J. Heyrovsky (1959) attended Ramsay's lectures during his studies.

Perhaps the most far reaching of collaborations links N. Bohr (1922) with E. Corey (1990) through R. Hoffmann (1981), where Pauling's link to both N. Bohr and E. Schrödinger was as a Guggenheim Fellow.

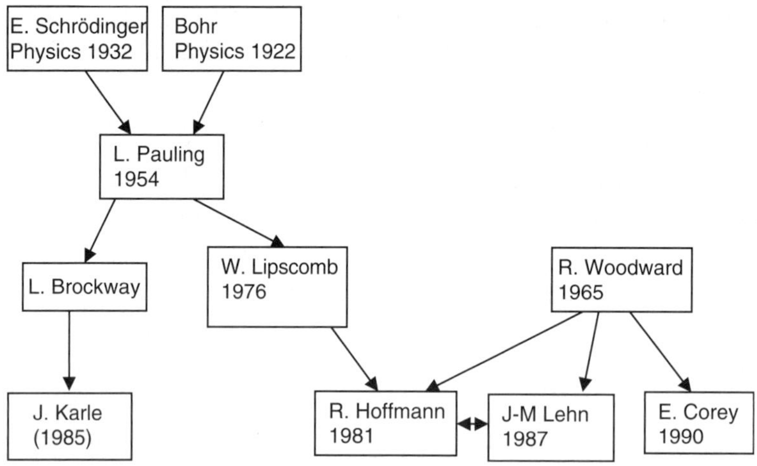

General Bibliography

Nobel Prize winners in Chemistry by Eduard Farber Aberlard-Schuman London revised 1962.
Eminent Chemists of our time by Benjamin Harrow pub: D. Van Nostrand Co. Inc. 1927.
British Chemists ed. Alexander Findlay and William Hobson Mills pub: The Chemical Society London 1947.
The Great Chemists by Eric John Holmyard Pub Methuen and Co. Ltd.
Makers of Chemistry by Eric John Holmyard Clarendon Press Oxford 1931.
Famous Chemists. The men and their work by Sir William A. Tilden FRS, pub: Routledge and Sons Ltd., 1921, RSC.
Nobel Laureates in Chemistry 1901–1992 ed. by Laylin K. James pub: American Chemical Society 1993, ISBN: 0-8412-2459-5 (cloth) 0-8412-2690-3 (pbk).
The International Who's who pub: Europa publications.
Biographical Encyclopedia of Science and Technology, by Isaac Asimov, Pan Books Ltd., 1972.
The scientific 100 – A ranking of the most Influential scientists past and present by John Simmons Citadel Press 1996.
Biographical Encyclopedia of Scientists, 2nd edition, pub: Institute of Physics Publishing, 1994.
Who's who of British Scientists.
Dictionary of Scientists by A. V. Howard, pub: W & R Chambers Ltd.
Dictionary of National Biography 1922–1930, pub: Oxford University Press 1953.
Chemical Who's who 1951 (3rd edition), Ed. Williams Haynes, pub: Lewis Historical Publishing Company Inc.
Nobel Prize Women in Science – Their lives, struggles and momentous discoveries by Sharon Bertsch McGrayne, pub: Birch Lane Press, 1993.
A Biographical Dictionary of Scientists ed. Trevor I. Williams, pub: Adam and Charles Black.
Biographical Dictionary of Scientists by Trevor Williams, Harper Collins, 1994 ISBN: 0 00 470109 7.

Concise Dictionary of Scientists ed. D Millar, I Millar, J Millar, and M Millar pub: Chambers Cambridge, 1989.
The Columbia Electronic Encyclopedia Copyright © 2003, Columbia University Press. Licensed from Columbia University Press.
The Road to Stockholm, Ista(acute)van Hargittai, Oxford University Press, 2002, ISBN: 019 850912X.
The Nobel Prize, Burton Feldman, Arcade Publishing, 2000, ISBN: 1-55970-537-X.
Nobel Prize winners in Chemistry, Eduard Farber, 1953 Abelard-Schuman. Library of congress card 62-17263.
Nobel the man and his prizes ed by the Nobel Foundation, 3rd ed. 1972, American Elsevier Publishing Company Inc, ISBN: 0-444-00117-4.
Nobel Prize Women in Science, Sharon Bertsch McGrayne, 1998, Joseph Henry Press, ISBN: 0-309-07270-0 (pbk).
The Nobel Prize ed Maj-Britt Neuman, 1983, Teknowledge Stockholm AB, ISBN: 91-970525-0-7.

Index